FINITE ELEMENT ANALYSIS OF STRUCTURES THROUGH UNIFIED FORMULATION

FINITE ELEMENT ANALYSIS OF STRUCTURES THROUGH UNIFIED FORMULATION

Erasmo Carrera

Department of Mechanical and Aerospace Engineering, Politecnico di Torino, Italy
School of Aerospace, Mechanical and Manufacturing Engineering,
RMIT University, Australia

Maria Cinefra
Enrico Zappino

Department of Mechanical and Aerospace Engineering, Politecnico di Torino, Italy

Marco Petrolo

School of Aerospace, Mechanical and Manufacturing Engineering,
RMIT University, Australia

WILEY

This edition first published 2014

Registered office
John Wiley & Sons Ltd, The Atrium, Southern Gate, Chichester, West Sussex, PO198SQ, United Kingdom

For details of our global editorial offices, for customer services and for information about how to apply for permission to reuse the copyright material in this book please see our website at www.wiley.com.

Library of Congress Cataloging-in-Publication Data

Carrera, Erasmo.
Finite element analysis of structures through unified formulation / Erasmo Carrera, Maria Cinefra, Marco Petrolo, Enrico Zappino.
pages cm
Includes bibliographical references and index.
ISBN 978-1-119-94121-7 (cloth)
1. Finite element method. 2. Numerical analysis. I. Cinefra, Maria. II. Petrolo, Marco. III. Zappino, Enrico. IV. Title.
QC20.7.F56C37 2014
518′.25–dc23

2014013805

A catalogue record for this book is available from the British Library.

ISBN: 9781119941217

Set in 10/12pt Times by Aptara Inc., New Delhi, India

1 2014

Contents

About the Authors

Erasmo Carrera

Erasmo Carrera graduated in Aeronautics in 1986 and in Space Engineering in 1988 from the Politecnico di Torino. He obtained a PhD in Aerospace Engineering in 1991 within the framework of a joint PhD programme between the Politecnico di Milano, the Politecnico di Torino and the Università di Pisa. He became assistant professor in 1992. He has continuously held courses at Bachelor, Master and PhD levels on Fundamentals of Theory of Structures, Aerospace Structures, Nonlinear Problems, Plates and Shells, Thermal Stress, Composite Materials, Multifield Problems and Computational Aeroelasticity. Currently he is a full professor in the Department of Mechanical and Aerospace Engineering. He has also been a visiting professor at the University of Stuttgart, Virginia Tech, Supmeca and the Centre of Research Public Henri Tudor.

His research topics cover: composite materials, nonlinear problems and the stability of structures, contact mechanics, multibody dynamics, finite elements, path-following methods in nonlinear finite element (FE) analysis, meshless methods, unconventional lifting systems, smart structures, thermal stress for coupled and uncoupled problems, multifield interaction, aeroelasticity, panel flutter, wind blades, explosion effects on flying aircraft, advanced theories for beams, plates and shells, mixed variational methods; zigzag, mixed and layer-wise modellings for multilayered beams, plates and shells; local–global methods and the Arlequin-type approach; advanced structural models for wings, fuselage and complete aircraft/spacecraft through the introduction of the so-called component-wise approach; failure and progressive failure analysis of laminated structures; inflatable structures for manned and unmanned space applications; and the design and analysis of full composite aircraft, including trikes and unmanned aerial vehicles (UAVs).

Professor Carrera developed the Reissner mixed variational theorem (RMVT) as a natural extension of the principle of virtual displacements to layered structure analysis. He introduced the unified formulation, or CUF (Carrera Unified Formulation), as a tool to establish a new framework in which beam, plate and shell theories can be developed for metallic and composite multilayered structures under mechanical, thermal, electrical and magnetic loadings. The CUF has been applied extensively to both strong and weak forms (FE and meshless solutions). The main feature of the CUF is that it permits any expansion of the unknown variables over the thickness/cross-section domain to be handled in a compact manner. Governing equations are in fact obtained in terms of a few fundamental nuclei whose forms do not depend on either the order of the expansion or the base functions used. As a result, the CUF allows the so-called best theory diagram (BTD) (which shows the minimum number of unknown variables vs the error on an assigned parameter) to be computed for a given problem. The BTD is a way of enhancing axiomatic and asymptotic approaches in the theory of structures.

Professor Carrera is the author and coauthor of about 500 papers on the above topics, most of which have been published in primary international journals, as well as of two recent books published by John Wiley & Sons, Ltd. His papers have received about 5000 citations with an h-index=39 (data from Scopus). He has held invited seminars in various European and North American universities, as well as plenary talks at international conferences. Professor Carrera serves as the Associate Editor for *Composite Structures*, *Journal of Thermal Stress*, *Mechanics of Advanced Structures*, *Computer and Structures* and the *International Journal of Aeronautical and Space Sciences*. He is founder and Editor-in-Chief of *Advances in Aircraft and Spacecraft Science*; acts as a reviewer for about 80 journals; and is on the Editorial Board of many international conferences. He is also in charge of the chapter on 'Shells' in the *Encyclopedia of Thermal Stress*, published by Springer. Professor Carrera is the founder of the non-profit international conference DeMEASS and the main organizer of ICMNMMCS (Turin, June 2012, co-chaired by Professor A. Ferreira), the ECCOMASS SMART 13 conference (Turin, June 2013) and ISVCS IX (Courmayeur, July 2013). He is member of the Distinguished Professor Board at King Abdulaziz University (Saudi Arabia). He has been a member of PhD and Habilitation committees in Germany, France, the Netherlands and Portugal. He is president of the Piedmont Section of AIDAA (Associazione Italiana di Aeronautica ed Astronautica).

Professor Carrera has been responsible for various research contracts granted by public and private national (including regional) and international institutions such as IVECO, the Italian Ministry of Education, the European Community, the European Space Agency, Alenia Spazio, Thales Alenia Space and Regione Piemonte. Among other projects, he has been responsible for the structural design and analysis of a full composite aircraft, named Sky-Y, by Alenia Aeronautica Torino, the first fully composite UAV made in Europe.

Professor Carrera is founder and leader of the MUL2 Group at the Politecnico di Torino. This group is considered one of the most active research teams in the Politecnico; it has acquired a significant international reputation in the field of *mul*tilayered structures subjected to *mul*tifield loadings; see also www.mul2.com. He is one of the Highly Cited Researchers by Thomson Reuters in both the Engineering and Materials Sections.

Maria Cinefra

Maria Cinefra is a research assistant at the Politecnico di Torino. She gained a BSc in Aerospace Engineering at the Politecnico di Torino in March 2007 with a thesis on the finite element method (FEM) in elliptic differential equations. Afterwards, she undertook an MSc in Aerospace Engineering at the Politecnico di Torino and gained her Master's degree, summa cum laude, in December 2008 from her work on the thermomechanical analysis of functionally graded material (FGM) shells. She began her PhD in January 2009, under the supervision of Professor Erasmo Carrera, on a research project related to the thermomechanical design of multilayered plates and shells embedding FGM layers. She was enrolled in a PhD with a foreign co-advisor, Professor Olivier Polit, at the University of Paris Ouest Nanterre. Her research project was funded by the Fonds National de la Recherche of Luxembourg and was performed in collaboration with the CRP Henri Tudor of Esch (Luxembourg). She was given the award for the best PhD paper (Ian Marshall's Award) at the 16th International Conference on Composite Structures (28–30 June 2011, Porto, Portugal). In January 2012, she was admitted to the final exam of her PhD and presented the defence of her thesis in April

2012. Since 2010, she has worked as a teaching assistant at the Politecnico di Torino on the courses Nonlinear Analysis of Structures, Structures for Space Vehicles and Fundamentals of Structural Mechanics. She is currently collaborating with the Department of Mathematics at Pavia University in order to develop a mixed shell FE based on the CUF for analysing composite structures. She has collaborated with Professor Ferreira, Editor of the *Composite Structures Journal*, on the radial basis functions method combined with the CUF. Dr Cinefra works as a reviewer for international journals such as *Composite Structures* and *Mechanics of Advanced Materials and Structures*. She is currently working on the STEPS regional project, in collaboration with Thales Alenia Space, and is also working on an extension of the shell FE, based on the CUF, to the analysis of multifield problems.

Marco Petrolo

Marco Petrolo is a Research Fellow at the School of Aerospace, Mechanical and Manufacturing Engineering, RMIT University, Melbourne, Australia. He was Post-Doc fellow at the Politecnico di Torino, Italy. He works in Professor Carrera's research group on various topics related to the development of refined structural models of composite structures. His research activity is connected with the structural analysis of composite lifting surfaces; refined beam, plate and shell models; component-wise approaches; and axiomatic/asymptotic analyses. He is the author and coauthor of some 50 publications, including 2 books and 25 articles that have been published in peer-reviewed journals.

Dr Petrolo gained his PhD in Aerospace Engineering at the Politecnico di Torino in April 2012, presenting a thesis on advanced aeroelastic models for the analysis of lifting surfaces made of composite materials. He also has an MSc in Aerospace Engineering from the Politecnico di Torino, an MSc in Aerospace Engineering from TU Delft (the Netherlands) and a BSc in Aerospace Engineering from the Politecnico di Torino. He has worked as an intern at EADS (Germany) and, as a Fulbright scholar, spent research periods at San Diego State University and the University of Michigan (USA). Dr Petrolo was appointed Adjunct Professor in Fundamentals of Strength of Materials (part of the BSc in Mechanical Engineering at the Turin Polytechnic University in Tashkent, Uzbekistan).

Enrico Zappino

Enrico Zappino is a post-doctoral fellow at the Politecnico di Torino. He has been in Professor Carrera's research group since 2010. His research activities concern structural analysis using classical and advanced models, multi-field analysis, and composite materials analysis. He is the coauthor of many works published in several international peer-reviewed journals. He obtained his PhD in April 2014, presenting a thesis on variable kinematic 1D, 2D, and 3D models for the analysis of aerospace structures. He also gained his BSc in Aerospace Engineering at the Politecnico di Torino in October 2007, presenting a thesis on advanced wing structures. He then obtained an MSc from the same university in July 2010, with a thesis on higher-order one-dimensional structural models applied to static, dynamic, and aeroelastic analysis. He was involved in many research programs supported by the European Space Agency and the European Union in cooperation with many European industrial and academic partners. From 2011, Dr. Zappino has worked as a teaching assistant at the Politecnico di Torino on the course of Aeroelasticity. In 2014, he was appointed as Adjunct Professor in Fundamentals of Strength of Materials at the Turin Polytechnic University in Tashkent, Uzbekistan.

Preface

This book deals with the finite element method (FEM) used for analysing the mechanics of structures in the case of linear elasticity. The novelty of this book is that the finite elements (FEs) are formulated on the basis of a class of theories of structures known as the Carrera Unified Formulation (CUF).

The CUF provides one-dimensional (beam) and two-dimensional (plate and shell) theories that go beyond classical theories (those of Euler, Kirchhoff, Reissner, Mindlin, Love) by exploiting a condensed notation and by expressing the displacement fields over the cross-section (the beam case) and along the thickness (plate and shell cases) in terms of base functions whose forms and orders are arbitrary. The condensed notation leads to the so-called *fundamental nucleus* (FN) of all the FEM matrices and vectors involved. The fundamental nuclei (FNs) and the related assembly technique are schematically shown in Table 1. The FNs consist of a few mathematical statements whose forms are independent of the theory of structures (TOS) employed. The FNs stem from the 3D elasticity equations via the principle of virtual displacements (PVD) and can be easily obtained for the 3D, 2D and 1D cases. This table will be reintroduced at the beginning of each chapter of this book that deals with 3D, 2D and 1D models to highlight the relevant fundamental nucleus.

The 1D and 2D FEs that stem from the CUF have enhanced capabilities since they can obtain results that are usually only provided by 3D elements with much lower computational costs. The 1D elements are particularly advantageous since they can deal with 2D and 3D problems in a proper manner.

The 1D and 2D CUF models are described in various chapters of this book. Particular attention has been paid to 1D and 2D FEs with only pure displacement degrees of freedom. The displacement unknowns of such FEs are defined over the physical surfaces of the real 3D body; this means that the definitions of mathematical reference axes (for beams) or reference surfaces (for plates and shells) are not needed. This capability is extremely important in an FEM/CAD coupling scenario. The modifications carried out in an FEM model can, in fact, be implemented directly in a CAD model (and vice versa) since physical surfaces are taken into account.

The concluding chapters of the book offer an overview of some of the most important features of the CUF models. In particular, the following topics are emphasized: multifield loads can be easily implemented; layered structures can be analysed; 1D, 2D and 3D models can be combined straightforwardly; and the CUF can lead to a definition of the BTD to evaluate the effectiveness of any structural theory. Numerical examples appear throughout the book on classical and non-classical TOS problems.

Table 1 A schematic description of the CUF and the related fundamental nucleus of the stiffness matrix for 3D, 2D and 1D models

---

Equilibrium equations in Strong Form $\rightarrow \quad \delta L_i = \int_V \delta \boldsymbol{u} \boldsymbol{k} \boldsymbol{u} dV + \int_S \ldots dS$

$$\underbrace{\begin{bmatrix} k_{xx} & k_{xy} & k_{xz} \\ k_{yx} & k_{yy} & k_{yz} \\ k_{zx} & k_{zy} & k_{zz} \end{bmatrix}}_{\boldsymbol{k}} \underbrace{\begin{Bmatrix} u_x \\ u_y \\ u_z \end{Bmatrix}}_{\boldsymbol{u}} = \underbrace{\begin{Bmatrix} p_x \\ p_y \\ p_z \end{Bmatrix}}_{\boldsymbol{p}}$$

$$k_{xx} = -(\lambda + 2G)\,\partial_{xx} - G\,\partial_{zz} - G\,\partial_{yy};$$
$$k_{xy} = -\lambda\,\partial_{xy} - G\,\partial_{yx};$$
$$k_{xz} = \ldots$$

$\lambda = (E\nu)/[(1+\nu)(1-2\nu)]; \quad G = E/[2(1+\nu)]$

$\boldsymbol{u} = \boldsymbol{u}(x, y, z)$

$\delta\boldsymbol{u} = \delta\boldsymbol{u}(x, y, z)$

The diagonal (e.g. k_{xx}) and the non-diagonal (e.g. k_{xy}) terms can be obtained through proper index permutations.

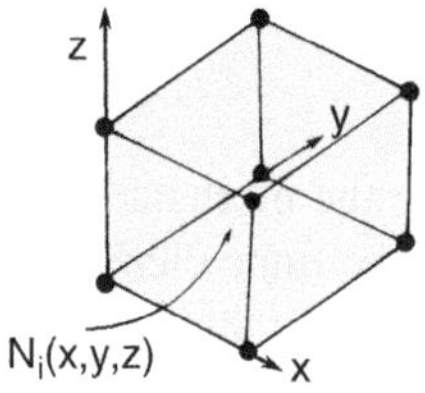

$\boldsymbol{u} = N_i(x, y, z)\boldsymbol{u}_i$

$\delta\boldsymbol{u} = N_j(x, y, z)\delta\boldsymbol{u}_j$

3D FEM Formulation $\rightarrow \quad \delta L_i = \delta\boldsymbol{u}_j \boldsymbol{k}^{ij} \boldsymbol{u}_i$

$$k^{ij}_{xx} = (\lambda + 2G)\int_V N_{j,x}N_{i,x}dV + G\int_V N_{j,z}N_{i,z}dV + G\int_V N_{j,y}N_{i,y}dV;$$

$$k^{ij}_{xy} = \lambda\int_V N_{j,y}N_{i,x}dV + G\int_V N_{j,x}N_{i,y}dV$$

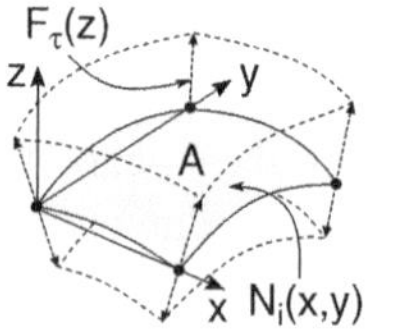

$\boldsymbol{u} = N_i(x, y)F_\tau(z)\boldsymbol{u}_{\tau i}$

$\delta\boldsymbol{u} = N_j(x, y)F_s(z)\delta\boldsymbol{u}_{sj}$

2D FEM Formulation $\rightarrow \quad \delta L_i = \delta\boldsymbol{u}_{sj} \boldsymbol{k}^{\tau sij} \boldsymbol{u}_{\tau i}$

$$k^{\tau sij}_{xx} = (\lambda + 2G)\int_\Omega N_{i,x}N_{j,x}d\Omega \int_h F_\tau F_s dz$$
$$+ G\int_\Omega N_i N_j d\Omega \int_h F_{\tau,z}F_{s,z}dz + G\int_V N_{i,y}N_{j,y}d\Omega \int_h F_\tau F_s dz;$$

$$k^{\tau sij}_{xy} = \lambda\int_\Omega N_{i,y}N_{j,y}d\Omega \int_h F_\tau F_s dz + G\int_\Omega N_{i,x}N_{j,y}d\Omega \int_h F_\tau F_s dz$$

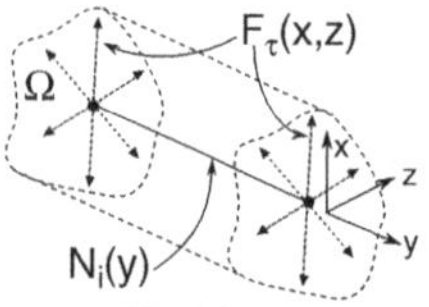

$\boldsymbol{u} = N_i(y)F_\tau(x, z)\boldsymbol{u}_{\tau i}$

$\delta\boldsymbol{u} = N_j(y)F_s(x, z)\delta\boldsymbol{u}_{sj}$

1D FEM Formulation $\rightarrow \quad \delta L_i = \delta\boldsymbol{u}_{sj} \boldsymbol{k}^{\tau sij} \boldsymbol{u}_{\tau i}$

$$k^{\tau sij}_{xx} = (\lambda + 2G)\int_l N_i N_j dy \int_A F_{\tau,x}F_{s,x}dA$$
$$+ G\int_l N_i N_j dy \int_A F_{\tau,z}F_{s,z}dA + G\int_l N_{i,y}N_{j,y}dy \int_A F_\tau F_s dA;$$

$$k^{\tau sij}_{xy} = \lambda\int_l N_{i,y}N_j dy \int_A F_\tau F_{s,x}dA + G\int_l N_i N_{j,y}dy \int_A F_{\tau,x}F_s dA$$

CUF leads to the automatic implementation of any theory of structures through 4 loops (i.e. 4 indexes):

- $\boldsymbol{\tau}$ and $\boldsymbol{s}$ deal with the functions that approximate the displacement field and its virtual variation along the plate/shell thickness ($F_\tau(z), F_s(z)$) or over the beam cross-section ($F_\tau(x, z), F_s(x, z)$);
- $\boldsymbol{i}$ and $\boldsymbol{j}$ deal with the shape functions of the FE model, (3D:$N_i(x, y, z), N_j(x, y, z)$; 2D:$N_i(x, y), N_j(x, y)$; 1D:$N_i(y), N_j(y)$).

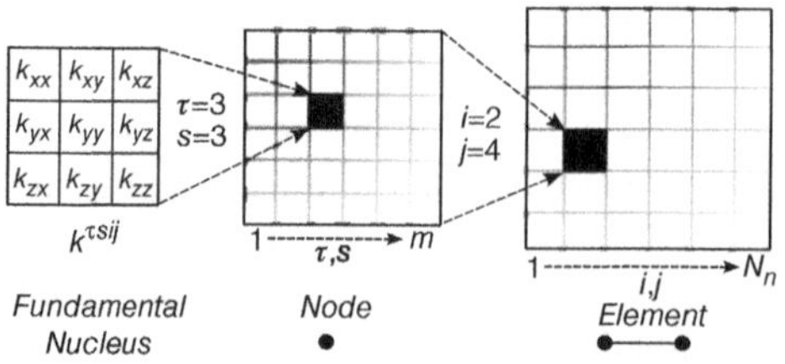

This table shows the essential features of the CUF. The strong form of the equilibrium equations allows one to derive a compact formulation for the fundamental nucleus. The nine elements of the FN can be written using only 2 terms. In this table, k_{xx} and k_{xy} are reported. All the remaining terms can be derived by a permutation of the indexes. This compact formulation is used to derive the 3D, 2D and 1D models in weak form.

This book follows on from two recent books where the CUF was applied to shell, plate and beam models: *Plates and Shells for Smart Structures: Classical and Advanced Theories for Modeling and Analysis* (E. Carrera, S. Brischetto and P. Nali, John Wiley & Sons, Ltd, 2011) deals with refined shell and plate models for smart structures; and *Beam Structures: Classical and Advanced Theories* (E. Carrera, G. Giunta and M. Petrolo, John Wiley & Sons, Ltd, 2011) deals with refined beam models. Analytical and FE formulations were introduced in both these books.

Nomenclature and Acronyms

The main symbols and acronyms that are defined in the book are listed below. Unless otherwise stated, the following definitions will be valid throughout the entire book.

Symbols

$\boldsymbol{B}, b$ Differential operator of the strain–displacements relations
B2, B3, B4 Beam elements with two, three and four nodes
$\boldsymbol{C}$ Hooke's law stiffness matrix
$C_{11}, C_{12}, C_{21}, C_{13}, C_{23}, C_{44}$ Hooke's law stiffness coefficients
E Young's modulus
F_τ, F_s Expansion functions
G Shear modulus
$\boldsymbol{g}$ Body forces per unit volume vector
g_x, g_y, g_z Body forces per unit volume components
H Metric factor
i, j Shape function indexes
k Layer index
$\boldsymbol{K}$ Stiffness matrix
$\mathbf{k}^{\tau sij}$ Fundamental nucleus of the stiffness matrix
$k_{xx}^{\tau sij}, k_{xy}^{\tau sij}, \ldots, k_{zz}^{\tau sij}$ Components of the stiffness matrix fundamental nucleus
L3, L4, L6, L9 Lagrange cross-section elements with three, four, six and nine nodes
L_{ext} Work of the external forces
L_{ine} Work of the inertial forces
L_{int} Work of internal forces
M Number of terms in an expansion
$\boldsymbol{M}$ Mass matrix
$\mathbf{m}^{\tau sij}$ Mass matrix fundamental nucleus
$m_{xx}^{\tau sij}, m_{xy}^{\tau sij}, \ldots, m_{zz}^{\tau sij}$ Components of the mass matrix fundamental nucleus
N Expansion order of F_τ, F_s
$\boldsymbol{n}$ Normal unit vector
N_a, N_b, N_c, N_d MITC4 interpolating shape functions for shear stresses
N_i, N_j Shape functions
N_m MITC9 interpolating shape functions for membrane stresses

$\boldsymbol{P}, \boldsymbol{p}$ Load vector
P_x, P_y, P_z Point load components
p_x, p_y, p_z Surface load components
q_x, q_y, q_z Line load components
R Radius of curvature
$\boldsymbol{U}, \boldsymbol{u}$ Displacement vector
u_x, u_y, u_z Displacement components
$\mathbf{u}_\tau$ Generalized displacement vector
$\mathbf{u}_{\tau i}$ Nodal displacement vector
$\ddot{\boldsymbol{U}}, \ddot{\boldsymbol{u}}$ Acceleration vector
V Volume
x, y, z Orthogonal Cartesian reference system
α, β, z Curvilinear coordinates
δ Virtual variation
$\boldsymbol{\epsilon}$ Strain vector
$\epsilon_{xx}, \epsilon_{yy}, \epsilon_{zz}$ Axial strain components
$\epsilon_{xy}, \epsilon_{yz}, \epsilon_{xz}, \gamma_{xy}, \gamma_{yz}, \gamma_{xz}$ Shear strain components
κ Shear correction factor
ν Poisson's ratio
ρ Material density
$\boldsymbol{\sigma}$ Stress vector
$\sigma_{xx}, \sigma_{yy}, \sigma_{zz}$ Axial stress components
$\sigma_{xy}, \sigma_{yz}, \sigma_{xz}, \tau_{xy}, \tau_{yz}, \tau_{xz}$ Shear stress components
τ, s Expansion function indexes
ϕ Rotation

Acronyms

1D/2D/3D One-/Two-/Three-Dimensional
BC Boundary Condition
BS Beam Semimonocoque
BTD Best Theory Diagram
CAD Computer-Aided Design
CLT Classical Lamination Theory
CNT Carbon Nanotube
CPT Classical Plate Theory
CST Classical Shell Theory
CUF Carrera Unified Formulation
CW Component-Wise
DOF Degree of Freedom
EBBT Euler–Bernoulli Beam Theory
ESL Equivalent Single Layer
ESLM ESL Model
FE Finite Element
FEA Finite Element Analysis

FEM Finite Element Method
FGM Functionally Graded Material
FN/FNs Fundamental Nucleus/Nuclei
FSDT First-order Shear Deformation Theory
HOT Higher-Order Theory
IC Interlaminar Continuity
LE Lagrange Expansion
LFAT Love First Approximation Theory
LM Lagrange Multiplier
LSAT Love Second Approximation Theory
LW Layer-Wise
LWM LW Model
MAAA Mixed Axiomatic–Asymptotic Approach
MAC Modal Assurance Criterion
MCS Multi-Component Structures
MFP Multifield Problem
MITC Mixed Interpolation of Tensorial Components
MLS Multilayered Structure
NRP Nanotube Reinforced Polymer
ODE/PDE Ordinary/Partial Differential Equations
PL Poisson Locking
PS Pure Semimonocoque
PVD Principle of Virtual Displacements
PVW Principle of Virtual Work
RMVT Reissner Mixed Variational Theorem
SDT Shear Deformation Theory
TBT Timoshenko Beam Theory
TE Taylor Expansion
TL Thickness Locking
TOS Theory of Structures
WRM Weight Residual Method
ZZ Zigzag

1

Introduction

1.1 What is in this Book

This book is devoted to the FE analysis of structures referring to linear elastic materials and small displacement assumptions. Attention is mainly focused on displacement formulations. This book is intended for two kinds of readers:

1. Those who are not familiar with FEs and would like to learn about them in a unified formulation framework.
2. Those who are familiar with FEs, but would like to learn how they are formulated with a unified formulation to overcome their limitations based on classical theories of structures.

Compared with other books on the subject, the present book offers the following novel features:

- It formulates 1D, 2D and 3D FEs on the basis of the same 'fundamental nucleus' that comes from geometrical relations and Hooke's law; the only difference between 3D elements and 1D and 2D elements is that cross-section and through-the-thickness integrals are introduced, respectively. The differential operators are the same in all three cases.
- It formulates refined 1D and 2D theories through formulae that remain invariant with respect to the variable order of the expansions used for the unknown displacement variables over the beam cross-section and plate/shell thickness.
- It shows that an appropriate refinement of 1D FEs can make them suitable for analysing shell structures.
- It presents both 1D and 2D refined FEs that only have displacement variables as in 3D elements. This is obtained using Lagrange polynomials to expand the displacement fields over the beam cross-section and plate/shell thickness.
- It presents 1D and 2D FEs that make use of 'real' physical surfaces rather than 'artificial' mathematical surfaces. Classical 1D and 2D elements in fact need lines and reference surfaces, both of which are artificial entities, to build mathematical models of a given structure. This means that the models from computer-aided design (CAD) tools have to be modified to obtain lines and reference surfaces for plate/shell elements. The use of real

Finite Element Analysis of Structures Through Unified Formulation, First Edition.
Erasmo Carrera, Maria Cinefra, Marco Petrolo and Enrico Zappino.

surface properties can facilitate the direct construction of FE mathematical models from CAD. On the other hand, modifications of the structure from FE analysis can easily be transferred to CAD, since physical surfaces are used.
- It shows how the described refined formulation can be easily and conveniently used to analyse laminated structures, such as sandwich and composite structures.
- It shows how refined elements are essential in the case of multifield loadings, e.g. thermal, electrical and magnetic loadings.
- It underlines that the use of refined FEs, unlike most available FEM codes, requires one to overcome the constraints of having only 6 DOFs for each node (3 displacements + 3 rotations, as in the Newtonian mechanics of a rigid body).
- It introduces an axiomatic/asymptotic approach that reduces the computational cost of the structural analysis without affecting the accuracy.
- It illustrates the performance of different FE models through the 'best theory diagram' (BTD), which allows different models to be compared in terms of accuracy and computational cost.

This book is devoted mostly to the development of FEs and not to procedures that can be used to solve FE problems since these procedures have already been covered in many other books on the FEM (Bathe 1996; Hughes 2000; Oñate 2009; Reddy 2005; Zienkiewicz *et al.* 2005).

1.2 The Finite Element Method

From a historical point of view, the FEM was introduced when the following two developments became available:

1. **Technological:** The development of computers capable of conducting mathematical operations very quickly.
2. **Mathematical:** The development of mathematical methods that can be used to solve differential equations in 'approximated' or 'weak' forms (both ordinary differential equations, ODEs, and partial differential equations, PDEs). The most significant of these methods is probably the weight residual method (WRM).

These two developments are common to all computational mechanics, and are discussed briefly in the following subsections.

1.2.1 Approximation of the Domain

'Technological' developments mean that automatic procedures are needed to solve the problem automatically in weak form. The problem can therefore be 'discretized' into a number of finite domains, or elements, in which mathematical tools work properly. The problem can then be 'assembled' and finally 'solved' using a computer.

In the case of the theory of structures, the problem is related to a 3D, 2D or 1D domain, which can be quite complex in terms of geometry, geometrical boundary conditions (conditions on generalized displacements in a given set of points) and mechanical boundary conditions

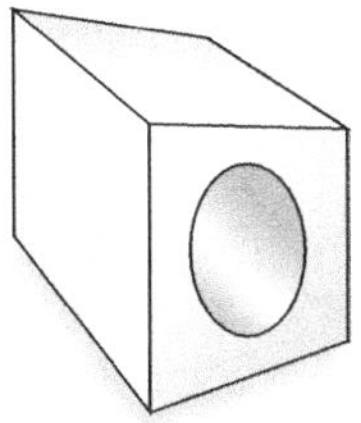

Figure 1.1 Example of 3D structure

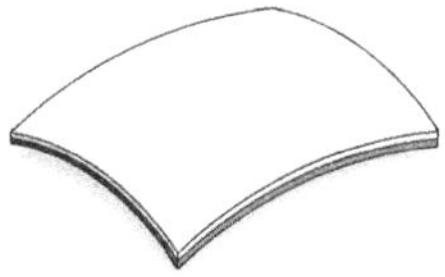

Figure 1.2 Example of 2D structure

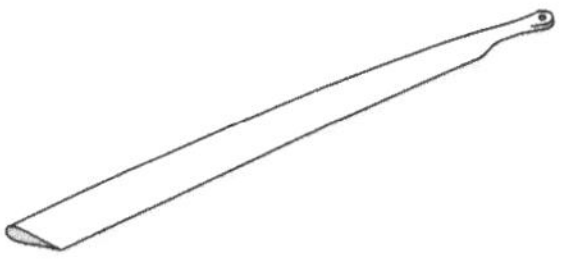

Figure 1.3 Example of 1D structure (helicopter blade)

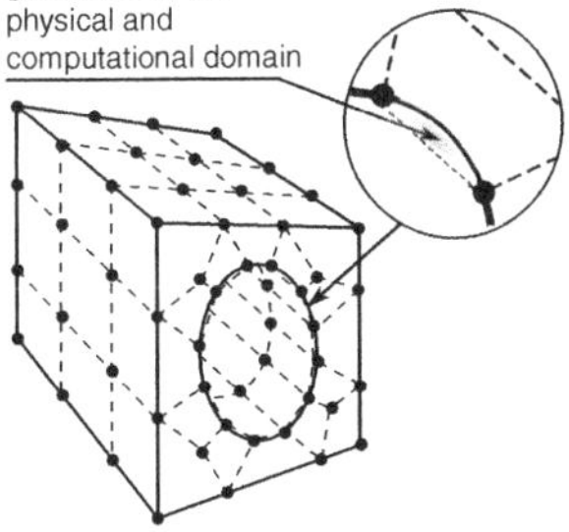

Figure 1.4 Example of a 3D FEM model

Figure 1.5 Example of 2D FEM model

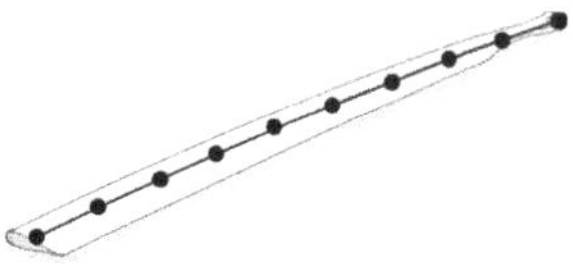

Figure 1.6 Example of 1D FEM model

(loadings). Geometrical domain examples are given in Figures 1.1, 1.2 and 1.3, where typical 3D, 2D and 1D domains are shown.

The key idea of the FEM is to 'discretize' complex domains into simpler ones, as in the following:

- Let us consider the 3D domain V in Figure 1.1 which can be discretized into a finite number of regular 3D solids (brick elements), as in Figure 1.4. It should be noted that the original domain can be violated slightly to correspond to its boundary surfaces. This cannot be avoided in the discretization process.
- Let us consider a 2D domain, Ω, which coincides with the reference surface of curved or flat panels, see Figure 1.2. The Ω domain can be discretized into a number of regular quadrilateral or triangular figures, or a mixture of these, as can be seen in Figure 1.5.
- Let us consider a 1D domain, l, which coincides with the reference line (axis) of a 'beam' structure, see Figure 1.3. The l domain can be discretized into a finite number of regular lines, as in Figure 1.6.

The above discretizations are common to all typical structures. One well-known example is the truss structure, shown in Figure 1.7. This kind of structure is significant from a historical point of view, since it represents one of the first applications of the FEM to the analysis of structures.

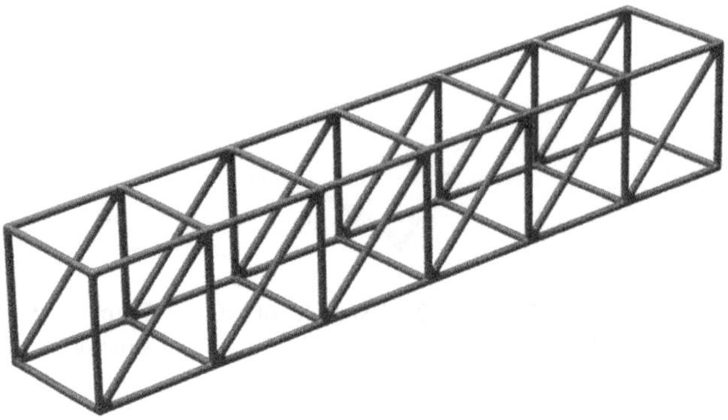

Figure 1.7 Example of truss structure

1.2.2 The Numerical Approximation

The solution of a structural problem consists of finding a solution to a given set of governing equations, which are defined in the V, Ω or l domain. The given set of governing equations of the unknown functions $\mathbf{\Phi}$

$$O(\Phi) = \Xi \tag{1.1}$$

can appear, from a mathematical point of view, as one of the following cases:

- O is a differential operator, for instance

$$a(x, y, z)\frac{\partial^2 \Phi}{\partial x^2} + \ldots = \Xi(x, y, z) \tag{1.2}$$

- O is an integral operator

$$\int_V a(x, y, z)\Phi(x, y, z)dV = \Xi(x, y, z) \tag{1.3}$$

- O is an algebraic operator

$$a(x, y, z)\Phi(x, y, z) = \Xi(x, y, z) \tag{1.4}$$

- O is given by any combination of the above.

The FEM solves the above equations, in a weak sense, at the element (or subdomain) level. The unknown function is usually assumed to be a combination of a finite number of $\mathbf{\Phi}$ in a given set of points, which are 'the nodes' of the element. In the 3D case

$$\Phi(x, y, z) = N_i(x, y, z)\Phi_i \tag{1.5}$$

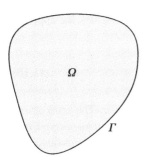

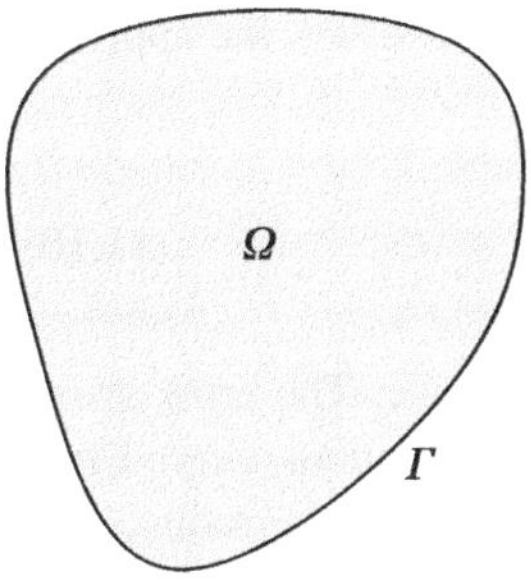

Figure 1.8 Surface Ω

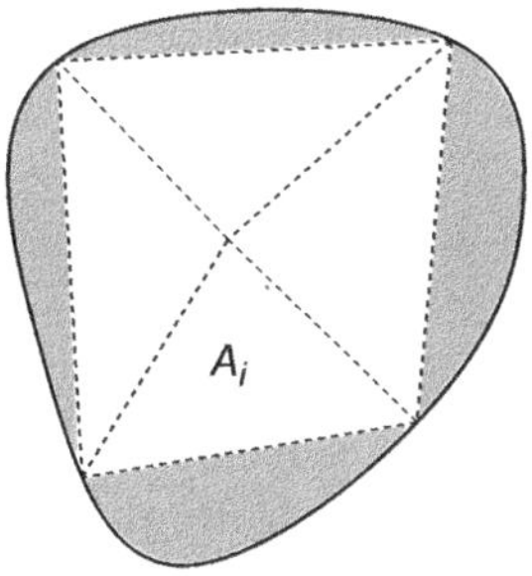

Figure 1.9 Coarse mesh

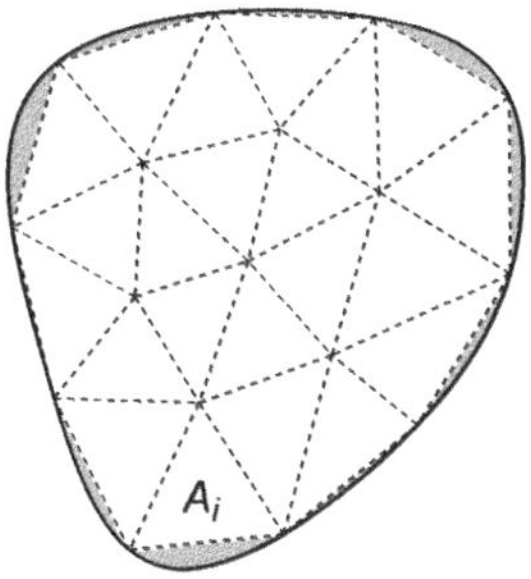

Figure 1.10 Refined mesh

The application of the WRM, for the governing equations (namely Equations (1.1–1.4)) of the element, leads to the following system of algebraic equations:

$$\boldsymbol{K}_E\boldsymbol{U}_E = \boldsymbol{P}_E \tag{1.6}$$

which is written for each element, E, and which is the governing equation for the element in the FEM sense. The governing FEM equation of the whole structure can be found, or 'assembled', by imposing compatibility/equilibrium conditions on the values of $\boldsymbol{\Phi}_i$ at the nodes:

$$\boldsymbol{K}_S\boldsymbol{U}_S = \boldsymbol{P}_S \tag{1.7}$$

Equation (1.6), the equilibrium equation, represents the result of a 'mathematical' problem. Writing and solving Equations (1.7) for a structure is a matter of automatic calculation. New 'technology' arising from the introduction of computers allows the problem to be solved easily.

1.3 Calculation of the Area of a Surface with a Complex Geometry via the FEM

Let us consider a surface Ω with a complex geometry (boundary Γ), such as that in Figure 1.8. The aim here is to calculate the area A of surface Ω. The complexity of the geometry of the surface does not allow a closed-form formula to be used to compute the area; that is, to express the area in terms of the geometrical parameters Ω and Γ:

$$A = A(\Omega, \Gamma) \tag{1.8}$$

A 'numerical' solution can be obtained by 'discretizing' the surface into simpler 'elements' or 'subdomains', such as quadrilaterals or triangles, whose area can easily be computed, see Figure 1.9. The number of elements, N_e, is finite. The words 'finite' and 'elements' give the name to the FEM.

The area, A, of a generic element i is denoted by A_i. If N_e is the number of elements, the unknown area is

$$\tilde{A} = \sum_{i}^{N_e} A_i \tag{1.9}$$

where $\tilde{A}$ is not the exact value of the area, but only an approximation. As usual, the approximation process leads to an error

$$E = A - \tilde{A} \tag{1.10}$$

where E is the grey area in Figure 1.9, which can be positive or negative. The error arises because of the approximation of the surface boundary, Γ, where a continuous curved line is approximated by the sum of the straight lines. However, such an error can be reduced by introducing smaller triangles or quadrilaterals, see Figure 1.10, which allow a better simulation of the boundary to be made.

This simple problem clearly illustrates the nature of 'discetrization' problems, and the need to conduct several calculations of the area for high N_e. When N_e is increased, an automatic tool, e.g. 'a computer', is needed to compute the area of a generic complex surface.

This simple example is not sufficient to show the mathematical difficulties involved in solving a problem at the element level. The area of a triangle does not in fact introduce any approximation, and can be computed exactly using well-known formulae.

Unfortunately, this is not the case for FEs of structures. This problem requires the application of a mathematical approximation process to solve the PDEs that usually govern elastic problems related to beams, plates, shells and solids.

1.4 Elasticity of a Bar

The simplest structural element is a bar. Let us consider a bar loaded by an axial loading, $q(y)$, as shown in Figure 1.11:

- it has a 1D behaviour, which means that the problem variables can be expressed in terms of only one coordinate, in this case the y coordinate;
- it can only carry loadings applied along its axis.

The only stress acting on the generic cross-section ($\sigma_{yy} = \sigma$) is assumed constant over the section itself. The stress resultant can be defined as

$$N = \int_A \sigma dx\, dz = \sigma A \tag{1.11}$$

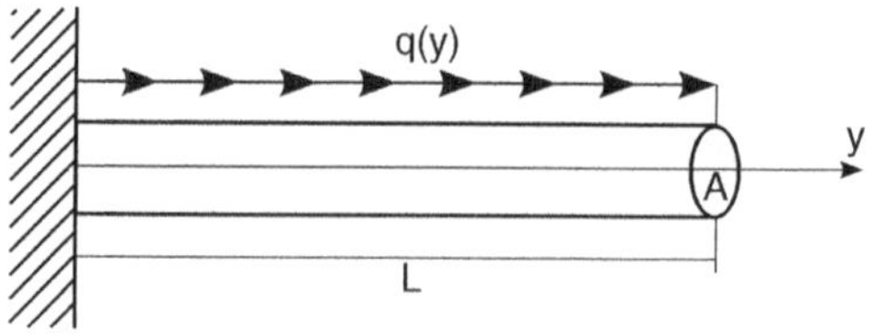

Figure 1.11 An axially loaded bar

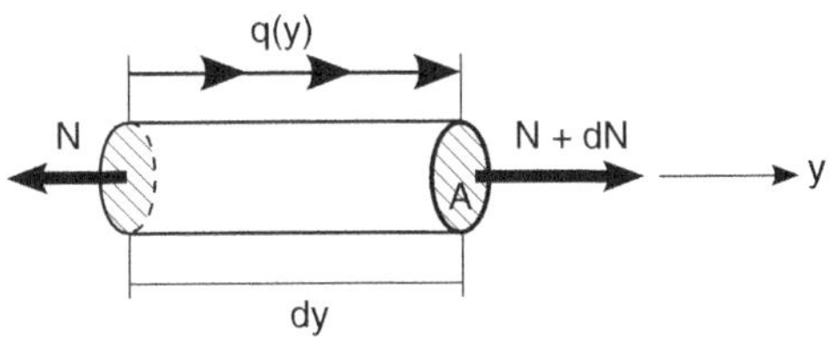

Figure 1.12 Equilibrium of an infinitesimal portion of a bar

If an infinitesimal portion of the bar, with length dy and cross-section A (see Figure 1.12), is considered, the equilibrium along the bar axis leads to

$$\frac{dN}{dy} = -q \tag{1.12}$$

which means that the variation in the axial forces (dN/dy) is balanced by the applied axial loading (q). It is clear that the solution of the previous equation, i.e. the distribution of N along the axis, is subordinate to the form of the applied loading q. If q is the only load applied on the bar, the following cases can be of particular interest:

- if $q = 0$,

$$N(y) = N_0 \tag{1.13}$$

  where N_0 is the value of N at $y = 0$.
- if q is constant,

$$N(y) = N_0 + yq \tag{1.14}$$

- in more complex cases, N can vary linearly or parabolically along the axis, and can be expressed as

$$N(y) = N_0 - \int_0^L q(y)dy \tag{1.15}$$

  where L is the length of the bar.

Hooke's law and the strain displacement relation are now introduced:

$$\varepsilon = \frac{\sigma}{E}, \quad \varepsilon = \frac{du_y}{dy} \tag{1.16}$$

where E is Young's modulus and A is the area of the cross-section of the bar. E and A are considered constant. The variation of $N(y)$, in terms of axial displacements, is

$$\frac{dN}{dy} = A\frac{d\sigma}{dy} = EA\frac{d\varepsilon}{dy} = EA\frac{d^2u_y}{dy^2} \tag{1.17}$$

The equilibrium equation can therefore be written in terms of displacements

$$EA\frac{d^2u_y}{dy^2} = -q \tag{1.18}$$

The derivative of the displacement at each point of the bar can be evaluated by integrating Equation (1.18),

$$\frac{du_y}{dy} = \left.\frac{du_y}{dy}\right|_{y=0} - \int_0^s \frac{q}{EA}ds \tag{1.19}$$

and, by integrating once more, the displacement becomes

$$u_y(y) = u_y|_{y=0} + \left.\frac{du_y}{dy}\right|_{y=0} y - \int_0^s \left(\int_0^r \frac{q}{EA}dr\right) ds \tag{1.20}$$

where s and r vary between 0 and L. The $q = 0$ case leads to a linear displacement field along the bar axis.

In the simplest case, $q = 0$ and $(du_y/dy)|_{y=0} = 0$, and if only a concentrated load is considered, one obtains

$$u_y(y) = u_y|_{y=0} + cy \tag{1.21}$$

where c is a constant that can be computed by imposing the given boundary condition. The linear form of u_y is consistent with the constant value of the axial stress resultant N.

1.5 Stiffness Matrix of a Single Bar

Let us consider the simplest case of an axially loaded bar of length L. The whole bar is the FE under investigation. Points $y = 0$ and $y = L$ are the nodes of the bar element. See Figure 1.13 for the notation.

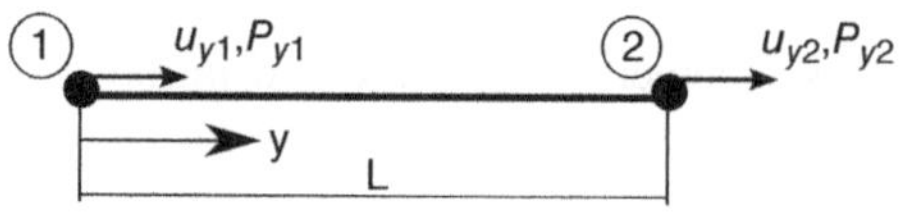

Figure 1.13 Local reference system and node numeration of a bar

If u_1 and u_2 are the values of the displacements at the nodes, then

$$u_y(y) = N_1(y)u_{y_1} + N_2(y)u_{y_2} \tag{1.22}$$

where $N_1(y)$ and $N_2(y)$ are the two linear Lagrange polynomials

$$N_1(y) = 1 - \frac{y}{L}, \quad N_2(y) = \frac{y}{L} \tag{1.23}$$

which satisfy the conditions

$$y = 0 \rightarrow u_y = u_{y_1}, \qquad y = L \rightarrow u_y = u_{y_2} \tag{1.24}$$

N_1 and N_2 are known as 'shape functions' in FE procedures.

The stress resultant, N, is

$$N = \sigma A = EA\varepsilon \tag{1.25}$$

where

$$\varepsilon = \frac{du_y}{dy} = -\frac{1}{L}u_{y_1} + \frac{1}{L}u_{y_2} = \frac{1}{L}(u_{y_2} - u_{y_1}) \tag{1.26}$$

Equation (1.25) represents the relationship between the stress resultant and the strain in the generic section that corresponds to y, see Figure 1.14. If a force P_{y_1} is applied at node 1, in the y-direction, or y direction equilibrium leads to

$$N = -P_{y_1} \tag{1.27}$$

therefore

$$P_{y_1} = \frac{EA}{L}(u_{y_1} - u_{y_2}) \tag{1.28}$$

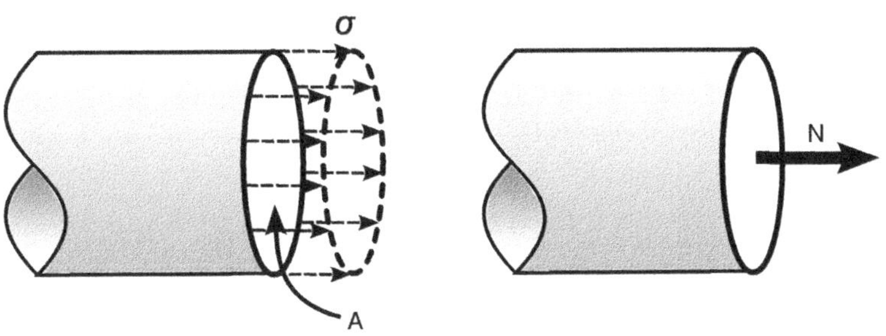

Figure 1.14 Relation between stress and resultant

while at node 2, one obtains

$$N = P_{y_2} \tag{1.29}$$

in terms of axial displacement

$$P_{y_2} = \frac{EA}{L}(-u_{y_1} + u_{y_2}) \tag{1.30}$$

These two equilibrium conditions can be expressed in matrix form:

$$\frac{EA}{L}\begin{bmatrix} 1 & -1 \\ -1 & 1 \end{bmatrix}\begin{Bmatrix} u_{y_1} \\ u_{y_2} \end{Bmatrix} = \begin{Bmatrix} P_{y_1} \\ P_{y_2} \end{Bmatrix} \tag{1.31}$$

By introducing the displacement vector (where T denotes vector/matrix transposition)

$$\boldsymbol{U}^T = (u_{y_1}, u_{y_2}) \tag{1.32}$$

and the force vector

$$\boldsymbol{P}^T = (P_{y_1}, P_{y_2}) \tag{1.33}$$

Equation (1.31) becomes

$$\boldsymbol{KU} = \boldsymbol{P} \tag{1.34}$$

where $\boldsymbol{K}$ is

$$\frac{EA}{L}\begin{bmatrix} 1 & -1 \\ -1 & 1 \end{bmatrix} \tag{1.35}$$

This matrix is known as the stiffness matrix in FE procedures. It should be noted that this matrix is symmetric and positive semidefinite.[1]

Example 1.5.1 *Let us consider the bar in Figure 1.15. The bar is axially loaded at node 2 and is clamped at node 1. The displacement vector has two contributions, the free displacements* $\bar{\boldsymbol{U}}^T$ *and the constrained displacements* $\tilde{\boldsymbol{U}}^T$*, which are*

$$\bar{\boldsymbol{U}}^T = u_{y_2}, \quad \tilde{\boldsymbol{U}}^T = 0 \tag{1.36}$$

[1] Equation (1.34) is a singular case of Equation (1.6). Furthermore, Equation (1.34) is only a weak solution of Equation (1.18). The solution of Equation (1.34) coincides with the solution of Equation (1.20) only in the particular case of $q = 0$.

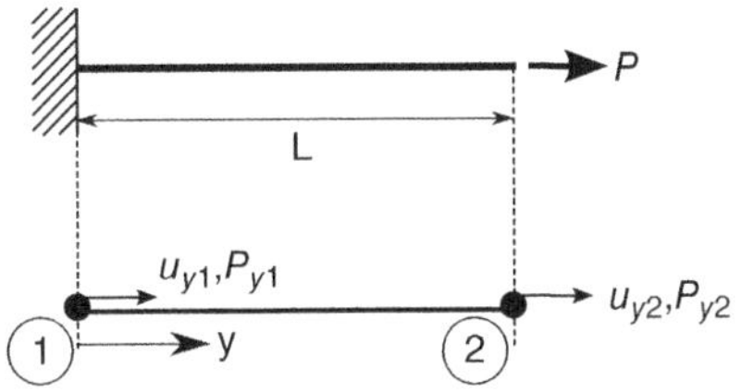

Figure 1.15 Example of an axially loaded bar, physical and FEM model

and the force vector

$$\bar{\boldsymbol{P}}^T = P_{y_2} = P, \quad \tilde{\boldsymbol{P}}^T = R_{y_1} \tag{1.37}$$

where $\bar{\boldsymbol{P}}^T$ *is the vector of the forces applied to the free nodes, while* $\tilde{\boldsymbol{P}}^T$ *is the reaction force in the constrained node, which is denoted by* R_{y_1}. *The problem that has to be solved can be written in the following form:*

$$\begin{Bmatrix} \bar{\boldsymbol{P}} \\ \tilde{\boldsymbol{P}} \end{Bmatrix} = \begin{bmatrix} \bar{\boldsymbol{K}} & \bar{\bar{\boldsymbol{K}}} \\ \tilde{\tilde{\boldsymbol{K}}} & \tilde{\boldsymbol{K}} \end{bmatrix} \begin{Bmatrix} \bar{\boldsymbol{U}} \\ \tilde{\boldsymbol{U}} \end{Bmatrix} \tag{1.38}$$

In this case, the problem becomes

$$\begin{Bmatrix} P_{y_2} \\ R_{y_1} \end{Bmatrix} = \frac{EA}{L} \begin{bmatrix} 1 & -1 \\ -1 & 1 \end{bmatrix} \begin{Bmatrix} u_{y_2} \\ 0 \end{Bmatrix} \tag{1.39}$$

The displacement vector is obtained by solving the first equation:

$$\begin{aligned} \bar{\boldsymbol{U}} &= \bar{\boldsymbol{K}}^{-1} \bar{\boldsymbol{P}} \\ u_{y_2} &= \left[\frac{EA}{L} 1 \right]^{-1} P_{y_2} \\ u_{y_2} &= P_{y_2} \frac{L}{EA} \end{aligned} \tag{1.40}$$

The reaction force at node 1 can be computed using the second equation:

$$\begin{aligned} \tilde{\boldsymbol{P}} &= \tilde{\tilde{\boldsymbol{K}}} \bar{\boldsymbol{U}} \\ R_{y_1} &= \frac{EA}{L} [-1] P_{y_2} \frac{L}{EA} \\ R_{y_1} &= -P_{y_2} \end{aligned} \tag{1.41}$$

1.6 Stiffness Matrix of a Bar via the PVD

The most powerful tool that can be used to derive FE equations in both static and dynamic cases and for linear and nonlinear problems is without doubt the principle of virtual work, PVW. In this book, PVW is used for FE problems with only displacement unknowns. In this case, it is referred to as the principle of virtual displacements (PVD). PVD is applied in this book to solids, beams, plates and shells for classical and refined FEM formulations.

Equations (1.31), which were introduced in Section 1.5, are now obtained using the PVD. For convenience, the displacement at point y is rewritten in the following rather universal notation:

$$u_y(y) = N_1(y)u_{y_1} + N_2(y)u_{y_2} = \boldsymbol{N}\boldsymbol{U} \tag{1.42}$$

where

$$\boldsymbol{N} = [N_1, N_2] \tag{1.43}$$

is the matrix of the shape functions and

$$\boldsymbol{U}^T = (u_{y_1}, u_{y_2}) \tag{1.44}$$

is the vector of the unknown displacements. In this case, N_1 and N_2 are linear Lagrange polynomials. These can also be derived by considering the conditions

$$u_y(y = 0) = u_{y_1} \tag{1.45}$$

$$u_y(y = L) = u_{y_2} \tag{1.46}$$

In order to verify these conditions, the shape functions have to fulfil the following requirements:

$$N_1(y = 0) = 1, \quad N_1(y = L) = 0, \quad N_2(y = 0) = 0, \quad N_2(y = L) = 1 \tag{1.47}$$

Because of the linearity of the functions, it is also possible to derive $N(y)$ in a simple geometrical manner. Figure 1.16 shows the application of the Thales theorem, which can be written as

$$\frac{u_y(y) - u_{y_1}}{y} = \frac{u_{y_2} - u_{y_1}}{L} \tag{1.48}$$

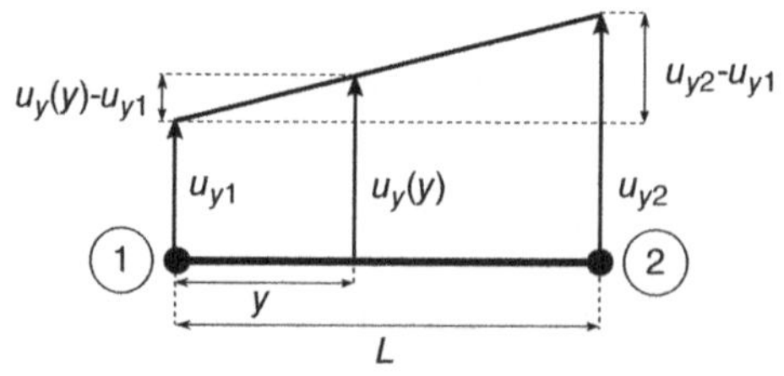

Figure 1.16 Linear Lagrange polynomials: triangle similarity

The explicit form of the displacement becomes

$$u_y(y) = \left(1 - \frac{y}{L}\right) u_{y_1} + \left(\frac{y}{L}\right) u_{y_2} \tag{1.49}$$

If matrix notation is introduced, the displacement can be rewritten as

$$u_y(y) = \left[\left(1 - \frac{y}{L}\right) \quad \left(\frac{y}{L}\right)\right] \begin{bmatrix} u_{y_1} \\ u_{y_2} \end{bmatrix} \tag{1.50}$$

and the explicit forms of the shape functions are

$$N_1(y) = \left(1 - \frac{y}{L}\right), \quad N_2(y) = \left(\frac{y}{L}\right) \tag{1.51}$$

which coincide with the linear Lagrange polynomials.

The axial strain is

$$\varepsilon_{yy} = \frac{du_y}{dy} = u_{y,y} \tag{1.52}$$

If the differential operator $\boldsymbol{b} = \left[\partial_y\right]$ is introduced and the displacement, u_y, is expressed in terms of nodal displacements, the axial strain becomes

$$\varepsilon_{yy} = \boldsymbol{b}u_y = \boldsymbol{bNU} = \boldsymbol{BU} \tag{1.53}$$

where matrix $\boldsymbol{B}$ is

$$\boldsymbol{B} = \boldsymbol{bN} \tag{1.54}$$

In explicit form, the strain becomes

$$\varepsilon_{yy} = \partial_y \left[\left(1 - \frac{y}{L}\right) \quad \left(\frac{y}{L}\right)\right] \begin{bmatrix} u_{y_1} \\ u_{y_2} \end{bmatrix} = \left[-\frac{1}{L} \quad \frac{1}{L}\right] \begin{bmatrix} u_{y_1} \\ u_{y_2} \end{bmatrix} \tag{1.55}$$

and matrix $\boldsymbol{B}$ is

$$\boldsymbol{B} = \left[-\frac{1}{L} \quad \frac{1}{L}\right] \tag{1.56}$$

In this case, matrix $\boldsymbol{B}$ is made up of constants.

The PVD states (in the static case) that

$$\delta L_{int} = \delta L_{ext} \tag{1.57}$$

where δL_{int} is the variation of the work done by the internal forces and δL_{ext} is the variation of the work done by the external forces. In order to derive Equation (1.34) for PVD application, it is necessary to express the variation of the internal work as

$$\delta L_{int} = \int_V \delta\varepsilon_{yy}\sigma_{yy} dV = \delta \boldsymbol{U}^T \boldsymbol{K}\boldsymbol{U} \tag{1.58}$$

and the variation of the external work as

$$\delta L_{ext} = \delta u_{y_1} P_{y_1} + \delta u_{y_2} P_{y_2} = \delta \boldsymbol{U}^T \boldsymbol{P} \tag{1.59}$$

Therefore

$$\delta \boldsymbol{U}^T \boldsymbol{K}\boldsymbol{U} = \delta \boldsymbol{U}^T \boldsymbol{P} \tag{1.60}$$

In order to obtain the explicit form of $\boldsymbol{K}$, Hooke's law is first applied to the internal work:

$$\delta L_{int} = \int_V \delta\varepsilon_{yy}^T \, E \, \varepsilon_{yy} \, dV \tag{1.61}$$

The strain and its virtual variation can be expressed as

$$\varepsilon_{yy} = \boldsymbol{B}\boldsymbol{U}, \quad \delta\varepsilon_{yy}^T = \delta\boldsymbol{U}^T\boldsymbol{B} \tag{1.62}$$

The virtual variation of the internal work can be written in terms of displacements as

$$\delta L_{int} = \int_V \delta\boldsymbol{U}^T \boldsymbol{B}^T \; \mathrm{E} \; \boldsymbol{B} \, \boldsymbol{U} \, dV \tag{1.63}$$

As $\boldsymbol{U}$ and $\delta\boldsymbol{U}$ are constant coefficients,

$$\delta L_{int} = \delta\boldsymbol{U}^T \left(\int_V \boldsymbol{B}^T \; \mathrm{E} \; \boldsymbol{B} \, dV \right) \boldsymbol{U} \tag{1.64}$$

The stiffness matrix can now be written as

$$\boldsymbol{K} = \int_V \boldsymbol{B}^T \; \mathrm{E} \; \boldsymbol{B} \, dV \tag{1.65}$$

and

$$\delta L_{int} = \delta\boldsymbol{U}^T \boldsymbol{K}\boldsymbol{U} \tag{1.66}$$

The explicit form of matrix $\boldsymbol{B}$, see Equation (1.56), can be inserted into Equation (1.65):

$$\boldsymbol{K} = \int_V \begin{bmatrix} -\frac{1}{L} \\ \frac{1}{L} \end{bmatrix} E \begin{bmatrix} -\frac{1}{L} & \frac{1}{L} \end{bmatrix} dV \tag{1.67}$$

Since $\boldsymbol{B}$ and E are constants, one has

$$\boldsymbol{K} = \frac{E}{L^2} \begin{bmatrix} -1 \\ 1 \end{bmatrix} \begin{bmatrix} -1 & 1 \end{bmatrix} \int_V dV \tag{1.68}$$

In the case of a constant section, $\int_V dV = AL$. The stiffness matrix of the element can therefore be written as

$$\boldsymbol{K} = \frac{EA}{L} \begin{bmatrix} 1 & -1 \\ -1 & 1 \end{bmatrix} \tag{1.69}$$

Stiffness matrix $\boldsymbol{K}$ is symmetric and singular. Moreover, $\boldsymbol{K}$ is a positive definite matrix, which is a common characteristic of the stiffness matrices of linear elastic problems. The equilibrium equation of a single bar can therefore be written as

$$\boldsymbol{KU} = \boldsymbol{P} \tag{1.70}$$

1.7 Truss Structures and Their Automatic Calculation by Means of the FEM

Bar structures are used in many applications to build the so-called truss structure. Unlike bars, which can only bear axial loads, truss structures can support any loadings, if these are applied at points where at least two bars are connected; these points are usually called nodes. An example is shown in Figure 1.17. This is a planar truss structure, which is very common in civil engineering.

The number of bars can be very large: dozens, hundreds or even thousands of bars can be usual in truss structures. It is clear that an automatic calculation is needed to solve practical problems.

It should be noted that, in most cases, only point loadings are applied to truss structures. These point loadings are applied to the nodes, and the loads are transferred to the bars that

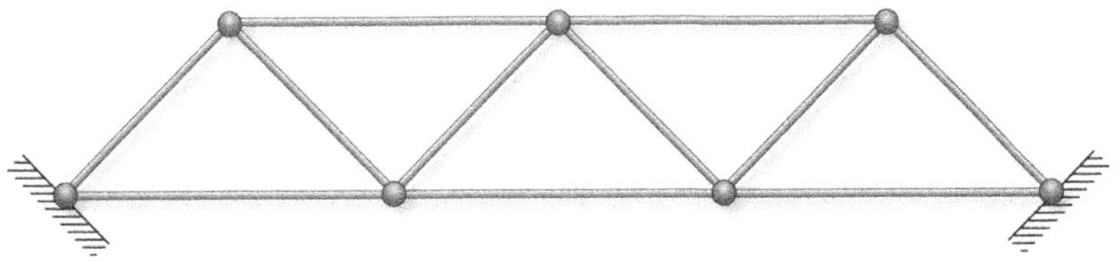

Figure 1.17 Example of a truss structure used in civil engineering

converge at those nodes, according to equilibrium and compatibility constraints. In order to solve truss structures, the equilibrium equations of the bar must be used, in terms of forces and displacements at the nodes. The equilibrium equation, which was derived in Section 1.5, has the following form:

$$\boldsymbol{K}\boldsymbol{U} = \boldsymbol{P} \tag{1.71}$$

It is important to point out that such an equation can be used to solve any truss structure. The following steps are required:

1. The nodes of the truss structure are depicted by numbers; that is, each bar is identified by two of these nodes. This is the so-called 'connectivity' of the various bar elements. The two nodes uniquely identify each bar of the truss structure. The number of nodes and the number of bars in the considered truss structure are denoted by N_{NE} and N_e, respectively.

2. Equation (1.71) is written for each bar of the 'element'

$$\boldsymbol{K}_i\boldsymbol{U}_i = \boldsymbol{P}_i, \quad i = 1, N_e \tag{1.72}$$

3. A global reference system is introduced for the whole truss structure, and the direction cosine of each bar is calculated in this reference system. The forces and displacements at the element level are introduced into the global reference system using a rotation matrix $\boldsymbol{R}$,

$$\boldsymbol{U}_{i_g} = \boldsymbol{R}_i\boldsymbol{U}_{i_e}, \quad \boldsymbol{P}_{i_g} = \boldsymbol{R}_i\boldsymbol{P}_{i_e} \tag{1.73}$$

4. The stiffness matrix of each bar is rotated according to the previous rotation matrix,

$$\boldsymbol{K}_i\,\boldsymbol{R}_i\boldsymbol{U}_{i_g} = \boldsymbol{R}_i\boldsymbol{P}_{i_g}, \quad i = 1, N_e \tag{1.74}$$

from which the stiffness matrix in the global reference system is obtained:

$$\boldsymbol{R}_i^{-1}\,\boldsymbol{K}_i\,\boldsymbol{R}_i\boldsymbol{U}_{i_g} = \boldsymbol{P}_{i_g}, \quad i = 1, N_e \tag{1.75}$$

that is,[2]

$$\boldsymbol{K}_{i_g}\boldsymbol{U}_{i_g} = \boldsymbol{P}_{i_g}, \quad i = 1, N_e \tag{1.76}$$

If the truss structure is in a plane, the size of the displacement and force vectors increases from two (in the local element reference system) to four; in the case of truss structures in space, these vectors have six components.

5. The stiffness matrices of the elements are 'summed' or 'assembled', according to the connectivity of each element, in a global stiffness matrix

$$\boldsymbol{K}_g\boldsymbol{U}_g = \boldsymbol{P}_g \tag{1.77}$$

[2] It should be noted that the inverse of the rotation matrix coincides with its transpose, $\boldsymbol{R}^T = \boldsymbol{R}^{-1}$, because it is orthogonal.

where U_g and P_g are the displacement and force vectors of the truss structure. Their dimension is $N = N_{NE} \times 2$, in the case of a planar truss structure, or $N = N_{NE} \times 3$, in the case of truss structures in 3D space.

K_g formally represents the sum of the stiffness matrices of each element introduced into the same global reference system,

$$K_g = \sum_{i=1}^{N_e} K_{i_g} \tag{1.78}$$

The way in which this matrix is built is a matter of assembling the FE technique.[3] It is possible to express the matrices of the elements in the global reference system and then sum them as shown in the previous formula. In practice, this is not convenient, since it increases both Central Processing Unit (CPU) time and memory allocation for computers. Specific FEM routines can be written to directly build local stiffness matrix K_{i_g} in the global stiffness matrix K_g.

6. Boundary conditions are imposed on constrained node displacements. In the most common case of homogeneous boundary conditions (rigid constraint), a number of displacements are fixed at zero. In other words, some components of the displacement vector U_g are known. As a consequence, the corresponding rows and lines of the governing systems can be removed, which leads to

$$\bar{K}_g \bar{U}_g = \bar{P}_g \tag{1.79}$$

This method, which is used to obtain the reduced matrix, can be considered as a partitioning technique of the complete stiffness matrix. If N_u is the number of imposed displacements, the displacement and force vectors can be split into two parts:

$$\bar{U}_g^T = (\bar{U}, \tilde{U}) \tag{1.80}$$

$$\bar{P}_g^T = (\bar{P}, \tilde{P}) \tag{1.81}$$

where $\bar{U}$ and $\bar{P}$ are the vectors that contain the displacements and the forces that correspond to the unconstrained degrees of freedom; $\tilde{U}$ and $\tilde{P}$ refer to the constrained nodes.

The dimension of $\bar{U}$ and $\bar{P}$ is

$$\bar{N} = N - N_u \tag{1.82}$$

while the dimension of $\tilde{U}$ and $\tilde{P}$ is

$$\tilde{N} = N_u \tag{1.83}$$

[3] The matrix has band properties, thus the elements that are different from zero are collected close to its diagonal for a length N_b; all the other elements are zero. The value of N_b is

$$N_b = Max(Node_i) \times Ndof_i$$

where $Ndof_i$ is equal to two or three in the case of planar or space truss structures, respectively.

It should be noted that $\tilde{N} = N_u$ displacements are known while $\bar{N}$ displacements need to be computed, the $\tilde{N} = N_u$ forces (the reactions at the constrained nodes) have to be computed and the $\bar{N}$ forces can be applied in each unconstrained node. As a consequence, matrix $\boldsymbol{K}_g$ can be partitioned into four matrices,

$$\boldsymbol{K}_g = \begin{bmatrix} \bar{\boldsymbol{K}}_g & \bar{\tilde{\boldsymbol{K}}}_g \\ \tilde{\bar{\boldsymbol{K}}}_g & \tilde{\boldsymbol{K}}_g \end{bmatrix} \tag{1.84}$$

$\bar{\boldsymbol{K}}$ and $\tilde{\boldsymbol{K}}$ are square matrices while $\tilde{\bar{\boldsymbol{K}}}_g$ and $\bar{\tilde{\boldsymbol{K}}}_g$ are rectangular matrices:

$$\tilde{\bar{\boldsymbol{K}}}_g = \bar{\tilde{\boldsymbol{K}}}_g^T \tag{1.85}$$

The system of equations can also be written as

$$\begin{bmatrix} \bar{\boldsymbol{K}}_g & \bar{\tilde{\boldsymbol{K}}}_g \\ \tilde{\bar{\boldsymbol{K}}}_g & \tilde{\boldsymbol{K}}_g \end{bmatrix} \begin{Bmatrix} \bar{\boldsymbol{U}} \\ \tilde{\boldsymbol{U}} \end{Bmatrix} = \begin{Bmatrix} \bar{\boldsymbol{P}} \\ \tilde{\boldsymbol{P}} \end{Bmatrix} \tag{1.86}$$

In practical FEM coding, the conditions on the displacements are imposed in a different way that does not modify the stiffness matrix.[4]

The unknown displacements are formally obtained by solving the algebraic system of equations related to $\bar{\boldsymbol{K}}$,[5]

$$\bar{\boldsymbol{U}}_g = \bar{\boldsymbol{K}}_g^{-1} \bar{\boldsymbol{P}}_g \tag{1.87}$$

while the unknown forces are

$$\tilde{\boldsymbol{P}}_g = \tilde{\bar{\boldsymbol{K}}}_g \bar{\boldsymbol{U}}_g \tag{1.88}$$

7. The stresses and strains can be obtained from the displacements using the relations that are described in Section 1.5.

An analysis of the previous points clearly shows that they can be automatized by writing appropriate computer codes. The number of trusses encountered in practical applications could make the method useless, unless a computer procedure is written.

It is important to note that the above points do not change if the FEs considered are changed: namely, beams, plates, shells or solids instead of bars. The only changes are in the number of DOFs and in the data of the structure. Such properties make it possible to combine bars,

---

[4] In most cases, a penalty technique is used. This consists of an increase in the stiffness in the diagonal terms related to the constrained displacements until the $\tilde{\boldsymbol{U}} = 0$ condition is obtained 'numerically'. Here, 'numerically' means that these displacements are at least 3–4 orders of magnitude lower than the other ones.

[5] In practice, the matrices are never inverted but factorized; one of the most popular methods is the Cholesky factorization method.

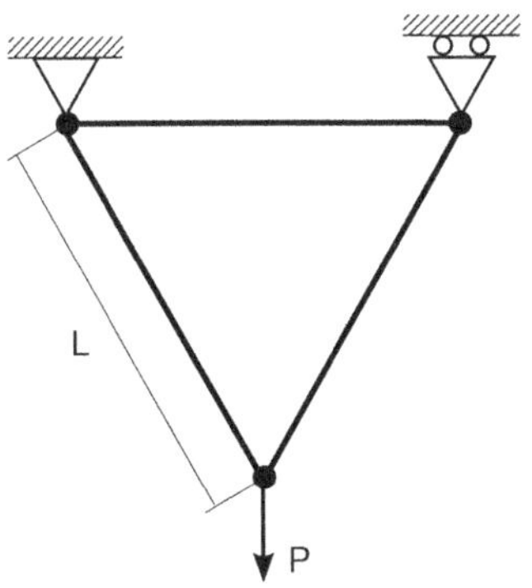

Figure 1.18 Three-bar truss structure

beams, plates, shells and solids in the analysis without any significant changes to the points listed above or in the related FE codes.

1.8 Example of a Truss Structure

In order to clarify points 1–7 introduced in Section 1.7, an example of a three-bar truss-structure is considered in this section. The structure, which is shown in Figure 1.18, consists of three bars and three nodes,

$$N_{NE} = 3, \quad N_e = 3 \tag{1.89}$$

Two nodes are constrained: one is hinged and the other one is simply supported. The third node is loaded with a force P.

As the first step, the nodes of the structure and the three bars have to be numbered. Figure 1.19 shows the global reference system with the numbering. The nodes are indicated by a number

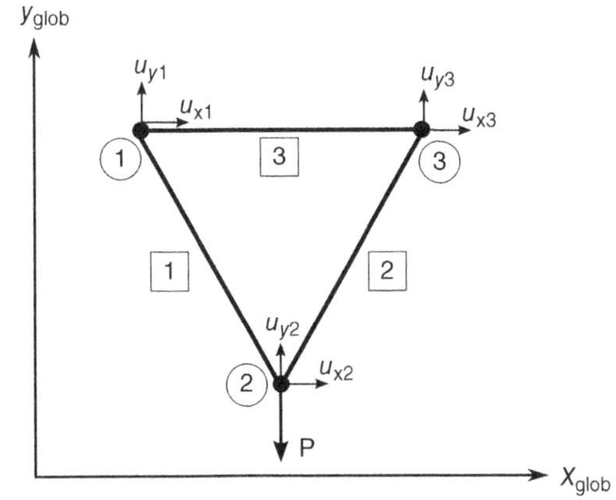

Figure 1.19 Global system of reference and node numbering of the truss structure

Table 1.1 Connectivity of the structure

Element	Local node 1	Local node 2	Rotation θ (deg)
1	1	2	−60
2	2	3	60
3	1	3	0

Connectivity between the local and the global nodes; the last column shows the rotation angle of each element.

in a circle, while the elements are indicated by a number in a square. The associated DOFs expressed in the global reference system are shown for each node.

1.8.1 Element Matrices in the Local Reference System

A bar element was depicted in Figure 1.13 in the local reference system. For convenience, it is assumed that the stiffness matrix of the bar assumes the following explicit form:

$$\boldsymbol{K}_i = \frac{E_i A_i}{L_i}\begin{bmatrix} 1 & -1 \\ -1 & 1 \end{bmatrix} \tag{1.90}$$

The displacement vector can be expressed as $\boldsymbol{U}_i^T = (u_{y_1}, u_{y_2})$, where the local force vector is $\boldsymbol{P}_i^T = (P_{y_1}, P_{y_2})$.

1.8.2 Element Matrices in the Global Reference System

The node numbers in the local reference system can be expressed in the global frame using the connectivity matrix given in Table 1.1. Connectivity links local nodes 1 and 2 to the global nodes for each element. For example, local node 1 of element 3 corresponds to global node 1, while local node 2 corresponds to global node 3. The rotation angle, θ, between the local and the global reference system is shown in the last column for each element.

Displacement vector $\boldsymbol{U}_{i_e}$, force vector $\boldsymbol{P}_{i_e}$ and stiffness matrix $\boldsymbol{K}_{i_e}$ can be expressed in the global reference system using the rotation matrix $\boldsymbol{R}_i$,

$$\boldsymbol{U}_{i_g} = \boldsymbol{R}_i \boldsymbol{U}_{i_e}, \quad \boldsymbol{P}_{i_g} = \boldsymbol{R}_i \boldsymbol{P}_{i_e}, \quad \boldsymbol{K}_{i_g} = \boldsymbol{R}_i \boldsymbol{K}_{i_e} \boldsymbol{R}_i^T \tag{1.91}$$

where $\boldsymbol{R}_i$ is

$$\boldsymbol{R}_i^T = \begin{bmatrix} \cos\theta_i & \sin\theta_i & 0 & 0 \\ 0 & 0 & \cos\theta_i & \sin\theta_i \end{bmatrix} \tag{1.92}$$

The global stiffness matrices of the three elements can be derived using Equation (1.91). The displacement and force vectors and the stiffness matrices in the global reference system become

$$U_{1_g} = \begin{Bmatrix} u_{x_1} \\ u_{y_1} \\ u_{x_2} \\ u_{y_2} \end{Bmatrix}; \quad P_{1_g} = \begin{Bmatrix} P_{x_1} \\ P_{y_1} \\ P_{x_2} \\ P_{y_2} \end{Bmatrix}; \quad K_{1_g} = \frac{E_1 I_1}{L_1} \begin{bmatrix} \frac{1}{4} & -\frac{\sqrt{3}}{4} & -\frac{1}{4} & \frac{\sqrt{3}}{4} \\ -\frac{\sqrt{3}}{4} & \frac{3}{4} & \frac{\sqrt{3}}{4} & -\frac{3}{4} \\ -\frac{1}{4} & \frac{\sqrt{3}}{4} & \frac{1}{4} & -\frac{\sqrt{3}}{4} \\ \frac{\sqrt{3}}{4} & -\frac{3}{4} & -\frac{\sqrt{3}}{4} & \frac{3}{4} \end{bmatrix} \quad (1.93)$$

$$U_{2_g} = \begin{Bmatrix} u_{x_2} \\ u_{y_2} \\ u_{x_3} \\ u_{y_3} \end{Bmatrix}; \quad P_{2_g} = \begin{Bmatrix} P_{x_2} \\ P_{y_2} \\ P_{x_3} \\ P_{y_3} \end{Bmatrix}; \quad K_{2_g} = \frac{E_2 I_2}{L_2} \begin{bmatrix} \frac{1}{4} & \frac{\sqrt{3}}{4} & -\frac{1}{4} & -\frac{\sqrt{3}}{4} \\ \frac{\sqrt{3}}{4} & \frac{3}{4} & -\frac{\sqrt{3}}{4} & \frac{3}{4} \\ -\frac{1}{4} & -\frac{\sqrt{3}}{4} & \frac{1}{4} & \frac{\sqrt{3}}{4} \\ -\frac{\sqrt{3}}{4} & -\frac{3}{4} & \frac{\sqrt{3}}{4} & \frac{3}{4} \end{bmatrix} \quad (1.94)$$

$$U_{3_g} = \begin{Bmatrix} u_{x_1} \\ u_{y_1} \\ u_{x_3} \\ u_{y_3} \end{Bmatrix}; \quad P_{3_g} = \begin{Bmatrix} P_{x_1} \\ P_{y_1} \\ P_{x_3} \\ P_{y_3} \end{Bmatrix}; \quad K_{3_g} = \frac{E_3 I_3}{L_3} \begin{bmatrix} 1 & 0 & -1 & 0 \\ 0 & 0 & 0 & 0 \\ -1 & 0 & 1 & 0 \\ 0 & 0 & 0 & 0 \end{bmatrix} \quad (1.95)$$

1.8.3 Global Structure Stiffness Matrix Assembly

Displacement vector U_g and loading vector P_g of the complete truss structure can be written as

$$U_g^T = (u_{x_1}, u_{y_1}, u_{x_2}, u_{y_2}, u_{x_3}, u_{y_3}) \quad (1.96)$$

$$P_g^T = (P_{x_1}, P_{y_1}, P_{x_2}, P_{y_2}, P_{x_3}, P_{y_3}) \quad (1.97)$$

The global stiffness matrix, K_g, can be assembled according to the connectivity. The stiffnesses of the shared nodes are 'summed'; in this way, the stiffness matrix of each element, see Equations (1.93), (1.94) and (1.95), can be assembled in the global stiffness matrix. The assembled global matrix becomes

$$K_g = \begin{bmatrix} K_{1_g}^{11} + K_{3_g}^{11} & K_{1_g}^{12} + K_{3_g}^{12} & K_{1_g}^{13} & K_{1_g}^{14} & K_{3_g}^{13} & K_{3_g}^{14} \\ K_{1_g}^{21} + K_{3_g}^{21} & K_{1_g}^{22} + K_{3_g}^{22} & K_{1_g}^{23} & K_{1_g}^{24} & K_{3_g}^{23} & K_{3_g}^{24} \\ K_{1_g}^{31} & K_{1_g}^{32} & K_{1_g}^{33} + K_{2_g}^{11} & K_{1_g}^{34} + K_{2_g}^{12} & K_{2_g}^{13} & K_{2_g}^{14} \\ K_{1_g}^{41} & K_{1_g}^{42} & K_{1_g}^{43} + K_{2_g}^{21} & K_{1_g}^{44} + K_{2_g}^{22} & K_{2_g}^{23} & K_{2_g}^{24} \\ K_{3_g}^{31} & K_{3_g}^{32} & K_{2_g}^{31} & K_{2_g}^{32} & K_{2_g}^{33} + K_{3_g}^{33} & K_{2_g}^{34} + K_{3_g}^{34} \\ K_{3_g}^{41} & K_{3_g}^{42} & K_{2_g}^{41} & K_{2_g}^{42} & K_{2_g}^{43} + K_{3_g}^{43} & K_{2_g}^{44} + K_{3_g}^{44} \end{bmatrix} \quad (1.98)$$

If the bars have the same geometry and material properties, it is possible to write

$$\frac{E_1 I_1}{L_1} = \frac{E_2 I_2}{L_2} = \frac{E_3 I_3}{L_3} = \frac{EI}{L} \tag{1.99}$$

The system that has to be solved becomes

$$\frac{EI}{L}\begin{bmatrix} \frac{5}{4} & -\frac{\sqrt{3}}{4} & -\frac{1}{4} & \frac{\sqrt{3}}{4} & -1 & 0 \\ -\frac{\sqrt{3}}{4} & \frac{3}{4} & \frac{\sqrt{3}}{4} & -\frac{3}{4} & 0 & 0 \\ -\frac{1}{4} & \frac{\sqrt{3}}{4} & \frac{1}{2} & 0 & -\frac{1}{4} & -\frac{\sqrt{3}}{4} \\ \frac{\sqrt{3}}{4} & -\frac{3}{4} & 0 & \frac{3}{2} & -\frac{\sqrt{3}}{4} & -\frac{3}{4} \\ -1 & 0 & -\frac{1}{4} & -\frac{\sqrt{3}}{4} & \frac{5}{4} & \frac{\sqrt{3}}{4} \\ 0 & 0 & -\frac{\sqrt{3}}{4} & -\frac{3}{4} & \frac{\sqrt{3}}{4} & \frac{3}{4} \end{bmatrix} \begin{Bmatrix} u_{x_1} \\ u_{y_1} \\ u_{x_2} \\ u_{y_2} \\ u_{x_3} \\ u_{y_4} \end{Bmatrix} = \begin{Bmatrix} P_{x_1} \\ P_{y_1} \\ P_{x_2} \\ P_{y_2} \\ P_{x_3} \\ P_{y_3} \end{Bmatrix} \tag{1.100}$$

1.8.4 Application of Boundary Conditions and the Numerical Solution

The boundary conditions and loads have to be imposed before the solution of the system is computed. The displacements at a number of points, N_u, are obtained from the boundary conditions. In the present problem, node 1 has both DOFs constrained, while only horizontal translation is allowed in node 3, thus N_u is equal to three. Because of these boundary conditions, it is possible to write

$$u_{x_1} = 0, \quad u_{y_1} = 0, \quad u_{y_3} = 0 \tag{1.101}$$

The displacement vector can be split into two contributions, the free displacement, $\bar{\boldsymbol{U}}_g$, and the constrained displacement, $\tilde{\boldsymbol{U}}_g$, see Equation (1.80), where

$$\bar{\boldsymbol{U}}_g^T = (u_{x_2}, u_{y_2}, u_{x_3}), \quad \tilde{\boldsymbol{U}}_g^T = (u_{x_1}, u_{y_1}, u_{y_3}) \tag{1.102}$$

Only one load is applied to the structure. It is placed on node 2 and has the u_{y_2}-direction or u_{y_2} direction. No forces are applied to the constrained nodes, but the boundary condition creates reaction forces. According to Equation (1.81), the force vector can be expressed as two contributions,

$$\bar{\boldsymbol{P}}_g^T = (0, -P, 0), \quad \tilde{\boldsymbol{P}}_g^T = (r_{x_1}, r_{y_1}, r_{y_3}) \tag{1.103}$$

where $\bar{\boldsymbol{P}}_g$ is the load vector of the free nodes and $\tilde{\boldsymbol{P}}_g$ is the vector of the reaction forces in the constrained nodes. The system can be written as

$$\begin{bmatrix} \bar{\boldsymbol{K}}_g & \bar{\tilde{\boldsymbol{K}}}_g \\ \tilde{\bar{\boldsymbol{K}}}_g & \tilde{\boldsymbol{K}}_g \end{bmatrix} \begin{Bmatrix} \bar{\boldsymbol{U}}_g \\ \tilde{\boldsymbol{U}}_g \end{Bmatrix} = \begin{Bmatrix} \bar{\boldsymbol{P}}_g \\ \tilde{\boldsymbol{P}}_g \end{Bmatrix} \tag{1.104}$$

The displacements are computed by solving the following system of equations:

$$\bar{\boldsymbol{U}}_g = \bar{\boldsymbol{K}}_g^{-1}\bar{\boldsymbol{P}}_g \tag{1.105}$$

and, for the present model, this system becomes

$$\begin{bmatrix} u_{x_2} \\ u_{y_2} \\ u_{x_3} \end{bmatrix} = \frac{PL}{EI}\begin{bmatrix} \frac{1}{2} & 0 & -\frac{1}{4} \\ 0 & \frac{3}{2} & -\frac{\sqrt{3}}{4} \\ -\frac{1}{4} & -\frac{\sqrt{3}}{4} & \frac{5}{4} \end{bmatrix}^{-1} \begin{Bmatrix} 0 \\ -1 \\ 0 \end{Bmatrix} \tag{1.106}$$

The solution leads to the following displacements:

$$u_{x_2} = -0.1443\,4\frac{PL}{EI}, \quad u_{y_2} = -0.7500\,0\frac{PL}{EI}, \quad u_{x_3} = -0.2886\,8\frac{PL}{EI} \tag{1.107}$$

Figure 1.20 shows the deformed configuration of the structure. The reaction forces in the constrained nodes can be evaluated by solving the following system:

$$\tilde{\boldsymbol{P}}_g = \tilde{\bar{\boldsymbol{K}}}_g\bar{\boldsymbol{U}}_g \tag{1.108}$$

In this case, Equation (1.108) becomes

$$\tilde{\boldsymbol{P}}_g = \frac{EI}{L}\begin{bmatrix} -\frac{1}{4} & \frac{\sqrt{3}}{4} & -1 \\ \frac{\sqrt{3}}{4} & -\frac{3}{4} & 0 \\ -\frac{\sqrt{3}}{4} & -\frac{3}{4} & \frac{\sqrt{3}}{4} \end{bmatrix} \frac{PL}{EI}\begin{bmatrix} -0.1443\,4 \\ -0.7500\,0 \\ -0.2886\,8 \end{bmatrix} \tag{1.109}$$

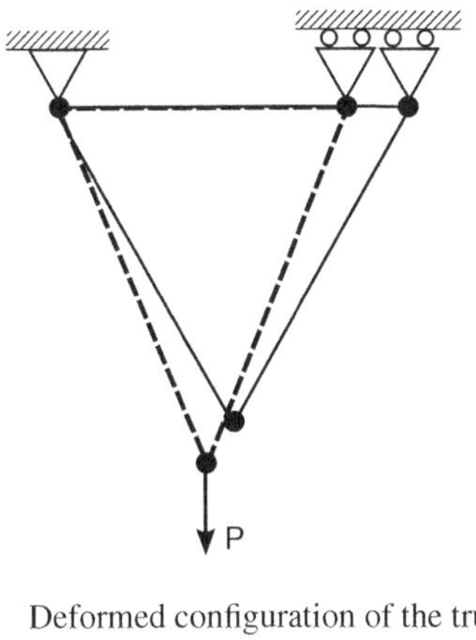

Figure 1.20 Deformed configuration of the truss structure

The solution leads to the following reaction forces:

$$u_{x_1} = 0, \quad u_{y_1} = 0.5P, \quad u_{y_3} = 0.5P \tag{1.110}$$

Axial loading on the bars can be computed in many ways, but the calculation has been omitted here.

1.9 Outline of the Book Contents

This book aims to introduce the use of the CUF in structural analysis. The subsequent chapters introduce the basis of structural engineering. The use of the CUF will be discussed in detail in the latter part of the book. Some appendices are reported to introduce some further details. The book is organized as follows:

CHAPTER 2- **Fundamental equations of three-dimensional elasticity**. The fundamental equations of continuous, deformable bodies are introduced briefly in this chapter. Particular attention is paid to those aspects that are of interest for this book. The geometrical and constitutive equations of continuum structural mechanics are given for the linear case.

CHAPTER 3- **From 3D problems to 2D and 1D problems: theories for beams, plates and shells**. The most common techniques that can be exploited to develop structural models are presented. The axiomatic method for 1D and 2D models is described. Examples of classical and advanced theories are provided. Further, the asymptotic method is briefly introduced.

CHAPTER 4- **Typical FE governing equations and procedures**. The governing equations of the main structural problem are introduced. The solutions of static, modal and dynamic problems are presented.

CHAPTER 5- **Introduction to the Unified Formulation**. A general method that can be used to derive refined FEM models is introduced. The main features of the CUF are presented step by step. The new formulation is derived from the classical FEM and extended to advanced models.

CHAPTER 6- **The displacement approach via the Principle of Virtual Displacements**. The strong form of the equilibrium equations is derived for a 3D problem. The strong form of the 'fundamental nucleus' is used as a unified formulation to derive the weak form of 1D, 2D and 3D models.

CHAPTER 7- **3D FEM formulation (solid elements)**. The formulation of the solid element is introduced in both matrix (classical) and index formulations. The FE matrices are derived in terms of fundamental nuclei.

CHAPTER 8- **1D models with N-order displacement field, the Taylor Expansion class (TE)**. The 1D (beam) models, based on Taylor-like expansions of the displacement field, are presented in this chapter. Classical models (Euler–Bernoulli and Timoshenko) are obtained as particular cases of the general *N*th-order model. The enhanced capabilities of 1D Taylor expansion (TE) models are shown via numerical examples. Static, modal and dynamic response problems are considered.

CHAPTER 9- **1D models with a physical volume/surface-based geometry and pure displacement variables, the Lagrange Expansion class (LE)**. The 1D (beam) models, based on Lagrange-like expansions of the displacement field, are presented in this chapter. The physics-based geometrical capabilities of 1D Lagrange expansion (LE) models are described and the component-wise approach is introduced. Numerical examples of static and modal analyses are provided.

CHAPTER 10- **2D plate models with N-order displacement field, the Taylor expansion class**. Here, 2D flat elements, based on Taylor expansions of the displacement variables, are discussed. Classical models (Kirchhoff and Reissner–Mindlin) are described briefly together with a more general complete linear expansion case. Higher-order models are presented and the unified formulation is introduced. The PVD is employed to derive governing equations. The shear locking phenomenon and its correction are discussed and numerical results are provided.

CHAPTER 11- **2D shell models with N-order displacement field, the Taylor expansion class**. Shell finite elements, based on Taylor expansions of the displacement variables, are presented here. The exact geometrical description of cylindrical shells is provided in order to obtain strain–displacement relations. Classical shell models and higher-order models are described briefly as extensions of the plate models. The governing equations are derived from the PVD for the FE analysis of shell structures. Membrane and shear locking phenomena and their corrections are discussed and numerical results are given.

CHAPTER 12- **2D models with physical volume/surface-based geometry and pure displacement variables, the Lagrange Expansion class (LE)**. In this chapter, 2D Lagrange expansion models – referred to as LE models – are described. It highlights how the variables and boundary conditions in LE models can be located above the physical surface of the structure, and the unknown variables of the problem are the pure displacement components. Numerical examples are provided in order to highlight the enhanced capabilities of 2D CUF LE models, in terms of 3D-like accuracies and very low computational costs.

CHAPTER 13- **Discussion on possible best beam, plate and shell diagrams**. The mixed axiomatic–asymptotic approach (MAAA) is described in this chapter. The MAAA is exploited to investigate the effectiveness of each unknown variable in 1D and 2D models. Reduced models that have the same accuracy as full expansion models, but fewer unknown variables, are obtained. Furthermore, the best theory diagram (BTD) is introduced as a tool to evaluate the accuracy of a model against the number of variables.

CHAPTER 14- **Mixing variable kinematic models**. This chapter introduces different approaches to combine variable kinematic models. Theoretical formulations and numerical results are given for each of them.

CHAPTER 15- **Extension to multilayered structures**. The capabilities of the CUF in the analysis of multilayered structures are shown. The CUF was in fact conceived by the first author to provide a powerful tool for the study of advanced new materials. For the sake of brevity, only 2D multilayered theories are discussed. However, the CUF can also be used for multilayered beam structures.

CHAPTER 16- **Extension to multifield problems**. Classical and mixed variational statements are discussed for the analysis of layered structures under the effect of four different fields (mechanical, thermal, electric and magnetic). Constitutive equations, in terms of coupled mechanical–thermal–electric–magnetic field variables, are obtained on the basis of a thermodynamics approach. The PVD and Reissner mixed variational theorem (RMVT) are employed. The analysis of multilayered plates is addressed to point out the effectiveness of the proposed approach.

APPENDIX A- **Numerical Integration**. A brief overview of the main techniques for numerical integration is given.

Appendix B- **CUF FE Models: Programming and Implementation Guidelines**. A comprehensive guide to the typical programming and implementation issues that are related to the CUF FEM is provided in this appendix. Numerical examples and data are given as benchmarks to assist the reader's implementation.

References

Bathe K 1996 *Finite Element Procedure*. Prentice Hall.

Hughes TJR 2000 *The Finite Element Method: Linear Static and Dynamic Finite Element Analysis*. Dover.

Oñate E 2009 *Structural Analysis with the Finite Element Method: Linear Statics*. Volume 1. *Basis and Solids*. Springer.

Reddy JN 2005 *An Introduction to the Finite Element Method*. Third Edition. McGraw-Hill Education.

Zienkiewicz OC, Taylor RL and Zhu JZ 2005 *The Finite Element Method: Its Basis and Fundamentals*. Sixth Edition. Elsevier.

2

Fundamental Equations of 3D Elasticity

The fundamental equations of continuous, deformable bodies are introduced briefly in this chapter. Particular attention is paid to those aspects that are of interest in this book. The geometrical and constitutive equations of continuum structural mechanics are given for the linear case. The main symbols and the reference systems that are used throughout the book are also introduced. For instance, x, y, z indicate the orthogonal Cartesian reference axes, P indicates point forces, q indicates line loads, p indicates surface loads, g indicates volume loads and S is the boundary of volume V.

2.1 Equilibrium Conditions

Let us consider a continuous body D (see Figure 2.1) of volume V and boundary S. The nine stress components[1]

$$\sigma_{xx}, \tau_{xy}, \tau_{xz}, \tau_{yx}, \sigma_{yy}, \tau_{yz}, \tau_{zx}, \tau_{zy}, \sigma_{zz}$$

at the generic point Q of volume V must fulfil the following differential indefinite dynamic equilibrium conditions along the three directions of an orthogonal Cartesian reference system x, y, z:

$$\begin{aligned} \frac{\partial \sigma_{xx}}{\partial x} + \frac{\partial \tau_{yx}}{\partial y} + \frac{\partial \tau_{zx}}{\partial z} &= g_x \\ \frac{\partial \tau_{xy}}{\partial x} + \frac{\partial \sigma_{yy}}{\partial y} + \frac{\partial \tau_{zy}}{\partial z} &= g_y \\ \frac{\partial \tau_{xz}}{\partial x} + \frac{\partial \tau_{yz}}{\partial y} + \frac{\partial \sigma_{zz}}{\partial z} &= g_z \end{aligned} \tag{2.1}$$

[1] In this book, shear stresses refer to both σ and τ, e.g. $\sigma_{xy} = \tau_{xy}$.

Finite Element Analysis of Structures Through Unified Formulation, First Edition.
Erasmo Carrera, Maria Cinefra, Marco Petrolo and Enrico Zappino.

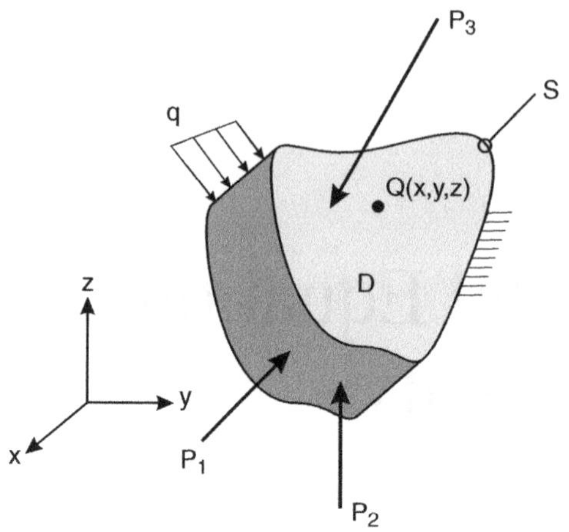

Figure 2.1 A continuous body and its boundary conditions

where g_x, g_y and g_z indicate the body forces per unit volume (e.g. inertial forces or weight). The equilibrium conditions related to rotations along the axes lead to the symmetry conditions or to the Cauchy theorem,

$$\sigma_{xz} = \sigma_{zx}, \quad \sigma_{yz} = \sigma_{zy}, \quad \sigma_{xy} = \sigma_{yx}$$

The equilibrium equations can be rewritten in vectorial form,

$$\boldsymbol{b}^T \boldsymbol{\sigma} = \boldsymbol{g} \tag{2.2}$$

where

$$\boldsymbol{\sigma}^T = \{\sigma_{xx}, \sigma_{yy}, \sigma_{zz}, \sigma_{xz}, \sigma_{yz}, \sigma_{xy}\} \tag{2.3}$$

and

$$\boldsymbol{b} = \begin{bmatrix} \partial/\partial x & 0 & 0 \\ 0 & \partial/\partial y & 0 \\ 0 & 0 & \partial/\partial z \\ \partial/\partial z & 0 & \partial/\partial x \\ 0 & \partial/\partial z & \partial/\partial y \\ \partial/\partial y & \partial/\partial x & 0 \end{bmatrix} \tag{2.4}$$

The loading vector $\boldsymbol{g}$ on the right-hand side is

$$\boldsymbol{g}^T = \{g_x, g_y, g_z\} \tag{2.5}$$

Mechanical boundary conditions must be fulfilled on S_m (the portion of S where the mechanical conditions on the loadings are given) with normal $\boldsymbol{n} = (n_x, n_y, n_z)$,

$$\begin{cases} \sigma_{xx}\, n_x + \sigma_{yx}\, n_y + \sigma_{zx}\, n_z = p_x \\ \sigma_{xy}\, n_x + \sigma_{yy}\, n_y + \sigma_{zy}\, n_z = p_y \\ \sigma_{xz}\, n_x + \sigma_{yz}\, n_y + \sigma_{zz}\, n_z = p_z \end{cases} \tag{2.6}$$

where $\boldsymbol{p} = (p_x, p_y, p_z)$ is the applied loading vector per unit area on S_m.

2.2 Geometrical Relations

In the case of linear problems, i.e. when the deformed configuration does not differ to any great extent from the undeformed one, the engineering strain components[2]

$$\varepsilon_{xx}, \varepsilon_{yy}, \varepsilon_{zz}, \gamma_{xy}, \gamma_{zx}, \gamma_{zy}$$

are related to the displacement components u_x, u_y, u_z of the displacement vector $\boldsymbol{u}$ through the following differential equations:

$$\begin{aligned} \varepsilon_{xx} &= \frac{\partial u_x}{\partial x} = u_{x,x} \\ \varepsilon_{yy} &= \frac{\partial u_y}{\partial y} = u_{y,y} \\ \varepsilon_{zz} &= \frac{\partial u_z}{\partial z} = u_{z,z} \\ \gamma_{xy} &= \frac{\partial u_x}{\partial y} + \frac{\partial u_y}{\partial x} = u_{x,y} + u_{y,x} \\ \gamma_{zx} &= \frac{\partial u_x}{\partial z} + \frac{\partial u_z}{\partial x} = u_{x,z} + u_{z,y} \\ \gamma_{zy} &= \frac{\partial u_y}{\partial z} + \frac{\partial u_z}{\partial y} = u_{y,z} + u_{z,y} \end{aligned} \tag{2.7}$$

where a compact notation to indicate derivatives is introduced (e.g. $u_{x,x}$ indicates the derivative of u_x with respect to x). Strains can be given in vectorial form,

$$\boldsymbol{\varepsilon} = \boldsymbol{b}\boldsymbol{u} \tag{2.8}$$

[2] γ_{xy}, γ_{zx} and γ_{zy} are the engineering shear strain components. Only engineering components are adopted in this book and they are referred to as γ_{xy}, γ_{zx} and γ_{zy} or ε_{xy}, ε_{zx} and ε_{zy}.

where

$$\boldsymbol{u}^T = \{u_x, u_y, u_z\} \tag{2.9}$$

$$\boldsymbol{\varepsilon}^T = \{\varepsilon_{xx}, \varepsilon_{yy}, \varepsilon_{zz}, \gamma_{xz}, \gamma_{yz}, \gamma_{xy}\} \tag{2.10}$$

Matrix $\boldsymbol{b}$ coincides with that of the equilibrium conditions in Equation (2.2).

The boundary conditions must be fulfilled at the boundary S_g (the portion of S where the geometrical boundary conditions are given),

$$\boldsymbol{u} = \bar{\boldsymbol{u}} \tag{2.11}$$

2.3 Hooke's Law

The physical relationship between stress and strain components, in terms of stiffness coefficients, is

$$\boldsymbol{\sigma} = \boldsymbol{C}\boldsymbol{\varepsilon} \tag{2.12}$$

Or, in terms of compliances,

$$\boldsymbol{\varepsilon} = \boldsymbol{S}\boldsymbol{\sigma} \tag{2.13}$$

The stiffness coefficients for isotropic materials are (with the usual notation)

$$\boldsymbol{C} = \begin{bmatrix} C_{11} & C_{12} & C_{12} & 0 & 0 & 0 \\ C_{21} & C_{11} & C_{12} & 0 & 0 & 0 \\ C_{21} & C_{21} & C_{11} & 0 & 0 & 0 \\ 0 & 0 & 0 & C_{44} & 0 & 0 \\ 0 & 0 & 0 & 0 & C_{44} & 0 \\ 0 & 0 & 0 & 0 & 0 & C_{44} \end{bmatrix} \tag{2.14}$$

where

$$C_{11} = 2G + \lambda, \quad C_{12} = C_{21} = \lambda, \quad C_{44} = G \tag{2.15}$$

and

$$G = \frac{E}{2(1+\nu)}, \quad \lambda = \frac{\nu E}{(1+\nu)(1-2\nu)} \tag{2.16}$$

where E is Young's modulus, G is the shear modulus and ν is Poisson's ratio. λ and G are also known as Lamé coefficients. The compliances are

$$\boldsymbol{S} = \begin{bmatrix} S_{11} & S_{12} & S_{12} & 0 & 0 & 0 \\ S_{21} & S_{11} & S_{12} & 0 & 0 & 0 \\ S_{21} & S_{21} & S_{11} & 0 & 0 & 0 \\ 0 & 0 & 0 & S_{44} & 0 & 0 \\ 0 & 0 & 0 & 0 & S_{44} & 0 \\ 0 & 0 & 0 & 0 & 0 & S_{44} \end{bmatrix} \tag{2.17}$$

where:

$$S_{11} = \frac{1}{E}, \quad S_{12} = S_{21} = -\frac{\upsilon}{E}, \quad S_{44} = \frac{1}{G} \tag{2.18}$$

2.4 Displacement Formulation

The governing equilibrium equations can be written in terms of displacements (u_x, u_y, u_z) by exploiting the previous equations. The stress–strain relations, written using Lamé coefficients, are

$$\begin{aligned} \sigma_{xx} &= (2G + \lambda)\, \varepsilon_{xx} + \lambda\, \varepsilon_{yy} + \lambda\, \varepsilon_{zz} \\ \sigma_{yy} &= \lambda\, \varepsilon_{xx} + (2G + \lambda)\, \varepsilon_{yy} + \lambda\, \varepsilon_{zz} \\ \sigma_{zz} &= \lambda\, \varepsilon_{xx} + \lambda\, \varepsilon_{yy} + (2G + \lambda)\, \varepsilon_{zz} \\ \sigma_{xz} &= G\, \gamma_{xz} \\ \sigma_{yz} &= G\, \gamma_{yz} \\ \sigma_{xy} &= G\, \gamma_{xy} \end{aligned} \tag{2.19}$$

And, by introducing the strain–displacement relations, the following equations hold:

$$\begin{aligned} \sigma_{xx} &= (2G + \lambda)\, u_{x,x} + \lambda\, u_{y,y} + \lambda\, u_{z,z} \\ \sigma_{yy} &= \lambda\, u_{x,x} + (2G + \lambda)\, u_{y,y} + \lambda\, u_{z,z} \\ \sigma_{zz} &= \lambda\, u_{x,x} + \lambda\, u_{y,y} + (2G + \lambda)\, u_{z,z} \\ \sigma_{xz} &= G\, (u_{x,z} + u_{z,x}) \\ \sigma_{yz} &= G\, (u_{y,z} + u_{z,y}) \\ \sigma_{xy} &= G\, (u_{x,x} + u_{x,y}) \end{aligned} \tag{2.20}$$

The previous relations can be substituted in the indefinite equilibrium equations,

$$\begin{aligned} (G + \lambda) I^{\star}_{1,x} + G \nabla^2\, u_x &= g_x \\ (G + \lambda) I^{\star}_{1,y} + G \nabla^2\, u_y &= g_y \\ (G + \lambda) I^{\star}_{1,z} + G \nabla^2\, u_z &= g_z \end{aligned} \tag{2.21}$$

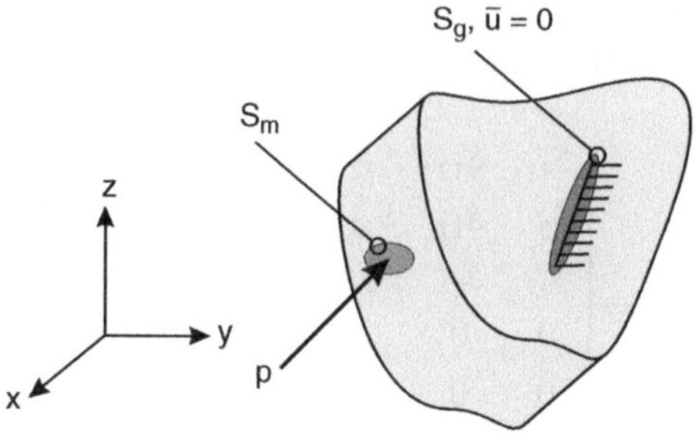

Figure 2.2 A continuous body with mechanical and geometrical boundary conditions

where $I_1^\star$ is the first-order invariant of the strain tensor,

$$I_1^\star = \varepsilon_{xx} + \varepsilon_{yy} + \varepsilon_{zz} \tag{2.22}$$

and ∇^2 is the the Laplacian or harmonic operator,

$$\nabla^2 = \frac{\partial^2}{\partial x^2} + \frac{\partial^2}{\partial y^2} + \frac{\partial^2}{\partial z^2} \tag{2.23}$$

Let us now consider a continuous body, as shown in Figure 2.2. The mechanical and geometrical boundary conditions must be defined to complete the set of equations related to the displacement formulation of the 3D problems. Let us suppose that they are defined on S_m and S_g, respectively.

S_m represents the portion of the boundary surface S at which forces are applied. If $\boldsymbol{n} = (n_x, n_y, n_z)$ is the vector normal to S_m and $\mathbf{p}$ is the surface force acting on an elementary surface dS, the following equilibrium boundary equations hold:

$$\begin{aligned} p_x &= \sigma_{xx}\, n_x + \sigma_{yx}\, n_y + \sigma_{zx}\, n_z \\ p_y &= \sigma_{xy}\, n_x + \sigma_{yy}\, n_y + \sigma_{zy}\, n_z \\ p_z &= \sigma_{xz}\, n_x + \sigma_{yz}\, n_y + \sigma_{zz}\, n_z \end{aligned} \tag{2.24}$$

By introducing Hooke's law and the strain–displacement relation, the following equations are obtained:

$$\begin{aligned} &\lambda I_1^\star\, n_x + G\,(u_{x,x}\, n_x + u_{x,y}\, n_y + u_{x,z}\, n_z) + G\,(u_{x,x}\, n_x + u_{y,x}\, n_y + u_{z,x}\, n_z) = p_x \\ &\lambda I_1^\star\, n_y + G\,(u_{y,x}\, n_x + u_{y,y}\, n_y + u_{y,z}\, n_z) + G\,(u_{x,y}\, n_x + u_{y,y}\, n_y + u_{z,y}\, n_z) = p_y \\ &\lambda I_1^\star\, n_z + G\,(u_{z,x}\, n_x + u_{z,y}\, n_y + u_{z,z}\, n_z) + G\,(u_{x,z}\, n_x + u_{y,z}\, n_y + u_{z,z}\, n_z) = p_z \end{aligned} \tag{2.25}$$

Conditions on the displacement variables (denoted by a bar) are instead imposed on S_g,

$$u_x(x,y,z) = \bar{u}_x(x,y,z), \quad u_y(x,y,z) = \bar{u}_y(x,y,z), \quad u_z(x,y,z) = \bar{u}_z(x,y,z) \tag{2.26}$$

These equations represent the *boundary-value problem* of the 3D elasticity problem whose solution leads to the calculation of the deformed configuration, i.e. to the calculation of the unknown displacements u_x, u_y and u_z.

The equations of elasticity and the related displacement approach can conveniently be formulated by means of the PVW or the PVD. Such a variational tool is in fact a powerful method which can deal with weak forms and the derivation of FE matrices, as will be shown in the following chapters of this book.

Further Reading

Washizu K 1968 *Variational Methods in Elasticity and Plasticity*. Oxford: Pergamon Press.

3

From 3D Problems to 2D and 1D Problems: Theories for Beams, Plates and Shells

The fundamental equations of continuum mechanics were introduced in the previous chapter. The exact analytical solution of these equations is generally available only for a few sets of geometries and boundary conditions. Approximated solutions of the general 3D problem are required in most cases. This has led, during the last two centuries, to the development of a significant number of structural theories that provide approximated solutions of the 3D problem. This chapter provides a short description of the two main methods adopted to derive structural theories, namely:

- *The asymptotic method*
- *The axiomatic method.*

These two approaches are usually exploited to reduce the 3D problem to a 2D or 1D problem. In a 3D problem, each variable $f(x, y, z)$ (displacements u, stresses σ and strains ε) is defined at each point $P(x, y, z)$ of volume V. In a 2D or 1D problem, each variable is defined by means of additional functions. These functions are defined above a surface Ω (2D problem) or along a line l (1D problem). In other words, in a 2D problem a generic variable (f) is defined as a function of two coordinates ($f = f(x, y)$), whereas in a 1D problem $f = f(y)$.[1]

The choice and the development of a 2D or 1D model are closely related to the structural component that has to be analysed. In practical applications, for instance, both 2D and 1D models are commonly adopted. In some cases, e.g. when local effects have to be accurately investigated, a direct solution of the 3D problem is often required.

[1] The choice of the coordinates (x, y, z) to be used depends on the body geometry, in this book we refer to $x - y$ for 2D problems and y for 1D cases.

Finite Element Analysis of Structures Through Unified Formulation, First Edition.
Erasmo Carrera, Maria Cinefra, Marco Petrolo and Enrico Zappino.

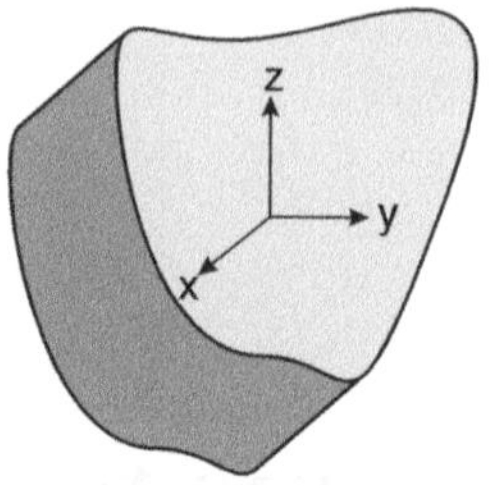

Figure 3.1 A 3D body

3.1 Typical Structures

The typical structures that are used in civil, automotive, mechanical, ship and aerospace constructions are, generally, composed of 3D, 2D or 1D structural elements.

3.1.1 Three-Dimensional Structures (Solids)

A structural component can be considered a 3D body if the characteristic lengths along three different directions have the same order of magnitude, see Figure 3.1.

3.1.2 Two-Dimensional Structures (Plates, Shells and Membranes)

Plates are 2D structural elements. They can be considered as special solids generated by segments of length h (see Figure 3.2), whose midpoints belong to a surface Ω. This mid-surface is composed of all the midpoints, and each h is perpendicular to Ω. h is the *thickness* of the plate and, in general, the thickness can vary from point to point. Ω is called the *mid-surface* or *reference surface* of the plate. The two surfaces generated by the top and bottom points of h are the top and bottom surfaces (or faces) of the plate. Γ is a closed line that represents the contour of Ω. The surface normal to Ω and passing through Γ is called the *edge* of the plate.

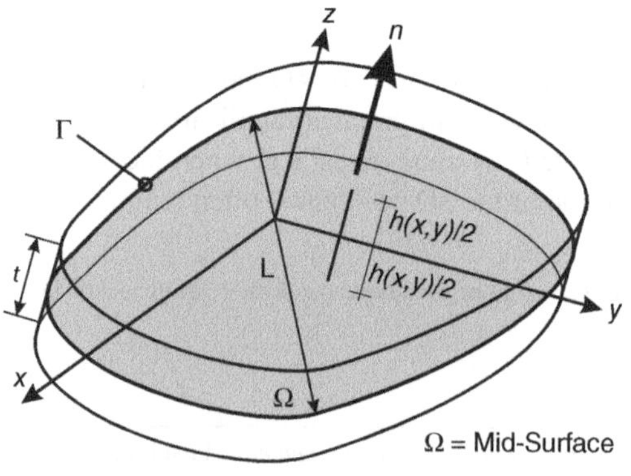

Figure 3.2 A plate element

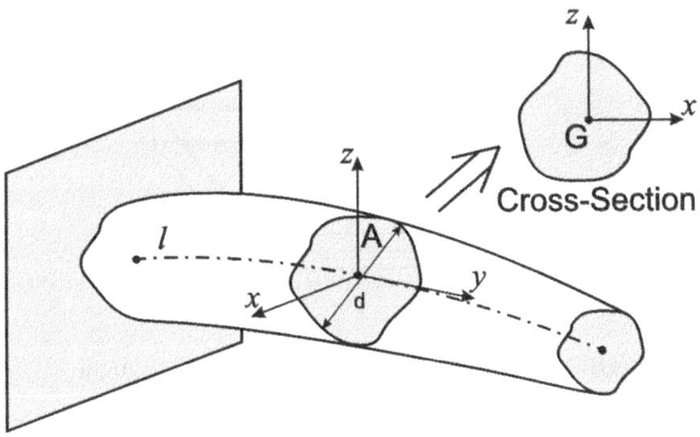

Figure 3.3 A beam element

Let us define t and L as the characteristic lengths of the plate, measured along the thickness and any direction above Ω, respectively. The two-dimensionality of the plate becomes more valid when L is much larger than t,

$$\frac{L}{t} \gg 1$$

Like plates, shells are 2D elements with a curved mid-surface. The nomenclature adopted for plates can also be used for shells.

Plates and shells are referred to as membranes if they are only loaded by forces that produce in-plane stresses which are constant along h.

3.1.3 One-Dimensional Structures (Beams and Bars)

Let us consider a section A whose centroid G moves along a line l where A is kept normal to l. The 1D solid generated in this manner is called *beam*, see Figure 3.3.

Let us define d as the characteristic length of A and let L be the length of l. The one-dimensionality of a solid body becomes more valid as

$$\frac{L}{d} \gg 1,$$

If a beam has no stiffness along the directions perpendicular to l, it is referred to as a *bar*.

3.2 Axiomatic Method

Axiomatic theories are developed on the basis of a number of hypotheses that *cannot* be mathematically proved. Despite this lack of mathematical strength, axiomatic structural theories have been exploited to design most of the structures built over the last few centuries.

In the development of an axiomatic theory, the starting point is the introduction of a number of *hypotheses* on the behaviour of the unknown functions. These hypotheses are introduced to reduce the mathematical complexity of the 3D differential equations mentioned in the previous

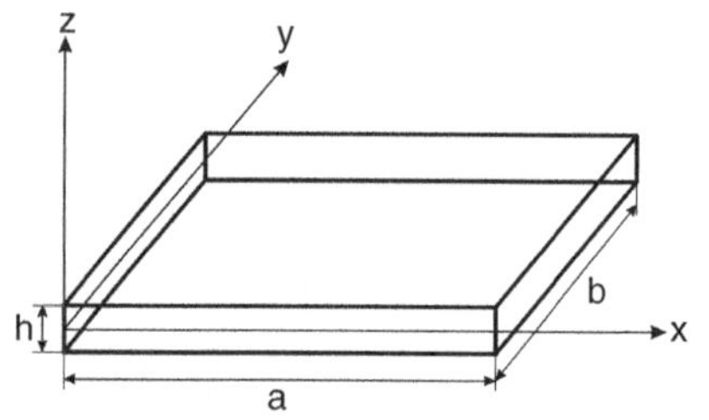

Figure 3.4 Reference frame for a plate element

chapters. In other words, these hypotheses allow us to find solutions for those equations and, in some cases, closed-form solutions can even be found.

Axiomatic theories have been introduced over the last few centuries by eminent scientists whose intuition allowed them to detect the most important aspects of a structural problem. The intuitions of these scientists represent the hypotheses of an axiomatic theory.

3.2.1 Two-Dimensional Case

Let us introduce an orthogonal Cartesian reference frame, as shown in Figure 3.4, where x and y lie on the mid-surface of the plate and z is defined along the thickness of the plate. Let $f(x, y, z)$ be an unknown variable of the 3D structural problem (e.g. a displacement u, stress σ or strain ϵ component). The 3D structural problem becomes a 2D problem when the unknown variables are defined above a reference surface (Ω). The 2D variables can be explicitly defined as $f_\Omega(x, y)$. In other words, it is necessary to define an explicit expression that relates variables (f) to the third coordinate (z). In the most general case, each variable f can require one or more f_Ω functions in order to reduce the 3D problem to a 2D one. The axiomatic approach can be seen as a tool that can be used to introduce an expansion of a 3D variable f along thickness z. The choice of the unknown variables that has to be expanded and the choice of the type of expansion are based on *intuition*. This intuition is axiomatic, i.e. it leads to a hypothesis and, in general, a relationship between the axiomatic expansion and the 3D solution cannot be found.

3.2.1.1 Membranal Behaviour

Let us consider the basic expansion case, i.e. an expansion in which the generic function f is expanded along z via a constant term,

$$f(x, y, z) = f_\Omega(x, y) \tag{3.1}$$

This means f is assumed independent of z, or, in other words, f is constant along the thickness of the plate (see Figure 3.5). If f represents a displacement component (along x or y, for instance), the previous expression leads to an axiomatic theory in which constant displacement distributions are assumed along the thickness of the plate. These kinds of theories (i.e. those with constant displacements along x and y) are referred to as *membranal*, since the mechanical behaviour of the plate is, in this case, equivalent to that of a membrane.

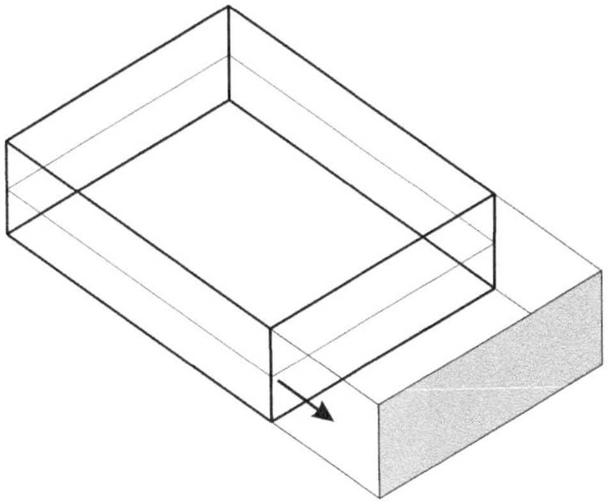

Figure 3.5 An axiomatic theory with only constant terms, the membranal case

3.2.1.2 Bending Behaviour

The membranal case can be extended to a more sophisticated plate model by adding a term that is linearly dependent on z,

$$f(x, y, z) = f_{\Omega}(x, y) + z\, f_{\Omega}^{*}(x, y) \tag{3.2}$$

The behaviour of a generic function f along z is now assumed to be given by the sum of a constant and a linear term (see Figure 3.6). Therefore, f at a generic point $P(x, y, z)$ is given by two functions (f_{Ω} and f_{Ω}^{*}) defined above the mid-surface of the plate. If f is a normal stress (e.g. σ_{xx}), the membranal (constant) term describes a constant stress distribution along the thickness of the plate, whereas the linear term introduces a variation of the stress distribution along the thickness. This implies the presence of compressed and stretched portions along the thickness, which are required to properly describe the *bending* behaviour of the plate.

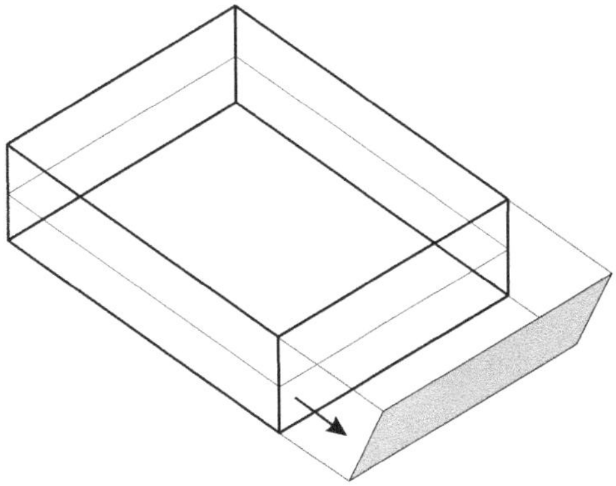

Figure 3.6 An axiomatic theory with constant and linear terms, the plate bending case

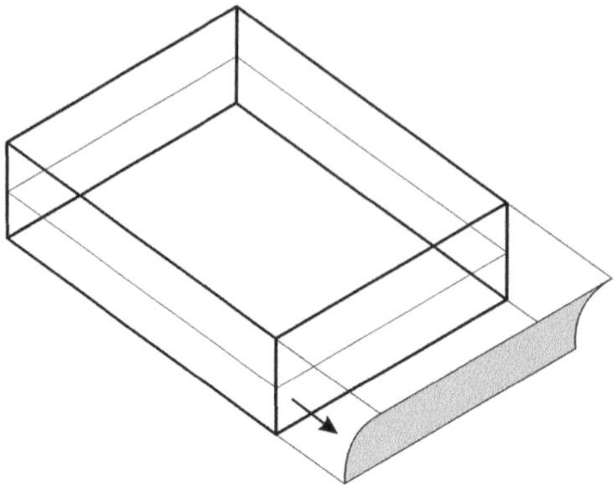

Figure 3.7 An axiomatic theory with constant, linear and higher-order terms

3.2.1.3 Higher-Order Behaviour

Further refinements can similarly be carried out in order to introduce more complicated expansions of f. For instance (see Figure 3.7),

$$f(x,y,z) = f_{\Omega}(x,y) + z\,f_{\Omega}^{*}(x,y) + z^{2}\,f_{\Omega}^{\star}(x,y) \tag{3.3}$$

where $f^{\star}$ is a new unknown variable. This expansion can also be considered as a series expansion of the unknown variables. However, it is important to emphasize that, from a practical point of view, the physical meaning of the 2D variables does not always coincide with the derivatives of f. This means that the new 2D variables, related to a certain expansion order, might only have a mathematical meaning. Most of the plate theories exploited in real applications are based on constant or linear distributions of the unknown variables (membranal and bending behaviour).

The procedure described above can easily be extended to shells. This extension does not imply any conceptual difficulties. However, some geometrical complications can arise due to the curvature of a shell.

3.2.1.4 The *N*th-order Case for 2D Theories

The displacement field refinement process can be extended to any expansion order. It is convenient to write the displacement expansion as

$$\boldsymbol{u} = F_{\tau}\,\boldsymbol{u}_{\tau}, \quad \tau = 1, \ldots, M \tag{3.4}$$

where F_{τ} is the expansion function, $\boldsymbol{u}_{\tau}$ is the vector of the unknown displacements and M is the number of expansion terms. According to the Einstein rule, repeated indexes denote summation. The following aspects are of fundamental importance:

- F_{τ} can be of any type. This means that one can assume polynomial expansions, Lagrange/Legendre polynomials, harmonics, exponentials, combinations of different expansion types, etc.

- M can be arbitrary. This means that the number of terms can be increased to any extent.

If, for instance, a Taylor-like polynomial expansion is adopted, Equation (3.4) can be rewritten as

$$\boldsymbol{u} = z^{\tau-1}\, \boldsymbol{u}_\tau, \quad \tau = 1, \ldots, M \tag{3.5}$$

A second-order model ($N = 2$, $M = 3$) is then described by the following displacement field:

$$\begin{aligned} u_x &= u_{x_1} + z\, u_{x_2} + z^2\, u_{x_3} \\ u_y &= u_{y_1} + z\, u_{y_2} + z^2\, u_{y_3} \\ u_z &= u_{z_1} + z\, u_{z_2} + z^2\, u_{z_3} \end{aligned} \tag{3.6}$$

In this case, the 2D model has nine unknown displacement variables.

3.2.2 One-Dimensional Case

The development of 1D structural models in an axiomatic framework requires a 2D expansion of the generic function f above the cross-section domain. Let y be the coordinate along the axis (l) of the beam.

3.2.2.1 Membranal Behaviour

The simplest expansion (constant term only) will therefore be

$$f(x, y, z) = f_s(y) \tag{3.7}$$

This expansion implies a constant distribution of f above the cross-section of the beam (see Figure 3.8). Such a distribution can vary from one section to another. If f is the axial stress σ_{yy}, a membranal stress state will be obtained. In this particular case, a beam is referred to as a *rod* or *bar*. The bending behaviour cannot be obtained in a bar and the axial tension is usually constant along the axis. In some cases, such as in stringers for aerospace applications, σ_{yy} is constant above the cross-section whereas it varies linearly along the axis.

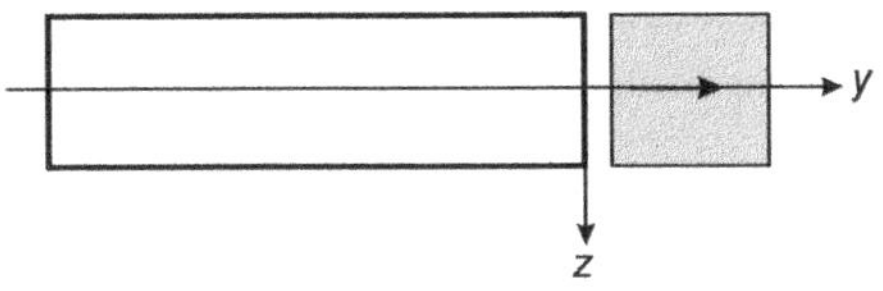

Figure 3.8 An axiomatic theory with constant terms, 1D case above the yz-plane

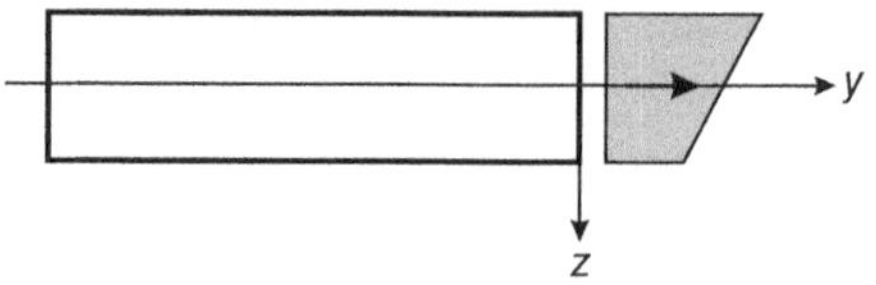

Figure 3.9 An axiomatic theory with constant and linear terms, 1D case above the yz-plane

3.2.2.2 Bending Behaviour

An improvement in the expansion can be obtained by adding a linear term (see Figure 3.9),

$$f(x, y, z) = f_s(y) + x f_s^a(y) \tag{3.8}$$

Or,

$$f(x, y, z) = f_s(y) + z f_s^b(y) \tag{3.9}$$

And, by combining the two cases,

$$f(x, y, z) = f_s(y) + x f_s^a(y) + z f_s^b(y) \tag{3.10}$$

These theories are usually adopted to include bending behaviour in a beam element.

3.2.2.3 Higher-Order Behaviour

Further improvements to the beam theory can be obtained through the addition of higher-order terms to the axiomatic expansion, for instance (see Figure 3.10)

$$f(x, y, z) = f_s(y) + x f^a(y) + z f_s^b(y) + x^2 f_s^c(y) + z^2 f_s^d(y) + xz f_s^e(y) \tag{3.11}$$

3.2.2.4 The *N*th-Order Case for 1D Theories

The displacement field refinement process can be extended to any expansion order even for the 1D case and the compact notation (see Equation (3.4)) does not change,

$$\boldsymbol{u} = F_\tau \boldsymbol{u}_\tau, \quad \tau = 1, \ldots, M \tag{3.12}$$

where, again, F_τ is the expansion function and it can be of any type (Taylor, Lagrange, harmonics, etc.), $\boldsymbol{u}_\tau$ is the vector of the unknown displacements and M is the number of

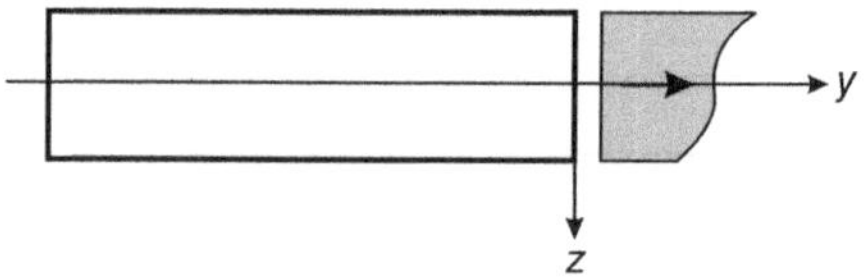

Figure 3.10 An axiomatic theory with higher-order terms, 1D case above the yz-plane

expansion terms and it can be arbitrary. A second-order model based on Taylor-like expansions ($N = 2, M = 6$) is described by the following displacement field:

$$\begin{aligned} u_x &= u_{x_1} + x\, u_{x_2} + z\, u_{x_3} + x^2\, u_{x_4} + xz\, u_{x_5} + z^2\, u_{x_6} \\ u_y &= u_{y_1} + x\, u_{y_2} + z\, u_{y_3} + x^2\, u_{y_4} + xz\, u_{y_5} + z^2\, u_{y_6} \\ u_z &= u_{z_1} + x\, u_{z_2} + z\, u_{z_3} + x^2\, u_{z_4} + xz\, u_{z_5} + z^2\, u_{z_6} \end{aligned} \tag{3.13}$$

In this case, the 1D model has 18 unknown displacement variables.

3.3 Asymptotic Method

As mentioned in the previous sections, axiomatic theories have been introduced by eminent scientists who were able to understand the most important mathematical terms that needed to be considered for a given structural problem. Further improvements are generally obtained by adding terms to the series expansions. An important drawback of axiomatic methods is the lack of information about the accuracy of the approximated theory with respect to the exact 3D solution. In other words, it is not usually possible to evaluate a priori the accuracy of an axiomatic theory. This lack of information has to be overcome by engineers who have to evaluate the validity of a theory on the basis of their knowledge and experience.

The asymptotic method can be seen as a step towards the development of approximated theories with known accuracy with respect to the 3D exact solution. Let us introduce two generic theories, A and B, and let B be a theory adopted to enhance A through additional terms in the expansion. An important issue is the effectiveness of the additional terms in B. In other words, do the additional terms improve the accuracy of A? In order to deal with this issue, let us assume an exact solution and its series expansion,

$$f_{exact} = f_0(x, y) + f_1(x, y)\, z + \frac{1}{2} f_2(x, y)\, z^2 + \frac{1}{6} f_3(x, y)\, z^3 + \cdots \tag{3.14}$$

Let us also assume that A contains all the terms that have the same effectiveness as $f_0(x, y)$ and $f_1(x, y)$ in the solution. It is necessary to evaluate whether B has all the terms with the same effectiveness as (for instance) $f_0(x, y)$, $f_1(x, y)$ and $f_2(x, y)$. During the last few decades, it has been shown that, in the development of axiomatic theories, some fundamental terms have erroneously been omitted from the series expansions. This means that many axiomatic theories are based on expansions that lack effective terms. The asymptotic method overcomes this drawback by introducing some controls on the order of magnitude of each term introduced into an expansion.

In order to build an asymptotic theory, a reference exact solution must first be introduced. This exact solution can usually be obtained by considering the limits of a function with respect to a characteristic feature of the problem (for instance, a characteristic length). This characteristic length is the *parameter* of the asymptotic method. For instance, in a 2D theory, this parameter could be the ratio between the thickness and the length of a plate,

$$\delta = \frac{t}{L} = \frac{\text{thickness}}{\text{reference length}} \tag{3.15}$$

When $\delta \to 0$, the 3D solid plate becomes a 2D surface and the 2D plate theory is exact.

A typical procedure that is followed to build an asymptotic theory is as follows:

1. An infinite expansion of an unknown function is introduced, for instance

$$f(x, y, z) = \sum_{1}^{\infty} f_i(x, y) z^i$$

2. These expansions are introduced into the governing equations of the problem and the thickness parameter is isolated.
3. The 3D equations are then written as a series expansion with respect to the thickness parameter δ (this step is usually extremely difficult).
4. All the terms in the equations that multiply δ by exponents that are lower than or equal to n are retrieved for a given value of the exponent.

The development of asymptotic theories is generally more difficult than the development of axiomatic ones. The main advantage of these theories is that they contain all the terms whose effectiveness is of the same order of magnitude. Moreover, these theories are exact as $\delta \to 0$.

Further Reading

Carrera E and Petrolo M 2010 Guidelines and recommendations to construct theories for metallic and composite plates. *AIAA Journal* **48**(12), 2852–2866.

Carrera E and Petrolo M 2011 On the effectiveness of higher-order terms in refined beam theories. *Journal of Applied Mechanics* **78**. DOI: 10.1115/1.4002207.

Cicala P 1965 *Systematic Approximation Approach to Linear Shell Theory*. Levrotto e Bella.

Gol'denweizer AL 1962 Derivation of an approximate theory of bending of a plate by the method of asymptotic integration of the equations of the theory of elasticity. *Prikladnaya Matematika i Mekhanika* **26**, 1000–1025.

4

Typical FE Governing Equations and Procedures

The analysis of structures has to face different problems. Different loading configurations, or boundary conditions, lead to different problems. The use of the PVD allows the problem to be described in terms of work. There are three main contributions to the work balance of a system in classical structural problems: internal work (we only refer to the elastic contribution), external work and inertial work.

The internal work originates from the deformation of the structure. The external work comes from the loads applied to the structure. Finally, the inertial work comes from the inertial forces that appear because the mass of the structure is subject to acceleration.

If only elastic and external forces are considered, the problem becomes a static response problem. When the system includes elastic and inertial forces, the problem coincides with free vibration analysis. If all three contributions are considered, i.e. inertial, elastic and external forces, a dynamic response analysis is necessary.

A short introduction to these problems is given in the following sections.

4.1 Static Response Analysis

A static response analysis includes the effects of elastic forces and external loads. The PVD in the static case states that

$$\delta L_{int} = \delta L_{ext} \tag{4.1}$$

The elastic work, or the internal work, can be expressed as

$$\delta L_{int} = \delta \boldsymbol{U}^T \boldsymbol{K} \boldsymbol{U} \tag{4.2}$$

Finite Element Analysis of Structures Through Unified Formulation, First Edition.
Erasmo Carrera, Maria Cinefra, Marco Petrolo and Enrico Zappino.

The external work can instead be written as

$$\delta L_{ext} = \delta \boldsymbol{U}^T \boldsymbol{P} \tag{4.3}$$

The PVD states

$$\delta \boldsymbol{U}^T \boldsymbol{K} \boldsymbol{U} = \delta \boldsymbol{U}^T \boldsymbol{P} \tag{4.4}$$

which can be reduced to the classic form

$$\boldsymbol{K}\boldsymbol{U} = \boldsymbol{P} \tag{4.5}$$

The solution of Equation (4.5) requires the calculation of vector $\boldsymbol{U}$. The solution is

$$\boldsymbol{U} = \boldsymbol{K}^{-1}\boldsymbol{P} \tag{4.6}$$

Matrix $\boldsymbol{K}$ can be inverted if and only if rigid body motions are removed. In practice, the matrices are never inverted but are factorized. One of the most popular methods is the Cholesky factorization method.

4.2 Free Vibration Analysis

A free vibration analysis investigates the equilibrium between elastic forces and inertial forces. The PVD in the dynamic case is written as

$$\delta L_{int} = -\delta L_{ine} \tag{4.7}$$

In order to introduce free vibration analysis, a description of the inertial forces is given briefly here. Details of this contribution will be given in the following chapters. The virtual variation of inertial work can be written as

$$\delta L_{ine} = \int_V \delta \boldsymbol{u} \rho \ddot{\boldsymbol{u}} dV \tag{4.8}$$

where ρ is the density of the material and $\ddot{u}$ is the acceleration. Introducing the FEM approximation, the variation of the inertial work assumes the form

$$\delta L_{ine} = \delta \boldsymbol{U}^T \boldsymbol{M} \ddot{\boldsymbol{U}} \tag{4.9}$$

It is possible to write the equilibrium equations

$$\delta L_{ine} + \delta L_{int} = 0 \tag{4.10}$$

$$\delta \boldsymbol{U}^T \boldsymbol{M} \ddot{\boldsymbol{U}} + \delta \boldsymbol{U}^T \boldsymbol{K} \boldsymbol{U} = 0 \tag{4.11}$$

$$\boldsymbol{M} \ddot{\boldsymbol{U}} + \boldsymbol{K} \boldsymbol{U} = 0 \tag{4.12}$$

The solution of Equation (4.12) gives the vector $\boldsymbol{U}$ that satisfies this equilibrium condition. The problem constitutes a homogeneous system and the solution must be calculated by solving an eigenvalue problem. The problem can be easily solved if the solution is considered to be harmonic. In this case, the displacement, the velocity and the acceleration become

$$\boldsymbol{U} = \bar{\boldsymbol{U}} e^{i\omega t} \tag{4.13}$$

$$\dot{\boldsymbol{U}} = i\omega \bar{\boldsymbol{U}} e^{i\omega t} \tag{4.14}$$

$$\ddot{\boldsymbol{U}} = -\omega^2 \bar{\boldsymbol{U}} e^{i\omega t} \tag{4.15}$$

where $\bar{\boldsymbol{U}}$ is the amplitude of the displacements and ω is the angular frequency. Equation (4.12) can be rewritten in the frequency domain, as

$$-\boldsymbol{M}\omega^2 \bar{\boldsymbol{U}} e^{i\omega t} + \boldsymbol{K}\bar{\boldsymbol{U}} e^{i\omega t} = 0 \tag{4.16}$$

This equation can be reduced to a standard eigenvalue problem,

$$\bar{\boldsymbol{U}} e^{i\omega t}(-\boldsymbol{M}\omega^2 + \boldsymbol{K}) = 0 \tag{4.17}$$

$$-\boldsymbol{M}\omega^2 + \boldsymbol{K} = 0 \tag{4.18}$$

The eigenvalue problem that has to be solved becomes

$$\boldsymbol{K}^{-1}\boldsymbol{M} - \frac{1}{\omega^2}\boldsymbol{I} = 0 \tag{4.19}$$

or

$$\omega^2 \boldsymbol{I} - \boldsymbol{M}^{-1}\boldsymbol{K} = 0 \tag{4.20}$$

The natural frequencies can be obtained from the eigenvalues,

$$f = \frac{\omega}{2\pi} \tag{4.21}$$

Each frequency gives an eigenvector that is the vector $\boldsymbol{U}$ which satisfies

$$\boldsymbol{K}^{-1}\boldsymbol{M}\boldsymbol{U} = \frac{1}{\omega^2}\boldsymbol{U} \tag{4.22}$$

4.3 Dynamic Response Analysis

If all inertial, elastic and external work contributions are considered, the problem that has to be solved can be written, through the PVD, in the form

$$\delta L_{ine} + \delta L_{int} = \delta L_{ext} \tag{4.23}$$

In FE form, the problem becomes

$$\delta \boldsymbol{U}^T \boldsymbol{M} \ddot{\boldsymbol{U}} + \delta \boldsymbol{U}^T \boldsymbol{K} \boldsymbol{U} = \delta \boldsymbol{U}^T \boldsymbol{P} \tag{4.24}$$

That is

$$\boldsymbol{M} \ddot{\boldsymbol{U}} + \boldsymbol{K} \boldsymbol{U} = \boldsymbol{P} \tag{4.25}$$

Equation (4.25) is written in the time domain. The solution of this equation in the time domain requires the use of a numerical technique. Three different numerical approaches can be used in a dynamic response analysis:

- the mode superposition method;
- the explicit direct integration method;
- the implicit direct integration method.

The first method uses the natural modes to approximate the structural behaviour and solve the uncoupled differential equations that correspond to the nodal DOFs. Direct integration techniques are widely used in FE analysis. These approaches solve the equations at a finite number of 'time points' and approximate the time variation of the solution. The equations are integrated step by step over time. Direct integration techniques can be explicit or implicit. The explicit method assumes the solution at time step $t + \Delta t$ as a function of the solution at previous time steps,

$$\boldsymbol{U}(t + \Delta t) = f(\boldsymbol{U}(t), \dot{\boldsymbol{U}}(t), \ddot{\boldsymbol{U}}(t), \boldsymbol{U}(t - \Delta t), \ldots) \tag{4.26}$$

The implicit method imposes the solution also to be a function of the velocities and the accelerations at the same time step,

$$\boldsymbol{U}(t + \Delta t) = f(\dot{\boldsymbol{U}}(t + \Delta t), \ddot{\boldsymbol{U}}(t + \Delta t), \boldsymbol{U}(t), \dot{\boldsymbol{U}}(t), \ddot{\boldsymbol{U}}(t), \boldsymbol{U}(t - \Delta t), \ldots) \tag{4.27}$$

Implicit methods make the computation more complex but increase the stability of the numerical approach and should therefore be preferred to explicit methods. Many of these approaches are used; in FE applications one of the most used is Newmark time integration. This is an implicit method that is widely used because it is intrinsically stable. The details and numerical formulations of these methods can be found in the books by Bathe (1996) and Volterra (1965). An application of Newmark time integration to FE models, in the CUF framework has been presented by Carrera and Varello (2012). The solution of the dynamic response problem is very time consuming, because the problem has to be solved for each time step. For this reason, reduced models are recommended to obtain solutions with lower computational effort.

References

Bathe K 1996 *Finite Element Procedure*. Prentice Hall.

Carrera E and Varello A 2012 Dynamic response of compact and thin-walled structures by variable kinematic one-dimensional models. *Journal of Sound and Vibration* **331**(24), 5268–5282.

Volterra E 1965 *Dynamics of Vibrations*. C.E. Merrill Books.

5

Introduction to the Unified Formulation

A new approach for the derivation of FE matrices is introduced in this chapter. This approach is based on the Carrera Unified Formulation which allows FE matrices/vectors to be derived in terms of fundamental nuclei. The CUF is introduced in this chapter by extending the index notation (indexes i and j), which is often used in FE procedures, to the theory of structures (indexes τ ans s). As a result, a fundamental nucleus (FN), expressed in terms of four indexes (τ, s, i and j), is obtained. This FN is a 3×3 array (3×1 in the case of a vector) and its form does not change for 1D, 2D or 3D problems, as anticipated in the Preface of this book (Table 1).

5.1 Stiffness Matrix of a Bar and the Related FN

In this section, the stiffness matrix of a bar is derived using a compact index approach which is quite common in FE procedures. In order to distinguish the displacements from their virtual variations, two different indexes or subscripts are introduced,

$$u_y(y) = N_i(y)\; u_{y_i} \tag{5.1}$$

$$\delta u_y(y) = N_j(y)\; \delta u_{y_j} \tag{5.2}$$

where i denotes the displacements and j the virtual variations. Repeated indexes denote a sum, therefore

$$u_y(y) = N_i(y)\; u_{y_i} = \sum_{i=1}^{N_{NE}} N_i(y) u_{y_i} = N_1 u_{y_1} + N_2 u_{y_2} + \ldots + N_{N_{NE}} u_{y_{N_{NE}}} \tag{5.3}$$

$$\delta u_y(y) = N_j(y)\; \delta u_j = \sum_{j=1}^{N_{NE}} N_i(y) \delta u_{y_j} = N_1 \delta u_{y_1} + N_2 \delta u_{y_2} + \ldots + N_{N_{NE}} \delta u_{y_{N_{NE}}} \tag{5.4}$$

Finite Element Analysis of Structures Through Unified Formulation, First Edition.
Erasmo Carrera, Maria Cinefra, Marco Petrolo and Enrico Zappino.

where N_{NE} is the number of nodes in the FE. The strains and the related virtual variations are

$$\varepsilon = N_{i,y}\ u_{y_i} \tag{5.5}$$

$$\delta\varepsilon = N_{j,y}\ \delta u_{y_j} \tag{5.6}$$

where $N_{i,y} = dN/dy$. The virtual variation of the internal work, according to the index notation, becomes

$$\begin{aligned}\delta L_{int} &= \int_V \delta\varepsilon^T\ \sigma\ dV \\ &= \int_V \delta\varepsilon^T\ E\ \varepsilon\ dV \\ &= \delta u_{y_j}\left(\int_V N_{j,y}\ \ E\ \ N_{i,y}\ dV\right) u_{y_i} \\ &= \delta u_{y_j}\ \ k^{ij}\ u_{y_i}\end{aligned} \tag{5.7}$$

where

$$k^{ij} = \int_V N_{j,y}\ \ E\ \ N_{i,y}\ dV \tag{5.8}$$

This is what we call the "fundamental nucleus" of a bar. It is invariant with respect to:

- the number of element nodes, N_{NE};
- the choice of the shape functions, N_i and N_j.

The explicit form of the nucleus, in the case of $N_{NE} = 2$ (linear shape functions), is

$$k^{11} = \int_V N_{1,y}\ \ E\ \ N_{1,y}\ dV = \frac{1}{L}\ E\ \frac{1}{L}\ \ AL\ \ = EA/L \tag{5.9}$$

$$k^{12} = \int_V N_{2,y}\ \ E\ \ N_{1,y}\ dV = -\frac{1}{L}\ E\ \frac{1}{L}\ \ AL\ \ = -EA/L \tag{5.10}$$

$$k^{21} = \int_V N_{1,y}\ \ E\ \ N_{2,y}\ dV = -\frac{1}{L}\ E\ \frac{1}{L}\ \ AL\ \ = -EA/L \tag{5.11}$$

$$k^{22} = \int_V N_{2,y}\ \ E\ \ N_{2,y}\ dV = \frac{1}{L}\ E\ \frac{1}{L}\ \ AL\ \ = EA/L \tag{5.12}$$

It is evident that the FN can easily be implemented in a computer program:

- two loops are made on i and j;
- the stiffness term for each i,j is calculated and assembled in the global matrix through appropriate identification of their position in the stiffness matrix.

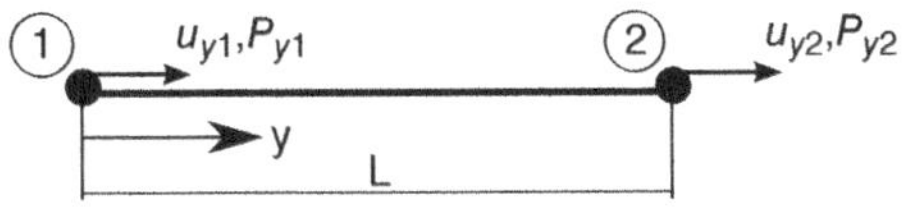

Figure 5.1 Bar element with two nodes in its local reference system

The stiffness matrix derived using the FN can be written as

$$\boldsymbol{K} = \begin{bmatrix} k^{11} & k^{12} \\ k^{21} & k^{22} \end{bmatrix} = \frac{EA}{L} \begin{bmatrix} 1 & -1 \\ -1 & 1 \end{bmatrix} \tag{5.13}$$

which coincides with Equation (1.69) derived using the classical approach. If the considered bar element has two nodes ($N_{NE} = 2$), as shown in Figure 5.1, the virtual variation of the internal virtual work can be written in explicit form, according to Equation (5.7),

$$\delta L_{int} = \delta u_{y_1} \; k^{11} \; u_{y_1} + \delta u_{y_2} \; k^{12} \; u_{y_1} + \delta u_{y_1} \; k^{21} \; u_{y_2} + \delta u_{y_2} \; k^{22} \; u_{y_2} \tag{5.14}$$

5.2 Case of a Bar Element with Internal Nodes

The index notation allows additional nodes to be introduced into the bar without any difficulty. Figures 5.2, 5.3 and 5.4 show different bar configurations. The first (Figure 5.2) has two nodes and the shape functions are linear. The second and third cases (Figures 5.3 and 5.4) have three and four nodes and the shape functions are quadratic and cubic, respectively. The stiffness matrix, for the linear case, was derived in Section 5.1. The stiffness matrix can be written for refined elements using the same approach. The displacement and its virtual variation can be written as

$$u_y(y) = N_i \; u_{y_i} \tag{5.15}$$

$$\delta u_y(y) = N_j \; \delta u_{y_j} \tag{5.16}$$

with $i,j = 1 \ldots N_{NE}$, where N_{NE} is the number of nodes of the bar element.

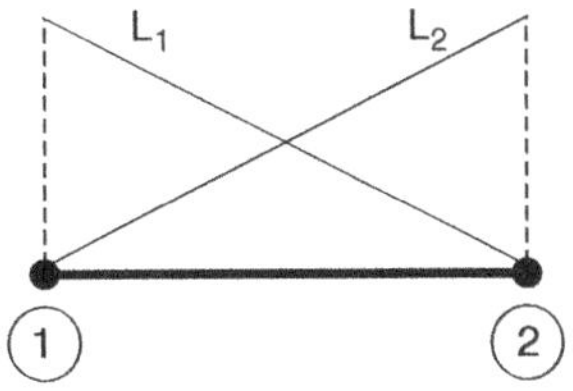

Figure 5.2 Example of two-node bar, $N_{NE} = 2$, linear shape functions

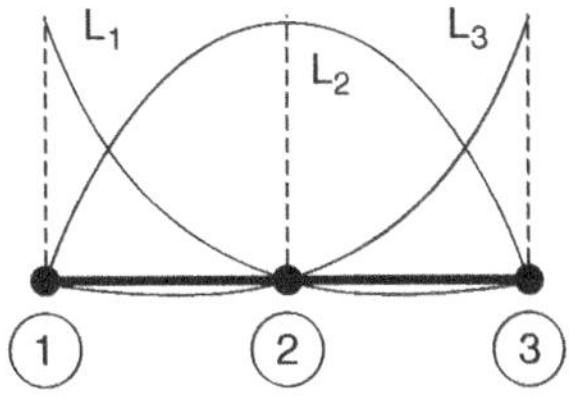

Figure 5.3 Example of three-node bar, $N_{NE} = 3$, parabolic shape functions

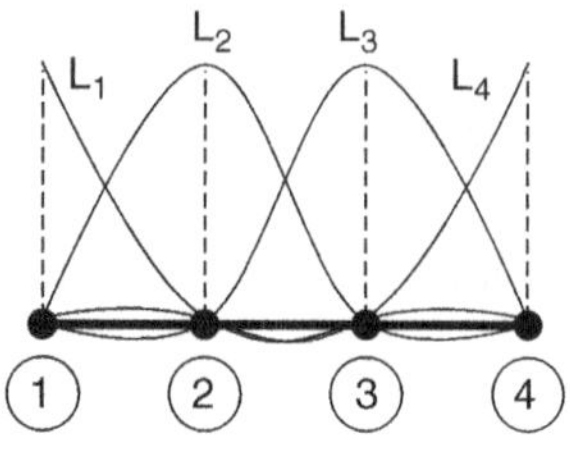

Figure 5.4 Example of four-node bar, $N_{NE} = 4$, cubic shape functions

5.2.1 The Case of Bar with Three Nodes

Let us consider the case in which $N_{NE} = 3$, the shape functions are quadratic and, in physical coordinates, their explicit expressions are

$$N_1(y) = 1 - 3\left(\frac{y}{L}\right) + 2\left(\frac{y}{L}\right)^2 \tag{5.17}$$

$$N_2(y) = 0 + 4\left(\frac{y}{L}\right) - 4\left(\frac{y}{L}\right)^2 \tag{5.18}$$

$$N_3(y) = 0 - 1\left(\frac{y}{L}\right) + 2\left(\frac{y}{L}\right)^2 \tag{5.19}$$

The index notation allows the stiffness matrix to be written independently of the number of nodes N_{NE},

$$k^{ij} = \int_V N_{j,y} \; E \; N_{i,y} \, dV \tag{5.20}$$

This is once again what we call the 'fundamental nucleus' of a bar as expressed in the notation of the CUF. Its form does not depend on N_i or N_j and it is exactly the same as the one computed in the $N_{NE} = 2$ case. Once again, this equation is invariant with respect to:

- the number of element nodes;
- the choice of the shape functions.

The derivatives of the shape functions are

$$N_1(y)_{,y} = -\left(\frac{3}{L}\right) + 4\left(\frac{y}{L^2}\right) \tag{5.21}$$

$$N_2(y)_{,y} = +\left(\frac{4}{L}\right) - 8\left(\frac{y}{L^2}\right) \tag{5.22}$$

$$N_3(y)_{,y} = -\left(\frac{1}{L}\right) + 4\left(\frac{y}{L^2}\right) \tag{5.23}$$

The components of the stiffness matrix can be calculated easily using an iterative procedure,

$$
\begin{aligned}
k^{11} &= EA \int_z N_{1,y}\, N_{1,y}\, dy = +\frac{7}{3}\frac{EA}{L} \\
k^{12} &= EA \int_z N_{2,y}\, N_{1,y}\, dy = -\frac{8}{3}\frac{EA}{L} \\
k^{13} &= EA \int_z N_{3,y}\, N_{1,y}\, dy = +\frac{1}{3}\frac{EA}{L} \\
k^{21} &= EA \int_z N_{1,y}\, N_{2,y}\, dy = -\frac{8}{3}\frac{EA}{L} \\
k^{22} &= EA \int_z N_{2,y}\, N_{2,y}\, dy = +\frac{16}{3}\frac{EA}{L} \\
k^{23} &= EA \int_z N_{3,y}\, N_{2,y}\, dy = -\frac{8}{3}\frac{EA}{L} \\
k^{31} &= EA \int_z N_{1,y}\, N_{3,y}\, dy = +\frac{1}{3}\frac{EA}{L} \\
k^{32} &= EA \int_z N_{2,y}\, N_{3,y}\, dy = -\frac{8}{3}\frac{EA}{L} \\
k^{33} &= EA \int_z N_{3,y}\, N_{3,y}\, dy = +\frac{7}{3}\frac{EA}{L}
\end{aligned}
\tag{5.24}
$$

Finally, the stiffness matrix is

$$
\boldsymbol{K} = \frac{EA}{3L}\begin{bmatrix} 7 & -8 & 1 \\ -8 & 16 & -8 \\ 1 & -8 & 7 \end{bmatrix} \tag{5.25}
$$

Example 5.2.1 *The $N_{NE} = 2$ case (two-node bar) does not allow a constant distribution of axial loading ($q = const$) to be applied. At least the $N_{NE} = 3$ case has to be adopted. In this example, a three-node bar loaded with a distributed load $q(y)$ is considered. Figure 5.5 shows*

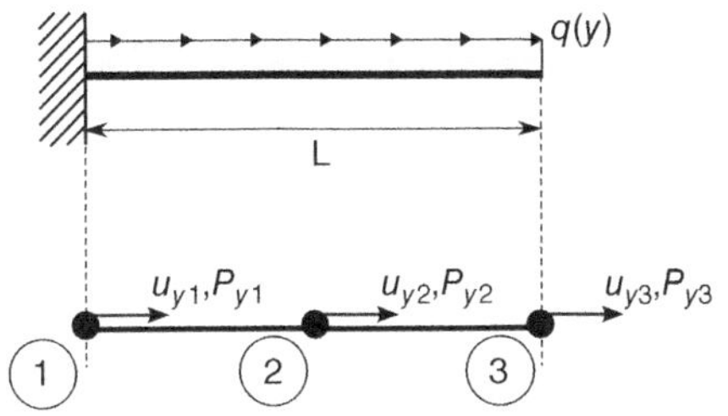

Figure 5.5 Example of axially loaded bar, physical and FEM model

the structure. Node 1 is clamped while load q(y) is constant and distributed along the whole axis of the bar. The external work done by load q(y) can be written as

$$\delta L_{ext} = \int_0^L \delta u_y q(y) dy = \bar{q} \int_0^L \delta u \ dy \tag{5.26}$$

Load q(y) is considered constant and it can therefore be removed from the integral. The index notation introduced by the CUF allows the external work to be written through the introduction of the shape functions,

$$\delta L_e = q \int_0^L N_j \delta u_j dy = \delta u_j \ q \int_0^L N_j dy \tag{5.27}$$

The integral of the shape functions is

$$\int_0^L N_1 dy = \frac{L}{6}, \quad \int_0^L N_2 dy = \frac{2L}{3}, \quad \int_0^L N_3 dy = \frac{L}{6} \tag{5.28}$$

therefore,

$$\delta L_e = q\frac{L}{6}\delta u_{y_1} + q\frac{2L}{3}u_{y_2} + q\frac{L}{6}u_{y_3} \tag{5.29}$$

The system that has to be solved becomes

$$\frac{EA}{3L}\begin{bmatrix} 7 & -8 & 1 \\ -8 & 16 & -8 \\ 1 & -8 & 7 \end{bmatrix} \begin{Bmatrix} u_{y_1} \\ u_{y_2} \\ u_{y_3} \end{Bmatrix} = \begin{Bmatrix} P_{y_1} \\ P_{y_2} \\ P_{y_3} \end{Bmatrix} \tag{5.30}$$

where

$$P_{y_1} = q\frac{L}{6}, \quad P_{y_2} = q\frac{2L}{3}, \quad P_{y_3} = q\frac{L}{6} \tag{5.31}$$

If node 1 is clamped and no other loads are applied, the boundary condition imposes the displacement equal to zero at node 1,

$$u_{y_1} = 0 \tag{5.32}$$

The system can be partitioned, as shown in the previous sections (see Equation (1.84)). The system that has to be solved becomes

$$\frac{EA}{3L}\left[\begin{array}{c|cc} 7 & -8 & 1 \\ \hline -8 & 16 & -8 \\ 1 & -8 & 7 \end{array}\right] \begin{Bmatrix} 0 \\ \hline u_{y_2} \\ u_{y_3} \end{Bmatrix} = \begin{Bmatrix} R_{y_1} \\ \hline q\,2L/3 \\ q\,L/6 \end{Bmatrix} \tag{5.33}$$

It is possible to compute the displacements u_{y_2} and u_{y_3} from the second and third equations,

$$\frac{EA}{3L}\begin{bmatrix} 16 & -8 \\ -8 & 7 \end{bmatrix}\begin{Bmatrix} u_{y_2} \\ u_{y_3} \end{Bmatrix} = Lq\begin{Bmatrix} 2/3 \\ 1/6 \end{Bmatrix} \tag{5.34}$$

The solution of the problem gives the displacements

$$u_{y_1} = \frac{3}{8}\frac{qL^2}{EA}, \quad u_{y_2} = \frac{1}{2}\frac{qL^2}{EA} \tag{5.35}$$

The solution of the first equation gives the reaction force at node 1, R_{y_1},

$$R_{y_1} = \frac{EA}{3L}[-8 \quad 1]\frac{qL^2}{EA}\begin{Bmatrix} 3/8 \\ 1/2 \end{Bmatrix} = -0.833\,33\, qL \tag{5.36}$$

The reaction force is not equal to $-qL$ (the applied load) because the load P_{y_1} acts on the constrained node and therefore is not taken into account in the solution. From these results, it is possible to recover the displacements along the whole axis,

$$u_y(y) = N_i u_{y_i} = \left(yL - \frac{y^2}{2}\right)\frac{q}{AE} \tag{5.37}$$

The axial strain is

$$\varepsilon_{yy}(y) = N_{i,y} u_{y_i} = (L - y)\frac{q}{AE} \tag{5.38}$$

Using Hooke's law, it is possible to derive the axial stress,

$$\sigma_{yy}(y) = E\varepsilon_y(y) = N_{i,y} u_{y_i} E = (L - y)\frac{q}{A} \tag{5.39}$$

Finally, the axial force becomes

$$N(y) = \int_A \sigma_{yy}(y) dA = \sigma_{yy}(y)\int_A 1 dA = (L - y)\, q \tag{5.40}$$

The results obtained in Equations (5.37)–(5.40) are shown in Figure 5.6. The displacement has quadratic behaviour, and, in the case of a distributed load, at least a three-node element

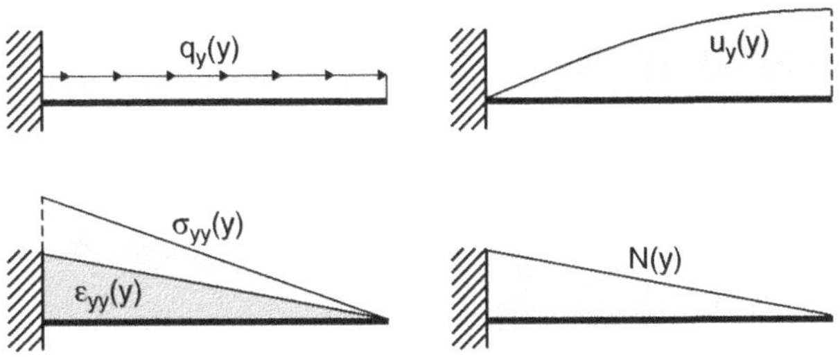

Figure 5.6 Qualitative solutions of a bar with a uniform axial load

is needed. The stresses, strains and resultant axial force are linear. They are equal to zero at $y = L$ and have the maximum value at $y = 0$, where the constraint was placed.

5.2.2 The Case of an Arbitrary Defined Number of Nodes

The results presented in Sections 5.1 and 5.2.1 show how to build stiffness matrices for a bar element with two or three nodes. The index formulation allows this approach to be extended to a bar with any number of nodes.

Because of the 1D nature of the bar problem, the FN has only one term,

$$k^{ij} = \int_V N_{j,y} \; E \; N_{i,y} \, dV \tag{5.41}$$

Indexes i and j represent the displacement and its virtual variation, respectively. There are no limitations on the choice of i and j, and the CUF therefore offers a tool that can be used to create the stiffness matrix of a bar element with any number of nodes[1]

$$\boldsymbol{K} = \begin{bmatrix} k^{11} & k^{12} & \dots & k^{1i} & \dots & k^{1\,N_{NE}} \\ k^{21} & k^{22} & & & & \\ \vdots & & \ddots & & & \vdots \\ k^{j1} & & & k^{ij} & & \vdots \\ \vdots & & & & \ddots & \\ k^{N_{NE}1} & & \dots & \dots & & k^{N_{NE}N_{NE}} \end{bmatrix} \tag{5.42}$$

Whatever the choice of N_{NE}, the stiffness matrix can be derived using only the FN, as shown in Equation (5.42). The same approach can be used to define the load vector $\boldsymbol{P}$. The nucleus of the load vector is

$$P_j = \int_0^L q(y) \, N_j \, dy \tag{5.43}$$

The vector P_j can be used to build the load vector for a bar element with any number of nodes,

$$\boldsymbol{P} = \begin{bmatrix} P_{y_1} \\ P_{y_2} \\ \vdots \\ P_{y_j} \\ \vdots \\ P_{y_{N_{NE}}} \end{bmatrix} \tag{5.44}$$

[1] Index notations are already known in FE applications. The CUF extends them to the theory of structures as will be shown in the following chapters.

5.3 Combination of the FEM and the Theory of Structure Approximations: A Four-Index FN and the CUF

In the previous section, the axial displacement, u_y, was considered to be constant over the cross-section and the transversal displacement was neglected. To overcome this limitations, the axial displacement should be considered as a 3D field,

$$u_y(y) \longrightarrow u_y(x, y, z) \tag{5.45}$$

The same must be done, in a mandatory sense, for the other displacement components,

$$u_y(x, y, z) \longrightarrow \boldsymbol{u}^T = (u_x, u_y, u_z) \tag{5.46}$$

The FNs for these cases are derived in the following sections.

5.3.1 FN for a 1D Element with a Variable Axial Displacement over the Cross-section

If the cross-section of a bar is considered rigid, the problem can be considered a 1D one. In other words, the value of the displacement on the axis is enough to describe the deformation of the whole cross-section. In order to overcome this assumption, the axial displacement, u_y, should not be considered constant over the cross-section,

$$u_y(y) \longrightarrow u_y(x, y, z) \tag{5.47}$$

The displacement field that was originally defined in a 1D domain now becomes a 3D field, as shown in Figure 5.7. A bar element with a variable displacement field over the cross-section does not have any physical applications in structural analysis but it is introduced here as an intermediate step. The next section, in fact, introduces the most appropriate case of a refined beam model.

The bar element introduced in Section 5.1 can be used to approximate this displacement field, thus:

$$u_y(x, y, z) = N_i(y) u_{y_i}(x, z) \tag{5.48}$$

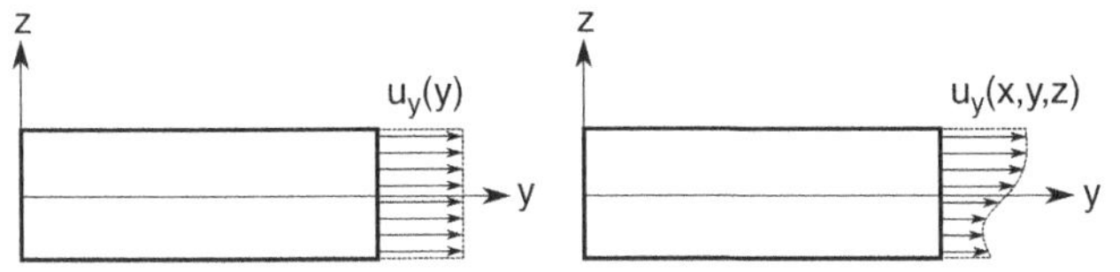

Figure 5.7 An example of a bar with a rigid cross-section, $u_y(y)$, on the left, and a bar with a variable axial displacement over the cross-section, $u_y\,(x, y, z)$, on the right

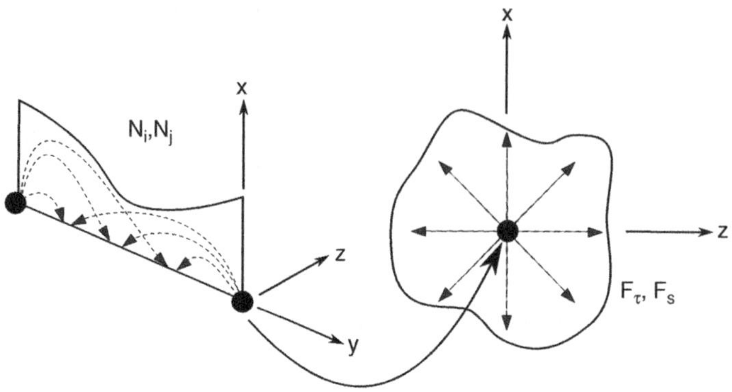

Figure 5.8 Representation of the axial approximation (using N_i, N_j), and the cross-section expansion (using F_τ, F_s)

The shape functions, $N_i(y)$, are used to approximate the displacements along the y-axis, while the u_{y_i} coefficients are now functions of the cross-section coordinates. The term $u_{y_i}(x,z)$ can be approximated by introducing a generic expansion for the cross-section. In general, u_{y_i} can be written as the sum of the generic functions $F_\tau(x,z)$,

$$\begin{aligned} u_{y_i}(x,z) &= F_\tau(x,z)u_{y_{\tau i}} \\ &= F_1(x,z)u_{y_{1i}} + F_2(x,z)u_{y_{2i}} + \ldots + F_\tau(x,z)u_{y_{\tau i}} + \ldots + F_m(x,z)u_{y_{Mi}} \end{aligned} \tag{5.49}$$

where M is the number of terms in the expansion. The complete displacement field can be obtained by inserting Equation (5.49) into Equation (5.48),

$$u_y(x,y,z) = N_i(y)\; u_{y_i}(x,z) = N_i(y)\; F_\tau(x,z)\; u_{y_{\tau i}} \tag{5.50}$$

It is possible to introduce the virtual variation using the s index instead of τ,

$$u_y(x,y,z) = N_i(y)\; F_\tau(x,z)\; u_{y_{\tau i}} \tag{5.51}$$

$$\delta u(x,y,z) = N_j(y)\; F_s(x,z)\; \delta u_{y_{sj}} \tag{5.52}$$

The introduction of the new expansion, indicated by the indexes τ and s, does not change the approach used to derive the FEM matrices introduced above.

Figure 5.8 shows the two approximations. The shape functions N_i and N_j expand the solution from the nodes to the axis. Expansions F_τ and F_s expand the solution from the nodes to the cross-section of the bar.

The stress and strain become

$$\boldsymbol{\varepsilon}(x,y,z) = \boldsymbol{b}\; N_i(y)\; F_\tau(x,z)\; u_{y_{\tau i}} \tag{5.53}$$

$$\boldsymbol{\sigma}(x,y,z) = \boldsymbol{C}\,\boldsymbol{b}\; N_i(y)\; F_\tau(x,z)\; u_{y_{\tau i}} \tag{5.54}$$

In this case, the stress and strain are vectors because the axial displacement is considered not to be constant over the cross-section. If u_y has a 3D formulation one has

$$\varepsilon_{xy} = \frac{\partial u_y}{\partial x} + \frac{\partial u_x}{\partial y} \neq 0, \quad \varepsilon_{zy} = \frac{\partial u_y}{\partial z} + \frac{\partial u_z}{\partial y} \neq 0 \tag{5.55}$$

therefore the stress and strain are expressed as vectors. These vectors are derived using matrix $\boldsymbol{b}$, which is introduced in Chapter 2. The virtual variation of the displacements is expressed using the j and s indexes,

$$\delta\varepsilon(x, y, z) = \boldsymbol{b}N_j(y)\ F_s(x, z)\ \delta u_{y_{sj}} \tag{5.56}$$

5.3.2 FN for a 1D Structure with a Complete Displacement Field: The Case of a Refined Beam Model

The formulation introduced in Section 5.3.1 considers a variable axial displacement on the cross-section of the bar. Engineering problems usually deal with 3D problems and the structural solution has the aim of studying the displacements in all three directions. If all the components of the displacement have to be considered, the displacement field becomes

$$u_y(x, y, z) \longrightarrow \boldsymbol{u}^T = (u_x, u_y, u_z) \tag{5.57}$$

If the approximations introduced into Equations (5.48) and (5.49) are applied to $\boldsymbol{u}$, the displacement field can be written in compact form as

$$\boldsymbol{u} = N_i(y)F_\tau(x, z)\boldsymbol{u}_{\tau i} \tag{5.58}$$

or

$$u_x(x, y, z) = N_i(y)\ F_\tau(x, z)\ u_{x_{\tau i}} \tag{5.59}$$

$$u_y(x, y, z) = N_i(y)\ F_\tau(x, z)\ u_{y_{\tau i}} \tag{5.60}$$

$$u_z(x, y, z) = N_i(y)\ F_\tau(x, z)\ u_{z_{\tau i}} \tag{5.61}$$

The introduction of the complete displacement field[2] does not require any change in the approach to build the FN or the stiffness matrix. The virtual variation can be introduced as

$$\delta\boldsymbol{u} = N_j(y)\ F_s(x, z)\ \delta\boldsymbol{u}_{sj} \tag{5.62}$$

[2] Functions N_i and F_τ can be different for the various displacement components, but for the sake of simplicity, they have here been assumed to be the same.

The stress and strain can be derived using the differential operator $\boldsymbol{b}$ and the material coefficient matrix $\boldsymbol{C}$,

$$\boldsymbol{\varepsilon}(x, y, z) = \boldsymbol{b} N_i(y)\, F_\tau(x, z)\, \boldsymbol{u}_{\tau i} \tag{5.63}$$

$$\boldsymbol{\sigma}(x, y, z) = \boldsymbol{C}\boldsymbol{b} N_i(y)\, F_\tau(x, z)\, \boldsymbol{u}_{\tau i} \tag{5.64}$$

The virtual variation becomes

$$\delta\boldsymbol{\varepsilon}(x, y, z) = \boldsymbol{b} N_j(y)\, F_s(x, z)\, \delta\boldsymbol{u}_{sj} \tag{5.65}$$

In this case, $\boldsymbol{\varepsilon}$ and $\boldsymbol{\sigma}$ are vectors that contain all the stress and strain components, $\boldsymbol{C}$ is a 6×6 matrix with all the material coefficients, $\boldsymbol{b}$ is a differential operator matrix of size 6×3 which contains the geometrical relation between the displacements and strains. All these quantities were introduced in the previous chapter of the book.

The virtual variation of the internal work can be derived in the same way as for the previous case,

$$\delta L_{int} = \int_V \delta\boldsymbol{\varepsilon}^T \boldsymbol{\sigma} dV = \delta\boldsymbol{u}_{sj} \boldsymbol{k}^{\tau sij} \boldsymbol{u}_{\tau s} \tag{5.66}$$

$$= \delta\boldsymbol{u}_{sj}^T \underbrace{\int_V \left[F_s(x, z) N_j(y) \underbrace{\overbrace{[3 \times 6]}^{\boldsymbol{b}^T} \overbrace{\begin{bmatrix} 6 \times 6 \end{bmatrix}}^{\boldsymbol{C}} \overbrace{\begin{bmatrix} 6 \times 3 \end{bmatrix}}^{\boldsymbol{b}}}_{\begin{bmatrix} 3 \times 3 \end{bmatrix}} N_i(y) F_\tau(x, z) \right] dV}_{FN\ \boldsymbol{k}^{\tau sij}} \boldsymbol{u}_{\tau i} \tag{5.67}$$

If all the components of the displacement are considered, the nucleus becomes a 3×3 matrix:

$$\boldsymbol{k}^{\tau sij} = \int_V F_s(x, z) N_j(y) \boldsymbol{b}^T \boldsymbol{C} \boldsymbol{b} N_i(y)\, F_\tau(x, z) dV \tag{5.68}$$

5.4 CUF Assembly Technique

The use of the CUF makes the assembly of the matrices a trivial operation that can be easily implemented in computer code. The assembly of the matrix consists of four loops on indexes i, j, τ and s, and an FN is calculated for each combination of these indexes. A representation of this procedure is shown in Figure 5.9. The diagram shows how it is possible to build a matrix of the node, of the element and, finally, of the global stiffness matrix by exploiting the nucleus.

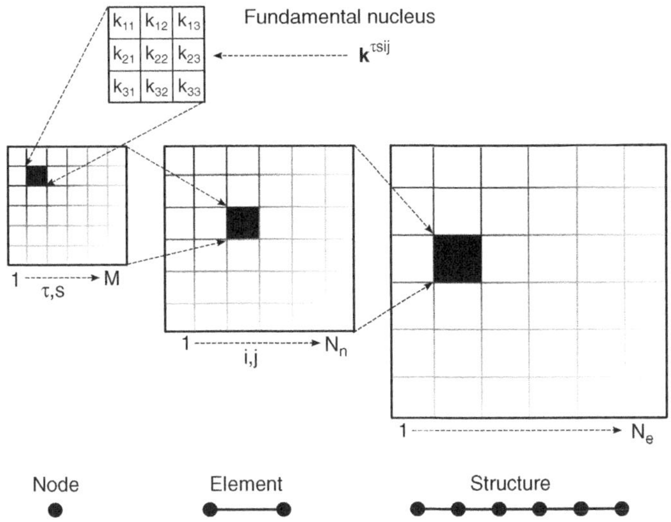

Figure 5.9 Representation of the assembly procedure: the FN is the core, the loops on τ and s build the matrix for a given pair of i and j, the loops on i and j give the matrix of the elements, and the loop on the elements gives the global stiffness matrix

The general form of the stiffness matrix is as follows:

$$
\begin{array}{cc|ccccccc}
 & & \multicolumn{3}{c}{\overbrace{\tau=1 \quad \cdots \quad \tau=M}^{i=1}} & & \multicolumn{3}{c}{\overbrace{\tau=1 \quad \cdots \quad \tau=M}^{i=N_{NE}}} \\
\multirow{3}{*}{$j=1$} & s=1 & \boldsymbol{k}^{1111} & \cdots & \boldsymbol{k}^{1M11} & & \boldsymbol{k}^{111N_n} & \cdots & \boldsymbol{k}^{1M1N_n} \\
 & \vdots & \vdots & & \vdots & \cdots & \vdots & & \vdots \\
 & s=M & \boldsymbol{k}^{M111} & \cdots & \boldsymbol{k}^{MM11} & & \boldsymbol{k}^{M11N_n} & \cdots & \boldsymbol{k}^{MM1N_n} \\
 & & & \vdots & & \ddots & & \vdots & \\
\multirow{3}{*}{$j=N_n$} & s=1 & \boldsymbol{k}^{11N_n1} & \cdots & \boldsymbol{k}^{1MN_n1} & & \boldsymbol{k}^{11N_nN_n} & \cdots & \boldsymbol{k}^{1MN_nN_n} \\
 & \vdots & \vdots & & \vdots & \cdots & \vdots & & \vdots \\
 & s=M & \boldsymbol{k}^{M1N_n1} & \cdots & \boldsymbol{k}^{MMN_n1} & & \boldsymbol{k}^{M1N_nN_n} & \cdots & \boldsymbol{k}^{MMN_nN_n}
\end{array}
\tag{5.69}
$$

Each FN is reported as $\boldsymbol{k}^{\tau sij}$ and it works as the core of the matrix construction. The indexes indicate the nucleus position in the global matrix.

5.5 CUF as a Unique Approach for 1D, 2D and 3D Structures

The displacement field of 1D models is described in Section 5.3.2 using the classical FE approach ($N_i(y)$) along the axis and functions $F_\tau(x,z)$ over the cross-section,

$$\boldsymbol{u} = N_i(y)F_\tau(x,z)\boldsymbol{u}_{\tau i} \tag{5.70}$$

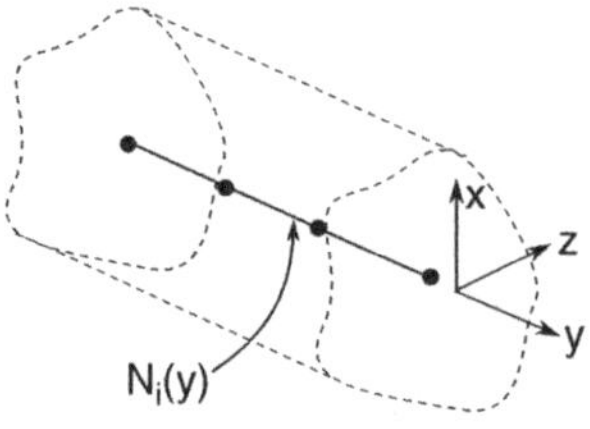

Figure 5.10 Example of 1D FEM model

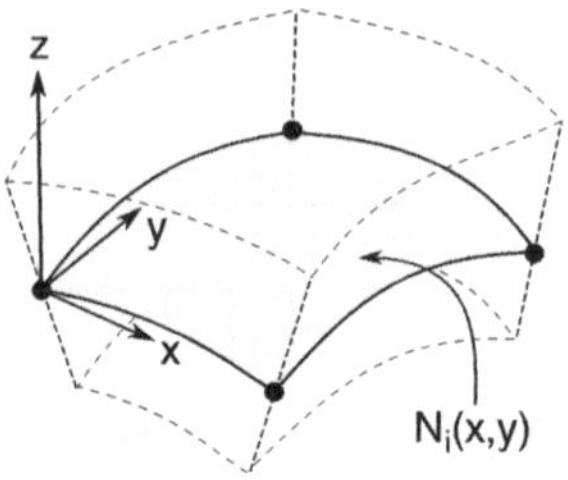

Figure 5.11 Example of 2D FEM model

The index formulation used in the CUF offers the possibility of extending the formulation introduced in the previous sections to any other structural theory, such as in plate and shell models. The shape functions, N_i, introduced by the FEM model, can be used to approximate 2D and 3D domains,

$$1\text{D} \longrightarrow N_i(y) \tag{5.71}$$

$$2\text{D} \longrightarrow N_i(x, y) \tag{5.72}$$

$$3\text{D} \longrightarrow N_i(x, y, z) \tag{5.73}$$

Figures 5.10, 5.11 and 5.12 show examples of 1D, 2D and 3D FE models, respectively. The introduction of expansions F_τ,

$$1\text{D} \longrightarrow F_\tau(y, z) \tag{5.74}$$

$$2\text{D} \longrightarrow F_\tau(z) \tag{5.75}$$

$$3\text{D} \longrightarrow 1 \tag{5.76}$$

completes the description of the theories of structures for the 1D, 2D and 3D cases. The combination of the FEM and the theory of structure approximations leads to

$$1\text{D}: \quad \boldsymbol{u}(x, y, z) = N_i(y)\, F_\tau(x, z) \tag{5.77}$$

$$2\text{D}: \quad \boldsymbol{u}(x, y, z) = N_i(x, y)\, F_\tau(z) \tag{5.78}$$

$$3\text{D}: \quad \boldsymbol{u}(x, y, z) = N_i(x, y, z) \tag{5.79}$$

Figures 5.13, 5.14 and 5.15 show how functions F_τ approximate the solution on the cross-section of a beam, or along the thickness of a plate/shell.

The 1D approach is described in Equation (5.77) and shown in Figure 5.13. The FE model is used to approximate the problem along the y-axis, while the expansion, $F_\tau(x, z)$, is used to approximate the displacement on the cross-section. Any beam model can be derived using this formulation and classical models, such as the Euler–Bernoulli or Timoshenko models, can be obtained as particular cases. The 2D model, see Equation (5.78), uses the FEM to solve the problem on a reference surface, while the expansion $F_\tau(z)$ is used to describe the

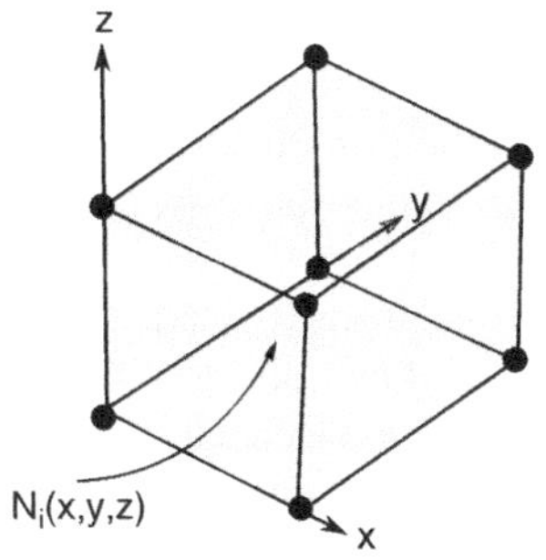

Figure 5.12 Example of 3D FEM model

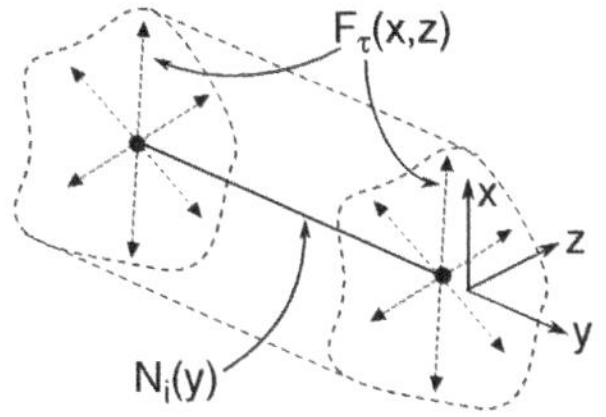

Figure 5.13 Example of beam model

displacements along the thickness of the element. An example is shown in Figure 5.14. This approach can be used to derive any plate or shell element. Finally, the 3D approach is shown in Equation (5.79). In this case, the whole domain is approximated using the FEM approach and the F_τ expansion is not used, see Figure 5.15; 3D FEM models can be derived using this approach.

Whatever model is used, the displacement field has a 3D formulation. Details of the 3D, 2D and 1D models will be introduced in the following chapters.

5.6 Literature Review of the CUF

The formulation that will be presented in the next few chapters has been developed by the first author of this book and his research group, the MUL2 team, over the last 20 years. Hundreds of works have been published on the CUF. A brief review of CUF works is presented in this section with the aim of highlighting the milestones in the development of such theories.

The CUF first appeared in the paper by Carrera (1995) where it was used to derive a class of 2D theories using a compact formulation. This work was the first of a number of papers that have dealt with several issues related to 2D problems. The CUF was used to derive Equivalent Single Layer (Carrera 1996) and Layer-Wise models (Carrera, 1999a,b). Many articles have been devoted to the importance of the inter laminar continuity of stresses (e.g. Carrera 1997),

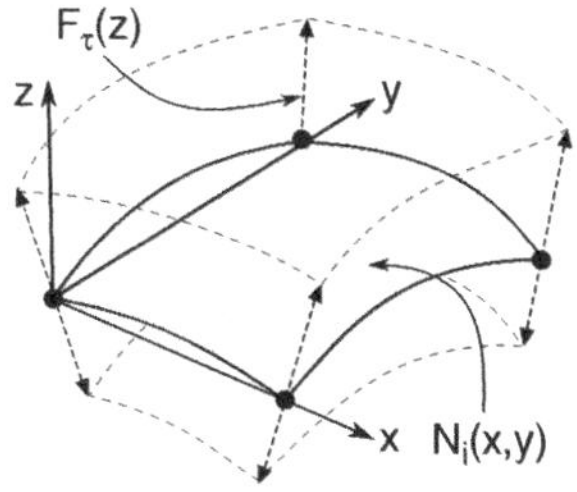

Figure 5.14 Example of plate/shell model

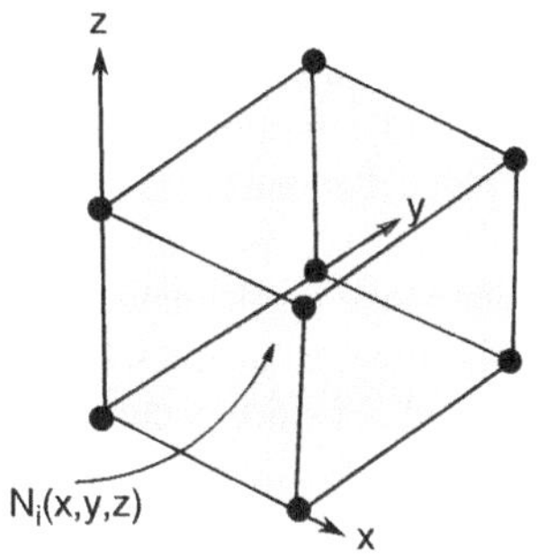

Figure 5.15 Example of solid model

the so-called C_z^0 requirement. The zigzag function proposed by Murakami (1986) was used in Carrera (2004). Classical 2D models were improved using the RMVT by Carrera and Demasi (2002a,b). The paper by Carrera (2003) presented the theoretical foundations of the unified method that has served as the basis of a large number of innovative applications and extensions of the CUF.

The features of the CUF allow the effects of thermal loads (Carrera 2002) to be included in the formulation. Thermoelastic coupling was presented by Carrera (2005) and Robaldo *et al.* (2005). After thermal loads, the multifield formulation was extended to piezoelectric structures by Robaldo *et al.* (2006) and Carrera *et al.* (2007). The complete multifield formulation was presented by Carrera *et al.* (2008a, 2009). The CUF has also been used in FGM analysis. The formulation was introduced by Carrera *et al.* (2008b) and extended to multifield problems in Brischetto *et al.* (2008).

In the last few years, the 2D formulation, which had previously been devoted to flat panels, was extended to shell FEs by Cinefra (2012). The model was presented by Cinefra *et al.* (2013) and its application to FGM shown by Cinefra *et al.* (2012).

Carrera *et al.* (2011c) presented a paper that considered the large number of models that can be derived using the CUF and attempted to offer some guidelines on the use of refined models. This activity was performed using a genetic algorithm and the results were presented by Carrera *et al.* (2011d) and Carrera and Miglioretti (2012).

The complete CUF formulation was recently presented for 2D problems in the book by Carrera *et al.* (2011a).

Since 2010, the CUF has been used to derive higher-order beam models. The first approach to 1D models was proposed by Carrera and Giunta (2010). This paper was the first of a number of papers that introduced different higher-order beam models based on the CUF. The works by Carrera *et al.* (2010), Carrera and Petrolo (2012) and Carrera *et al.* (2013a) have introduced different classes of 1D models, although they all refer to the CUF. These models were used to perform static analyses (Carrera *et al.*, 2012e) and dynamic analyses (Carrera and Varello 2012) and to show that these models are able to provide reliable results for complex problems that cannot be addressed with classical beam models. Carrera and Petrolo (2011) and Carrera *et al.* (2011c,d, 2012b) proposed axiomatic/asymptotic approaches for beam formulations.

The large number of new theories introduced by the CUF has made it necessary to derive a new approach to investigate the accuracy of each model. Carrera *et al.* (2012b) used a genetic algorithm to investigate the influence of the higher-order terms on the solution.

The enhanced capabilities of the 1D models introduced using the CUF suggested the use of such models for more challenging analyses. One of the first applications concerned the study of the stability of structures. The problem of buckling was investigated by Ibrahim *et al.* (2012), in which compact and thin-walled beams were considered. Advanced beam models have been employed to analyse complex structures, such as thin-walled structures with reinforcements. Some results have been presented by Carrera *et al.* (2013d,h). The use of higher-order models allows a "quasi-3D" solution to be obtained. This allows a failure analysis to be accurately carried out since accurate stresses and strains can be recovered. Carrera *et al.* (2012a) used a 1D model, based on the CUF, to investigate the failure of composite materials. Refined beam models, like the advanced plate/shell models introduced above, are suitable for multifield applications. The main research activities have focused on the aeroelasticity of lifting surfaces and panels. The works by Carrera *et al.* (2012d) and Varello *et al.* (2011) deal with static aeroelasticity. Petrolo (2012a) introduced a classical unsteady flow formulation that extends the use of advanced beam models to flutter analysis; some examples are given in Petrolo (2012b, 2013). Aeroelastic phenomena have been investigated in the supersonic range by Carrera and Zappino (2013a).

The use of a 1D advanced model is currently being extended to biomechanical applications (Varello and Carrera 2012), electro mechanical analysis (Miglioretti *et al.* 2014) and rotor dynamics (Carrera *et al.* 2013b,c). A complete review of all the 1D models derived by means of the CUF can be found in the book by Carrera *et al.* (2011b).

The analysis of complex structures requires the use of different numerical models in the same analysis. Different techniques can be used to join together different theories. Biscani *et al.* (2011, 2012) used the Arlequin method to combine different models. Lagrange multipliers have been used by Carrera *et al.* (2013e) to connect variable kinematic models together. A "component-wise" approach was proposed by Carrera *et al.* (2013f,g). This method allows different structural components to be investigated using a refined 1D model. Carrera and Zappino (2013b) introduced a 3D model based on the CUF. This model has been used as a connection between different 1D and 2D structural models.

References

Biscani F, Giunta G, Belouettar S, Carrera E and Hud H. 2011 Variable kinematic beam elements coupled via Arlequin method. *Composite Structures* **93**(2), 697–708.

Biscani F, Nali P, Carrera E, and Belouettar S. 2012 Coupling of hierarchical piezoelectric plate finite elements via Arlequin method. *Journal of Intelligent Material Systems and Structures* **23**(7), 749–764.

Brischetto S, Leetsch R, Carrera E, Wallmersperger T and Kroplin B 2008 Variable kinematic model for the analysis of functionally graded material plates. *Journal of thermal Stress* **31**(3), 286–308.

Carrera E 1995 A class of two dimensional theories for multilayered plates analysis. *Atti Accademia delle Scienze di Torino, Memorie Scienze Fisiche* **19–20**, 49–87.

Carrera E 1996 C_z^0 Reissner-Mindlin multilayered plate elements including zig-zag and interlaminar stresses continuity. *International Journal for Numerical Methods in Engineering* **39**, 1797–1820.

Carrera E 1997 C_z^0 requirements – models for the two dimensional analysis of multilayered structures. *Composite Structure* **37**, 373–384.

Carrera E 1999a Multilayered shell theories that account for a layer-wise mixed description. Part I. Governing equations. *AIAA Journal* **37**, 1107–1116.

Carrera E 1999b Multilayered shell theories that account for a layer-wise mixed description. Part II. Numerical evaluations. *AIAA Journal* **37**, 1117–1124.

Carrera E 2002 Temperature profile influence on layered plates response considering classical and advanced theories. *AIAA Journal* **40**, 1885–1896.

Carrera E 2003 Theories and finite elements for multilayered plates and shells: a unified compact formulation with numerical assessment and benchmarking. *Archives of Computational Methods in Engineering* **10**, 215–297.

Carrera E 2004 On the use of the Murakami's zig-zag function in the modeling of layered plates and shells. *Computers & Structures* **82**(7–8), 541–554.

Carrera E 2005 Transverse normal strain effects on thermal stress analysis of homogeneous and layered plates. *AIAA Journal* **43**, 2232–2242.

Carrera E, Boscolo M and Robaldo A 2007 Hierarchic multilayered plate elements for coupled multifield problems of piezoelectric adaptive structures: formulation and numerical assessment. *Archives of Computational Methods in Engineering* **4**(14), 383–430.

Carrera E, Brischetto S, Fagiano C and Nali P 2009 Mixed multilayered plate elements for coupled magneto-electro-elastic problems. *Multidiscipline Modeling in Materials and Structures* **5**(3), 251–256.

Carrera E, Brischetto S and Nali P 2008a Variational statements and computational models for multifield problems and multilayered structures. *Mechanics of Advanced Materials and Structures* **15**(3–4), 182–198.

Carrera E, Brischetto S, and Nali P 2011a Plates and shells for smart structures: classical and advanced theories for modeling and analysis. John Wiley & Sons, Ltd.

Carrera E, Brischetto S and Robaldo A 2008b Variable kinematic model for the analysis of functionally graded material plates. *AIAA Journal* **46**(1), 194–203.

Carrera E and Demasi L 2002a Classical and advanced multilayered plate element based upon PVD and RMVT. Part 1: Derivation of finite element matrices. *International Journal for Numerical Methods in Engineering* **55**, 191–231.

Carrera E and Demasi L 2002b Classical and advanced multilayered plate element based upon PVD and RMVT. Part II: Numerical implementations. *International Journal for Numerical Methods in Engineering* **55**, 253–291.

Carrera E, Filippi M and Zappino E 2013a Laminated beam analysis by polynomial, trigonometric, exponential and zig-zag theories. *European Journal of Mechanics – A/Solids* **41**, 58–69.

Carrera E, Filippi M and Zappino E 2013b Analysis of rotor dynamic by one-dimensional variable kinematic theories. *Journal of Engineering for Gas Turbines and Power* **135**(9), 092501.

Carrera E, Filippi M and Zappino E 2013c Free vibration analysis of rotating composite blades via Carrera unified formulation. *Composite Structures* **106**, 317–325.

Carrera E and Giunta G 2010 Refined beam theories based on a unified formulation. *International Journal of Applied Mechanics and Engineering* **2**(1), 117–143.

Carrera E, Giunta G, Nali P, and Petrolo M 2010 Refined beam elements with arbitrary cross-section geometries. *Computers & Structures* **88**, 283–293.

Carrera E, Giunta G, and Petrolo M 2011b *Beam Structures: Classical and Advanced Theories*. John Wiley & Sons, Ltd.

Carrera E, Maiarù M and Petrolo M 2012a A component-wise approach for the failure analysis of composite structures. *53rd AIAA/ASME/ASCE/AHS/ASC Structures, Structural Dynamics, and Materials Conference (SDM)*, Honolulu, Hawaii, USA, 23–26 April 2012.

Carrera E and Miglioretti F 2012 Selection of appropriate multilayered plate theories by using a genetic like algorithm. *Composite Structures* **94**, 1175–1186.

Carrera E, Miglioretti F and Petrolo M 2011c Guidelines and recommendations on the use of higher order finite elements for bending analysis of plates. *International Journal for Computational Methods in Engineering Science and Mechanics* **12**, 303–324.

Carrera E, Miglioretti F and Petrolo M 2011d Accuracy of refined finite elements for laminated plate analysis. *Composite Structures* **93**(5), 1311–1327.

Carrera E, Miglioretti M and Petrolo M 2012b Computations and evaluations of higher-order theories for free vibration analysis of beams. *Journal of Sound and Vibration* **331**, 4269–4284.

Carrera E, Pagani A and Petrolo M 2013d Classical, refined and component-wise analysis of reinforced-shell structures. *AIAA Journal* **51**(5), 1255–1268.

Carrera E, Pagani A and Petrolo M 2013e Use of Lagrange multipliers to combine 1D variable kinematic finite elements. *Computers & Structures* **129**, 194–206. DOI:10.1016/j.compstruc.2013.07.005.

Carrera E, Pagani A and Petrolo M 2013f Classical, refined and component-wise analysis of reinforced-shell structures. *AIAA Journal* **51**(5), 1255–1268.

Carrera E, Pagani A and Petrolo M 2013g Component-wise method applied to vibration of wing structures. *Journal of Applied Mechanics* **88**(4), 041012.

Carrera E and Petrolo M 2011 On the effectiveness of higher-order terms in refined beam theories. *Journal of Applied Mechanics* **78**. DOI: 10.1115/1.4002207.

Carrera E and Petrolo M 2012 Refined beam elements with only displacement variables and plate/shell capabilities. *Meccanica* **47**, 537–556.

Carrera E, Petrolo M and Varello A 2012d Advanced beam formulations for free vibration analysis of conventional and joined wings. *Journal of Aerospace Engineering* **25**(2), 282–293.

Carrera E, Petrolo M and Zappino E 2012e Performance of CUF approach to analyze the structural behavior of slender bodies. *Journal of Structural Engineering* **138**(2), 285–297.

Carrera E and Varello A 2012 Dynamic response of thin-walled structures by variable kinematic one-dimensional models. *Journal of Sound and Vibration* **331**(24), 5268–5282.

Carrera E and Zappino E 2013a Aeroelastic analysis of pinched panels in a variable supersonic flow changing with altitude. *Journal of Spacecraft and Rockets* **51**(1), 187–199.

Carrera E and Zappino E 2013b Full aircraft dynamic response by simplified structural models. *54rd AIAA/ASME/ASCE/AHS/ASC Structures, Structural Dynamics, and Materials Conference (SDM)*, Boston, Massachusetts, USA, 8–11 April 2013.

Carrera E, Zappino E and Petrolo M 2013h Analysis of thin-walled structures with longitudinal and transversal stiffeners. *Journal of Applied Mechanics* **80**, 011006.

Cinefra M 2012 Refined and advanced shell models for the analysis of advanced structures. PhD thesis, Politecnico di Torino – Université Paris Ouest – Nanterre La Défense (France).

Cinefra M, Carrera E, Della Croce L, and Chinosi C 2012 Refined shell elements for the analysis of functionally graded structures. *Composite Structures* **94**, 415–422.

Cinefra M, Chinosi C and Della Croce L 2013 MITC9 shell elements based on refined theories for the analysis of isotropic cylindrical structures. *Mechanics of Advanced Materials ans Structures* **20**, 91–100.

Ibrahim SM, Carrera E, Petrolo M and Zappino E 2012 Buckling of composite thin walled beams by refined theory. *Composite Structures* **94**, 563–570.

Miglioretti F, Carrera E and Petrolo M 2014 Variable kinematic beam elements for electro-mechanical analysis. *DeMEASS V, Special Issue.*

Murakami H 1986 Laminated composite plate theory with improved in-plane responses. *Journal of Applied Mechanics* **53**, 661–666.

Petrolo M 2012a Advanced aeroelastic models for the analysis of lifting surfaces made of composite materials. PhD thesis, Politecnico di Torino.

Petrolo M 2012b Advanced 1D structural models for flutter analysis of lifting surfaces. *International Journal of Aeronautical and Space Sciences* **13**(2), 199–209.

Petrolo M 2013 Flutter analysis of composite lifting surfaces by the 1D Carrera unified formulation and the doublet lattice method. *Composite Structures* **95**, 539–546.

Robaldo A, Carrera E and Benjeddou A 2005 Unified formulation for finite element thermoelastic analysis of multilayered anisotropic composite plates. *Journal of Thermal Stress* **28**, 1031–1064.

Robaldo A, Carrera E and Benjeddou A 2006 A unified formulation for finite element analysis of piezoelectric plates. *Computers & Structures* **84**, 1494–1505.

Varello A and Carrera E 2012 Accurate 1D structural models for the analysis of non-homogeneous biomechanical structures. *Fifth International Symposium on Design, Modelling and Experiments of Advanced Structures and Systems*, Ulrichsberg, Austria, 28–31 October 2012.

Varello A, Carrera E and Demasi L 2011 Vortex lattice method coupled with advanced one-dimensional structural models. *Journal of Aeroelasticity and Structural Dynamics* **2**, 53–78.

6

The Displacement Approach via the PVD and FN for 1D, 2D and 3D Elements

In this chapter, the PVD (Principle of Virtual Displacements) is used to derive the displacement formulation for a 3D structural problem. In the first part, some general considerations are made on the PVD and its use in the theory of elasticity. The relationship between FNs, which were introduced in Chapter 5, and the strong formulation of the equilibrium equations is highlighted. The weak form of the 3D structural problem is introduced in the second part, where the CUF approach is used to derive a compact form and the related FNs for 3D, 2D and 1D problems.

6.1 Strong Form of the Equilibrium Equations via the PVD

Indefinite equilibrium equations are well known in the theory of elasticity and were introduced in Chapter 2. The use of the PVD allows one to derive the same equations, both weak and strong forms. The PVD can be written in its static case as

$$\delta L_{int} = \delta L_{ext} \tag{6.1}$$

where L_{int} is the internal elastic work, L_{ext} is the work done by the external forces and δ indicates the virtual variation. The internal work can be expressed in explicit form as

$$\delta L_{int} = \int_V \left(\sigma_{xx}\delta\varepsilon_{xx} + \sigma_{yy}\delta\varepsilon_{yy} + \sigma_{zz}\delta\varepsilon_{zz} + \sigma_{xz}\delta\varepsilon_{xz} + \sigma_{yx}\delta\varepsilon_{yz} + \sigma_{xy}\delta\varepsilon_{xy}\right) dV \tag{6.2}$$

The same equation can be written in compact form using matrix notation:

$$\delta L_{int} = \int_V \delta\boldsymbol{\varepsilon}^T \boldsymbol{\sigma} dV \tag{6.3}$$

Finite Element Analysis of Structures Through Unified Formulation, First Edition.
Erasmo Carrera, Maria Cinefra, Marco Petrolo and Enrico Zappino.

The external work on a general body D is expressed as a sum of four contributions: volume forces, $\boldsymbol{g}$, on volume V, surface forces, $\boldsymbol{p}$, on surface S, line forces, $\boldsymbol{q}$, on line l and the concentrated force, $\boldsymbol{P}$, at point Q. The formulation of the external work, introduced in Chapter 2, becomes

$$\delta L_{ext} = \int_V \delta \boldsymbol{u}^T \boldsymbol{g} dV + \int_S \delta \boldsymbol{u}^T \boldsymbol{p} dS + \int_L \delta \boldsymbol{u}^T \boldsymbol{q} dy + \delta \boldsymbol{u}^T|_Q \boldsymbol{P} \tag{6.4}$$

The relationship between the displacement vector, $\boldsymbol{u}$, and the strain vector, $\boldsymbol{\varepsilon}$, is obtained from the geometrical relation

$$\boldsymbol{\varepsilon} = \boldsymbol{b}\boldsymbol{u} \tag{6.5}$$

where matrix $\boldsymbol{b}$ is a differential operator. In explicit form, the equation becomes

$$\begin{Bmatrix} \varepsilon_{xx} \\ \varepsilon_{yy} \\ \varepsilon_{zz} \\ \gamma_{xz} \\ \gamma_{yz} \\ \gamma_{xy} \end{Bmatrix} = \boldsymbol{b}\boldsymbol{u} = \begin{bmatrix} \partial_x & 0 & 0 \\ 0 & \partial_y & 0 \\ 0 & 0 & \partial_z \\ \partial_z & 0 & \partial_x \\ 0 & \partial_z & \partial_y \\ \partial_y & \partial_x & 0 \end{bmatrix} \begin{Bmatrix} u_x \\ u_y \\ u_z \end{Bmatrix} \tag{6.6}$$

The internal work can be written in terms of displacements as

$$\delta L_{int} = \int_V \delta(\boldsymbol{b}\boldsymbol{u})^T \boldsymbol{\sigma} dV = \int_V (\delta \boldsymbol{u}^T \boldsymbol{b}^T) \boldsymbol{\sigma} dV \tag{6.7}$$

In order to obtain strong form equations, it is possible to move the differential operator from the displacements to the strains by integrating by parts,[1]

$$\int_V (\delta \boldsymbol{u}^T \boldsymbol{b}^T) \boldsymbol{\sigma} dV = -\int_V \delta \boldsymbol{u}^T (\boldsymbol{b}^T \boldsymbol{\sigma}) dV + \int_S \delta \boldsymbol{u}^T (\boldsymbol{I_n}^T \boldsymbol{\sigma}) dS \tag{6.10}$$

[1] Integration by parts of a volume integral can be expressed as

$$\int_V (\delta \boldsymbol{u}^T \boldsymbol{D}^T) \boldsymbol{\sigma} dV = -\int_V \delta \boldsymbol{u}^T (\boldsymbol{D}^T \boldsymbol{\sigma}) dV + \int_S \delta \boldsymbol{u}^T (\boldsymbol{I_n}^T \boldsymbol{\sigma}) dS \tag{6.8}$$

where $\boldsymbol{I_n}$ is a matrix with cosine directors,

$$\boldsymbol{I}_n = \begin{bmatrix} n_x & 0 & 0 \\ 0 & n_y & 0 \\ 0 & 0 & n_z \\ n_z & 0 & n_x \\ 0 & n_z & n_x \\ n_y & n_x & 0 \end{bmatrix} \tag{6.9}$$

n_x, n_y, n_z are the cosine directors of the point P above the boundary surface S, see Washizu (1968).

where $\boldsymbol{I_n}$ is a matrix with cosine directors. In the first term on the right-hand side, operator $\boldsymbol{b}$ acts on the stress vector. The PVD can be written as

$$-\int_V \delta\boldsymbol{u}^T(\boldsymbol{b}^T\boldsymbol{\sigma})dV + \int_S \delta\boldsymbol{u}^T(\boldsymbol{I_n}^T\boldsymbol{\sigma})dS = \int_V \delta\boldsymbol{u}^T\boldsymbol{g}dV + \int_S \delta\boldsymbol{u}^T\boldsymbol{p}dS + \int_L \delta\boldsymbol{u}^T\boldsymbol{q}dy + \delta\boldsymbol{u}^T|_Q\boldsymbol{P} \tag{6.11}$$

From this equation, and using the virtual variation definition, it is possible to derive the equilibrium equation at a generic point P on volume V of body D,

$$\delta\boldsymbol{u}: \quad -\boldsymbol{b}^T\boldsymbol{\sigma} = \boldsymbol{g} \tag{6.12}$$

From this equation, it is clear that the differential operator $\boldsymbol{b}$ must be the same for both the equilibrium and geometrical equations.

The integrals on the surface give the boundary conditions, which can be expressed as

$$\delta\boldsymbol{u}: \quad \boldsymbol{I_n}^T\boldsymbol{\sigma} = \boldsymbol{p} \tag{6.13}$$

The equilibrium equations can be derived in explicit form by expanding Equation (6.12)

$$\begin{aligned}
\delta u_x: &\quad \frac{\partial\sigma_{xx}}{\partial x} + \frac{\partial\sigma_{xz}}{\partial z} + \frac{\partial\sigma_{xy}}{\partial y} = g_x \\
\delta u_y: &\quad \frac{\partial\sigma_{yy}}{\partial y} + \frac{\partial\sigma_{yz}}{\partial z} + \frac{\partial\sigma_{yx}}{\partial x} = g_y \\
\delta u_z: &\quad \frac{\partial\sigma_{zz}}{\partial z} + \frac{\partial\sigma_{zx}}{\partial x} + \frac{\partial\sigma_{zy}}{\partial y} = g_z
\end{aligned} \tag{6.14}$$

as in Chapter 2. The same can be done for the boundary conditions, see Equation (6.13). This expansion is omitted for the sake of brevity.

Hooke's law allows the equilibrium equations to be written in terms of displacements,

$$\delta\boldsymbol{u}: \quad -\boldsymbol{b}^T\boldsymbol{C}\boldsymbol{b}\boldsymbol{u} = \boldsymbol{g} \tag{6.15}$$

If the material is isotropic, matrix $\boldsymbol{C}$ can be written using Lamé coefficients, see Chapter 3:

$$\boldsymbol{C} = \begin{bmatrix} \lambda+2G & \lambda & \lambda & 0 & 0 & 0 \\ \lambda & \lambda+2G & \lambda & 0 & 0 & 0 \\ \lambda & \lambda & \lambda+2G & 0 & 0 & 0 \\ 0 & 0 & 0 & G & 0 & 0 \\ 0 & 0 & 0 & 0 & G & 0 \\ 0 & 0 & 0 & 0 & 0 & G \end{bmatrix} \tag{6.16}$$

where

$$\lambda = \frac{E\nu}{(1+\nu)(1-2\nu)}, \quad G = \frac{E}{2(1+\nu)} \tag{6.17}$$

The equilibrium equations can be written, in strong form, by introducing a matrix $\boldsymbol{k}$ that originates from the previous matrix multiplication,

$$\delta \boldsymbol{u} : \quad \boldsymbol{k}\boldsymbol{u} = \boldsymbol{g} \tag{6.18}$$

where

$$\boldsymbol{k} = -\boldsymbol{b}^T \boldsymbol{C} \boldsymbol{b} \tag{6.19}$$

This book deals, above all, with isotropic and homogeneous materials and, for the sake of simplicity, matrix $\boldsymbol{C}$ is assumed to be constant in V. Matrix $\boldsymbol{k}$ is a 3×3 matrix and it contains nine differential operators,

$$\boldsymbol{k} = \begin{bmatrix} k_{xx} & k_{xy} & k_{xz} \\ k_{yx} & k_{yy} & k_{yz} \\ k_{zx} & k_{zy} & k_{zz} \end{bmatrix} \tag{6.20}$$

which, in explicit form, become (see also Chapter 2)

$$\begin{aligned}
k_{xx} &= -(\lambda + 2G)\ \partial_x \partial_x - G\ \partial_y \partial_y - G\ \partial_z \partial_z \\
k_{xy} &= -\lambda\ \partial_x \partial_y - G\ \partial_y \partial_x \\
k_{xz} &= -\lambda\ \partial_x \partial_z - G\ \partial_z \partial_x \\
k_{yx} &= -\lambda\ \partial_y \partial_x - G\ \partial_x \partial_y \\
k_{yy} &= -(\lambda + 2G)\ \partial_y \partial_y - G\ \partial_x \partial_x - G\ \partial_z \partial_z \\
k_{yz} &= -\lambda\ \partial_y \partial_z - G\ \partial_z \partial_y \\
k_{zx} &= -\lambda\ \partial_z \partial_x - G\ \partial_x \partial_z \\
k_{zy} &= -\lambda\ \partial_z \partial_y - G\ \partial_y \partial_z \\
k_{zz} &= -(\lambda + 2G)\ \partial_z \partial_z - G\ \partial_x \partial_x - G\ \partial_y \partial_y
\end{aligned} \tag{6.21}$$

The symbol ∂_x means partial differentiation with respect to x. The derivatives in Equation (6.21) appear in pairs, where the first derivative is due to a virtual variation of the strains, while the second is due to the stresses. Since the displacements are continuous functions, it is possible to state that

$$\partial_y \partial_x = \partial_x \partial_y = \partial_{yx}, \quad \partial_z \partial_x = \partial_x \partial_z = \partial_{zx}, \quad \partial_z \partial_y = \partial_y \partial_z = \partial_{zy} \tag{6.22}$$

Therefore Equation (6.21) can be written as

$$\begin{aligned}
k_{xx} &= -(\lambda + 2G)\ \partial_{xx} - G\ \partial_{yy} - G\ \partial_{zz}\\
k_{xy} &= -(\lambda + G)\ \partial_{xy}\\
k_{xz} &= -(\lambda + G)\ \partial_{xz}\\
k_{yx} &= -(\lambda + G)\ \partial_{yx}\\
k_{yy} &= -(\lambda + 2G)\ \partial_{yy} - G\ \partial_{xx} - G\ \partial_{zz}\\
k_{yz} &= -(\lambda + G)\ \partial_{yz}\\
k_{zx} &= -(\lambda + G)\ \partial_{zx}\\
k_{zy} &= -(\lambda + G)\ \partial_{zy}\\
k_{zz} &= -(\lambda + 2G)\ \partial_{zz} - G\ \partial_{xx} - G\ \partial_{yy}
\end{aligned} \tag{6.23}$$

Finally, the equilibrium equations can be written in terms of displacements,

$$\begin{aligned}
\delta u_x :\ & -(\lambda + 2G)\left(\frac{\partial^2 u_x}{\partial x^2}\right) - G\left(\frac{\partial^2 u_z}{\partial y^2} + \frac{\partial^2 u_x}{\partial z^2}\right)\\
& -(\lambda + G)\left(\frac{\partial^2 u_y}{\partial x \partial y} + \frac{\partial^2 u_z}{\partial x \partial z}\right) = g_x
\end{aligned} \tag{6.24}$$

$$\begin{aligned}
\delta u_y :\ & -(\lambda + 2G)\left(\frac{\partial^2 u_y}{\partial y^2}\right) - G\left(\frac{\partial^2 u_y}{\partial x^2} + \frac{\partial^2 u_y}{\partial z^2}\right)\\
& -(\lambda + G)\left(\frac{\partial^2 u_x}{\partial y \partial x} + \frac{\partial^2 u_z}{\partial y \partial z}\right) = g_y
\end{aligned} \tag{6.25}$$

$$\begin{aligned}
\delta u_z :\ & -(\lambda + 2G)\left(\frac{\partial^2 u_z}{\partial z^2}\right) - G\left(\frac{\partial^2 u_z}{\partial x^2} + \frac{\partial^2 u_z}{\partial y^2}\right)\\
& -(\lambda + G)\left(\frac{\partial^2 u_z}{\partial z \partial x} + \frac{\partial^2 u_y}{\partial z \partial y}\right) = g_z
\end{aligned} \tag{6.26}$$

6.1.1 The Two Fundamental Terms of the FN

Although there are nine terms in matrix $\boldsymbol{k}$, only two have a different structure. Let us consider the following two terms:

$$k_{xx} = -(\lambda + 2G)\ \partial_{xx} - \lambda\ \partial_{zz} - \lambda\ \partial_{yy} \tag{6.27}$$

$$k_{xy} = -\lambda\ \partial_{xy} - G\ \partial_{yx} \tag{6.28}$$

It is evident that the other components of matrix $\boldsymbol{k}$ can be obtained in a similar form as k_{xx} and k_{xy}. The elements on the diagonal have the form of k_{xx}, and the terms k_{yy} and k_{zz} therefore have the same form as k_{xx} with the indexes permuted. The non-diagonal terms are given by a permutation of the indexes of k_{xy}, and k_{xz}, k_{yz}, k_{yx}, k_{zx} and k_{zy} can in fact be obtained by permuting the indexes in k_{xy}.

6.2 Weak Form of the Solid Model Using the PVD

In the previous section, the PVD was used to derive the equilibrium equations of a generic point, *Q*, in a body *D*. This form is called a 'strong' formulation, as it provides the exact equations of the problem in terms of displacements, stresses and strains at each point of *D*. However, it is not an easy task to find solutions of such equations. They can usually only be solved for simple geometries and boundary conditions.

A 'weak' form could be used to solve the problem in most applications. The solution of such a weak form only satisfies the previously introduced equations according to a fixed criterion that is given in integral form. The weak form is used extensively in real applications of structural analysis, and it is the basis of computational mechanics.

The PVD is a tool that is able to derive the weak form of structural mechanics problems. A weak form assumption allows one directly to approximate the derivatives of the virtual variations without using integration by parts.

A cubic solid, *D*, is considered.[2] The displacement field can be written using Lagrange polynomials. These polynomials allow the displacement field of the solid *D* to be described as a function of the displacements at a finite number of points, i.e. the 'nodes':[3]

$$u_x = N_i(x,y,z)u_{x_i}, \quad u_y = N_i(x,y,z)u_{y_i}, \quad u_z = N_i(x,y,z)u_{z_i}, \quad i = 1, N_{NE} \tag{6.30}$$

where N_i is the *i*th Lagrange polynomial, and u_{x_i}, u_{y_i} and u_{z_i} are the nodal displacements at the *i*th node in the *x*, *y* and *z* directions, respectively. Figure 6.1 shows a 3D element with eight nodes, $N_{NE} = 8$. The approach shown hereafter can be considered valid whatever the number of nodes in the element.

The PVD in the static case is,

$$\delta L_{int} = \delta L_{ext} \tag{6.31}$$

where L_{int} is the internal work and L_{ext} is the external work. The internal work can be written in terms of the stress and strain,

$$\delta L_{int} = \int_V \delta \epsilon^T \sigma dV \tag{6.32}$$

[2] Cubic geometry is a strong assumption, but the approach introduced in this section is valid for any geometry.

[3] The Lagrange polynomials can be written in a form that is convenient for computational calculation. One of the best ways is to express the polynomials in 'natural' coordinates: ξ, η, ζ. If the cube has dimensions $2a$, $2b$ and $2c$, in the x, y and z directions, and G is the centre of gravity with positions x_G, y_G and z_G, then the natural coordinates can be written as

$$\xi = \frac{x - x_G}{a}, \quad \eta = \frac{y - y_G}{b}, \quad \zeta = \frac{z - z_G}{c} \tag{6.29}$$

The *i*th node has ξ_i, η_i, ζ_i coordinates that are combinations of: $\xi = \pm 1, \eta = \pm 1, \zeta = \pm 1$. The *i*th Lagrange polynomial has a value of 1 at the *i*th node and of 0 at all the other nodes.

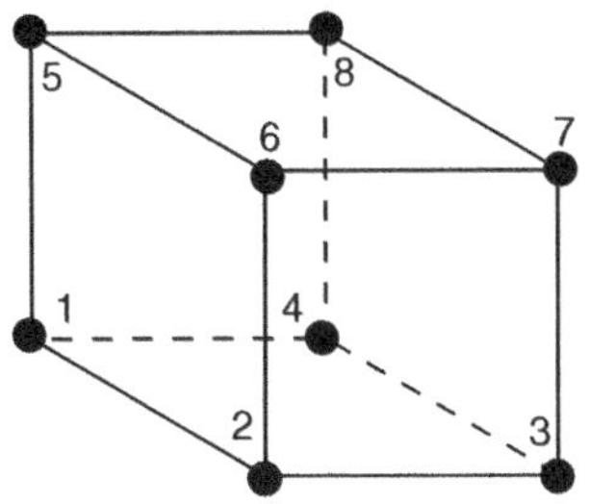

Figure 6.1 Eight-node cubic element

The external work is derived from the external forces that are known and, for the sake of simplicity, are only considered as forces acting at the nodes of the cube; in this case, the external work becomes

$$\delta L_{ext} = \delta \boldsymbol{U}^T \boldsymbol{P} \tag{6.33}$$

where $\boldsymbol{P}$ is the vector of the nodal forces

$$\boldsymbol{P}^T = \{P_{x_1}, P_{y_1}, P_{z_1}, P_{x_2}, P_{y_2}, P_{z_2}, \dots, P_{x_8}, P_{y_8}, P_{z_8}\} \tag{6.34}$$

and $\boldsymbol{U}$ is the vector of the nodal displacements

$$\boldsymbol{U}^T = \{u_{x_1}, u_{y_1}, u_{z_1}, u_{x_2}, u_{y_2}, u_{z_2}, \dots, u_{x_8}, u_{y_8}, u_{z_8}\} \tag{6.35}$$

Both vectors have $3 \times N_{NE}$ components, since it is considered that three displacements are allowed at each node. The internal work has to be written in terms of displacements. In order to write the PVD in terms of displacements, the following relations can be used:

1. Geometrical relations: matrix $\boldsymbol{b}$ is used to express the strain as a function of the displacements.
2. Constitutive equations: matrix $\boldsymbol{C}$ allows the stresses to be derived from the strains.

The displacement field can be expressed, in terms of nodal displacements, using a matrix notation:

$$\boldsymbol{u} = \boldsymbol{N}\boldsymbol{U} \tag{6.36}$$

where $\boldsymbol{N}$ is the matrix that contains Lagrange polynomials that act as shape functions.

If Hooke's law is applied to the PVD equation (see Equation (6.32)), one has

$$\delta L_{int} = \int_V \delta \boldsymbol{\varepsilon}^T \boldsymbol{C} \boldsymbol{\varepsilon} dV \tag{6.37}$$

The geometrical relations allow the strains to be expressed, in terms of displacements, through the introduction of matrix $\boldsymbol{b}$:

$$\boldsymbol{\varepsilon} = \boldsymbol{b}\boldsymbol{u} = \boldsymbol{b}\boldsymbol{N}\boldsymbol{U} = \boldsymbol{B}\boldsymbol{U} \tag{6.38}$$

where $\boldsymbol{B}$ represents the results of the application of the differential operator $\boldsymbol{b}$ to the matrix of shape functions $\boldsymbol{N}$. The virtual variation can be written in the same way:

$$\delta\boldsymbol{u} = \boldsymbol{N}\delta\boldsymbol{U} \tag{6.39}$$

The internal work can be written in terms of nodal displacements as

$$\delta L_{int} = \int_V \delta\boldsymbol{U}^T\boldsymbol{B}^T\ \boldsymbol{C}\ \boldsymbol{B}\boldsymbol{U}dV \tag{6.40}$$

The nodal displacements are constant and can therefore be removed from the integral,

$$\delta L_{int} = \delta\boldsymbol{U}^T\left(\int_V \boldsymbol{B}^T\boldsymbol{C}\boldsymbol{B}dV\right)\boldsymbol{U} \tag{6.41}$$

The same equation can be written in the form

$$\delta L_{int} = \delta\boldsymbol{U}^T\ \boldsymbol{K}\ \boldsymbol{U} \tag{6.42}$$

where

$$\boldsymbol{K} = \int_V \boldsymbol{B}^T\boldsymbol{C}\boldsymbol{B}dV \tag{6.43}$$

is the stiffness matrix for a 3D problem derived using the PVD. It is possible to substitute Equations (6.42) and (6.33) in the equilibrium equation, Equation (6.1), the PVD becomes

$$\delta\boldsymbol{U}^T\boldsymbol{K}\boldsymbol{U} = \delta\boldsymbol{U}^T\boldsymbol{P} \tag{6.44}$$

that is,

$$\boldsymbol{K}\boldsymbol{U} = \boldsymbol{P} \tag{6.45}$$

Equation (6.45) is the equilibrium equation in 'weak form' for a 'solid' element.

6.3 Weak Form of a Solid Element Using Index Notation

The matrices introduced in Section 6.2 can be written in compact form using the index notation introduced in Chapter 5. The displacement can be written as

$$\boldsymbol{u} = N_i \boldsymbol{u}_i \tag{6.46}$$

where index i ranges from 1 to the number of nodes of the elements N_{NE}. The virtual variation is indicated through index j,

$$\delta \boldsymbol{u} = N_j \delta \boldsymbol{u}_j \tag{6.47}$$

The strains and their virtual variations can be written in index form as

$$\boldsymbol{\varepsilon} = \boldsymbol{b}\boldsymbol{u} = \boldsymbol{b}N_i\boldsymbol{u}_i = \boldsymbol{B}_i\boldsymbol{u}_i \tag{6.48}$$

$$\delta\boldsymbol{\varepsilon} = \boldsymbol{b}\delta\boldsymbol{u} = \boldsymbol{b}N_j\delta\boldsymbol{u}_j = \boldsymbol{B}_j\delta\boldsymbol{u}_j \tag{6.49}$$

Matrices $\boldsymbol{B}_i$ and $\boldsymbol{B}_j$ can be written in the explicit form

$$\boldsymbol{B}_i = \begin{bmatrix} N_{i,x} & 0 & 0 \\ 0 & N_{i,y} & 0 \\ 0 & 0 & N_{i,z} \\ N_{i,z} & 0 & N_{i,x} \\ 0 & N_{i,z} & N_{i,y} \\ N_{i,y} & N_{i,x} & 0 \end{bmatrix}, \quad \boldsymbol{B}_j = \begin{bmatrix} N_{j,x} & 0 & 0 \\ 0 & N_{j,y} & 0 \\ 0 & 0 & N_{j,z} \\ N_{j,z} & 0 & N_{j,x} \\ 0 & N_{j,z} & N_{j,y} \\ N_{j,y} & N_{j,x} & 0 \end{bmatrix} \tag{6.50}$$

The expression for the internal work can be written accordingly:

$$\delta L_{int} = \delta\boldsymbol{u}_j^T \left(\int_V \boldsymbol{B}_j^T \boldsymbol{C} \boldsymbol{B}_i dV \right) \boldsymbol{u}_i \tag{6.51}$$

In this case, $\boldsymbol{k}^{ij}$ is written as

$$\boldsymbol{k}^{ij} = \int_V \boldsymbol{B}_j^T \boldsymbol{C} \boldsymbol{B}_i dV \tag{6.52}$$

The matrix in Equation (6.52) is of size 3×3,

$$\boldsymbol{k}^{ij} = \begin{bmatrix} k_{xx}^{ij} & k_{xy}^{ij} & k_{xz}^{ij} \\ k_{yx}^{ij} & k_{yy}^{ij} & k_{yz}^{ij} \\ k_{zx}^{ij} & k_{zy}^{ij} & k_{zz}^{ij} \end{bmatrix} \tag{6.53}$$

which is the 'fundamental nucleus' of a solid element. Indexes i and j range from 1 to the number of nodes N_{NE}. In explicit form, the terms of the nucleus of the stiffness matrix are

$$
\begin{aligned}
k_{xx}^{ij} &= (\lambda+2G)\int_V N_{i,x}N_{j,x}dV + G\int_V N_{i,y}N_{j,y}dV + G\int_V N_{i,z}N_{j,z}dV \\
k_{xy}^{ij} &= \lambda\int_V N_{i,y}N_{j,x}dV + G\int_V N_{i,x}N_{j,y}dV \\
k_{xz}^{ij} &= \lambda\int_V N_{i,z}N_{j,x}dV + G\int_V N_{i,x}N_{j,z}dV \\
k_{yx}^{ij} &= \lambda\int_V N_{i,x}N_{j,y}dV + G\int_V N_{i,y}N_{j,x}dV \\
k_{yy}^{ij} &= (\lambda+2G)\int_V N_{i,y}N_{j,y}dV + G\int_V N_{i,x}N_{j,x}dV + G\int_V N_{i,z}N_{j,z}dV \\
k_{yz}^{ij} &= \lambda\int_V N_{i,z}N_{j,y}dV + G\int_V N_{i,y}N_{j,z}dV \\
k_{zx}^{ij} &= \lambda\int_V N_{i,x}N_{j,z}dV + G\int_V N_{i,z}N_{j,x}dV \\
k_{zy}^{ij} &= \lambda\int_V N_{i,y}N_{j,z}dV + G\int_V N_{i,z}N_{j,y}dV \\
k_{zz}^{ij} &= (\lambda+2G)\int_V N_{i,z}N_{j,z}dV + G\int_V N_{i,x}N_{j,x}dV + G\int_V N_{i,y}N_{j,y}dV
\end{aligned}
\tag{6.54}
$$

6.4 FN for 1D, 2D and 3D Problems in Unique Form

The fundamental fact that Equation (6.54) can be derived from Equation (6.23) should be noted. The weak form can be derived, in the case of a solid, according to the following steps:

- the derivative is moved from the displacements to the shape functions;
- interpolation over the volume V is introduced;

and, as a result, the sign changes, since there is no need to integrate by parts (and no boundary terms are obtained).

In the case of the weak form, Equation (6.22) is not verified and each derivative should therefore be applied to the proper term; consequently, Equation (6.21) is used to derive the weak form instead of Equation (6.23). Let us compare the first two terms of Equation (6.21) and Equation (6.54):

Strong:

$$k_{xx} = -(\lambda+2G) \quad \partial_x \; \partial_x \quad - G \quad \partial_y \; \partial_y \quad - G \quad \partial_z \; \partial_z$$

Weak: (6.55)

$$k_{xx}^{ij} = +(\lambda+2G) \int_V N_{i,x}N_{j,x}dV \quad + G \int_V N_{i,y}N_{j,y}dV \quad + G \int_V N_{i,z}N_{j,z}dV$$

Once again the weak form can be derived from the strong form by introducing the shape functions: one for the displacements, N_i, and one for the virtual variations, N_j.

The integral over the volume, V, is a consequence of the introduction of the weak form. Both strong and weak forms are used to write an equilibrium equation but the former is satisfied at each point of volume V, while the latter is therefore written in integral form, and equilibrium is satisfied for mean quantities that come from an integral of V.

It should be noted that each term of the nucleus element includes a second derivative. In the case of the weak form, each derivative should be applied to the proper term. Let us compare the k_{xy} and k_{xy}^{ij} elements:

$$\begin{aligned} k_{xy} &= \; -\lambda \qquad \partial_x \; \partial_y \qquad - G \qquad \partial_y \; \partial_x \\ k_{xy}^{ij} &= \; +\lambda \int_V N_{j,x} N_{i,y} dV \; + G \int_V N_{j,y} N_{i,x} dV \end{aligned} \tag{6.56}$$

The first derivation should be applied to the shape function of the virtual variation of the displacement and it therefore acts on N_j, while the second derivation acts on the shape function of displacement, N_i.

The fundamental nucleus can thus be derived in weak form from the strong form by imposing an approximated formulation of the displacements. Since the compact formulation of the operators in strong form is

$$k_{xx} = -(\lambda + 2G)\ \partial_x \partial_x - \lambda\ \partial_z \partial_z - \lambda\ \partial_y \partial_y \tag{6.57}$$

$$k_{xy} = -\lambda\ \partial_x \partial_y - G\ \partial_y \partial_x \tag{6.58}$$

it is possible to derive the FN for any structural model introduced in Section 5.5. It is important to remember that the terms k_{yy} and k_{zz} in Equation (6.58) have the same form as k_{xx} with the indexes permuted, while k_{xz}, k_{yz}, k_{yx}, k_{zx} and k_{zy} can be obtained by permuting the indexes of k_{xy}.

6.4.1 *Three-Dimensional Models*

In the case of 3D models, the displacement field is approximated using only the shape functions introduced by the FEM. As a result,

$$\boldsymbol{u}(x, y, z) = N_i(x, y, z)\boldsymbol{u}_i \tag{6.59}$$

In compact notation, the weak form of the FN becomes

$$k_{xx}^{ij} = (\lambda + 2G) \int_V N_{i,x} N_{j,x} dV + G \int_V N_{i,z} N_{j,z} dV + G \int_V N_{i,y} N_{j,y} dV \tag{6.60}$$

$$k_{xy}^{ij} = \lambda \int_V N_{i,y} N_{j,x} dV + G \int_V N_{i,x} N_{j,y} dV \tag{6.61}$$

6.4.2 Two-Dimensional Models

The displacement field of a 2D model can be written as the product of the FE approximation on the reference surface, $N_i(x, y)$, and an expansion on the thickness direction, $F_\tau(z)$,

$$\boldsymbol{u}(x, y, z) = N_i(x, y)F_\tau(z)\boldsymbol{u}_{\tau i} \tag{6.62}$$

In compact form, the weak form of the FN becomes

$$\begin{aligned} k_{xx}^{\tau sij} &= (\lambda + 2G)\int_V (N_iF_\tau)_{,x}(N_jF_s)_{,x}dV + G\int_V (N_iF_\tau)_{,z}(N_jF_s)_{,z}dV \\ &+G\int_V (N_iF_\tau)_{,y}(N_jF_s)_{,y}dV \end{aligned} \tag{6.63}$$

$$k_{xy}^{\tau sij} = \lambda\int_V (N_iF_\tau)_{,y}(N_jF_s)_{,x}dV + G\int_V (N_iF_\tau)_{,x}(N_jF_s)_{,y}dV \tag{6.64}$$

The integral over volume V can be written as the product of two contributions: the integral over the reference surface, Ω, and the integral over thickness, h. Thus one obtains

$$dV = d\Omega\, dz \tag{6.65}$$

Using the rule for the derivative of a product, Equations (6.63) and (6.64) become

$$\begin{aligned} k_{xx}^{\tau sij} &= (\lambda + 2G)\int_\Omega\int_h (N_{i,x}F_\tau + N_i\cancel{F_{\tau,x}})(N_{j,x}F_s + N_j\cancel{F_{s,x}})dz\, d\Omega \\ &+G\int_\Omega\int_h (N_{i,y}F_\tau + N_i\cancel{F_{\tau,y}})(N_{j,y}F_s + N_j\cancel{F_{s,y}})dz\, d\Omega \\ &+G\int_\Omega\int_h (\cancel{N_{i,z}}F_\tau + N_iF_{\tau,z})(\cancel{N_{j,z}}F_s + N_jF_{s,z})dz\, d\Omega \end{aligned} \tag{6.66}$$

$$\begin{aligned} k_{xy}^{\tau sij} &= \lambda\int_\Omega\int_h (N_{i,y}F_\tau + N_i\cancel{F_{\tau,y}})(N_{j,x}F_s + N_j\cancel{F_{s,x}})dz\, d\Omega \\ &+G\int_\Omega\int_h (N_{i,x}F_\tau + N_i\cancel{F_{\tau,x}})(N_{j,y}F_s + N_j\cancel{F_{s,y}})dz\, d\Omega \end{aligned} \tag{6.67}$$

Equations (6.66) and (6.67) can be written as

$$\begin{aligned} k_{xx}^{\tau sij} &= (\lambda + 2G)\int_\Omega N_{i,x}N_{j,x}d\Omega\, dy \int_h F_\tau F_s dz + G\int_\Omega N_iN_jd\Omega \int_h F_{\tau,z}F_{s,z}dz \\ &+G\int_V N_{i,y}N_{j,y}d\Omega\, dy \int_h F_\tau F_s dz \end{aligned} \tag{6.68}$$

$$k_{xy}^{\tau sij} = \lambda\int_\Omega N_{i,y}N_{j,x}d\Omega\, dy \int_h F_\tau F_s dz + G\int_\Omega N_{i,x}N_{j,y}d\Omega\, dy \int_h F_\tau F_s dz \tag{6.69}$$

The explicit form of the FN related to a 2D problem is given in Chapters 10–12.

6.4.3 One-Dimensional Models

The 1D models are characterized by an FE approximation on the axis, $N_i(y)$, and an expansion on the cross-section, $F_\tau(x, z)$,

$$\boldsymbol{u}(x, y, z) = N_i(y)F_\tau(x, z)\boldsymbol{u}_{\tau i} \tag{6.70}$$

The compact formulation of the nucleus becomes

$$\begin{aligned} k_{xx}^{\tau sij} &= (\lambda + 2G)\int_V (N_iF_\tau)_{,x}(N_jF_s)_{,x}dV + G\int_V (N_iF_\tau)_{,z}(N_jF_s)_{,z}dV \\ &\quad + G\int_V (N_iF_\tau)_{,y}(N_jF_s)_{,y}dV \end{aligned} \tag{6.71}$$

$$k_{xy}^{\tau sij} = \lambda\int_V (N_iF_\tau)_{,y}(N_jF_s)_{,x}dV + G\int_V (N_iF_\tau)_{,x}(N_jF_s)_{,y}dV \tag{6.72}$$

The integral over volume V can be split into the integral over the cross-section, A, and the integral along the beam axis, y. Thus one has

$$dV = dA\ dy \tag{6.73}$$

Using the rule for the derivative of a product, Equations (6.71) and (6.72) become

$$\begin{aligned} k_{xx}^{\tau sij} &= (\lambda + 2G)\int_l\int_A (\cancel{N_{i,x}}F_\tau + N_iF_{\tau,x})(\cancel{N_{j,x}}F_s + N_jF_{s,x})dA\,dy \\ &\quad + G\int_l\int_A (\cancel{N_{i,z}}F_\tau + N_iF_{\tau,z})(\cancel{N_{j,z}}F_s + N_jF_{s,z})dA\,dy \end{aligned} \tag{6.74}$$

$$\begin{aligned} k_{xy}^{\tau sij} &= \lambda\int_l\int_A (N_{i,y}F_\tau + (N_i\cancel{F_{\tau,y}})(\cancel{N_{j,x}}F_s + N_jF_{s,x})dA\,dy \\ &\quad + G\int_l\int_A (\cancel{N_{i,x}}F_\tau + (N_iF_{\tau,x})(N_{j,y}F_s + N_j\cancel{F_{s,y}})dA\,dy \\ &\quad + G\int_l\int_A (N_{i,y}F_\tau + N_i\cancel{F_{\tau,y}})(N_{j,y}F_s + N_j\cancel{F_{s,y}})dA\,dy \end{aligned} \tag{6.75}$$

Equations (6.74) and (6.75) can be written as

$$\begin{aligned} k_{xx}^{\tau sij} &= (\lambda + 2G)\int_l N_iN_j\,dy\int_A F_{\tau,x}F_{s,x}dA + G\int_l N_iN_j\,dy\int_A F_{\tau,z}F_{s,z}dA \\ &\quad + G\int_l N_{i,y}N_{j,y}\,dy\int_A F_\tau F_sdA \end{aligned} \tag{6.76}$$

$$k_{xy}^{\tau sij} = \lambda\int_l N_{i,y}N_jdy\int_A F_\tau F_{s,x}dA + G\int_l N_iN_{j,y}dy\int_A F_{\tau,x}F_sdA \tag{6.77}$$

The extended form of the FN for a 1D problem is given in Chapters 8 and 9.

6.5 CUF at a Glance

The CUF consists of a unified framework which can be used to derive any FEM model for any structural element. The use of the FNs allows a compact form to be obtained (see also Carrera 2014). Table 6.1 summarizes the main features of the CUF in only one page. The upper part of the table shows the strong form of the equilibrium equation, written in terms of displacements. No assumptions are made about the displacements. Matrix $\boldsymbol{k}$ can be used to obtain the formulation of the FN for the stiffness matrix of a general structural problem. Two representative terms of the nucleus are reported and the others can be derived through a permutation of the indexes. The problem can be solved for any arbitrary geometry and boundary condition using the FEM. The displacement field can be assumed 'a priori' and it can be 1D, 2D or 3D. The solid case is reported. The displacements are expressed in terms of shape functions, and no expansion is introduced in this model. The 2D model introduces an expansion, F_τ, along the thickness direction, while the solution on the reference surface is obtained via FEM. The 1D model approximates the cross-section behaviour using the F_τ functions. The FEM is used to investigate a structure along the beam axis. Whatever the nature of the model, the FN can be derived from the formulation of the virtual variation of the internal work. It can be derived from the strong form, as differential operators are applied to the shape functions, N_i or N_j, and to the approximation functions, F_τ or F_s. The integrals appear because, in the weak formulation, equilibrium is not satisfied at each point, but only at some points, called nodes. The FN can be easily implemented in an FEM code and the global matrix can be assembled using a different loop for each index. The assembly procedure, from the nucleus to the element matrix, is reported at the bottom of the table.

6.5.1 *Choice of N_i, N_j, F_τ and F_s*

The displacement field is composed of three displacement components: u_x, u_y and u_z. These three components can be approximated using the same expansion, but they can also be approximated using different function expansions. The compact formulation introduced by the CUF does not place any constraint on the choice of the functions that have to be used in the approximation. The displacement and its virtual variation can be written in index form as

$$\boldsymbol{u} = F_\tau N_i \boldsymbol{u}_{\tau i}, \quad \delta\boldsymbol{u} = F_s N_j \delta\boldsymbol{u}_{sj} \tag{6.78}$$

If each displacement component is approximated using a different expansion, each component could be written as

$$u_x = F_\tau^x N_i^x u_{x_{\tau i}}, \quad \delta u_x = F_s^x N_j^x \delta u_{x_{sj}} \tag{6.79}$$

$$u_y = F_\tau^y N_i^y u_{y_{\tau i}}, \quad \delta u_y = F_s^y N_j^y \delta u_{y_{sj}} \tag{6.80}$$

$$u_z = F_\tau^z N_i^z u_{z_{\tau i}}, \quad \delta u_z = F_s^z N_j^z \delta u_{z_{sj}} \tag{6.81}$$

where the indexes x, y or z indicate the expansions used for each component. The use of different expansions for each displacement component introduces a different number of terms

Table 6.1 A schematic description of the CUF and the related fundamental nucleus of the stiffness matrix for 3D, 2D and 1D models

Equilibrium equations in Strong Form $\rightarrow \quad \delta L_i = \int_V \delta \boldsymbol{u} \boldsymbol{k} \boldsymbol{u} dV + \int_S \ldots dS$

$$\underbrace{\begin{bmatrix} k_{xx} & k_{xy} & k_{xz} \\ k_{yx} & k_{yy} & k_{yz} \\ k_{zx} & k_{zy} & k_{zz} \end{bmatrix}}_{\boldsymbol{k}} \underbrace{\begin{Bmatrix} u_x \\ u_y \\ u_z \end{Bmatrix}}_{\boldsymbol{u}} = \underbrace{\begin{Bmatrix} p_x \\ p_y \\ p_z \end{Bmatrix}}_{\boldsymbol{p}}$$

$$k_{xx} = -(\lambda + 2G)\, \partial_{xx} - G\, \partial_{zz} - G\, \partial_{yy};$$
$$k_{xy} = -\lambda\, \partial_{xy} - G\, \partial_{yx};$$
$$k_{xz} = \ldots$$
$$\lambda = (E\nu)/[(1+\nu)(1-2\nu)]; \quad G = E/[2(1+\nu)]$$

$\boldsymbol{u} = \boldsymbol{u}(x, y, z)$

$\delta \boldsymbol{u} = \delta \boldsymbol{u}(x, y, z)$

The diagonal (e.g. k_{xx}) and the non-diagonal (e.g. k_{xy}) terms can be obtained through proper index permutations.

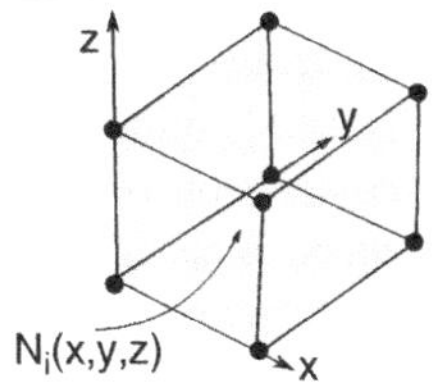

$\boldsymbol{u} = N_i(x, y, z)\boldsymbol{u}_i$

$\delta \boldsymbol{u} = N_j(x, y, z)\delta \boldsymbol{u}_j$

3D FEM Formulation $\rightarrow \quad \delta L_i = \delta \boldsymbol{u}_j \boldsymbol{k}^{ij} \boldsymbol{u}_i$

$$k_{xx}^{ij} = (\lambda + 2G)\int_V N_{j,x} N_{i,x} dV + G \int_V N_{j,z} N_{i,z} dV + G \int_V N_{j,y} N_{i,y} dV;$$

$$k_{xy}^{ij} = \lambda \int_V N_{j,y} N_{i,x} dV + G \int_V N_{j,x} N_{i,y} dV$$

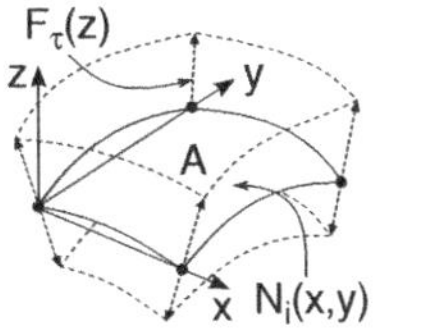

$\boldsymbol{u} = N_i(x, y) F_\tau(z) \boldsymbol{u}_{\tau i}$

$\delta \boldsymbol{u} = N_j(x, y) F_s(z) \delta \boldsymbol{u}_{sj}$

2D FEM Formulation $\rightarrow \quad \delta L_i = \delta \boldsymbol{u}_{sj} \boldsymbol{k}^{\tau sij} \boldsymbol{u}_{\tau i}$

$$k_{xx}^{\tau sij} = (\lambda + 2G)\int_\Omega N_{i,x} N_{j,x} d\Omega \int_h F_\tau F_s dz$$
$$+ G \int_\Omega N_i N_j d\Omega \int_h F_{\tau,z} F_{s,z} dz + G \int_V N_{i,y} N_{j,y} d\Omega \int_h F_\tau F_s dz;$$

$$k_{xy}^{\tau sij} = \lambda \int_\Omega N_{i,y} N_{j,y} d\Omega \int_h F_\tau F_s dz + G \int_\Omega N_{i,x} N_{j,y} d\Omega \int_h F_\tau F_s dz$$

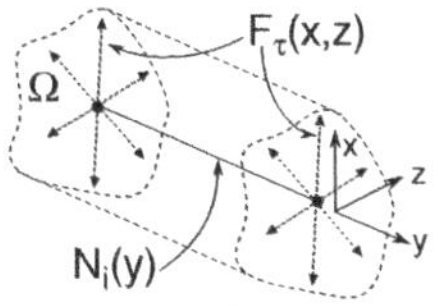

$\boldsymbol{u} = N_i(y) F_\tau(x, z) \boldsymbol{u}_{\tau i}$

$\delta \boldsymbol{u} = N_j(y) F_s(x, z) \delta \boldsymbol{u}_{sj}$

1D FEM Formulation $\rightarrow \quad \delta L_i = \delta \boldsymbol{u}_{sj} \boldsymbol{k}^{\tau sij} \boldsymbol{u}_{\tau i}$

$$k_{xx}^{\tau sij} = (\lambda + 2G)\int_l N_i N_j dy \int_A F_{\tau,x} F_{s,x} dA$$
$$+ G \int_l N_i N_j dy \int_A F_{\tau,z} F_{s,z} dA + G \int_l N_{i,y} N_{j,y} dy \int_A F_\tau F_s dA;$$

$$k_{xy}^{\tau sij} = \lambda \int_l N_{i,y} N_j dy \int_A F_\tau F_{s,x} dA + G \int_l N_i N_{j,y} dy \int_A F_{\tau,x} F_s dA$$

CUF leads to the automatic implementation of any theory of structures through 4 loops (i.e. 4 indexes):

- $\boldsymbol{\tau}$ and $\boldsymbol{s}$ deal with the functions that approximate the displacement field and its virtual variation along the plate/shell thickness ($F_\tau(z), F_s(z)$) or over the beam cross-section ($F_\tau(x, z), F_s(x, z)$);
- $\boldsymbol{i}$ and $\boldsymbol{j}$ deal with the shape functions of the FE model, (3D:$N_i(x, y, z), N_j(x, y, z)$; 2D:$N_i(x, y), N_j(x, y)$; 1D:$N_i(y), N_j(y)$).

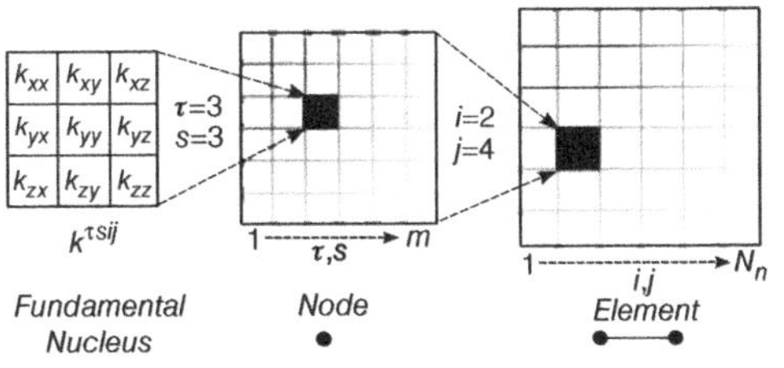

This Table shows the essential features of the CUF. The strong form of the equilibrium equations allows one to derive a compact formulation for the fundamental nucleus. The nine elements of the FN can be written using only 2 terms. In this table, k_{xx} and k_{xy} are reported. All the remaining terms can be derived by a permutation of the indexes. This compact formulation is used to derive the 3D, 2D and 1D models in weak form for each expansion, therefore, instead of M, one has M^x, M^y and M^z, which are the number of terms in the expansion of u_x, u_y and u_z, respectively.

The choice of these expansions is arbitrary. The shape functions used in the FEM solution, N_i and N_j, are the same shape functions that are used in classical models. They usually come from Lagrange polynomials, but many other functions could be used, such as Hermite polynomials. The books by Hughes (2000), Oñate (2009), Reddy (2005), Zienkiewicz *et al.* (2005) and Bathe (1996) offer an exhaustive list of classical FEM formulations.

While shape functions have a classical formulation, the approximation introduced through the use of expansions F_τ and F_s can consider a large number of functions. Three expansions are considered in this book: Taylor, Lagrange and Legendre expansions. The details of these expansions and their applications to refined 1D and 2D models are given in the following chapters. However, the CUF allows any expansion to be used in the approximation of the displacement field. Trigonometric, exponential and logarithmic expansions can be of particular interest in this context. The paper by Carrera *etal.* (2013) discusses the use of such expansions for the case of 1D models.

References

Bathe K 1996 *Finite Element Procedure*. Prentice Hall.

Carrera E 2014 A modern approach to theory of beams, plates, shells and solids based on fundamental nuclei.

Carrera E, Filippi M and Zappino E 2013 Laminated beam analysis by polynomial, trigonometric, exponential and zig-zag theories. *European Journal of Mechanics – A/Solids*, **41** 58–69.

Hughes TJR 2000 *The Finite Element Method: Linear Static and Dynamic Finite Element Analysis*. Dover.

Oñate E 2009 *Structural Analysis with the Finite Element Method: Linear Statics*. Volume 1. *Basis and Solids*. Springer.

Reddy JN 2005 *An Introduction to the Finite Element Method*. Third Edition. McGraw-Hill Education.

Washizu K 1968 *Variational Methods in Elasticity and Plasticity*: Pergamon Press.

Zienkiewicz OC, Taylor RL and Zhu JZ 2005 *The Finite Element Method: Its Basis and Fundamentals*. Sixth Edition. Elsevier.

7

Three-Dimensional FEM Formulation (Solid Elements)

The subject of this chapter is the 3D FE model. Table 7.1 shows the FN for 3D models that is discussed in this chapter. In the 3D cases FE approximations are introduced in the three directions (namely, x, y and z) and the related FNs are only described by indexes i and j. That is, the theory of structure approximation is taken out of the derivation and the CUF is not used. However, it is useful to point out that the use of FE index notation i and j is formally the same as index notation τ and s in the theory of structures.

7.1 An Eight-Node Element Using Classical Matrix Notation

The displacement field of a 3D FE is not affected by any 'a priori' assumptions. The FEM is used to solve the problem over the full 3D domain, thus the shape functions are 3D. The displacement field can be written as follows:

$$u(x, y, z) = \boldsymbol{N}\boldsymbol{u} \tag{7.1}$$

where matrix $\boldsymbol{N}$ contains the shape functions and vector $\boldsymbol{u}$ contains the nodal displacements. An eight-node element is considered in this section. This element has a hexahedral geometry, which can be seen in Figure 7.1, and the shape functions have a linear formulation.

Finite Element Analysis of Structures Through Unified Formulation, First Edition.
Erasmo Carrera, Maria Cinefra, Marco Petrolo and Enrico Zappino.

Table 7.1 A schematic description of the CUF and the related FN of the stiffness matrix for 3D models

Equilibrium equations in Strong Form $\rightarrow \quad \delta L_i = \int_V \delta \boldsymbol{u} \boldsymbol{k} \boldsymbol{u} dV + \int_S \ldots dS$

$$\underbrace{\begin{bmatrix} k_{xx} & k_{xy} & k_{xz} \\ k_{yx} & k_{yy} & k_{yz} \\ k_{zx} & k_{zy} & k_{zz} \end{bmatrix}}_{\boldsymbol{k}} \underbrace{\begin{Bmatrix} u_x \\ u_y \\ u_z \end{Bmatrix}}_{\boldsymbol{u}} = \underbrace{\begin{Bmatrix} p_x \\ p_y \\ p_z \end{Bmatrix}}_{\boldsymbol{p}}$$

$$k_{xx} = -(\lambda + 2G)\,\partial_{xx} - G\,\partial_{zz} - G\,\partial_{yy};$$
$$k_{xy} = -\lambda\,\partial_{xy} - G\,\partial_{yx};$$
$$k_{xz} = \ldots$$

$\lambda = (E\nu)/[(1+\nu)(1-2\nu)]; \quad G = E/[2(1+\nu)]$

$\boldsymbol{u} = \boldsymbol{u}(x,y,z)$
$\delta\boldsymbol{u} = \delta\boldsymbol{u}(x,y,z)$

The diagonal (e.g. k_{xx}) and the non-diagonal (e.g. k_{xy}) terms can be obtained through proper index permutations.

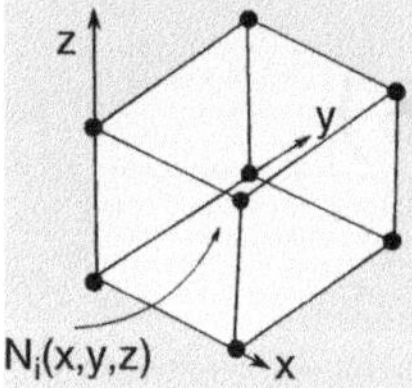

$\boldsymbol{u} = N_i(x,y,z)\boldsymbol{u}_i$
$\delta\boldsymbol{u} = N_j(x,y,z)\delta\boldsymbol{u}_j$

3D FEM Formulation $\rightarrow \quad \delta L_i = \delta \boldsymbol{u}_j \boldsymbol{k}^{ij} \boldsymbol{u}_i$

$$k_{xx}^{ij} = (\lambda + 2G) \int_V N_{j,x} N_{i,x} dV + G \int_V N_{j,z} N_{i,z} dV + G \int_V N_{j,y} N_{i,y} dV;$$
$$k_{xy}^{ij} = \lambda \int_V N_{j,y} N_{i,x} dV + G \int_V N_{j,x} N_{i,y} dV$$

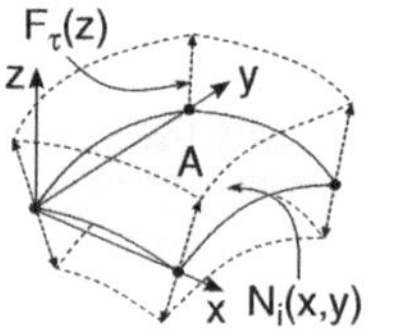

$\boldsymbol{u} = N_i(x,y)F_\tau(z)\boldsymbol{u}_{\tau i}$
$\delta\boldsymbol{u} = N_j(x,y)F_s(z)\delta\boldsymbol{u}_{sj}$

2D FEM Formulation $\rightarrow \quad \delta L_i = \delta \boldsymbol{u}_{sj} \boldsymbol{k}^{\tau sij} \boldsymbol{u}_{\tau i}$

$$k_{xx}^{\tau sij} = (\lambda + 2G) \int_\Omega N_{i,x} N_{j,x} d\Omega \int_h F_\tau F_s dz$$
$$+ G \int_\Omega N_i N_j d\Omega \int_h F_{\tau,z} F_{s,z} dz + G \int_V N_{i,y} N_{j,y} d\Omega \int_h F_\tau F_s dz;$$
$$k_{xy}^{\tau sij} = \lambda \int_\Omega N_{i,y} N_{j,y} d\Omega \int_h F_\tau F_s dz + G \int_\Omega N_{i,x} N_{j,y} d\Omega \int_h F_\tau F_s dz$$

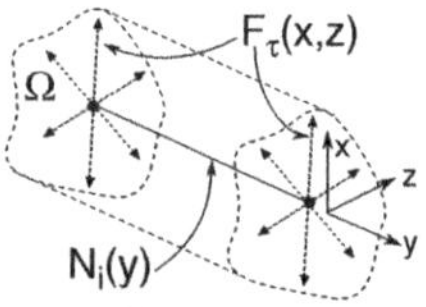

$\boldsymbol{u} = N_i(y)F_\tau(x,z)\boldsymbol{u}_{\tau i}$
$\delta\boldsymbol{u} = N_j(y)F_s(x,z)\delta\boldsymbol{u}_{sj}$

1D FEM Formulation $\rightarrow \quad \delta L_i = \delta \boldsymbol{u}_{sj} \boldsymbol{k}^{\tau sij} \boldsymbol{u}_{\tau i}$

$$k_{xx}^{\tau sij} = (\lambda + 2G) \int_l N_i N_j dy \int_A F_{\tau,x} F_{s,x} dA$$
$$+ G \int_l N_i N_j dy \int_A F_{\tau,z} F_{s,z} dA + G \int_l N_{i,y} N_{j,y} dy \int_A F_\tau F_s dA;$$
$$k_{xy}^{\tau sij} = \lambda \int_l N_{i,y} N_j dy \int_A F_\tau F_{s,x} dA + G \int_l N_i N_{j,y} dy \int_A F_{\tau,x} F_s dA$$

CUF leads to the automatic implementation of any theory of structures through 4 loops (i.e. 4 indexes):

- τ and s deal with the functions that approximate the displacement field and its virtual variation along the plate/shell thickness ($F_\tau(z), F_s(z)$) or over the beam cross-section ($F_\tau(x,z), F_s(x,z)$);
- $\boldsymbol{i}$ and $\boldsymbol{j}$ deal with the shape functions of the FE model, (3D:$N_i(x,y,z), N_j(x,y,z)$; 2D:$N_i(x,y), N_j(x,y)$; 1D:$N_i(y), N_j(y)$).

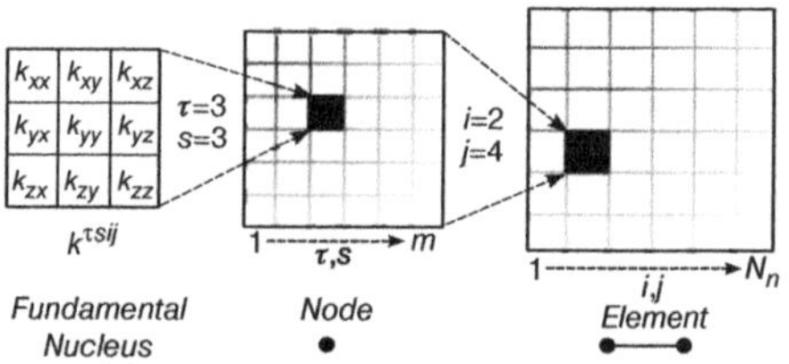

This table shows the essential features of the CUF for 3D models.

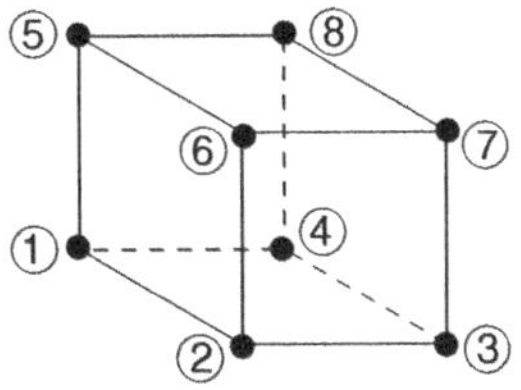

Figure 7.1 Example of an eight-node element

The displacement field can be written in explicit form as

$$u = \begin{bmatrix} N_1 & 0 & 0 & \dots & N_i & 0 & 0 & \dots & N_8 & 0 & 0 \\ 0 & N_1 & 0 & \dots & 0 & N_i & 0 & \dots & 0 & N_8 & 0 \\ 0 & 0 & N_1 & \dots & 0 & 0 & N_i & \dots & 0 & 0 & N_8 \end{bmatrix} \begin{Bmatrix} u_{x_1} \\ u_{y_1} \\ u_{z_1} \\ \vdots \\ u_{x_i} \\ u_{y_i} \\ u_{z_i} \\ \vdots \\ u_{x_8} \\ u_{y_8} \\ u_{z_8} \end{Bmatrix} \tag{7.2}$$

7.1.1 Stiffness Matrix

Let us consider a static response problem. The PVD states that

$$\delta L_{int} = \delta L_{ext} \tag{7.3}$$

The formulation of the internal work was derived in Equation (6.41) as

$$\delta L_i = \delta \boldsymbol{U}^T \left(\int_V \boldsymbol{B}^T \boldsymbol{C} \boldsymbol{B} dV \right) \boldsymbol{U} \tag{7.4}$$

Matrix $\boldsymbol{B}$ is the product of the differential operator $\boldsymbol{b}$, see Equation (2.4), and the matrix of the shape functions, $\boldsymbol{N}$,

$$\boldsymbol{B} = \boldsymbol{b}\boldsymbol{N} = \begin{bmatrix} N_{1,x} & 0 & 0 & \dots & N_{i,x} & 0 & 0 & \dots & N_{8,x} & 0 & 0 \\ 0 & N_{1,y} & 0 & \dots & 0 & N_{i,y} & 0 & \dots & 0 & N_{8,y} & 0 \\ 0 & 0 & N_{1,z} & \dots & 0 & 0 & N_{i,z} & \dots & 0 & 0 & N_{8,z} \\ N_{1,z} & 0 & N_{1,x} & \dots & N_{i,z} & 0 & N_{i,x} & \dots & N_{8,z} & 0 & N_{8,x} \\ 0 & N_{1,z} & N_{1,y} & \dots & 0 & N_{i,z} & N_{i,y} & \dots & 0 & N_{8,z} & N_{8,y} \\ N_{1,y} & N_{1,x} & 0 & \dots & N_{i,y} & N_{i,x} & 0 & \dots & N_{8,y} & N_{8,x} & 0 \end{bmatrix} \tag{7.5}$$

Matrix $\boldsymbol{B}$ has a dimension of 6×24 while matrix $\boldsymbol{C}$ has a dimension of 6×6, as shown in Equation (2.14). The internal work thus can be written as

$$\delta L_{int} = \delta \boldsymbol{U}^T \underbrace{\int_V \underbrace{\Big[\overbrace{[6 \times 24]}^{\boldsymbol{B}^T} \overbrace{[6 \times 6]}^{\boldsymbol{C}} \overbrace{[\quad 24 \times 6 \quad]}^{\boldsymbol{B}} \Big]}_{[24 \times 24]} dV \, \boldsymbol{U}}_{\boldsymbol{K}} \tag{7.6}$$

Equation (7.6) shows the form of the stiffness matrix, which has a dimension of 24×24.

7.1.2 Load Vector

The load vector, $\boldsymbol{P}$, originates from the work done by the external forces. Different loads can be applied to the structure, namely volume loads (g), surface loads (p), line loads (q) and concentrated loads (P).

The external work δL_{ext} can be expressed using the PLV,

$$\begin{aligned} \delta L_{ext} = & \int_V \delta \boldsymbol{u}^T \boldsymbol{g} dV + \int_S \delta \boldsymbol{u}^T \boldsymbol{p} dS \\ & + \int_L \delta \boldsymbol{u}^T \boldsymbol{q} dy + \delta \boldsymbol{u}^T \boldsymbol{P} \end{aligned} \tag{7.7}$$

If the displacements are expressed in the FE formulation, the equation becomes

$$\begin{aligned} \delta L_{ext} = & \delta \boldsymbol{U}^T \int_V \boldsymbol{N}^T \boldsymbol{g} dV + \delta \boldsymbol{U}^T \int_S \boldsymbol{N}^T \boldsymbol{p} dS \\ & + \delta \boldsymbol{U}^T \int_L \boldsymbol{N}^T \boldsymbol{q} dy + \delta \boldsymbol{U}^T \boldsymbol{N}^T \boldsymbol{P} \end{aligned} \tag{7.8}$$

The external work in the case of a concentrated load, P, acting at a point with coordinates x_P, y_P and z_P, is given by

$$\delta L^P_{ext} = \delta \boldsymbol{u}(x_P, y_P, z_P)^T P \tag{7.9}$$

and, in terms of nodal variables,

$$\delta L^P_{ext} = \delta \boldsymbol{U}^T \boldsymbol{N}^T |_{(x_P, y_P, z_P)} P = \delta \boldsymbol{U} \boldsymbol{P} \tag{7.10}$$

where the matrix of the shape functions, evaluated at the loading point, is denoted by $\boldsymbol{N}|_{(x_p, y_p, z_p)}$. Load vector $\boldsymbol{P}$ has a dimension of 24×1,

$$\boldsymbol{P}^T = \{P_{x_1}, P_{y_1}, P_{z_1}, \dots, P_{x_i}, P_{y_i}, P_{z_i}, \dots, P_{x_8}, P_{y_8}, P_{z_8}\} \tag{7.11}$$

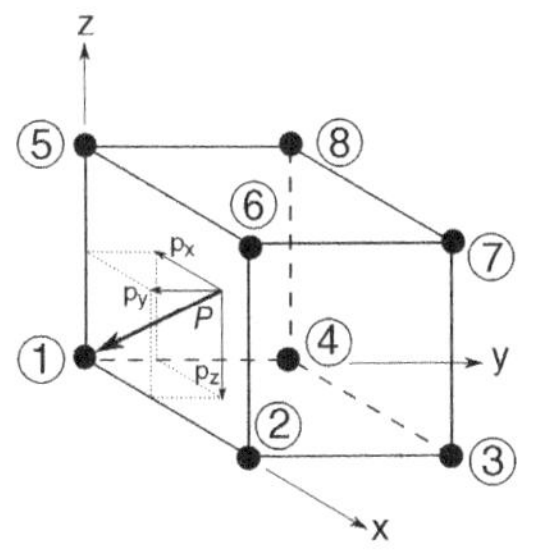

Figure 7.2 Eight-node element loaded by a concentrated load at node 1

The dimension of the load vector depends on the number of nodes of the element.

Example 7.1.1 *Let us consider an eight-node element with a concentrated load, P, applied to node 1. Figure 7.2 shows this configuration. Vector load* $\boldsymbol{P}$ *can be derived using the formulation given in Equation (7.10). On the basis of the definition of the shape functions, all the functions in node 1 are zero, except* N_1. *Matrix* $\boldsymbol{N}|_{(x_P,y_P,z_P)}$ *becomes*

$$\boldsymbol{N}|_{(x_P,y_P,z_P)} = \begin{bmatrix} 1 & 0 & 0 & \dots & 0 & 0 & 0 \\ 0 & 1 & 0 & \dots & 0 & 0 & 0 \\ 0 & 0 & 1 & \dots & 0 & 0 & 0 \end{bmatrix} \tag{7.12}$$

Thus, the load vector is

$$\boldsymbol{P} = \boldsymbol{N}^T|_{(x_P,y_P,z_P)} P = \begin{bmatrix} 1 & 0 & 0 \\ 0 & 1 & 0 \\ 0 & 0 & 1 \\ 0 & 0 & 0 \\ \vdots & \vdots & \vdots \\ 0 & 0 & 0 \end{bmatrix} \begin{Bmatrix} P_x \\ P_y \\ P_z \end{Bmatrix} = \begin{Bmatrix} P_{x_1} \\ P_{y_1} \\ P_{z_1} \\ 0 \\ \vdots \\ 0 \end{Bmatrix} \tag{7.13}$$

7.2 Derivation of the Stiffness Matrix Using the Index Notation

Section 7.1 introduced the FE formulation for an eight-node element. The form of the stiffness matrix and the load vector depends on the element that is used. This means that if the element is changed, the form of the stiffness matrix and of the load vector changes. Therefore, an 'ad hoc' form is needed for each different element that is considered. The use of the index notation allows the matrices to be derived for a solid element, regardless of the number and nature of the shape functions used in the element.

7.2.1 Governing Equations

The equations that describe the dynamic motion of a deformable body are addressed. They can be derived using the PVD, which in dynamic form can be written as

$$\delta L_{int} = \delta L_{ext} - \delta L_{ine} \tag{7.14}$$

where L_{int} is the internal work, L_{ext} is the work done by the external forces, L_{ine} is the inertial work and δ is the virtual variation. The different work contributions can be expressed in terms of stresses, strains and displacements. The virtual internal work is

$$\delta L_{int} = \int_V \delta \boldsymbol{\varepsilon}^T \boldsymbol{\sigma} dV \tag{7.15}$$

The inertial virtual work can be written as

$$\delta L_{ine} = \int_V \delta \boldsymbol{u} \rho \ddot{\boldsymbol{u}} dV \tag{7.16}$$

where ρ is the density of the material.

Finally, the virtual work done by the external loads is

$$\begin{aligned} \delta L_{ext} = & \int_V \delta \boldsymbol{u}^T \boldsymbol{g} dV + \int_S \delta \boldsymbol{u}^T \boldsymbol{p} dS \\ & + \int_L \delta \boldsymbol{u}^T \boldsymbol{q} dy + \delta \boldsymbol{u}^T \boldsymbol{P} \end{aligned} \tag{7.17}$$

The stresses and strains used in the expression of the virtual work can be expressed in terms of displacements. By recalling the equations introduced in Chapter 2, the strains can be written, in terms of displacements, using the geometric relation

$$\boldsymbol{\varepsilon} = \boldsymbol{b}\boldsymbol{u} \tag{7.18}$$

where $\boldsymbol{b}$ is a differential operator that was introduced in Equation (2.4). The stress can be derived using Hooke's law:

$$\boldsymbol{\sigma} = \boldsymbol{C}\boldsymbol{\varepsilon} \tag{7.19}$$

where $\boldsymbol{C}$ contains the coefficients of the material. If these equations are used, it is possible to write the virtual work in terms of displacements.

7.2.2 FE Approximation in the CUF

Use of the FEM allows the displacement field to be written as the sum of the known functions multiplied by a constant. In the simple case introduced in Section 7.1, an eight-node element was considered, and the displacement field can therefore be written as

$$\boldsymbol{u} = u_1 N_1 + u_2 N_2 + u_3 N_3 + u_4 N_4 + u_5 N_5 + u_6 N_6 + u_7 N_7 + u_8 N_8 \tag{7.20}$$

The same displacement field can be written in index form, as shown in Section 6.3. If i is the index used for the displacement,

$$\boldsymbol{u} = N_i \boldsymbol{u}_i \tag{7.21}$$

The virtual variation of the displacements can be written in the same form using index j,

$$\delta\boldsymbol{u} = N_j \delta\boldsymbol{u}_j \tag{7.22}$$

The strains, and their virtual variations, can also be written in this compact form,

$$\boldsymbol{\varepsilon} = \boldsymbol{b} N_i \boldsymbol{u}_i \tag{7.23}$$

$$\delta\boldsymbol{\varepsilon} = \boldsymbol{b} N_j \delta\boldsymbol{u}_j \tag{7.24}$$

In the same way, the stresses become

$$\boldsymbol{\sigma} = \boldsymbol{C}\boldsymbol{b} N_i \boldsymbol{u}_i \tag{7.25}$$

Indexes i and j can vary according to the number of nodes of the element.

7.2.3 Stiffness Matrix

The stresses and strains can be expressed in the compact form shown in Equations (7.24) and (7.25). The virtual variation of the internal work, in compact form, becomes

$$\delta L_{int} = \delta\boldsymbol{u}_j \left(\int_V N_j \boldsymbol{b}^T \boldsymbol{C}\boldsymbol{b} N_i dV \right) \boldsymbol{u}_i \tag{7.26}$$

It is now possible to introduce the 3×3 matrix $\boldsymbol{k}^{ij}$,

$$\delta L_{int} = \delta\boldsymbol{u}_j \boldsymbol{k}^{ij} \boldsymbol{u}_i \tag{7.27}$$

Matrix $\boldsymbol{k}^{ij}$ is the fundamental nucleus of the stiffness matrix and is a 3×3 matrix, as shown in the following formula:

$$\boldsymbol{k}^{ij} = \int_V \underbrace{\left[N_j \overbrace{[3\times 6]}^{\boldsymbol{b}^T} \overbrace{[6\times 6]}^{\boldsymbol{C}} \overbrace{[6\times 3]}^{\boldsymbol{b}} N_i \right]}_{[3\times 3]} dV = \begin{bmatrix} k^{ij}_{xx} & k^{ij}_{xy} & k^{ij}_{xz} \\ k^{ij}_{yx} & k^{ij}_{yy} & k^{ij}_{yz} \\ k^{ij}_{zx} & k^{ij}_{zy} & k^{ij}_{zz} \end{bmatrix} \tag{7.28}$$

In explicit form, the nine terms are

$$
\begin{aligned}
k_{xx}^{ij} &= (\lambda + 2G)\int_V N_{i,x}N_{j,x}dV + G\int_V N_{i,y}N_{j,y}dV + G\int_V N_{i,z}N_{j,z}dV \\
k_{xy}^{ij} &= \lambda\int_V N_{i,y}N_{j,x}dV + G\int_V N_{i,x}N_{j,y}dV \\
k_{xz}^{ij} &= \lambda\int_V N_{i,z}N_{j,x}dV + G\int_V N_{i,x}N_{j,z}dV \\
k_{yx}^{ij} &= \lambda\int_V N_{i,x}N_{j,y}dV + G\int_V N_{i,y}N_{j,x}dV \\
k_{yy}^{ij} &= (\lambda + 2G)\int_V N_{i,y}N_{j,y}dV + G\int_V N_{i,x}N_{j,x}dV + G\int_V N_{i,z}N_{j,z}dV \\
k_{yz}^{ij} &= \lambda\int_V N_{i,z}N_{j,y}dV + G\int_V N_{i,y}N_{j,z}dV \\
k_{zx}^{ij} &= \lambda\int_V N_{i,x}N_{j,z}dV + G\int_V N_{i,z}N_{j,x}dV \\
k_{zy}^{ij} &= \lambda\int_V N_{i,y}N_{j,z}dV + G\int_V N_{i,z}N_{j,y}dV \\
k_{zz}^{ij} &= (\lambda + 2G)\int_V N_{i,z}N_{j,z}dV + G\int_V N_{i,x}N_{j,x}dV + G\int_V N_{i,y}N_{j,y}dV
\end{aligned}
\tag{7.29}
$$

The FN has an invariant form which does not change for any change in the shape functions or the number of nodes in the element. It is straightforward to build the stiffness matrix using the FN. Figure 7.3 shows a graphical representation of the assembly procedure.

The loops on indexes i and j allow the matrix of the single element to be built. More elements can be assembled as in the classical FEM by superimposing the stiffness of the shared nodes

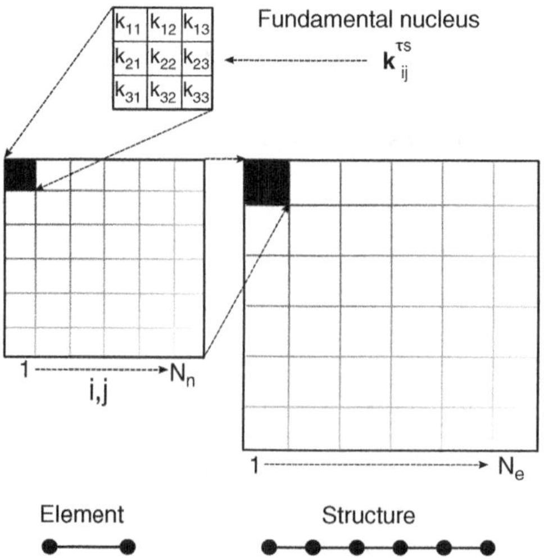

Figure 7.3 Assembly procedure for the stiffness matrix of a solid model

according to the PVD statements. Equation (7.29) clearly shows that the FNs of the 3D FE are the same as those of the weak form in 3D elasticity problems, see Chapter 6. If the strong form of element k_{xx} is compared with the same term in Equation (7.29), we obtain

$$\begin{array}{lcccccc} k_{xx} = -(\lambda + 2G) & \partial_x \ \partial_x & -G & \partial_y \ \partial_y & -G & \partial_z \ \partial_z \\ k_{xx}^{ij} = +(\lambda + 2G) & \int_V N_{i,x} N_{j,x} dV & +G & \int_V N_{i,y} N_{j,y} dV & +G & \int_V N_{i,z} N_{j,z} dV \end{array} \tag{7.30}$$

The weak form of the FN can be easily derived from the strong form in just a few steps:

- The derivative is moved from the displacements to the shape functions; the second derivatives must be split into a derivative of the displacement and one for its virtual variation.
- The sign changes since there is no need to integrate by parts.

It is clear that the nine components of the FN in Equation (7.29) can be derived from those in Table 7.1.

7.2.4 Mass Matrix

The virtual variation of the inertial work, as well as the internal work, can be expressed in terms of displacements. Equation (7.16) can be used for this purpose. If the displacements are expressed in compact form, the inertial work becomes

$$\delta L_{ine} = \delta \boldsymbol{u}_j \left(\int_V N_j \boldsymbol{I} \rho \boldsymbol{I} N_i dV \right) \ddot{\boldsymbol{u}}_i \tag{7.31}$$

Identity matrix $\boldsymbol{I}$ is introduced and the FN of the mass matrix is

$$\delta L_{ine} = \delta \boldsymbol{u}_j \boldsymbol{m}^{ij} \ddot{\boldsymbol{u}}_i \tag{7.32}$$

where

$$\boldsymbol{m}^{ij} = \int_V N_j \boldsymbol{I} \rho \boldsymbol{I} N_i dV = \begin{bmatrix} m_{xx}^{ij} & 0 & 0 \\ 0 & m_{yy}^{ij} & 0 \\ 0 & 0 & m_{zz}^{ij} \end{bmatrix} \tag{7.33}$$

Matrix $\boldsymbol{m}^{ij}$ is a 3×3 matrix. It consist of the FN of the mass matrix. It has only three elements on the diagonal that are different from zero,

$$\begin{aligned} m_{xx}^{ij} &= \int_V N_j \rho N_i dV \\ m_{yy}^{ij} &= \int_V N_j \rho N_i dV \\ m_{zz}^{ij} &= \int_V N_j \rho N_i dV \end{aligned} \tag{7.34}$$

while the elements outside the diagonal are null,

$$m_{yz}^{ij} = m_{zx}^{ij} = m_{zy}^{ij} = m_{xy}^{ij} = m_{xz}^{ij} = m_{yx}^{ij} = 0 \tag{7.35}$$

Assembly of the global mass matrix follows the same rules as those for the stiffness matrix. The loops on indexes i and j give the mass matrix of the elements. The mass matrix of the structure can be assembled by superimposing the masses of the shared nodes.

7.2.5 Loading Vector

The loading vector can be derived using the formulation of the virtual variation of the external work. The virtual variation of the displacements in Equation (7.22) can be used to express the virtual variation of the external work in the CUF, see Equation (7.17). The virtual variation of the external work can be written as

$$\delta L_{ext} = \int_V \delta \boldsymbol{u}^T \boldsymbol{g} dV + \int_S \delta \boldsymbol{u}^T \boldsymbol{p} dS + \int_L \delta \boldsymbol{u}^T \boldsymbol{q} dy + \delta \boldsymbol{u}_j^T P \tag{7.36}$$

where $\boldsymbol{g}$ are the volume forces, $\boldsymbol{p}$ are the surface forces, $\boldsymbol{q}$ are the line forces and P are the concentrated loads. The external loads are usually applied as surface loads (a pressure), or as a concentrated load. Each contribution of the external load can be written in index form. The volume loads become

$$\delta L_{ext} = \int_V \delta \boldsymbol{u}^T \boldsymbol{g} dV = \delta \boldsymbol{u}_j^T \int_V N_j \boldsymbol{g} dV \tag{7.37}$$

The surface loads are

$$\delta L_{ext} = \int_S \delta \boldsymbol{u}^T \boldsymbol{p} dS = \delta \boldsymbol{u}_j^T \int_S N_j \boldsymbol{p} dS \tag{7.38}$$

The line loads become

$$\delta L_{ext} = \int_l \delta \boldsymbol{u}^T \boldsymbol{q} dl = \delta \boldsymbol{u}_j^T \int_l N_j \boldsymbol{q} dl \tag{7.39}$$

where l is the line on which the load is applied. Finally, the concentrated loads are

$$\delta L_{ext} = \delta \boldsymbol{u}^T P = \delta \boldsymbol{u}_j^T N_j P \tag{7.40}$$

The load vector can be written as the sum of the previous contributions

$$\boldsymbol{P} = \int_V N_j \boldsymbol{g} dV + \int_S N_j \boldsymbol{p} dS + \int_l N_j \boldsymbol{q} dl + N_j P \tag{7.41}$$

The load vector of the element can be assembled following the same procedure that was introduced for the stiffness matrix. In this case, only a loop on j gives the load vector of the element. The global vector can be derived by summing the loads in the shared nodes.

7.3 Three-Dimensional Numerical Integration

The calculation of the FNs (for both the stiffness and mass matrices) requires the calculation of integrals. The elements of the nucleus have a standard form that can be expressed as

$$\mathrm{FN} = C \int_V f(x, y, z) dV \tag{7.42}$$

where C is a constant that can be either a mass parameter or a material stiffness constant, and $f(x, y, z)$ is a generic function that involves the product of the derivatives of the shape functions. This integral can only be calculated in exact form for elements that have a simple geometry. The integral must be evaluated by a numerical approach for elements with an arbitrary shape. The method presented here uses the Gauss quadrature formulation. In order to use this technique, it is convenient to introduce a new reference system that switches from the 'physical' to the 'natural' reference system.

7.3.1 Three-Dimensional Gauss–Legendre Quadrature

A function can be integrated in a finite domain using several techniques. The use of numerical solutions, instead of analytical ones, allows the integration approach to be generalized and used for any domain. The most frequently used quadrature technique in the FEM is the Gauss–Legendre quadrature formulation, as it allows an integral to be calculated as the sum of a finite number of terms. The Gauss–Legendre quadrature formulation is performed in a normalized reference system with coordinates ξ, η, μ. In the *natural* reference system, the coordinates can vary between -1 and 1. Details of the integration, using the Gauss–Legendre quadrature formulas, are presented in Appendix A.

The equation for the calculation of an integral I_V over volume V in the natural plane is

$$I_V = \int_{-1}^{1} \int_{-1}^{1} \int_{-1}^{1} f(\xi, \eta, \zeta) d\xi \, d\eta \, d\zeta \approx \sum_{l=1}^{N_G} \sum_{m=1}^{N_G} \sum_{k=1}^{N_G} f(\xi_l, \eta_m, \zeta_k) w_l \, w_m w_k \tag{7.43}$$

Equation (7.43) shows that the integral of a 3D function, defined in the natural reference system, can be calculated as the sum of the product of the values of the function at a finite number of points (Gauss points), N_G, with coordinates (ξ_l, η_m, ζ_k) and some weights $(w_l \, w_m w_k)$.

In the case of the Gauss–Legendre quadrature formulation, the coordinates of the Gauss points and weights are given in tabular form in Appendix A.

The position of the Gauss points does not depend on the geometry of the 3D domain that has to be integrated, because the points are written in normalized coordinates ξ, η and ζ.

In the FEM solution, the integrals of the shape functions must be expressed in the real reference system, while the integrals evaluated using the Gauss–Legendre quadrature formulation are normalized. The isoparametric formulation allows the integral to be transformed from the natural to the real reference system.

7.3.2 Isoparametric Formulation

The 3D isoparametric formulation has been adopted here for the definition of 3D elements. The word 'isoparametric' means that the same formulation has been used to express the displacement field and the geometry of the 3D element,

$$\boldsymbol{u} = N_i \boldsymbol{u}_i \tag{7.44}$$

$$x = N_i x_i, \quad y = N_i y_i, \quad z = N_i z_i \tag{7.45}$$

An integral in the real reference system can be expressed in the natural reference system using the following formula:

$$\int_V N_{i_{,x}}(x,y,z) N_{j_{,x}}(x,y,z) dx\, dy\, dz = \int_{-1}^{1}\int_{-1}^{1}\int_{-1}^{1} N_{i_{,x}}(\xi,\eta,\zeta) N_{j_{,x}}(\xi,\eta,\zeta) |\boldsymbol{J}(\xi,\eta,\zeta)| d\xi\, d\eta\, d\zeta \tag{7.46}$$

where $|\boldsymbol{J}(\xi,\eta,\zeta)|$ is the determinant of the Jacobian and, from a physical point of view, it represents the ratio between the volume of the element in the real reference system and the volume of the element in the normalized formulation.

In order to calculate the integral in Equation (7.46), the derivatives of the shape functions have to be written in normalized coordinates, with respect to real coordinates,

$$\begin{aligned}
\frac{\partial N_i}{\partial \xi} &= \frac{\partial N_i}{\partial x}\frac{\partial x}{\partial \xi} + \frac{\partial N_i}{\partial y}\frac{\partial y}{\partial \xi} + \frac{\partial N_i}{\partial z}\frac{\partial z}{\partial \xi} \\
\frac{\partial N_i}{\partial \eta} &= \frac{\partial N_i}{\partial x}\frac{\partial x}{\partial \eta} + \frac{\partial N_i}{\partial y}\frac{\partial y}{\partial \eta} + \frac{\partial N_i}{\partial z}\frac{\partial z}{\partial \eta} \\
\frac{\partial N_i}{\partial \zeta} &= \frac{\partial N_i}{\partial x}\frac{\partial x}{\partial \zeta} + \frac{\partial N_i}{\partial y}\frac{\partial y}{\partial \zeta} + \frac{\partial N_i}{\partial z}\frac{\partial z}{\partial \zeta}
\end{aligned} \tag{7.47}$$

Equation (7.47) can be rewritten in matrix form using the Jacobian matrix, $\boldsymbol{J}$:

$$\begin{Bmatrix} \dfrac{\partial N_i}{\partial \xi} \\ \dfrac{\partial N_i}{\partial \eta} \\ \dfrac{\partial N_i}{\partial \zeta} \end{Bmatrix} = \underbrace{\begin{bmatrix} \dfrac{\partial x}{\partial \xi} & \dfrac{\partial y}{\partial \xi} & \dfrac{\partial z}{\partial \xi} \\ \dfrac{\partial x}{\partial \eta} & \dfrac{\partial y}{\partial \eta} & \dfrac{\partial z}{\partial \eta} \\ \dfrac{\partial x}{\partial \zeta} & \dfrac{\partial y}{\partial \zeta} & \dfrac{\partial z}{\partial \zeta} \end{bmatrix}}_{\boldsymbol{J}} \begin{Bmatrix} \dfrac{\partial N_i}{\partial x} \\ \dfrac{\partial N_i}{\partial y} \\ \dfrac{\partial N_i}{\partial z} \end{Bmatrix} \tag{7.48}$$

Therefore, the derivatives of the shape functions in the physical reference system are

$$\begin{Bmatrix} \dfrac{\partial N_i}{\partial x} \\ \dfrac{\partial N_i}{\partial y} \\ \dfrac{\partial N_i}{\partial z} \end{Bmatrix} = \underbrace{\begin{bmatrix} \dfrac{\partial x}{\partial \xi} & \dfrac{\partial y}{\partial \xi} & \dfrac{\partial z}{\partial \xi} \\ \dfrac{\partial x}{\partial \eta} & \dfrac{\partial y}{\partial \eta} & \dfrac{\partial z}{\partial \eta} \\ \dfrac{\partial x}{\partial \zeta} & \dfrac{\partial y}{\partial \zeta} & \dfrac{\partial z}{\partial \zeta} \end{bmatrix}^{-1}}_{\boldsymbol{J}^{-1}} \begin{Bmatrix} \dfrac{\partial N_i}{\partial \xi} \\ \dfrac{\partial N_i}{\partial \eta} \\ \dfrac{\partial N_i}{\partial \zeta} \end{Bmatrix} \tag{7.49}$$

The Jacobian can therefore be evaluated easily since the geometry is expressed in terms of shape functions. For example, in the case of an eight-node element, the derivative of x with respect to ξ becomes

$$\frac{\partial x}{\partial \xi} = \frac{N_1}{\partial \xi}x_1 + \frac{N_2}{\partial \xi}x_2 + \ldots + \frac{N_8}{\partial \xi}x_8 \tag{7.50}$$

Equation (7.49) allows the derivatives of the shape function to be expressed in the physical reference system. The Jacobian must be not singular to be invertible; this requirement is only fulfilled if the geometry of the element does not show any major distortions. In this sense, the aspect ratio of a 3D element is a critical parameter. Equation (7.46) can be solved using the Gauss–Legendre quadrature and the integrals of the shape functions become

$$\int_V N_{i,_x}(x,y,z)N_{j,_x}(x,y,z)dx\,dy\,dz = \sum_{l,m,k} w_l\, w_m w_k N_{i,_x}(\xi_l,\eta_m,\zeta_k)N_{j,_x}(\xi_l,\eta_m,\zeta_k)|\boldsymbol{J}(\xi_l,\eta_m,\zeta_k)| \tag{7.51}$$

7.3.3 Reduced Integration: Shear Locking Correction

The use of 'a priori' assumed kinematic models in the FEM gives rise to locking phenomena that can have an important effect on the convergence of the solution of the problem. The shear locking phenomenon is well known and is caused by the linear formulation of 3D elements. A solid structure with a pure bending load can be seen on the left of Figure 7.4. If the problem is solved with an eight-node element, which has linear shape functions, the displacements appear as in the central figure. If a 20-node quadratic element is used, the solution appears as in the figure on the right.

Since a pure bending moment is applied, the solution should show no shear strains. This is true for the quadratic element. If we look at points a, b and c, the vertical lines on the front face are perpendicular to the horizontal line. In other cases, the linear element is not able to ensure this solution. While the horizontal and vertical lines are perpendicular at point b, the lines are not perpendicular at points a and c, and the angle γ is due to shear deformation. The development of shear deformations in a linear element under a bending load increases the stiffness of the element, and the solution will therefore not converge to the correct value.

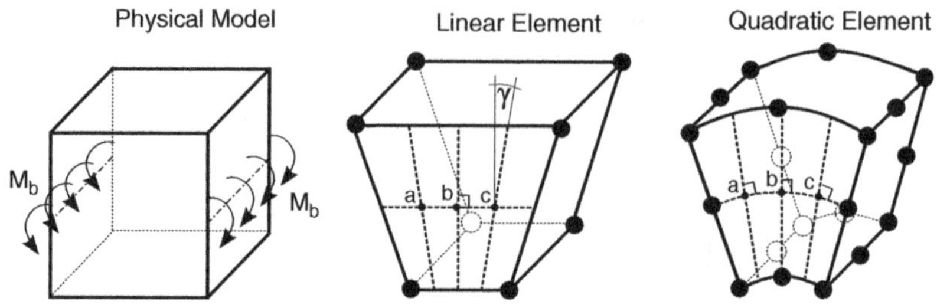

Figure 7.4 Physical demonstration of shear locking. Structure under a pure bending load (left). Deformation of an eight-node element under pure bending load (centre). Deformation of a 20-node element under pure bending load (right)

An excess of shear stiffness can be corrected during integration. In a linear element, only point b fulfils the requirement of no shear under a bending load. If we try to reproduce the same procedure over all three directions, it can be seen that only the central point of the solid element does not show shear strains. Since the solution is only correct at this point, only this point can be used to integrate the terms related to the shear strains. Figure 7.5 shows the Gauss points for full integration (on the left) and in the case of reduced integration (on the right). It is important to select the terms that have to be treated with full integration and those that have to be treated with reduced integration. The strain vector can be split into two contributions, i.e. normal strains, $\boldsymbol{\varepsilon}_n$, and shear strains, $\boldsymbol{\varepsilon}_t$,

$$\boldsymbol{\varepsilon}_n^T = \{\varepsilon_{xx}, \varepsilon_{yy}, \varepsilon_{zz}\}, \quad \boldsymbol{\varepsilon}_t^T = \{\varepsilon_{xz}, \varepsilon_{yz}, \varepsilon_{xy}\} \tag{7.52}$$

The stresses can be written using the same approach

$$\boldsymbol{\sigma}_n^T = \{\sigma_{xx}, \sigma_{yy}, \sigma_{zz}\}, \quad \boldsymbol{\sigma}_t^T = \{\sigma_{xz}, \sigma_{yz}, \sigma_{xy}\} \tag{7.53}$$

The relations between the stresses and strains become

$$\boldsymbol{\sigma}_n = \boldsymbol{C}_{nn}\boldsymbol{\varepsilon}_n + \boldsymbol{C}_{nt}\boldsymbol{\varepsilon}_t \tag{7.54}$$

$$\boldsymbol{\sigma}_t = \boldsymbol{C}_{tn}\boldsymbol{\varepsilon}_n + \boldsymbol{C}_{tt}\boldsymbol{\varepsilon}_t \tag{7.55}$$

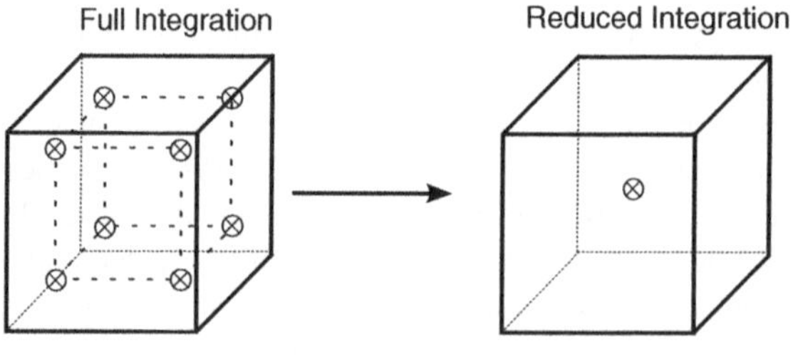

Figure 7.5 From full to reduced integration

where

$$\boldsymbol{C}_{nn} = \begin{bmatrix} C_{11} & C_{12} & C_{13} \\ C_{21} & C_{22} & C_{23} \\ C_{31} & C_{32} & C_{33} \end{bmatrix}, \quad \boldsymbol{C}_{nt} = \boldsymbol{C}_{tn}^T = 0, \quad \boldsymbol{C}_t = \begin{bmatrix} C_{11} & 0 & 0 \\ 0 & C_{22} & 0 \\ 0 & 0 & C_{33} \end{bmatrix} \tag{7.56}$$

$\boldsymbol{C}_{nt}$ and $\boldsymbol{C}_{tn}$ are equal to zero because the material is considered isotropic. The internal work can be written as the contribution of the work from the normal strains plus the work from the shear strains

$$\delta L_i = \delta L_{i_n} + \delta L_{i_t} = \int_V \delta \varepsilon_n \sigma_n dV + \int_V \delta \varepsilon_t \sigma_t dV \tag{7.57}$$

The integral of the normal quantities has to be computed using full integration, while the integral of the tangential quantities should be computed using reduced integration. Full integration must be applied to

$$\begin{aligned} k_{xx}^{ij} &= (2G+\lambda)\int_V N_{i,x}N_{j,x}dV + G\int_V N_{i,y}N_{j,y}dV + G\int_V N_{i,z}N_{j,z}dV \\ k_{yy}^{ij} &= (2G+\lambda)\int_V N_{i,y}N_{j,y}dV + G\int_V N_{i,x}N_{j,x}dV + G\int_V N_{i,z}N_{j,z}dV \\ k_{zz}^{ij} &= (2G+\lambda)\int_V N_{i,z}N_{j,z}dV + G\int_V N_{i,x}N_{j,x}dV + G\int_V N_{i,z}N_{j,z}dV \end{aligned} \tag{7.58}$$

while reduced integration affects the following terms:

$$\begin{aligned} k_{xy}^{ij} &= \lambda\int_V N_{i,y}N_{j,x}dV + G\int_V N_{i,x}N_{j,y}dV \\ k_{xz}^{ij} &= \lambda\int_V N_{i,z}N_{j,x}dV + G\int_V N_{i,x}N_{j,z}dV \\ k_{yx}^{ij} &= \lambda\int_V N_{i,x}N_{j,y}dV + G\int_V N_{i,y}N_{j,x}dV \\ k_{yz}^{ij} &= \lambda\int_V N_{i,z}N_{j,y}dV + G\int_V N_{i,y}N_{j,z}dV \\ k_{zx}^{ij} &= \lambda\int_V N_{i,x}N_{j,z}dV + G\int_V N_{i,z}N_{j,x}dV \\ k_{zy}^{ij} &= \lambda\int_V N_{i,y}N_{j,z}dV + G\int_V N_{i,z}N_{j,y}dV \end{aligned} \tag{7.59}$$

Alternative and more reliable techniques to contrast shear locking should be based on the use of different shape functions for different contributions to the stiffness matrix. This approach is used in the case of the plate/shell structures in Chapters 10–12.

Table 7.2 The eight-node element

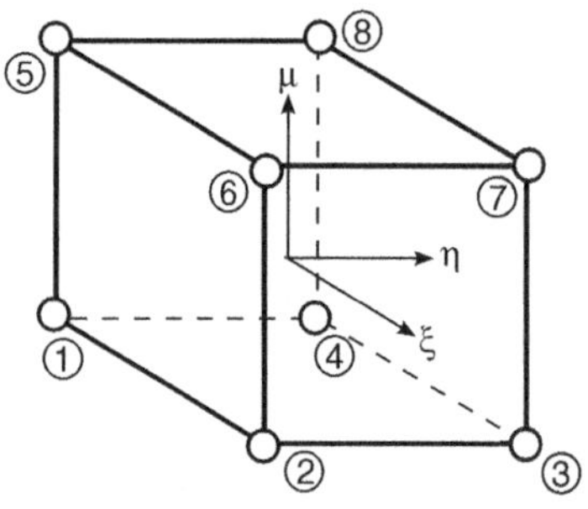

Node ID	ξ	η	ζ
1	−1	−1	−1
2	1	−1	−1
3	1	1	−1
4	−1	1	−1
5	−1	−1	1
6	1	−1	1
7	1	1	1
8	−1	1	1

Shape functions:

$$N_i = \frac{(1+\xi_i\xi)(1+\eta_i\eta)(1+\zeta_i\zeta)}{8}$$

where ξ_i, η_i and ζ_i are the coordinates of the $i-th$ node in the normalized reference system

This table describes the eight-node element. The coordinates of the nodes and the shape functions are also presented.

Table 7.3 The 20-node element

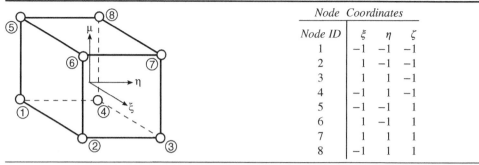

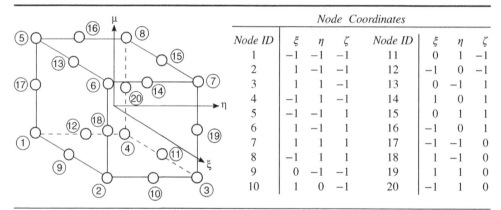

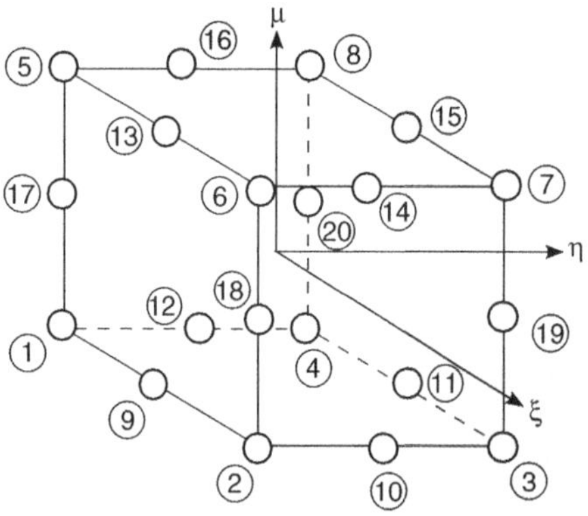

Node ID	ξ	η	ζ	Node ID	ξ	η	ζ
1	−1	−1	−1	11	0	1	−1
2	1	−1	−1	12	−1	0	−1
3	1	1	−1	13	0	−1	1
4	−1	1	−1	14	1	0	1
5	−1	−1	1	15	0	1	1
6	1	−1	1	16	−1	0	1
7	1	1	1	17	−1	−1	0
8	−1	1	1	18	1	−1	0
9	0	−1	−1	19	1	1	0
10	1	0	−1	20	−1	1	0

Shape functions:

$$N_i = \frac{(1+\xi_i\xi)(1+\eta_i\eta)(1+\zeta_i\zeta)(\xi_i\xi+\eta_i\eta+\zeta_i\zeta-2)}{8};\ \xi_i=\pm1, \eta_i=\pm1, \zeta_i=\pm1$$

$$N_i = \frac{(1-\xi_i^2)(1+\eta_i\eta)(1+\zeta_i\zeta)}{4};\ \xi_i=0, \eta_i=\pm1, \zeta_i=\pm1$$

$$N_i = \frac{(1+\xi_i\xi)(1-\eta_i^2)(1+\zeta_i\zeta)}{4};\ \xi_i=\pm1, \eta_i=0, \zeta_i=\pm1$$

$$N_i = \frac{(1+\xi_i\xi)(1+\eta_i\eta)(1-\zeta_i^2)}{4};\ \xi_i=\pm1, \eta_i=\pm1, \zeta_i=0$$

where ξ_i, η_i and ζ_i are the coordinates of the ith node in the normalized reference system.

This table describes the 20-node element. The coordinates of the nodes and the shape functions are also presented.

7.4 Shape Functions

This section presents some classical 3D FE formulations. Many FEs, with different geometries and shape functions, can be found in the literature. In this section, only two hexahedral elements are introduced, the first with 8 nodes and the second with 20 nodes. A complete derivation of these and other elements can be found in many classical books on FEs, such as those by Oñate (2009) and Zienkiewicz *et al.* (2005).

Table 7.2 lists the coordinates of the nodes and the shape functions of a hexahedral element with eight nodes. The shape functions are defined in normalized coordinates. The element has a linear formulation and therefore shows shear locking. Reduced integration must be used to achieve accurate results. The element described in Table 7.3 has a hexahedral form, but with 20 nodes. The displacement and geometry are expressed using quadratic shape functions, and the element is therefore free of shear locking. However, it is much more computationally expensive than a linear element.

References

Oñate E 2009 *Structural Analysis with the Finite Element Method: Linear Statics.* Volume 1. *Basis and Solids.* Springer.

Zienkiewicz OC, Taylor RL and Zhu JZ 2005 *The Finite Element Method: Its Basis and Fundamentals.* Sixth Edition. Elsevier.

8

One-Dimensional Models with *N*th-Order Displacement Field, the Taylor Expansion Class

The unknown variables of a 1D model depend on one coordinate which is generally the axial coordinate of the structure. In this book, the axial coordinate of a 1D model is y, while x and z are the so-called cross-section coordinates. The behaviour of the unknowns of a 1D structural problem can be axiomatically assumed. For instance, polynomial expansions of the unknown variables can be adopted above the cross-section; this means that for a given y the distribution of the unknowns above the cross-section will be given by a 2D polynomial in x and z. The choice of a particular expansion characterizes the capabilities of a structural model: the richer the expansion, the more accurate the results (see Washizu 1968). The number of expansion terms to be retained to fulfil a given accuracy requirement is strictly related to the characteristics of the structural problem to be analysed (e.g. geometry, boundary conditions, material, etc.). The main feature of the present unified formulation is due to the possibility of arbitrarily choosing the number of terms of the expansion. Therefore, the number of terms to be retained can be evaluated through a convergence analysis.

This chapter presents 1D elements based on Taylor-like expansions of the displacement variables, these elements are hereafter referred to as TE elements. Table 8.1 shows the FN for 1D models that is introduced in this chapter. These expressions are valid in a Cartesian orthogonal reference frame. If, for instance, a curvilinear system is adopted, the explicit expressions of the FN components can vary as will be shown in Chapter 11. Classical models (Euler–Bernoulli and Timoshenko models) are briefly presented together with the more general complete linear expansion case. Higher-order models are also presented and the unified formulation introduced. The PVD is employed to derive the governing equations and the FE formulation. Locking phenomena and their corrections are discussed and numerical results provided.

Finite Element Analysis of Structures Through Unified Formulation, First Edition.
Erasmo Carrera, Maria Cinefra, Marco Petrolo and Enrico Zappino.

Table 8.1 A schematic description of the CUF and the related FN of the stiffness matrix for 1D models

---

Equilibrium equations in Strong Form $\longrightarrow \quad \delta L_i = \int_V \delta \boldsymbol{u} \boldsymbol{k} \boldsymbol{u} dV + \int_S \ldots dS$

$$\underbrace{\begin{bmatrix} k_{xx} & k_{xy} & k_{xz} \\ k_{yx} & k_{yy} & k_{yz} \\ k_{zx} & k_{zy} & k_{zz} \end{bmatrix}}_{\boldsymbol{k}} \underbrace{\begin{Bmatrix} u_x \\ u_y \\ u_z \end{Bmatrix}}_{\boldsymbol{u}} = \underbrace{\begin{Bmatrix} p_x \\ p_y \\ p_z \end{Bmatrix}}_{\boldsymbol{p}}$$

$$k_{xx} = -(\lambda + 2G)\, \partial_{xx} - G\, \partial_{zz} - G\, \partial_{yy};$$
$$k_{xy} = -\lambda\, \partial_{xy} - G\, \partial_{yx};$$
$$k_{xz} = \ldots$$
$$\lambda = (E\nu)/[(1+\nu)(1-2\nu)]; \quad G = E/[2(1+\nu)]$$

$\boldsymbol{u} = \boldsymbol{u}(x, y, z)$
$\delta \boldsymbol{u} = \delta \boldsymbol{u}(x, y, z)$

The diagonal (e.g. k_{xx}) and the non-diagonal (e.g. k_{xy}) terms can be obtained through proper index permutations.

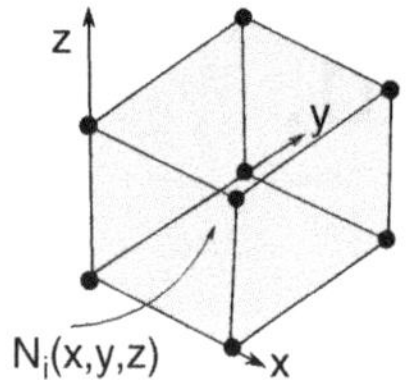

$\boldsymbol{u} = N_i(x, y, z)\boldsymbol{u}_i$
$\delta \boldsymbol{u} = N_j(x, y, z)\delta \boldsymbol{u}_j$

3D FEM Formulation $\longrightarrow \quad \delta L_i = \delta \boldsymbol{u}_j \boldsymbol{k}^{ij} \boldsymbol{u}_i$

$$k^{ij}_{xx} = (\lambda + 2G) \int_V N_{j,x} N_{i,x} dV + G \int_V N_{j,z} N_{i,z} dV + G \int_V N_{j,y} N_{i,y} dV;$$

$$k^{ij}_{xy} = \lambda \int_V N_{j,y} N_{i,x} dV + G \int_V N_{j,x} N_{i,y} dV$$

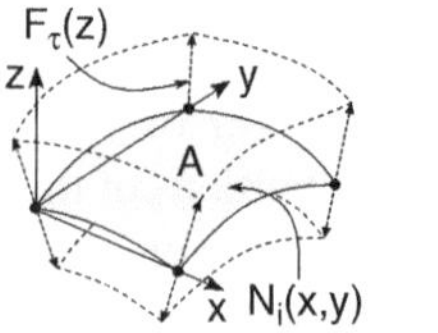

$\boldsymbol{u} = N_i(x, y) F_\tau(z) \boldsymbol{u}_{\tau i}$
$\delta \boldsymbol{u} = N_j(x, y) F_s(z) \delta \boldsymbol{u}_{sj}$

2D FEM Formulation $\longrightarrow \quad \delta L_i = \delta \boldsymbol{u}_{sj} \boldsymbol{k}^{\tau sij} \boldsymbol{u}_{\tau i}$

$$k^{\tau sij}_{xx} = (\lambda + 2G) \int_\Omega N_{i,x} N_{j,x} d\Omega \int_h F_\tau F_s dz$$
$$+ G \int_\Omega N_i N_j d\Omega \int_h F_{\tau,z} F_{s,z} dz + G \int_V N_{i,y} N_{j,y} d\Omega \int_h F_\tau F_s dz;$$

$$k^{\tau sij}_{xy} = \lambda \int_\Omega N_{i,y} N_{j,y} d\Omega \int_h F_\tau F_s dz + G \int_\Omega N_{i,x} N_{j,y} d\Omega \int_h F_\tau F_s dz$$

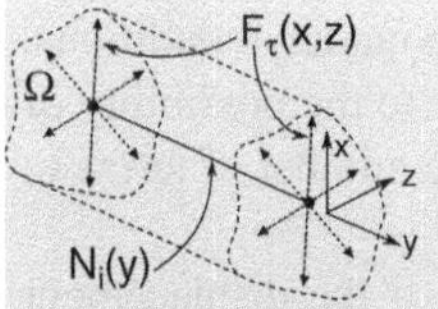

$\boldsymbol{u} = N_i(y) F_\tau(x, z) \boldsymbol{u}_{\tau i}$
$\delta \boldsymbol{u} = N_j(y) F_s(x, z) \delta \boldsymbol{u}_{sj}$

1D FEM Formulation $\longrightarrow \quad \delta L_i = \delta \boldsymbol{u}_{sj} \boldsymbol{k}^{\tau sij} \boldsymbol{u}_{\tau i}$

$$k^{\tau sij}_{xx} = (\lambda + 2G) \int_l N_i N_j dy \int_A F_{\tau,x} F_{s,x} dA$$
$$+ G \int_l N_i N_j dy \int_A F_{\tau,z} F_{s,z} dA + G \int_l N_{i,y} N_{j,y} dy \int_A F_\tau F_s dA;$$

$$k^{\tau sij}_{xy} = \lambda \int_l N_{i,y} N_j dy \int_A F_\tau F_{s,x} dA + G \int_l N_i N_{j,y} dy \int_A F_{\tau,x} F_s dA$$

CUF leads to the automatic implementation of any theory of structures through 4 loops (i.e. 4 indexes):

- τ and s deal with the functions that approximate the displacement field and its virtual variation along the plate/shell thickness ($F_\tau(z), F_s(z)$) or over the beam cross-section ($F_\tau(x, z), F_s(x, z)$);
- i and j deal with the shape functions of the FE model, (3D:$N_i(x, y, z), N_j(x, y, z)$; 2D:$N_i(x, y), N_j(x, y)$; 1D:$N_i(y), N_j(y)$).

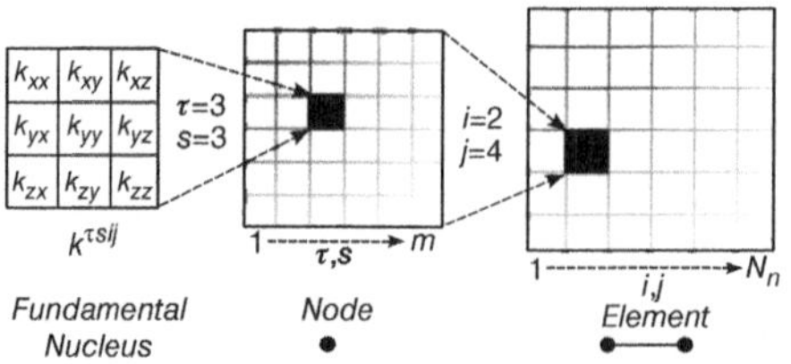

This table shows the essential features of the CUF for 1D models.

8.1 Classical Models and the Complete Linear Expansion Case

The mechanics of a beam under bending was first understood and described by Leonardo da Vinci, as stated by Reti (1974) and Ballarini (2003). The mathematical formulations of beams under bending were provided by Euler–Bernoulli (Euler, 1744) and Timoshenko (1921, 1922), whose models represent the classical beam theories. They are the reference models for analysing slender homogeneous structures under bending loads or for computing bending natural modes. These theories will be briefly described below; more details can be found in Carrera *et al.* (2011).

8.1.1 The Euler–Bernoulli Beam Model

The Euler–Bernoulli beam theory, hereafter referred to as EBBT, was derived from the following a priori assumptions (see Figure 8.1):

1. The cross-section is rigid on its plane.
2. The cross-section rotates around a neutral surface, remaining plane.
3. The cross-section remains perpendicular to the neutral surface during deformation.

According to the first assumption, no in-plane deformations are accounted for and, therefore, the in-plane displacements u_x and u_z depend upon the axial coordinate y only,

$$\begin{cases} \varepsilon_{xx} = u_{x,x} = 0 \\ \varepsilon_{zz} = u_{z,z} = 0 \\ \gamma_{xz} = u_{x,z} + u_{z,x} = 0 \end{cases} \Rightarrow \begin{cases} u_x(x,y,z) = u_{x_1}(y) \\ u_z(x,y,z) = u_{z_1}(y) \end{cases} \tag{8.1}$$

On the basis of the second assumption, the out-of-plane (or axial displacement) u_y is linear versus the in-plane coordinates,

$$u_y(x,y,z) = u_{y_1}(y) + \phi_z(y)\,x + \phi_x(y)\,z \tag{8.2}$$

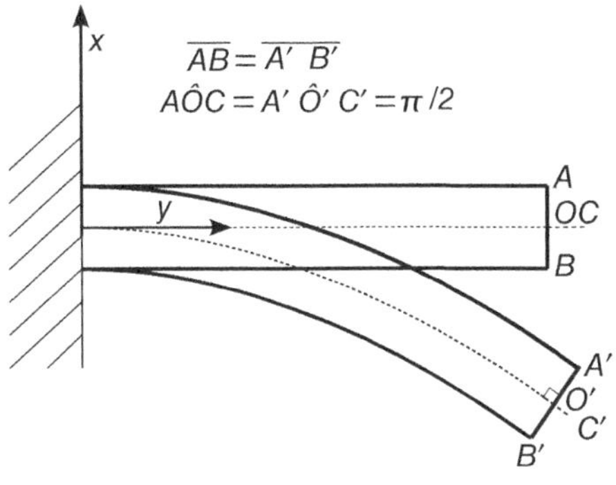

Figure 8.1 Euler–Bernoulli beam model

where ϕ_z and ϕ_x are the rotation angles along the z- and the x-axis, respectively. ϕ_z is positive when, according to the 'right-hand grip rule', the thumb points in the positive direction of the z-axis, whereas the thumb points in the negative direction of the x-axis for positive values of ϕ_x. On the basis of the third assumption and according to the definition of shear strains, shear deformations γ_{yz} and γ_{yx} are disregarded,

$$\gamma_{yz} = \gamma_{yx} = 0 \tag{8.3}$$

Equations (8.2), (8.3) and (8.1) allow the rotation angles to be obtained as functions of the derivatives of the in-plane displacements,

$$\begin{cases} \gamma_{xy} = u_{y,x} + u_{x,y} = \phi_z + u_{x_1,y} = 0 \\ \gamma_{yz} = u_{y,z} + u_{z,y} = \phi_x + u_{z_1,y} = 0 \end{cases} \Rightarrow \begin{cases} \phi_z = -u_{x_1,y} \\ \phi_x = -u_{z_1,y} \end{cases} \tag{8.4}$$

The displacement field of EBBT is then

$$\begin{aligned} u_x &= u_{x_1} \\ u_y &= u_{y_1} - u_{x_1,y}\, x - u_{z_1,y}\, z \\ u_z &= u_{z_1} \end{aligned} \tag{8.5}$$

From a mathematical point of view, the EBBT displacement field can be seen as a Maclaurin-like series expansion in which a zeroth-order approximation is used for the in-plane components and an expansion order (N) equal to one is adopted for the axial displacement. The relations among the unknowns have been derived from kinematic considerations. EBBT presents three unknown variables.

According to the kinematic hypotheses, EBBT accounts for the axial strain only. On the basis of its definition and of the EBBT displacement field, the axial strain ε_{yy} is

$$\varepsilon_{yy} = u_{y,y} = \underbrace{\frac{\partial u_{y_1}}{\partial y}}_{k_y^y} \underbrace{-\frac{\partial^2 u_{x_1}}{\partial y^2}}_{k_{yy}^x} x \underbrace{-\frac{\partial^2 u_{z_1}}{\partial y^2}}_{k_{yy}^z} z = k_y^y + k_{yy}^x x + k_{yy}^z z \tag{8.6}$$

k_y^y has the physical meaning of membrane deformation, whereas k_{yy}^x and k_{yy}^x are the second-order derivatives of the transverse displacements; they represent the curvatures in the case of infinitesimal deformations and small rotations. The axial stress, σ_{yy}, is obtained from the axial strain by means of the reduced constitutive equations:

$$\sigma_{yy} = E\varepsilon_{yy} = E\left(k_y^y + k_{yy}^x x + k_{yy}^z z\right) \tag{8.7}$$

8.1.2 *The Timoshenko Beam Theory (TBT)*

In a TBT model, the cross-section is still rigid on its plane, it rotates around a neutral surface, remaining plane, but it is no longer constrained to remain perpendicular to it (see Figure 8.2).

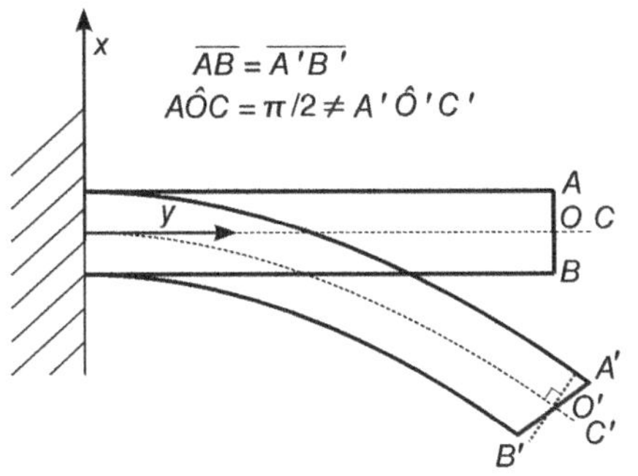

Figure 8.2 Timoshenko beam model

Shear deformations γ_{xy} and γ_{yz} are now accounted for. According to the previous a priori kinematic assumptions, the displacement field of TBT is

$$\begin{aligned} u_x(x,y,z) &= u_{x_1}(y) \\ u_y(x,y,z) &= u_{y_1}(y) + \phi_z(y)\,x + \phi_x(y)\,z \\ u_z(x,y,z) &= u_{z_1}(y) \end{aligned} \tag{8.8}$$

The strain components are obtained by substituting the displacement field in Equation (8.8) into the geometrical relations. Only the non-null strain components are reported,

$$\begin{aligned} \varepsilon_{yy} &= u_{y,y} = u_{y_1,y} + \phi_{z,y}\,x + \phi_{x,y}\,z \\ \gamma_{xy} &= u_{y,x} + u_{x,y} = \phi_z + u_{x_1,y} \\ \gamma_{yz} &= u_{y,z} + u_{z,y} = \phi_x + u_{z_1,y} \end{aligned} \tag{8.9}$$

The constitutive relations are used to obtain the axial stress and the shear stress components,

$$\begin{aligned} \sigma_{yy} &= E\epsilon_{yy} = E\left(u_{y_1,y} + \phi_{z,y}\,x + \phi_{x,y}\,z\right) \\ \sigma_{xy} &= \kappa G\left(\phi_z + u_{x_1,y}\right) \\ \sigma_{yz} &= \kappa G\left(\phi_x + u_{z_1,y}\right) \end{aligned} \tag{8.10}$$

where κ is the shear correction factor. The shear predicted by TBT should be corrected since the model yields a constant shear distribution above the cross-section, whereas the shear distribution has to be at least parabolic in order to satisfy the stress-free boundary conditions on the unloaded edges of the cross-section. The shear correction factor is mainly related to the cross-section geometry. In the literature, there are many methods for computing κ: see, for instance, Carrera *et al.* (2012c); Cowper (1966); Gruttmann and Wagner (2001); Gruttmann *et al.* (1999); Pai and Schulz (1999); and Timoshenko (1921). A discussion of the shear correction factor is beyond the scope of this book. It will be shown that the adoption of higher-order models represents a general approach to avoid the introduction of shear correction factors.

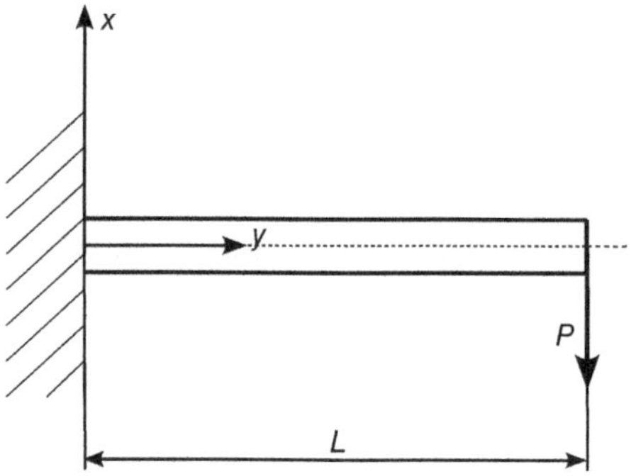

Figure 8.3 Cantilever beam bent by a vertical force

Example 8.1.1 *A cantilever beam is considered to highlight the differences between EBBT and TBT solutions. Figure 8.3 shows the structure that is loaded by a vertical force applied at the free tip. Let us consider the elastica equation,*

$$\frac{\partial^2 u_{x_1}}{\partial y^2} = \frac{M_z(y)}{EI_z} = -\frac{P\,L}{E\,I_z}\left(\frac{L-y}{L}\right) \tag{8.11}$$

By integrating Equation (8.11), the following relations are obtained:

$$\begin{aligned} u_{x_1,y} &= -\frac{P\,L}{E\,I_z}\left(y-\frac{y^2}{2\,L}\right)+C_1 \\ u_{x_1} &= -\frac{P\,L}{E\,I_z}\left(\frac{y^2}{L}-\frac{y^3}{6\,L}\right)+C_1\,y+C_2 \end{aligned} \tag{8.12}$$

The following boundary conditions (BCs) are applied in order to evaluate C_1 and C_2:

$$\begin{aligned} u_{x_1,y}\Big|_{y=0} &= 0 \Rightarrow C_1 = 0 \\ u_{x_1}\Big|_{y=1} &= 0 \Rightarrow C_2 = 0 \end{aligned} \tag{8.13}$$

The value of the maximum vertical displacement is then given by

$$u_{x_1}\Big|_{y=1} = -\frac{P\,L^3}{3\,E\,I_z} \tag{8.14}$$

In the case of TBT, the elastica equation becomes

$$\begin{cases} \phi_{z,y} = -\dfrac{M_z(y)}{EI_z} = \dfrac{P\,L}{E\,I_z}\left(\dfrac{L-y}{L}\right) \\ u_{x_1,y} = -\dfrac{P}{\kappa\,G\,A} - \phi_z \end{cases} \tag{8.15}$$

By integrating the first relation of Equation (8.15), the following relations are obtained:

$$\begin{cases} \phi_z = \dfrac{P\,L}{E\,I_z}\left(y - \dfrac{y^2}{2\,L}\right) + C_1 \\ u_{x_1,y} = -\dfrac{P}{\kappa\,G\,A} - \dfrac{P\,L}{E\,I_z}\left(y - \dfrac{y^2}{2\,L}\right) - C_1 \end{cases} \tag{8.16}$$

Further integration leads to

$$\begin{cases} \phi_z = \dfrac{P\,L}{E\,I_z}\left(y - \dfrac{y^2}{2\,L}\right) + C_1 \\ u_{x_1} = -\dfrac{P\,y}{\kappa\,G\,A} - \dfrac{P\,L}{E\,I_z}\left(\dfrac{y^2}{2\,L} - \dfrac{y^3}{6\,L}\right) - C_1\,y + C_2 \end{cases} \tag{8.17}$$

The following BCs are applied in order to evaluate C_1 and C_2:

$$\begin{aligned} \phi_z\big|_{y=0} &= 0 \Rightarrow C_1 = 0 \\ u_{x_1}\big|_{y=0} &= 0 \Rightarrow C_2 = 0 \end{aligned} \tag{8.18}$$

The value of the maximum vertical displacement is then given by

$$u_{x_1}\big|_{y=0} = -\underbrace{\frac{P\,L}{\kappa\,G\,A}}_{TBT,\ shear\ contribution} - \underbrace{\frac{P\,L^3}{3\,E\,I_z}}_{EBBT,\ bending\ contribution} \tag{8.19}$$

By comparing this last equation with Equation (8.14) it is clear that the TBT solution provides the EBBT solution enhanced by the shear deformation contribution. The TBT contribution is significant whenever short beams or composite materials are considered. In an orthotropic material, for instance, the E/G ratio is much larger than in isotropic materials, which leads to higher shear deformability (Kapania and Raciti 1989). Similar considerations are still valid when the Classical Laminated Theory (CLT) and the First-Order Shear Deformability Theory (FSDT) for plates are considered (Carrera and Petrolo 2010).

8.1.3 The Complete Linear Expansion Case

The complete linear expansion model involves a first-order ($N = 1$) Taylor-like polynomial to describe the cross-section displacement field,

$$
\begin{aligned}
u_x &= \underbrace{u_{x_1}}_{N=0} + \underbrace{x\, u_{x_2} + z\, u_{x_3}}_{N=1} \\
u_y &= u_{y_1} + x\, u_{y_2} + z\, u_{y_3} \\
u_z &= u_{z_1} + x\, u_{z_2} + z\, u_{z_3}
\end{aligned}
\tag{8.20}
$$

The beam model given in Equation (8.20) has nine displacement variables (or unknowns): three constant ($N = 0$) and six linear ($N = 1$). Strain components are given by

$$
\begin{aligned}
\varepsilon_{xx} &= u_{x,x} = u_{x_2} \\
\varepsilon_{yy} &= u_{y,y} = u_{y_1,y} + x\, u_{y_2,y} + z\, u_{y_3,y} \\
\varepsilon_{zz} &= u_{z,z} = u_{z_3} \\
\gamma_{xy} &= u_{x,y} + u_{y,x} = u_{x_1,y} + x\, u_{x_2,y} + z\, u_{x_3,y} + u_{y_2} \\
\gamma_{xz} &= u_{x,z} + u_{z,x} = u_{x_3} + u_{z_2} \\
\gamma_{yz} &= u_{y,z} + u_{z,y} = u_{y_3} + u_{z_1,y} + x\, u_{z_2,y} + z\, u_{z_3,y}
\end{aligned}
\tag{8.21}
$$

The linear model leads to a constant distribution of ε_{xx}, ε_{zz} and γ_{xz} above the cross-section, and a linear distribution of ε_{yy}, γ_{xy} and γ_{yz}.

The adoption of the $N = 1$ model is necessary to introduce the in-plane stretching of the cross-section. A simple example is presented in the following to emphasize the importance of the in-plane stretching terms (u_{x_2}, u_{x_3}, u_{z_2} and u_{z_3}). Figure 8.4 shows a rectangular cross-section loaded by two point forces which provide a pure torsion load. The adoption of EBBT or TBT to analyse this problem would provide null displacements since only constant in-plane translations can be detected. The $N = 1$ model can provide a linear distribution of the in-plane stretching as shown in Figure 8.5.

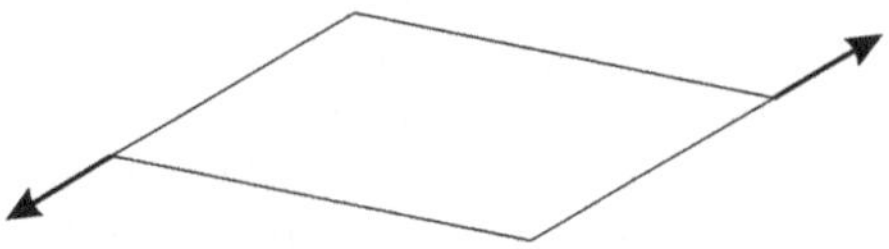

Figure 8.4 Torsion of a square beam

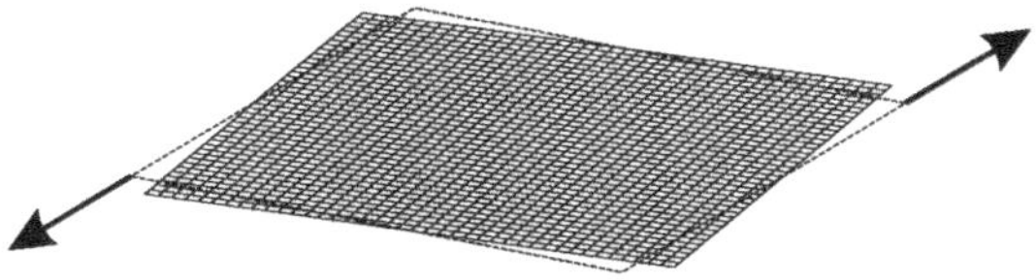

Figure 8.5 Deformed cross-section due to torsion, $N = 1$ model

8.1.4 A Finite Element Based on $N = 1$

The $N = 1$ model has nine displacement variables; this implies that, in an FE formulation, each node has nine generalized displacement variables. The aim of this section is to provide the FE formulation for $N = 1$. The derivation of the governing FE equations begins with the definition of the nodal displacement vector

$$\mathbf{u}_i = \left\{ u_{x1} \quad u_{y1} \quad u_{z1} \quad u_{x2} \quad u_{y2} \quad u_{z2} \quad u_{x3} \quad u_{y3} \quad u_{z3} \right\}^T \tag{8.22}$$

The displacement variables are interpolated along the beam axis by means of the shape functions, N_i:

$$\mathbf{u} = N_i \mathbf{u}_i \tag{8.23}$$

Beam elements with two (B2) nodes are considered here, with the following shape functions:

$$N_1 = \frac{1}{2}(1 - r), \;\; N_2 = \frac{1}{2}(1 + r), \;\; \begin{cases} r_1 = -1 \\ r_2 = +1 \end{cases} \tag{8.24}$$

where the natural coordinate, r, varies from -1 and $+1$ and r_i indicates the position of the node within the natural boundaries of the beam. The total number of DOFs of the structural model will be given by

$$\text{DOFs} = \underbrace{3 \times 3}_{\text{number of DOFs per node}} \times [(\underbrace{2}_{\text{number of nodes per element}} - 1) \times \underbrace{N_{BE}}_{\text{total number of beam elements}} + 1] \tag{8.25}$$

The PVD is employed to compute the FE matrices

$$\delta L_{int} = \delta L_{ext} \tag{8.26}$$

where

$$\delta L_{int} = \int_V \delta \epsilon^T \sigma dV \tag{8.27}$$

L_{int} stands for the internal work, L_{ext} is the work of the external loadings, and δ stands for the virtual variation. Using Equation (8.24), a compact form of the virtual variation of the internal work can be obtained, as known from the FE procedure,

$$\delta L_{int} = \delta \mathbf{u}_j^T \mathbf{k}^{ij} \mathbf{u}_i \tag{8.28}$$

where $\mathbf{k}^{ij}$ is the stiffness matrix. For a given i, j pair, the stiffness matrix has the form

$$\left[\begin{array}{ccc|ccc|ccc}
k_{xx}^{1,1} & k_{xy}^{1,1} & k_{xz}^{1,1} & k_{xx}^{1,x} & k_{xy}^{1,x} & k_{xz}^{1,x} & k_{xx}^{1,z} & k_{xy}^{1,z} & k_{xz}^{1,z} \\
k_{yx}^{1,1} & k_{yy}^{1,1} & k_{yz}^{1,1} & k_{yx}^{1,x} & k_{yy}^{1,x} & k_{yz}^{1,x} & k_{yx}^{1,z} & k_{yy}^{1,z} & k_{yz}^{1,z} \\
k_{zx}^{1,1} & k_{zy}^{1,1} & k_{zz}^{1,1} & k_{zx}^{1,x} & k_{zy}^{1,x} & k_{zz}^{1,x} & k_{zx}^{1,z} & k_{zy}^{1,z} & k_{zz}^{1,z} \\
\hline
k_{xx}^{x,1} & k_{xy}^{x,1} & k_{xz}^{x,1} & k_{xx}^{x,x} & k_{xy}^{x,x} & k_{xz}^{x,x} & k_{xx}^{x,z} & k_{xy}^{x,z} & k_{xz}^{x,z} \\
k_{yx}^{x,1} & k_{yy}^{x,1} & k_{yz}^{x,1} & k_{yx}^{x,x} & k_{yy}^{x,x} & k_{yz}^{x,x} & k_{yx}^{x,z} & k_{yy}^{x,z} & k_{yz}^{x,z} \\
k_{zx}^{x,1} & k_{zy}^{x,1} & k_{zz}^{x,1} & k_{zx}^{x,x} & k_{zy}^{x,x} & k_{zz}^{x,x} & k_{zx}^{x,z} & k_{zy}^{x,z} & k_{zz}^{x,z} \\
\hline
k_{xx}^{z,1} & k_{xy}^{z,1} & k_{xz}^{z,1} & k_{xx}^{z,x} & k_{xy}^{z,x} & k_{xz}^{z,x} & k_{xx}^{z,z} & k_{xy}^{z,z} & k_{xz}^{z,z} \\
k_{yx}^{z,1} & k_{yy}^{z,1} & k_{yz}^{z,1} & k_{yx}^{z,x} & k_{yy}^{z,x} & k_{yz}^{z,x} & k_{yx}^{z,z} & k_{yy}^{z,z} & k_{yz}^{z,z} \\
k_{zx}^{z,1} & k_{zy}^{z,1} & k_{zz}^{z,1} & k_{zx}^{z,x} & k_{zy}^{z,x} & k_{zz}^{z,x} & k_{zx}^{z,z} & k_{zy}^{z,z} & k_{zz}^{z,z}
\end{array}\right]_{ij} \tag{8.29}$$

where the superscripts indicate the expansion functions that are involved in each component of the stiffness matrix, i.e. 1, x or z. For the sake of clarity, the explicit expression of two components is reported here:

$$\begin{aligned}
k_{xx}^{1,1} &= C_{44} \int_A 1 \cdot 1 \, dx\, dz \int_l N_{i,y} N_{j,y} dy \\
k_{yx}^{x,z} &= C_{23} \int_A \frac{\partial x}{\partial x} \cdot z \, dx\, dz \int_l N_i N_{j,y} dy + C_{44} \int_A x \cdot \frac{\partial z}{\partial x} \, dx\, dz \int_l N_{i,y} N_j dy \\
&= C_{23} \int_A 1 \cdot z \, dx\, dz \int_l N_i N_{j,y} dy
\end{aligned} \tag{8.30}$$

where A indicates the cross-section area and l the element length.

8.2 EBBT, TBT and $N = 1$ in Unified Form

EBBT, TBT and $N = 1$ models can be obtained in a unified manner via a condensed notation that represents the basic step towards the CUF. In this section, the $N = 1$ model will be reformulated by means of a new notation, then TBT and EBBT will be obtained as particular cases.

8.2.1 Unified Formulation of $N = 1$

The $N = 1$ stiffness matrix given in Equation (8.29) can be considered as being composed of nine 3×3 sub-matrices,

$$
\begin{array}{cccc}
 & 1 & x & z \\
1 & \begin{bmatrix} k_{xx}^{1,1} & k_{xy}^{1,1} & k_{xz}^{1,1} \\ k_{yx}^{1,1} & k_{yy}^{1,1} & k_{yz}^{1,1} \\ k_{zx}^{1,1} & k_{zy}^{1,1} & k_{zz}^{1,1} \end{bmatrix} & \cdots & \cdots \\
x & \cdots & \cdots & \begin{bmatrix} k_{xx}^{x,z} & k_{xy}^{x,z} & k_{xz}^{x,z} \\ k_{yx}^{x,z} & k_{yy}^{x,z} & k_{yz}^{x,z} \\ k_{zx}^{x,z} & k_{zy}^{x,z} & k_{zz}^{x,z} \end{bmatrix} \\
z & \cdots & \begin{bmatrix} k_{xx}^{z,x} & k_{xy}^{z,x} & k_{xz}^{z,x} \\ k_{yx}^{z,x} & k_{yy}^{z,x} & k_{yz}^{z,x} \\ k_{zx}^{z,x} & k_{zy}^{z,x} & k_{zz}^{z,x} \end{bmatrix} & \cdots
\end{array}
\tag{8.31}
$$

Each sub-matrix has a fixed pair of expansion functions that are used in the explicit computation of the integrals, as shown in Equation (8.30). It is extremely important to note that the formal expression of each component of the sub-matrices does not depend on the expansion functions. That is, the corresponding components of different sub-matrices have the same formal expression, as shown here:

$$
\begin{aligned}
k_{xx}^{1,1} &= C_{44} \int_A 1 \cdot 1 \, dx\, dz \int_l N_{i,y} N_{j,y} dy \\
k_{xx}^{x,z} &= C_{44} \int_A x \cdot z \, dx\, dz \int_l N_{i,y} N_{j,y} dy \\
k_{xx}^{1,z} &= C_{44} \int_A 1 \cdot z \, dx\, dz \int_l N_{i,y} N_{j,y} dy
\end{aligned}
\tag{8.32}
$$

This implies that the sub-matrix can be considered as a fundamental invariant nucleus which can be used to build the global stiffness matrix. Let us introduce the following notation for the expansion functions, F_τ:

$$
\begin{aligned}
F_{\tau=1} &= 1 \\
F_{\tau=2} &= x \\
F_{\tau=3} &= z
\end{aligned}
\tag{8.33}
$$

The displacement field in Equation (8.20) becomes

$$\begin{aligned} u_x &= F_1\ u_{x_1} + F_2\ u_{x_2} + F_3\ u_{x_3} = F_\tau\ u_{x_\tau} \\ u_y &= F_1\ u_{y_1} + F_2\ u_{y_2} + F_3\ u_{y_3} = F_\tau\ u_{y_\tau} \\ u_z &= F_1\ u_{z_1} + F_2\ u_{z_2} + F_3\ u_{z_3} = F_\tau\ u_{z_\tau} \end{aligned} \tag{8.34}$$

where the repeated indexes indicate summation according to the Einstein notation. The displacement vector can be written as

$$\mathbf{u} = F_\tau \mathbf{u}_\tau, \qquad \tau = 1, 2, 3 \tag{8.35}$$

If an FE formulation is introduced and two-node elements are adopted, the nodal unknown vector is given by

$$\mathbf{u}_{\tau i} = \left\{ u_{x_{\tau i}} \quad u_{y_{\tau i}} \quad u_{z_{\tau i}} \right\}^T \qquad \tau = 1, 2, 3, \qquad i = 1, 2 \tag{8.36}$$

The compact form of the internal work in Equation (8.28) becomes

$$\delta L_{int} = \delta \mathbf{u}_{sj}^T \mathbf{k}^{\tau sij} \mathbf{u}_{\tau i} \tag{8.37}$$

where:

- τ and s are the expansion function indexes;
- i and j are the shape function indexes.

Coherently with the notation introduced, the matrix in Equation (8.31) can be expressed as

$$\begin{array}{cccc} & \tau = 1 & \tau = 2 & \tau = 3 \\ s = 1 & \begin{bmatrix} k_{xx}^{11} & k_{xy}^{11} & k_{xz}^{11} \\ k_{yx}^{11} & k_{yy}^{11} & k_{yz}^{11} \\ k_{zx}^{11} & k_{zy}^{11} & k_{zz}^{11} \end{bmatrix} & \dots & \dots \\ s = 2 & \dots & \dots & \begin{bmatrix} k_{xx}^{32} & k_{xy}^{32} & k_{xz}^{32} \\ k_{yx}^{32} & k_{yy}^{32} & k_{yz}^{32} \\ k_{zx}^{32} & k_{zy}^{32} & k_{zz}^{32} \end{bmatrix} \\ s = 3 & \dots & \begin{bmatrix} k_{xx}^{23} & k_{xy}^{23} & k_{xz}^{23} \\ k_{yx}^{23} & k_{yy}^{23} & k_{yz}^{23} \\ k_{zx}^{23} & k_{zy}^{23} & k_{zz}^{23} \end{bmatrix} & \dots \end{array} \tag{8.38}$$

Each 3×3 block is the FN of the stiffness matrix. A component of the nucleus is given for different combinations of the expansion functions,

$$
\begin{aligned}
k_{xx}^{ij11} &= C_{44} \int_A F_1 \cdot F_1 \, dx\, dz \int_l N_{i,y} N_{j,y} dy \\
k_{xx}^{ij23} &= C_{44} \int_A F_2 \cdot F_3 \, dx\, dz \int_l N_{i,y} N_{j,y} dy \\
k_{xx}^{ij13} &= C_{44} \int_A F_1 \cdot F_3 \, dx\, dz \int_l N_{i,y} N_{j,y} dy
\end{aligned}
\tag{8.39}
$$

The unified formulation introduced is of particular interest when a computational implementation is considered. Exploitation of the four indexes in the formal expression of the FN makes it possible to compute the stiffness matrix by means of four nested *FOR* cycles.

8.2.2 EBBT and TBT as Particular Cases of $N = 1$

EBBT and TBT are particular cases of the $N = 1$ formulation and they can be obtained from the full linear expansion. As far as TBT is concerned, the displacement field is given by

$$
\begin{aligned}
u_x &= u_{x_1} \\
u_y &= u_{y_1} + x\, u_{y_2} + z\, u_{y_3} \\
u_z &= u_{z_1}
\end{aligned}
\tag{8.40}
$$

A linear out-of-plane warping distribution is considered and constant in-plane displacement distributions are accounted for. Starting from the $N = 1$ case, two possible techniques can be used to obtain TBT: (1) the rearranging of rows and columns of the stiffness matrix; (2) penalization of the stiffness terms related to u_{x_2}, u_{x_3}, u_{z_2} and u_{z_3} (the latter is preferred here in the numerical applications). The main diagonal terms have to be considered, namely $i = j$ and $\tau = s$; moreover, only the component with $\tau, s = 2, 3$ has to be penalized, therefore

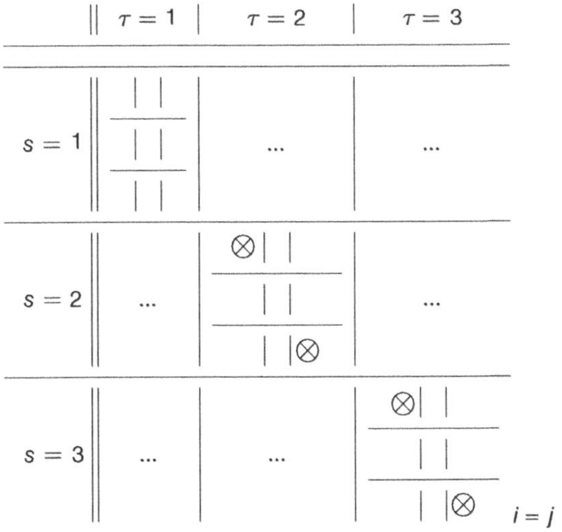

(8.41)

EBBT can be obtained through the penalization of γ_{xy} and γ_{zy}. This condition can be imposed using a penalty value χ in the following constitutive equations:

$$\begin{aligned}\tau_{xy} &= \chi C_{55}\gamma_{xy} + \chi C_{45}\gamma_{zy}\\ \tau_{zy} &= \chi C_{45}\gamma_{xy} + \chi C_{44}\gamma_{zy}\end{aligned} \tag{8.42}$$

8.3 CUF for Higher-Order Models

Classical 1D models can deal properly with the bending of a compact cross-section beam, but on the other hand, refined models are mandatory to describe the mechanical response of more complex boundary (e.g. torsion) or geometrical (e.g. thin walls) conditions. Refinements of classical theories are possible as shown by many authors; however, most of the time, refinements are problem dependent. This means that, as the structural problem changes, a new model has to be developed and adopted.

This section presents a novel unified approach to deal with structural models of any order. In the present unified formulation, in fact, displacement fields are obtained through a single formal expression in a unified manner regardless of the order of the theory (N) which is considered as an input of the analysis.

The unified formulation of the cross-section displacement field is described by an expansion of generic functions (F_τ),

$$\mathbf{u} = F_\tau \mathbf{u}_\tau, \qquad \tau = 1, 2, \ldots, M \tag{8.43}$$

where F_τ are functions of the cross-section coordinates x and z, $\mathbf{u}_\tau$ is the displacement vector and M stands for the number of terms in the expansion. According to the Einstein notation, the repeated subscript τ indicates summation. The choice of F_τ and M is arbitrary; different base functions of any order can be taken into account to model the displacement field of a structure above its cross-section. One possible choice deals with the adoption of Taylor-like polynomials consisting of the 2D base $x^i\, z^j$, where i and j are positive integers. Table 8.2 presents M and F_τ as functions of the order of the beam model, N. Each row shows the expansion terms of an Nth-order theory. The second-order model ($N = 2$, $M = 6$) exploits a parabolic expansion of

Table 8.2 Taylor-like polynomials

N	M	F_τ
0	1	$F_1 = 1$
1	3	$F_2 = x \;\; F_3 = z$
2	6	$F_4 = x^2 \;\; F_5 = xz \;\; F_6 = z^2$
3	10	$F_7 = x^3 \;\; F_8 = x^2 z \;\; F_9 = xz^2 \;\; F_{10} = z^3$
$\vdots$	$\vdots$	$\vdots$
N	$(N+1)(N+2)/2$	$F_{(N^2+N+2)/2} = x^N \;\; \ldots \;\; F_{(N+1)(N+2)/2} = z^N$

This table presents the compact form of the Taylor-like polynomials.

the Taylor-like polynomials,

$$
\begin{array}{llll}
u_x = & \underbrace{u_{x_1}}_{N=0} & \underbrace{+x\,u_{x_2} + z\,u_{x_3}}_{N=1} & \underbrace{+x^2\,u_{x_4} + xz\,u_{x_5} + z^2\,u_{x_6}}_{N=2} \\
u_y = & u_{y_1} & +x\,u_{y_2} + z\,u_{y_3} & +x^2\,u_{y_4} + xz\,u_{y_5} + z^2\,u_{y_6} \\
u_z = & u_{z_1} & +x\,u_{z_2} + z\,u_{z_3} & +x^2\,u_{z_4} + xz\,u_{z_5} + z^2\,u_{z_6}
\end{array}
\tag{8.44}
$$

The 1D model given by Equation (8.44) has 18 displacement variables: three constant ($N = 0$), six linear ($N = 1$) and nine parabolic ($N = 2$). The strain components can be obtained through partial derivations; for instance, the normal components are

$$
\begin{aligned}
\varepsilon_{xx} = \; & u_{x,x} = u_{x_2} + 2\,x\,u_{x_4} + z\,u_{x_5} \\
\varepsilon_{yy} = \; & u_{y,y} = u_{y_1,y} + x\,u_{y_2,y} + z\,u_{y_3,y} + x^2\,u_{y_4,y} + xz\,u_{y_5,y} + z^2\,u_{y_6,y} \\
\varepsilon_{zz} = \; & u_{z,z} = u_{z_3} + z\,u_{z_5} + 2\,z\,u_{z_6}
\end{aligned}
\tag{8.45}
$$

Example 8.3.1 *Let us consider a two-node 1D element (B2) modelled via the $N = 2$ theory. The normal strain components have to be computed at node 1 (i.e. $y_p = 0$) in a generic point of coordinates (x_p, z_p). For an $N = 2$ model the expansion functions (and their partial derivatives) are*

$$\tau, s = 1,2,3,4,5,6 \Rightarrow F_1 = 1, F_2 = x, F_3 = z, F_4 = x^2, F_5 = x\,z, F_6 = z$$

$$\tau, s = 1,2,3,4,5,6 \Rightarrow F_{1,x} = 0, F_{2,x} = 1, F_{3,x} = 0, F_{4,x} = 2\,x, F_{5,x} = z, F_{6,x} = 0$$

$$\tau, s = 1,2,3,4,5,6 \Rightarrow F_{1,z} = 0, F_{2,z} = 0, F_{3,z} = 1, F_{4,z} = 0, F_{5,z} = x, F_{6,z} = 2\,z$$

The use of a B2 element implies that

$$
\begin{aligned}
i,j = 1,2 \Rightarrow N_1 = 1 - \frac{y}{L}, \quad N_2 = \frac{y}{L} \\
i,j = 1,2 \Rightarrow N_{1,y} = -\frac{1}{L}, \quad N_{2,y} = \frac{1}{L}
\end{aligned}
$$

The strain components are given by

$$
\begin{aligned}
\varepsilon_{xx} = \; & u_{x,x} = (F_\tau\,N_i)_x\,u_{x_{\tau i}} = F_{\tau,x}(x_p, z_p)\,N_i(y_p)\,u_{x_{\tau i}} \\
\varepsilon_{yy} = \; & u_{y,y} = (F_\tau\,N_i)_y\,u_{y_{\tau i}} = F_\tau(x_p, z_p)\,N_{i,y}(y_p)\,u_{y_{\tau i}} \\
\varepsilon_{zz} = \; & u_{z,z} = (F_\tau\,N_i)_z\,u_{z_{\tau i}} = F_{\tau,z}(x_p, z_p)\,N_i(y_p)\,u_{z_{\tau i}}
\end{aligned}
$$

Thus

$$
\begin{aligned}
\varepsilon_{xx}(x_p, y_p, z_p) &= \qquad\qquad 1\; u_{x_{21}} + 2\, x_p\; u_{x_{41}} + z_p\; u_{x_{51}} \\
\varepsilon_{yy}(x_p, y_p, z_p) &= -\frac{1}{L}\left(1\; u_{y_{11}} + x_p\; u_{y_{21}} + z_p\; u_{y_{31}} + x_p^2\; u_{y_{41}} + x_p\, z_p\; u_{y_{51}} + z_p^2\; u_{y_{61}}\right) \\
&\quad + \frac{1}{L}\left(1\; u_{y_{12}} + x_p\; u_{y_{22}} + z_p\; u_{y_{32}} + x_p^2\; u_{y_{42}} + x_p\, z_p\; u_{y_{52}} + z_p^2\; u_{y_{62}}\right) \\
\varepsilon_{zz}(x_p, y_p, z_p) &= \qquad\qquad 1\; u_{z_{31}} + x_p\; u_{z_{51}} + 2\, z_p\; u_{z_{61}}
\end{aligned}
$$

The CUF can also deal with *reduced* models by exploiting a lower number of variables; see Carrera *et al.* (2012a) and Carrera and Petrolo (2010, 2011). An example of reduced models is given by

$$
\begin{aligned}
u_x &= u_{x_1} + x\, u_{x_2} + \qquad + x^2\, u_{x_4} + xz\, u_{x_5} \\
u_y &= u_{y_1} + x\, u_{y_2} + z\, u_{y_3} + x^2\, u_{y_4} + \qquad + z^2\, u_{y_6} \\
u_z &= u_{z_1} + \qquad + z\, u_{z_3} + \qquad + xz\, u_{z_5} + z^2\, u_{z_6}
\end{aligned}
\tag{8.46}
$$

The 1D model given in Equation (8.46) has thirteen displacement variables: three constant, four linear and six parabolic. More details about this kind of reduced model will be provided in the following chapters.

8.3.1 N = 3 and N = 4

The third-order model ($N = 3$, $M = 10$) exploits a cubic expansion of the Taylor-like polynomials,

$$
\begin{aligned}
u_x &= \underbrace{\quad\dots\quad}_{} \Big\| + x^3\, u_{x_7} + x^2\, z\, u_{x_8} + x\, z^2\, u_{x_9} + z^3\, u_{x_{10}} \\
u_y &= \quad\dots\quad \Big\| + x^3\, u_{y_7} + x^2\, z\, u_{y_8} + x\, z^2\, u_{y_9} + z^3\, u_{y_{10}} \\
u_z &= \underbrace{\quad\dots\quad}_{N\leq 2} \Big\| \underbrace{+ x^3\, u_{z_7} + x^2\, z\, u_{z_8} + x\, z^2\, u_{z_9} + z^3\, u_{z_{10}}}_{N=3}
\end{aligned}
\tag{8.47}
$$

The 1D model given by Equation (8.47) has 30 displacement variables: 12 cubic ($N = 3$) and 18 lower-order terms ($N \leq 2$). The shear strain components are

$$
\begin{aligned}
\gamma_{xy} = u_{x,y} + u_{y,x} = \overbrace{\ldots}^{N\leq 2} \quad & +x^3\, u_{x_7,y} + x^2\, z\, u_{x_8,y} + x\, z^2\, u_{x_9,y} + z^3\, u_{x_{10},y} \\
& +3\, x^2\, u_{y_7} + 2\, x\, z\, u_{y_8} + z^2\, u_{y_9} \\
\gamma_{yz} = u_{y,z} + u_{z,y} = \ldots \quad & +x^2\, u_{y_8} + 2\, x\, z\, u_{y_9} + 3\, z^2\, u_{y_{10}} \\
& +x^3\, u_{z_7,y} + x^2\, z\, u_{z_8,y} + x\, z^2\, u_{z_9,y} + z^3\, u_{z_{10},y} \\
\gamma_{xz} = u_{x,z} + u_{z,x} = \ldots \quad & +x^2\, u_{x_8} + 2\, x\, z\, u_{x_9} + 3\, z^2\, u_{x_{10}} \\
& +3\, x^2\, u_{z_7} + 2\, x\, z\, u_{z_8} + z^2\, u_{z_9}
\end{aligned}
\tag{8.48}
$$

The fourth-order model ($N = 4$) exploits a quartic expansion of the Taylor-like polynomials

$$
\begin{aligned}
u_x = \ldots \quad & +x^4\, u_{x_{11}} + x^3\, z\, u_{x_{12}} + x^2\, z^2\, u_{x_{13}} + x\, z^3\, u_{x_{14}} + z^4\, u_{x_{15}} \\
u_y = \ldots \quad & +x^4\, u_{y_{11}} + x^3\, z\, u_{y_{12}} + x^2\, z^2\, u_{y_{13}} + x\, z^3\, u_{y_{14}} + z^4\, u_{y_{15}} \\
u_z = \underbrace{\ldots}_{N\leq 3} \quad & \underbrace{+x^4\, u_{z_{11}} + x^3\, z\, u_{z_{12}} + x^2\, z^2\, u_{z_{13}} + x\, z^3\, u_{z_{14}} + z^4\, u_{z_{15}}}_{N=4}
\end{aligned}
\tag{8.49}
$$

The beam model given by Equation (8.49) has 45 displacement variables: 15 quartic ($N = 4$) and 30 lower-order terms ($N \leq 3$).

8.3.2 Nth-Order

The present 1D formulation is able to implement theory of any order by setting the order (N) as an input. An arbitrary refined model can be obtained by the following compact expression:

$$
\begin{aligned}
u_x = \ldots \quad & + \sum_{M=0}^{N} x^{N-M}\, z^{M}\, u_{x_{\frac{N(N+1)}{2}+M+1}} \\
u_y = \ldots \quad & + \sum_{M=0}^{N} x^{N-M}\, z^{M}\, u_{y_{\frac{N(N+1)}{2}+M+1}} \\
u_z = \underbrace{\ldots}_{1,\ldots,N-1} \quad & \underbrace{+ \sum_{M=0}^{N} x^{N-M}\, z^{M}\, u_{z_{\frac{N(N+1)}{2}+M+1}}}_{N\text{th-order}}
\end{aligned}
\tag{8.50}
$$

The total number of displacement variables of the model (N_{DV}) is related to N,

$$N_{DV} = 3 \times \frac{(N+1)(N+2)}{2} \tag{8.51}$$

In the case of the FE formulation, N_{DV} indicates the number of DOFs per node. The strain components can be expressed in a compact manner,

$$\begin{array}{rl||l}
\varepsilon_{xx} = & \ldots & + \sum_{M=0}^{N-1} (N-M)\, x^{N-M-1}\, z^{M}\, u_{x\,\frac{N(N+1)+M+1}{2}} \\
\varepsilon_{yy} = & \ldots & + \sum_{M=0}^{N} x^{N-M}\, z^{M} \left(u_{y\,\frac{N(N+1)+M+1}{2}} \right)_{,y} \\
\varepsilon_{zz} = & \ldots & + \sum_{M=1}^{N} M\, x^{N-M}\, z^{M-1}\, u_{z\,\frac{N(N+1)+M+1}{2}} \\
\gamma_{xy} = & \ldots & + \sum_{M=0}^{N} x^{N-M}\, z^{M} \left(u_{x\,\frac{N(N+1)+M+1}{2}} \right)_{,y} \\
 & & + \sum_{M=0}^{N-1} (N-M)\, x^{N-M-1}\, z^{M}\, u_{y\,\frac{N(N+1)+M+1}{2}} \\
\gamma_{yz} = & \ldots & + \sum_{M=0}^{N} x^{N-M}\, z^{M} \left(u_{z\,\frac{N(N+1)+M+1}{2}} \right)_{,y} \\
 & & + \sum_{M=1}^{N} M\, x^{N-M}\, z^{M-1}\, u_{y\,\frac{N(N+1)+M+1}{2}} \\
\gamma_{xz} = & \ldots & + \sum_{M=0}^{N-1} (N-M)\, x^{N-M-1}\, z^{M}\, u_{z\,\frac{N(N+1)+M+1}{2}} \\
 & \underbrace{\ldots}_{1,\ldots,N-1} & \underbrace{+ \sum_{M=1}^{N} M\, x^{N-M}\, z^{M-1}\, u_{x\,\frac{N(N+1)+M+1}{2}}}_{N\text{th-order}}
\end{array} \tag{8.52}$$

8.4 Governing Equations, FE Formulation and the FN

The governing equations are derived by means of the PVD. Starting from the unified form of the displacement field in Equation (8.43), stiffness, mass and loading arrays are obtained in terms of the FNs whose form is independent of the order of the model. The weak form of the governing equations is obtained by means of the FEM, which can easily handle arbitrary geometries, loading and boundary conditions. Closed-form solutions can be found in Carrera *et al.* (2011).

8.4.1 Governing Equations

According to the PVD, the following equation holds:

$$\delta L_{iin} = \delta L_{ext} - \delta L_{ine} \tag{8.53}$$

Stress (σ) and strain (ε) components are grouped as follows:

$$\begin{aligned} \boldsymbol{\sigma}_p &= \left\{ \sigma_{zz} \quad \sigma_{xx} \quad \sigma_{zx} \right\}^T, \qquad & \boldsymbol{\varepsilon}_p &= \left\{ \varepsilon_{zz} \quad \varepsilon_{xx} \quad \varepsilon_{zx} \right\}^T \\ \boldsymbol{\sigma}_n &= \left\{ \tau_{zy} \quad \tau_{xy} \quad \tau_{yy} \right\}^T, \qquad & \boldsymbol{\varepsilon}_n &= \left\{ \gamma_{zy} \quad \gamma_{xy} \quad \gamma_{yy} \right\}^T \end{aligned} \tag{8.54}$$

Linear strain–displacement relations can be rewritten as

$$\begin{aligned} \boldsymbol{\varepsilon}_p &= \boldsymbol{b}_p \mathbf{u} \\ \boldsymbol{\varepsilon}_n &= \boldsymbol{b}_n \mathbf{u} = (\boldsymbol{b}_{n\Omega} + \boldsymbol{b}_{ny}) \mathbf{u} \end{aligned} \tag{8.55}$$

where

$$\boldsymbol{b}_p = \begin{bmatrix} 0 & 0 & \partial/\partial z \\ \partial/\partial x & 0 & 0 \\ \partial/\partial z & 0 & \partial/\partial x \end{bmatrix}, \quad \boldsymbol{b}_{n\Omega} = \begin{bmatrix} 0 & \partial/\partial z & 0 \\ 0 & \partial/\partial x & 0 \\ 0 & 0 & 0 \end{bmatrix}, \quad \boldsymbol{b}_{ny} = \begin{bmatrix} 0 & 0 & \partial/\partial y \\ \partial/\partial y & 0 & 0 \\ 0 & \partial/\partial y & 0 \end{bmatrix} \tag{8.56}$$

Hooke's law becomes

$$\boldsymbol{\sigma} = \boldsymbol{C}\boldsymbol{\varepsilon} \tag{8.57}$$

According to Equation (8.54), the previous equation becomes

$$\begin{aligned} \boldsymbol{\sigma}_p &= \boldsymbol{C}_{pp}\boldsymbol{\varepsilon}_p + \boldsymbol{C}_{pn}\boldsymbol{\varepsilon}_n \\ \boldsymbol{\sigma}_n &= \boldsymbol{C}_{np}\boldsymbol{\varepsilon}_p + \boldsymbol{C}_{nn}\boldsymbol{\varepsilon}_n \end{aligned} \tag{8.58}$$

In the case of isotropic material, the matrices $\boldsymbol{C}_{pp}$, $\boldsymbol{C}_{nn}$, $\boldsymbol{C}_{pn}$ and $\boldsymbol{C}_{np}$ are

$$\boldsymbol{C}_{pp} = \begin{bmatrix} C_{11} & C_{12} & 0 \\ C_{12} & C_{22} & 0 \\ 0 & 0 & C_{66} \end{bmatrix}, \quad \boldsymbol{C}_{nn} = \begin{bmatrix} C_{55} & 0 & 0 \\ 0 & C_{44} & 0 \\ 0 & 0 & C_{33} \end{bmatrix}, \quad \boldsymbol{C}_{pn} = \boldsymbol{C}_{np}^T = \begin{bmatrix} 0 & 0 & C_{13} \\ 0 & 0 & C_{23} \\ 0 & 0 & 0 \end{bmatrix} \tag{8.59}$$

where $C_{13} = C_{23} = C_{12}$ and $C_{44} = C_{55} = C_{66}$, see also Equation (2.15). According to Equation (8.54), the virtual variation of the internal work is considered as the sum of two contributions,

$$\delta L_i = \int_V \delta\epsilon_n^T \sigma_n \, dV + \int_V \delta\epsilon_p^T \sigma_p \, dV \tag{8.60}$$

The virtual variation of the work of the inertial loadings is

$$\delta L_{ine} = \int_V \rho \ddot{u} \delta u^T dV \tag{8.61}$$

where ρ is the density of the material and $\ddot{u}$ is the acceleration vector.

The application of surface, line and point loads is now discussed to derive the variation of the external work. A generic surface load acting on a lateral face of the structure is first considered, i.e. $p_{\alpha\beta}$ where α can be equal to x or z and β can be equal to x, y or z. The first subscript (α) indicates the axis perpendicular to the surface (A_α) where the load is applied, whereas the second one (β) indicates the direction of the load. The virtual variation of the external work due to $p_{\alpha\beta}$ is given by

$$\delta L_{ext}^{p_{\alpha\beta}} = \int_{A_\alpha} \delta u_\beta \, p_{\alpha\beta} \, dA \tag{8.62}$$

A generic line load, $q_{\alpha\beta}$, can be treated similarly. The variation of the external work due to $q_{\alpha\beta}$ is given by

$$\delta L_{ext}^{q_{\alpha\beta}} = \int_l \delta u_\beta \, q_{\alpha\beta} \, dy \tag{8.63}$$

The loading vector in the case of a generic concentrated load $\boldsymbol{P}$ is

$$\boldsymbol{P} = \left\{ P_x \quad P_y \quad P_z \right\}^T \tag{8.64}$$

and the work due to $\boldsymbol{P}$ is

$$\delta L_{ext} = \boldsymbol{P} \delta \mathbf{u}^T \tag{8.65}$$

It is important to note that the loading vector changes according to the order of the expansion.

8.4.2 FE Formulation

The nodal displacement vector is introduced, i.e.

$$\mathbf{u}_{\tau i} = \left\{ u_{x_{\tau i}} \quad u_{y_{\tau i}} \quad u_{z_{\tau i}} \right\}^T, \qquad \tau = 1, 2, \ldots, M, \; i = 1, 2, \ldots, N_{EN} \tag{8.66}$$

where the index i indicates the element node and N_{EN} stands for the number of nodes per element. If a linear model is considered ($N = 1$, $M = 3$), and a two-node element is adopted, the element unknowns will be

$$\mathbf{u}_{\tau i} = \begin{Bmatrix} u_{x_{11}} & u_{y_{11}} & u_{z_{11}} & u_{x_{21}} & u_{y_{21}} & u_{z_{21}} & u_{x_{31}} & u_{y_{31}} & u_{z_{31}} \\ u_{x_{12}} & u_{y_{12}} & u_{z_{12}} & u_{x_{22}} & u_{y_{22}} & u_{z_{22}} & u_{x_{32}} & u_{y_{32}} & u_{z_{32}} \end{Bmatrix}^T \tag{8.67}$$

The displacement variables are interpolated along the axis of the beam by means of the shape functions (N_i),

$$\mathbf{u} = N_i F_\tau \mathbf{u}_{\tau i} \tag{8.68}$$

Beam elements with two (B2), three (B3) and four (B4) nodes are considered whose shape functions are

$$\begin{aligned}
&N_1 = \tfrac{1}{2}(1-r),\ N_2 = \tfrac{1}{2}(1+r), \quad \begin{cases} r_1 = -1 \\ r_2 = +1 \end{cases} \\
&N_1 = \tfrac{1}{2}r(r-1),\ N_2 = \tfrac{1}{2}r(r+1),\ N_3 = -(1+r)(1-r), \quad \begin{cases} r_1 = -1 \\ r_2 = +1 \\ r_3 = 0 \end{cases} \\
&\begin{matrix} N_1 = -\tfrac{9}{16}\left(r+\tfrac{1}{3}\right)\left(r-\tfrac{1}{3}\right)(r-1),\ N_2 = \tfrac{9}{16}\left(r+\tfrac{1}{3}\right)\left(r-\tfrac{1}{3}\right)(r+1), \\ N_3 = +\tfrac{27}{16}(r+1)\left(r-\tfrac{1}{3}\right)(r-1),\ N_4 = -\tfrac{27}{16}(r+1)\left(r+\tfrac{1}{3}\right)(r-1), \end{matrix} \quad \begin{cases} r_1 = -1 \\ r_2 = +1 \\ r_3 = -\tfrac{1}{3} \\ r_4 = +\tfrac{1}{3} \end{cases}
\end{aligned} \tag{8.69}$$

The natural coordinate (r) ranges from -1 to $+1$. r_i indicates the position of the node within the natural beam boundaries. The beam model order is given by the expansion above the cross-section. The number of nodes per element is related to the approximation along the longitudinal axis. An Nth-order beam model is therefore a theory that exploits an Nth-order Taylor-like polynomial to describe the displacement field of the cross-section. The choice of the beam model, the beam element and the mesh (i.e. the number of beam elements) determines the total number of DOFs of the structural model,

$$\text{DOFs} = \underbrace{3 \times M}_{\text{number of DOFs per node}} \times [(\underbrace{N_{NE}}_{\text{number of nodes per element}} - 1) \times \underbrace{N_{BE}}_{\text{total number of beam elements}} + 1] \tag{8.70}$$

8.4.3 Stiffness Matrix

In the CUF, FE matrices are formulated in terms of the FN. A compact form of the stiffness matrix can be obtained through Equations (8.55), (8.58) and (8.68),

$$\delta L_{int} = \delta \mathbf{u}_{sj}^T \mathbf{k}^{\tau sij} \mathbf{u}_{\tau i} \tag{8.71}$$

where $\mathbf{k}^{\tau sij}$ is the stiffness matrix written in the form of the FN. The FN is a 3×3 array which is formally independent of the order of the structural model. The nine components of the nucleus were introduced in Table 8.1, their explicit expressions are as follows:

$$
\begin{aligned}
k_{xx}^{\tau sij} &= C_{22} \int_A F_{\tau,x} F_{s,x} dx\, dz \int_l N_i N_j dy + C_{66} \int_A F_{\tau,z} F_{s,z} dx\, dz \int_l N_i N_j dy \\
&\quad + C_{44} \int_A F_\tau F_s dx\, dz \int_l N_{i,y} N_{j,y} dy \\
k_{xy}^{\tau sij} &= C_{23} \int_A F_\tau F_{s,x} dx\, dz \int_l N_{i,y} N_j dy + C_{44} \int_A F_{\tau,x} F_s dx\, dz \int_l N_i N_{j,y} dy \\
k_{xz}^{\tau sij} &= C_{12} \int_A F_{\tau,z} F_{s,x} dx\, dz \int_l N_i N_j dy + C_{66} \int_A F_{\tau,x} F_{s,z} dx\, dz \int_l N_i N_j dy \\
k_{yx}^{\tau sij} &= C_{44} \int_A F_\tau F_{s,x} dx\, dz \int_l N_{i,y} N_j dy + C_{23} \int_A F_{\tau,x} F_s dx\, dz \int_l N_i N_{j,y} dy \\
k_{yy}^{\tau sij} &= C_{55} \int_A F_{\tau,z} F_{s,z} dx\, dz \int_l N_i N_j dy + C_{44} \int_A F_{\tau,x} F_{s,x} dx\, dz \int_l N_i N_j dy \\
&\quad + C_{33} \int_A F_\tau F_s dx\, dz \int_l N_{i,y} N_{j,y} dy \\
k_{yz}^{\tau sij} &= C_{55} \int_A F_\tau F_{s,z} dx\, dz \int_l N_{i,y} N_j dy + C_{13} \int_A F_{\tau,z} F_s dx\, dz \int_l N_i N_{j,y} dy \\
k_{zx}^{\tau sij} &= C_{12} \int_A F_{\tau,x} F_{s,z} dx\, dz \int_l N_i N_j dy + C_{66} \int_A F_{\tau,z} F_{s,x} dx\, dz \int_l N_i N_j dy \\
k_{zy}^{\tau sij} &= C_{13} \int_A F_\tau F_{s,z} dx\, dz \int_l N_{i,y} N_j dy + C_{55} \int_A F_{\tau,z} F_s dx\, dz \int_l N_i N_{j,y} dy \\
k_{zz}^{\tau sij} &= C_{11} \int_A F_{\tau,z} F_{s,z} dx\, dz \int_l N_i N_j dy + C_{66} \int_A F_{\tau,x} F_{s,x} dx\, dz \int_l N_i N_j dy \\
&\quad + C_{55} \int_A F_\tau F_s dx\, dz \int_l N_{i,y} N_{j,y} dy
\end{aligned}
\tag{8.72}
$$

Four indexes (τ, s, i and j) are exploited to assemble the stiffness matrix. τ and s are related to the expansion functions (F_τ and F_s) and the FN is computed by varying τ and s, as shown in Figure 8.6 where the construction of the so-called τs-block, which coincides with the node stiffness matrix, is described. The position of each τs-block in the element stiffness matrix is

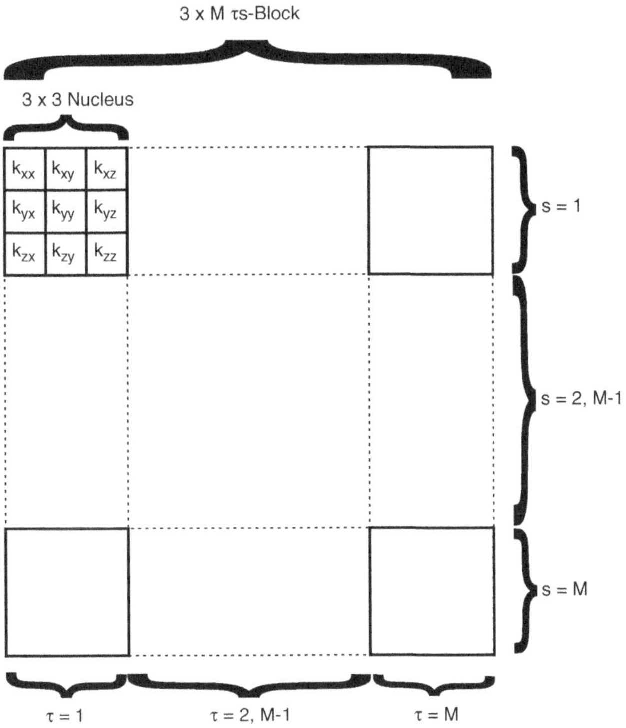

Figure 8.6 Assembly scheme of the nodal stiffness matrix

shown in Figure 8.7. The element stiffness matrix is obtained through the assembly of all the ij-blocks, as shown in Figure 8.8. Structural models of any order can be computed since the definition of the order acts on the τs-loop.

Example 8.4.1 *Let us consider a rectangular beam element, with length equal to L. A B2 element is used and a linear theory ($N = 1$) is adopted:*

$$\begin{aligned} \tau, s = 1, 2, 3 &\Rightarrow F_1 = 1, \quad F_2 = x, \quad F_3 = z \\ i, j = 1, 2 &\Rightarrow N_1 = 1 - \tfrac{y}{L}, \quad N_2 = \tfrac{y}{L} \end{aligned}$$

The cross-section coordinates range from $-a$ to $+a$ along the x direction and from $-b$ to $+b$ along the z direction. The k_{xx} component has to be computed for τ, $s = 2$, $i = 1$ and $j = 2$:

$$\begin{aligned} k_{xx}^{2212} = C_{22} &\int_{-a}^{+a}\int_{-b}^{+b} F_{2,x}\, F_{2,x} dx\, dz \int_0^L N_1\, N_2 dy \\ +C_{66} &\int_{-a}^{+a}\int_{-b}^{+b} F_{2,z}\, F_{2,z} dx\, dz \int_0^L N_1\, N_2 dy \\ +C_{44} &\int_{-a}^{+a}\int_{-b}^{+b} F_2\, F_2 dx\, dz \int_0^L N_{1,y}\, N_{2,y} dy \end{aligned}$$

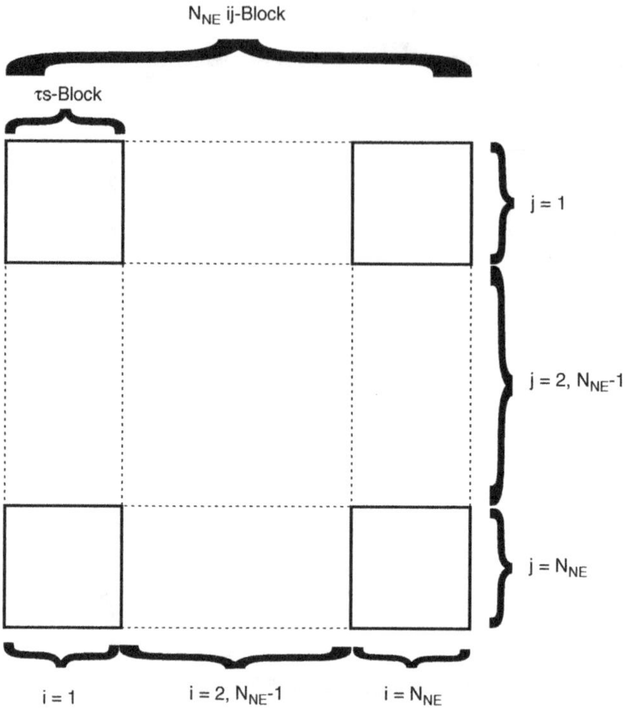

Figure 8.7 Assembly scheme of the element stiffness matrix

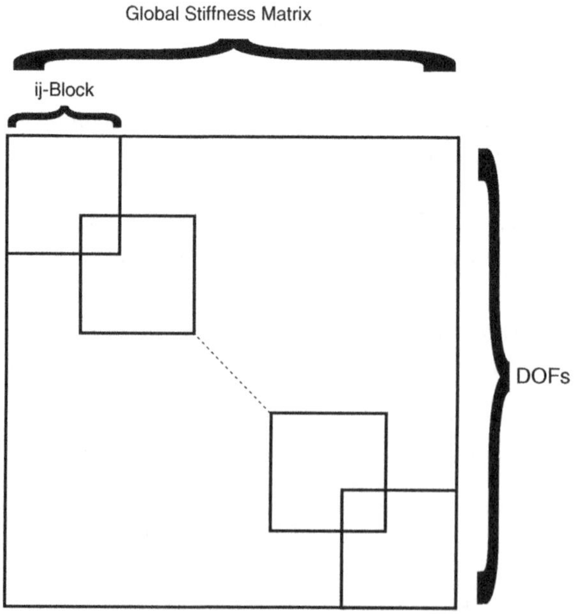

Figure 8.8 Assembly scheme of the global stiffness matrix

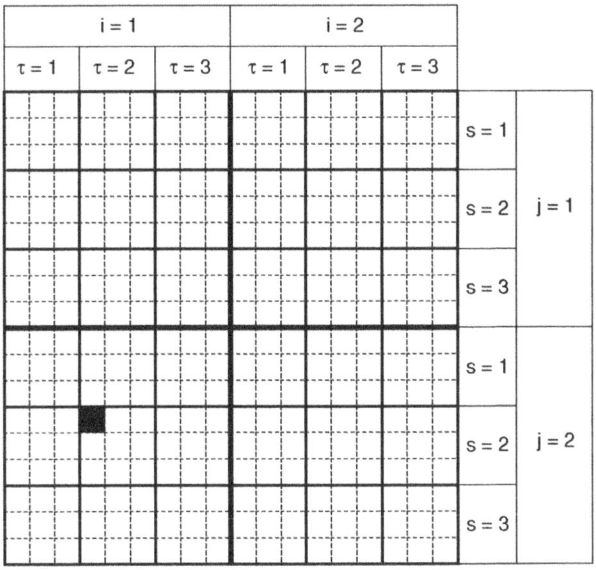

Figure 8.9 Position of k_{xx}^{2212} within the element stiffness matrix

By substituting the explicit expressions of the functions, the integrals become

$$
\begin{aligned}
k_{xx}^{2212} = C_{22} & \int_{-a}^{+a}\int_{-b}^{+b} 1\cdot 1\,dx\,dz \int_0^L \left(1-\frac{y}{L}\right)\frac{y}{L}\,dy \\
+ C_{66} & \int_{-a}^{+a}\int_{-b}^{+b} 0\cdot 0\,dx\,dz \int_0^L \left(1-\frac{y}{L}\right)\frac{y}{L}\,dy \\
+ C_{44} & \int_{-a}^{+a}\int_{-b}^{+b} x\cdot x\,dx\,dz \int_0^L \left(-\frac{1}{L}\right)\frac{1}{L}\,dy
\end{aligned}
$$

The final result is

$$k_{xx}^{2212} = \frac{2}{3}\,C_{22}\,a\,b\,L - \frac{4}{3}\,C_{44}\,\frac{b\,a^3}{L}$$

Figure 8.9 shows the position of k_{xx}^{2212} *within the element stiffness matrix.*

8.4.4 Mass Matrix

The FN assembly technique can be exploited for every FE matrix or vectors. In order to obtain the mass matrix, Equation (8.61) is rewritten using Equations (8.55) and (8.68):

$$\delta L_{ine} = \int_l \delta \mathbf{u}_{sj}^T N_j \left[\int_\Omega \rho (F_s \mathbf{I})(F_\tau \mathbf{I}) d\Omega\right] N_i \ddot{\mathbf{u}}_{\tau i} dy \tag{8.73}$$

where $\ddot{\boldsymbol{u}}$ is the nodal acceleration vector. The last equation can be rewritten in the following compact manner:

$$\delta L_{ine} = \delta \mathbf{u}_{sj}^T \mathbf{m}^{\tau sij} \ddot{\boldsymbol{u}}_{\tau i} \tag{8.74}$$

where $\mathbf{m}^{\tau sij}$ is the mass matrix in the form of the FN. Its components are

$$\begin{aligned} m_{xx}^{\tau sij} &= m_{yy}^{\tau sij} = m_{zz}^{\tau sij} = \rho \int_A F_\tau F_s dx\, dz \int_l N_i N_j dy \\ m_{xy}^{\tau sij} &= m_{xz}^{\tau sij} = m_{yx}^{\tau sij} = m_{yz}^{\tau sij} = m_{zx}^{\tau sij} = m_{zy}^{\tau sij} = 0 \end{aligned} \tag{8.75}$$

The undamped dynamic problem can be written as follows:

$$\boldsymbol{M}\ddot{\boldsymbol{A}} + \boldsymbol{K}\boldsymbol{A} = \boldsymbol{P} \tag{8.76}$$

where $\mathbf{A}$ is the vector of the nodal unknowns and $\boldsymbol{P}$ is the loading vector. Introducing harmonic solutions, it is possible to compute the natural frequencies (ω_i) by solving an eigenvalue problem:

$$(-\omega_i^2 \boldsymbol{M} + \boldsymbol{K})\mathbf{A}_i = 0 \tag{8.77}$$

where $\mathbf{A}_i$ is the ith eigenvector.

8.4.5 Loading Vector

Surface loads are addressed first. The F_τ expansions and the nodal displacements were introduced in Equation (8.62),

$$\delta L_{ext}^{p_{\alpha\beta}} = \int_{A_\alpha} F_s\, N_j\, \delta u_{\beta_{sj}}\, p_{\alpha\beta}\, dA \tag{8.78}$$

For the line load case, Equation (8.63) becomes

$$\delta L_{ext}^{q_{\alpha\beta}} = \int_l F_s\, N_j\, \delta u_{\beta_{sj}}\, q_{\alpha\beta}\, dy \tag{8.79}$$

Equation (8.65) is exploited for the point load case and becomes

$$\delta L_{ext} = F_s(x_p, z_p) N_j(y_p) \boldsymbol{P} \delta \mathbf{u}_{sj}^T \tag{8.80}$$

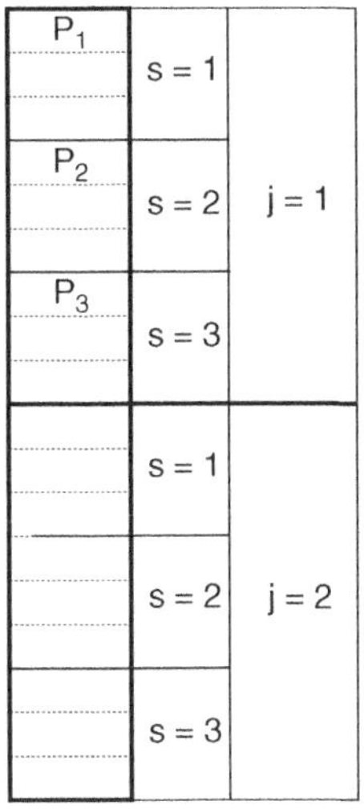

Figure 8.10 Position of the loading components within the nodal force vector

where it is assumed that P is applied at $[x_p, y_p, z_p]$. This equation is used to assemble the loading vector by detecting the displacement variables that have to be loaded.

Example 8.4.2 *Let us consider the same beam element as in Example 8.4.1. A point load* **P** *acts on node 1 in the x direction only*

$$\boldsymbol{P} = \{P_x \quad 0 \quad 0\}^T \tag{8.81}$$

The virtual variation of the external work is

$$\delta L_{ext} = \underbrace{P_x}_{P_1} \delta u_{x1} + \underbrace{x_p P_x}_{P_2} \delta u_{x2} + \underbrace{z_p P_x}_{P_3} \delta u_{x3} \tag{8.82}$$

where $[x_p, z_p]$ are the coordinates on the cross-section of the loading application point. Figure 8.10 shows the position of the load components within the force vector.

8.5 Locking Phenomena

Two types of locking phenomena are described in this section: Poisson and shear locking. The former is due to deformation coupling phenomena and affects EBBT, TBT and $N = 1$ models. The latter is a numerical phenomenon related to the FE formulation. The effects of both lockings can be extremely detrimental; however, a number of correction techniques have been proposed, as will be shown in the following.

8.5.1 Poisson Locking and its Correction

Poisson locking (PL) is a phenomenon that arises from the coupling among the normal deformations ε_{xx}, ε_{yy} and ε_{zz} as stated by the Poisson coefficients,

$$\nu_{ij} = -\frac{\varepsilon_{jj}}{\varepsilon_{ii}}, \qquad i,j = x,y,z \tag{8.83}$$

PL affects the accuracy of EBBT, TBT and $N = 1$ models.

8.5.1.1 Kinematic Considerations on Strains

Let us first consider strain distributions derived from the EBBT, TBT and $N = 1$ displacement models,

$$\begin{aligned}
\varepsilon_{xx}^{kin} &= u_{x,x} = u_{x_2} && \text{if} \quad N = 1\\
&= 0 && \text{if} \quad \text{EBBT or TBT}\\
\varepsilon_{yy}^{kin} &= u_{y,y} = u_{y_1,y} + x\, u_{y_2,y} + z\, u_{y_3,y} && \text{if} \quad N = 1 \ \text{ or EBBT or TBT}\\
\varepsilon_{zz}^{kin} &= u_{z,z} = u_{z_3} && \text{if} \quad N = 1\\
&= 0 && \text{if} \quad \text{EBBT or TBT}
\end{aligned} \tag{8.84}$$

The following can be stated:

- the out-of-plane axial strain, ε_{yy}^{kin}, is linear;
- the in-plane axial strains, ε_{xx}^{kin} and ε_{zz}^{kin}, are null in classical models;
- $N = 1$ provides a constant geometrical distribution of in-plane axial strains;
- to avoid constant distributions of ε_{xx}^{kin} and ε_{zz}^{kin}, the order of the expansion above the cross-section must be greater than two, $N \geq 2$.

8.5.1.2 Physical Considerations on Strains

Starting from the constitutive relations between the in-plane and out-of-plane strains, some physical considerations can be made,

$$\varepsilon_{xx}^{con}, \varepsilon_{zz}^{con} \propto \nu \varepsilon_{yy}^{kin} \tag{8.85}$$

This means that the orders of magnitude of the in-plane and out-of-plane normal strains are the same; that is, since EBBT, TBT and $N = 1$ give linear distributions of ε_{yy}^{kin}, then ε_{xx}^{con} and ε_{zz}^{con} will also be linear. It is clear that a contradiction exists between kinematics and the constitutive laws, and while the former does not account for in-plane strain distributions ($\varepsilon_{xx}^{kin}, \varepsilon_{zz}^{kin} = 0$) the latter provides linear distributions of ε_{xx}^{con} and ε_{zz}^{con}. This contradiction generates the PL that is responsible for the poor convergence rates of EBBT, TBT and $N = 1$. Two possible remedies will now be described to contrast PL (Carrera and Brischetto, 2008a,b): that is, to contrast ε_{xx}^{con} and ε_{zz}^{con}.

First Remedy: Use of Higher-order Kinematics

The most 'natural' way of overcoming PL is based on the use of refined displacement models. At least second-order terms to describe u_x and u_z should be employed,

$$\begin{aligned} u_x &= u_{x_1} + x\, u_{x_2} + z\, u_{x_3} + x^2\, u_{x_4} + \ldots \\ u_z &= u_{z_1} + x\, u_{z_2} + z\, u_{z_3} + x^2\, u_{z_4} + \ldots \end{aligned} \tag{8.86}$$

This kind of solution is particularly convenient in the framework of the present unified formulation.

Second Remedy: Modification of Elastic Coefficients

It is clear that PL originates from constitutive laws which state the intrinsic coupling between in- and out-of-plane strain components. The second remedy illustrated here is based on a proper modification of these laws.

The 3D constitutive relations between stresses (σ) and strains (ε) are given by Hooke's law,

$$\{\sigma\} = [\mathbf{C}]\{\varepsilon\} \tag{8.87}$$

For isotropic materials, its explicit form is

$$\begin{Bmatrix} \sigma_{xx} \\ \sigma_{yy} \\ \sigma_{zz} \\ \sigma_{yz} \\ \sigma_{xz} \\ \sigma_{xy} \end{Bmatrix} = \begin{bmatrix} C_{11} & C_{12} & C_{13} & 0 & 0 & 0 \\ C_{12} & C_{22} & C_{23} & 0 & 0 & 0 \\ C_{13} & C_{23} & C_{33} & 0 & 0 & 0 \\ 0 & 0 & 0 & C_{44} & 0 & 0 \\ 0 & 0 & 0 & 0 & C_{55} & 0 \\ 0 & 0 & 0 & 0 & 0 & C_{66} \end{bmatrix} \begin{Bmatrix} \varepsilon_{xx} \\ \varepsilon_{yy} \\ \varepsilon_{zz} \\ \gamma_{yz} \\ \gamma_{xz} \\ \gamma_{xy} \end{Bmatrix} \tag{8.88}$$

The elastic coefficients are related to the engineering constants (Poisson's ν, Young's E and the shear G moduli) as follows:

$$\begin{aligned} C_{11} &= C_{22} = C_{33} = \frac{E\,(1-\nu)}{(1+\nu)(1-2\nu)} \\ C_{12} &= C_{13} = C_{23} = \frac{\nu E}{(1+\nu)(1-2\nu)} \\ C_{44} &= C_{55} = C_{66} = G \end{aligned} \tag{8.89}$$

These are the coefficients that have to be modified in order to prevent PL. Classical plate theories correct PL by imposing that the out-of-plane normal stress is zero. This hypothesis yields reduced material stiffness coefficients which have to be accounted for in Hooke's law for the in-plane stress and strain components. PL correction is obtained for the beam theory in the same manner, where σ_{xx} and σ_{zz} are assumed to be zero under Hooke's law, i.e.

$$\begin{cases} \sigma_{xx} = C_{11}\varepsilon_{xx} + C_{12}\varepsilon_{yy} + C_{13}\varepsilon_{zz} = 0 \\ \sigma_{zz} = C_{13}\varepsilon_{xx} + C_{23}\varepsilon_{yy} + C_{33}\varepsilon_{zz} = 0 \end{cases} \tag{8.90}$$

This system has to be solved with respect to ε_{xx} and ε_{zz}

$$\begin{cases} \varepsilon_{xx} = \dfrac{C_{13}\,C_{23} - C_{12}\,C_{33}}{C_{11}\,C_{33} - C_{13}^2}\,\varepsilon_{yy} \\ \varepsilon_{zz} = \dfrac{C_{13}\,C_{12} - C_{23}\,C_{11}}{C_{11}\,C_{33} - C_{13}^2}\,\varepsilon_{yy} \end{cases} \tag{8.91}$$

These expressions have to be inserted into Equation (8.88) in the σ_{yy} line,

$$\sigma_{yy} = \underbrace{\left(C_{22} - C_{12}\,\frac{C_{33}\,C_{12} - C_{13}\,C_{23}}{C_{11}\,C_{33} - C_{13}^2} - C_{23}\,\frac{C_{23}\,C_{11} - C_{13}\,C_{12}}{C_{11}\,C_{33} - C_{13}^2} \right)}_{C'_{22}} \varepsilon_{yy} \tag{8.92}$$

C'_{22} is the reduced elastic coefficient that has to be used to contrast PL. The above procedure is also valid in the case of orthotropic materials; in this case, not only will C_{22} be reduced, but also the elastic coefficient that relates σ_{yy} to the in-plane shear strain γ_{xz}. In the particular case of isotropic materials, the reduced elastic coefficient is equal to Young's modulus,

$$C'_{22} = E \tag{8.93}$$

Some final remarks are necessary to highlight the role of the PL correction:

- The correction of the material coefficients does not have a consistent theoretical proof.
- This means that the adoption of reduced material coefficients does not necessarily lead to the exact 3D solution, as shown in Carrera *et al.* (2012b), where the correction of PL led to a model with less bending stiffness than the correct one.
- A consistent way of preventing PL consists of adopting higher-order models, as will be shown in subsequent chapters.
- The correction of the material coefficients becomes detrimental as nonlinear terms are included in the displacement field, as will be shown in the next chapter.

Table 8.3 Effect of the PL correction on u_z (Carrera *et al.* 2010a)

Model	*PL corrected*	*PL uncorrected*
	$L/h = 100,\ u_z \times 10^2\, m,\ u_{z_b} \times 10^2 = -1.333\, m$	
EBBT	-1.333	-0.901
TBT	-1.333	-0.901
$N = 1$	-1.333	-0.901
	$L/h = 10,\ u_z \times 10^5\, m,\ u_{z_b} \times 10^5 = -1.333\, m$	
EBBT	-1.333	-0.901
TBT	-1.343	-0.909
$N = 1$	-1.343	-0.909

This table shows the beneficial effect of the PL correction for the case of the EBBT, TBT and $N = 1$ models.

Example 8.5.1 *Let us consider a cantilever square cross-section beam loaded with a force ($P_z = -50\, N$) applied at the centre point of the free-tip cross-section. The cross-section edge, h, is 0.2 m long and two slenderness ratios, L/h, are considered: 100 and 10. The material is isotropic with $E = 75$ GPa and $\nu = 0.33$. The results are obtained via an FE model having a 40 B4 mesh. A benchmark solution is obtained by means of EBBT,*

$$u_{z_b} = \frac{1}{3}\frac{P_z L^3}{EI} \tag{8.94}$$

where I indicates the moment of inertia of the cross-section. Table 8.3 shows the displacement values at the loading point for the EBBT, TBT and $N = 1$ models. The second column indicates the results obtained with the PL correction activated, while the third column reports the results with the PL correction deactivated. It can be observed how the correction of PL enhances the flexibility of the structure and the convergence rate. Similar results are also obtained for free vibration analyses (Carrera et al.*, 2012b); in this case, the correction of PL makes the natural bending frequencies decrease.*

8.5.2 Shear Locking

Shear locking is a numerical phenomenon that can occur as the thickness of beams or plates decreases (Cinefra *et al.* 2010; Reddy, 1997). Locking is due to an overestimation of the shear stiffness of the structures, which tends to be infinite as the thickness tends to zero. Many techniques are able to attenuate this effect. Among these, the one adopted for the present formulation is the reduced selective integration. This method is based on a reduced Gauss integration of the terms of the stiffness matrix that are related to shear. 'Reduced' here means that a lower number of Gauss points are used and this results in a reduction in the shear stiffness of the structure. It is important to point out that refined beam theories obtained by means of the CUF can lead to models which are hardly affected by shear locking as shown in Example 8.5.2. A more exhaustive description of shear locking and related correction techniques can be found in the plate and shell chapters of this book (Section 10.5.2) where the MITC4 technique is also introduced.

Table 8.4 Effect of the selective integration on u_z for different meshes and beam elements via an N = 4 model (Carrera *et al.* 2010a)

No. of elements	*B2*	*B2**	*B3*	*B3**	*B4*	*B4**
		$L/h = 100$, $u_y \times 10^2\ m$				
1	−0.893	−0.0004	−1.158	−0.905	−1.240	−1.227
2	−1.065	−0.001	−1.255	−1.191	−1.290	−1.278
3	−1.165	−0.003	−1.275	−1.247	−1.302	−1.297
4	−1.210	−0.006	−1.289	−1.271	−1.310	−1.310
5	−1.236	−0.009	−1.298	−1.285	−1.320	−1.312
10	−1.287	−0.035	−1.316	−1.312	−1.325	−1.324
20	−1.311	−0.129	−1.325	−1.324	−1.329	−1.329
40	−1.323	−0.399	−1.330	−1.329	−1.332	−1.332
		$L/h = 10$, $u_y \times 10^5\ m$				
1	−0.904	−0.035	−1.168	−0.988	−1.250	−1.241
2	−1.076	−0.128	−1.255	−1.223	−1.296	−1.293
3	−1.176	−0.255	−1.285	−1.274	−1.311	−1.310
4	−1.220	−0.392	−1.299	−1.294	−1.319	−1.318
5	−1.246	−0.524	−1.307	−1.305	−1.323	−1.322
10	−1.296	−0.954	−1.324	−1.323	−1.330	−1.330
20	−1.318	−1.208	−1.330	−1.330	−1.333	−1.333
40	−1.328	−1.298	−1.333	−1.333	−1.333	−1.333

*Full integration adopted.
This table shows the effect of selective integration on the solution given by a fourth-order beam model for different meshes.

Example 8.5.2 *Let us consider the cantilever beam of Example 8.5.1, in which the effect of the reduced integration has to be evaluated by using a fourth-order model. Table 8.4 shows the results for different meshes. The following points are evident:*

- *The use of a fourth-order model makes selective integration necessary only if beams are meshed with B2 elements.*
- *In all the other cases, full integration is able to detect the right solution.*
- *The detrimental effect of shear locking due to low thickness values is confirmed.*
- *The use of higher-order models leads to shear locking free models.*

8.6 Numerical Applications

Refined 1D models are particularly powerful tools when thin-walled structures are considered. Since in- and out-of-plane warping phenomena must be accounted for, classical models cannot be exploited. On the other hand, the use of shell or solid FEs can result in very cumbersome numerical models. The aim of this section is to show the enhanced capabilities of the present 1D formulation in detecting 3D-like results with very low computational costs.

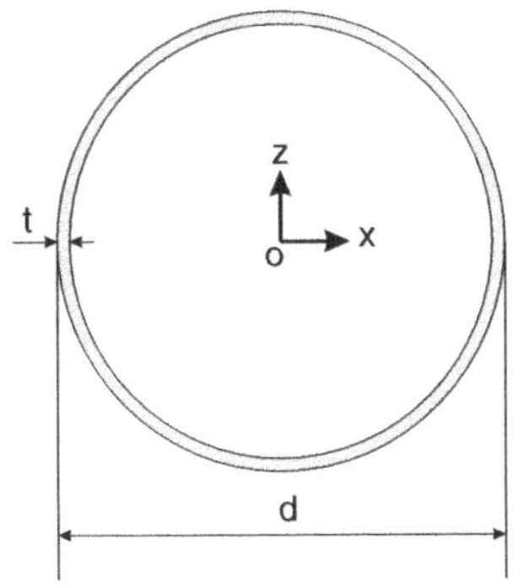

Figure 8.11 Cross-section of the hollow cylinder

8.6.1 Structural Analysis of a Thin-Walled Cylinder

A thin-walled cylinder is considered with the cross-section geometry shown in Figure 8.11 where the diameter, d, is equal to 2 m, and the thickness, t, is equal to 0.02 m. The length of the cylinder, L, is equal to 20 m. The structure is modelled as a clamped–clamped beam made of isotropic material ($E = 75$ GPa and $\nu = 0.33$).

8.6.1.1 Detection of Local Effects Due to a Point Load

A downward point load (P_z) is applied at $[0, L/2, d/2]$. P_z is parallel to the z-axis and is equal to -5 MN. A 10 B4 mesh is adopted since it leads to convergent results in terms of vertical displacement of the loaded point. An MSC Nastran shell model is used for comparison purposes. Figure 8.12 shows the deformed cross-section at $y = L/2$. Results from different 1D models are shown as well as those from the shell model. The following comments hold:

- TBT detects the bending behaviour of the structure (EBBT gives a similar result, but it is not reported on the plot for the sake of clarity). Since local effects are predominant, the TBT result is extremely inaccurate.

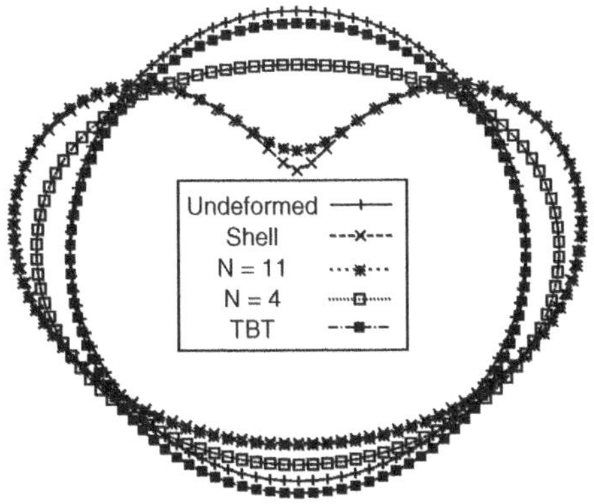

Figure 8.12 Deformed cross-section of the hollow cylinder

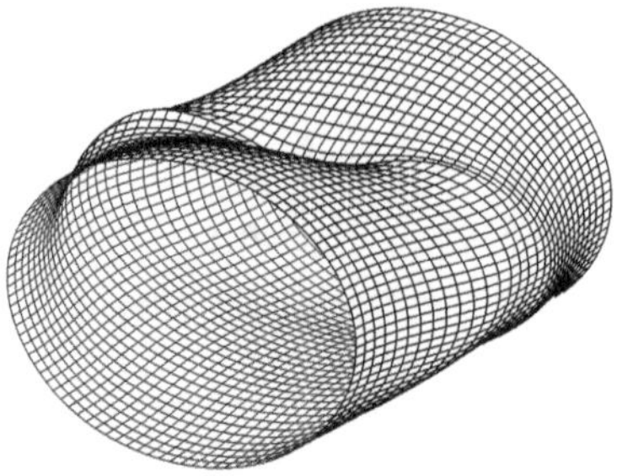

Figure 8.13 The 3D deformation field of the thin-walled cylinder from the $N = 11$ beam model

- A fourth-order model ($N = 4$) is able to model roughly the in-plane distortion of the cross-section.
- An 11th-order ($N = 11$) beam model ensures a good overall accuracy of the result.

Figure 8.13 shows the deformed $N = 11$ configuration of the whole cylinder. Table 8.5 presents the u_z values of the loading point from different 1D models; an indication of the total number of DOFs for each model is also given. It can be stated that:

- Theories up to second order ($N \leq 2$) detect only the bending behaviour.
- Significant improvements in the solution can be observed by increasing the order of the theory. The adoption of higher-order models is effective in detecting the shell-like solution.
- The computational costs of the 1D models are far lower than those required for the shell model.

Table 8.5 u_z at the loading point of the thin-walled cylinder (Carrera *et al.* 2010b)

Theory	DOFs	u_z (m)
Detection of the bending behaviour only		
EBBT	155	−0.046
TBT	155	−0.053
$N = 1$	279	−0.053
$N = 2$	558	−0.052
Distortion of the cross-section		
$N = 3$	930	−0.114
$N = 4$	1395	−0.229
$N = 5$	1953	−0.335
$N = 6$	2604	−0.386
$N = 7$	3348	−0.486
$N = 8$	4185	−0.535
$N = 9$	5115	−0.564
$N = 10$	6138	−0.584
$N = 11$	7254	−0.597
Shell	49 500	−0.670

This table presents the vertical displacement of the loading point of the thin-walled cylinder; results from different higher-order 1D theories and shell models are reported.

Table 8.6 First and second bending natural frequency of the thin-walled cylinder (Carrera *et al.* 2010b)

Theory	DOFs	f_1 (Hz)	f_2 (Hz)
EBBT	155	32.598	88.072
TBT	155	30.304	76.447
$N = 1$	279	30.304	76.447
$N = 2$	558	30.730	77.338
$N = 3$	930	28.754	69.448
$N = 4$	1395	28.747	69.402
$N = 5$	1953	28.745	69.397
$N = 6$	2604	28.745	69.397
Shell	49 500	28.489	68.940
Solid	174 000	28.369	68.687

This table presents the frequency values related to the first two bending modes of the thin-walled cylinder.

8.6.1.2 Detection of Shell-Like Natural Modes

A free vibration analysis was conducted to investigate the role of higher-order theories in detecting the natural modes and frequencies of a thin-walled structure. A 10 B4 mesh is adopted since it leads to convergent results in terms of natural frequencies. MSC Nastran shell and solid models are used for comparison purposes. Table 8.6 lists the first and second frequencies related to the first and second bending modes, respectively. Figure 8.14 shows the position of the modes within the eigenvector matrix. For instance, in the case of a third-order model ($N = 3$), the first bending mode corresponds to the first two natural modes, whereas the second bending mode corresponds to the fifth and sixth natural modes. Each bending mode

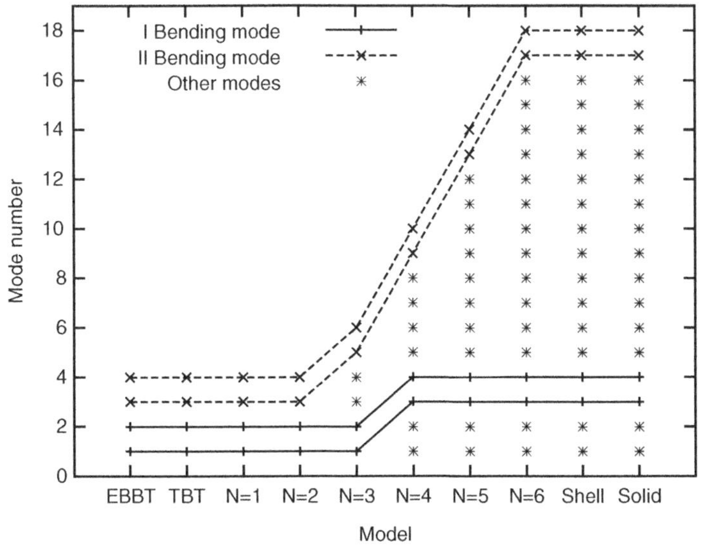

Figure 8.14 Mode-type distribution of the thin-walled cylinder for different structural models

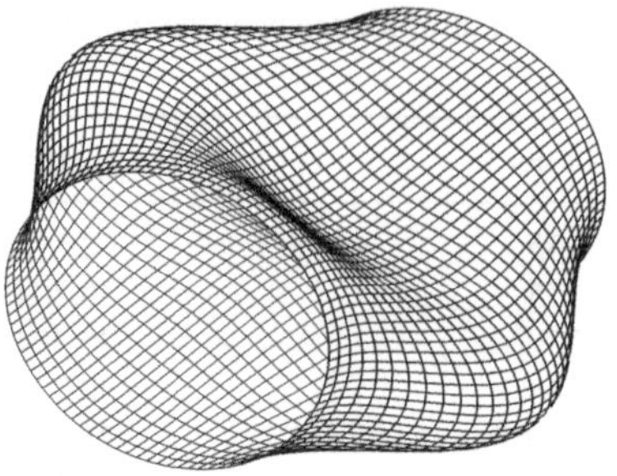

Figure 8.15 A two-lobe mode of the thin-walled cylinder obtained with $N = 3$

appears twice in the eigenvector since the structure is symmetric. The following statements hold:

- At least a third-order 1D model is needed to obtain accurate flexural frequencies. However, the higher the mode number, the more inaccurate the classical model is.
- Through the adoption of refined 1D theories, modal shapes that are different from the bending ones can be detected. In particular, a sixth-order ($N = 6$) 1D model is able to detect correctly the first two bending frequencies and also to detect all the natural modes which lie in between the flexural ones.
- The computational costs of the refined 1D models are far lower than those of shell and solid elements.

Typical modal shapes of a thin-walled cylinder have lobes in the circumferential direction of the cylinder. Figures 8.15 and 8.16 show a two- and three-lobe mode, respectively. The modal shapes above the cross-section are reported in Figures 8.17 and 8.18. The frequency values of the first two- and three-lobe frequencies are given in Tables 8.7 and 8.8 and compared with those from shell and solid models. The following considerations hold:

- The two-lobe mode requires $N = 3$ to be detected, whereas the three-lobe mode needs $N = 4$.
- The correct frequency is obtained with $N = 7$. In the case of the two-lobe model, $N = 8$ must be used to compute the exact frequency.

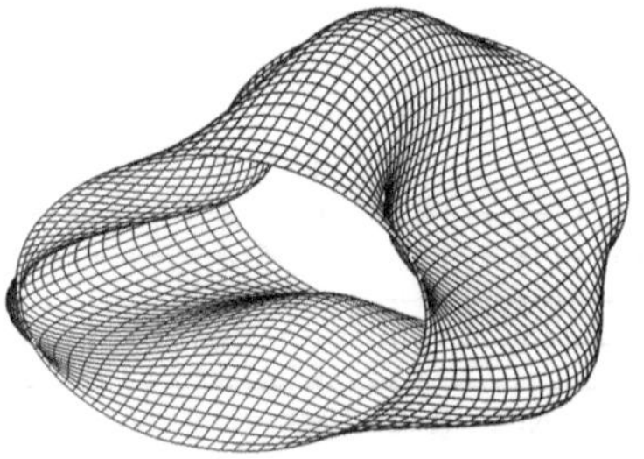

Figure 8.16 A three-lobe mode of the thin-walled cylinder obtained with $N = 4$

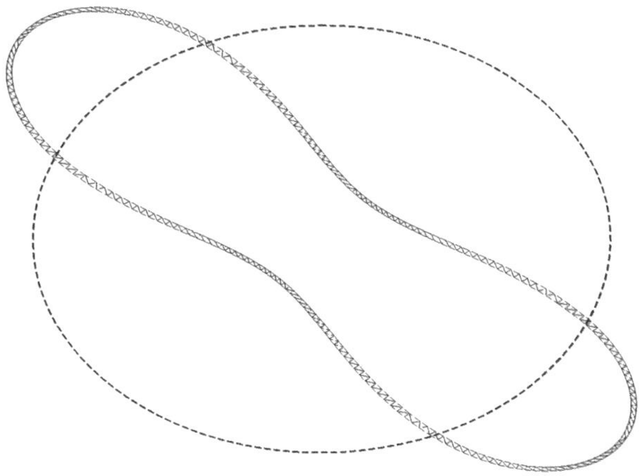

Figure 8.17 A 2D view of the two-lobe mode of the thin-walled cylinder obtained with $N = 3$

8.6.2 *Dynamic Response of Compact and Thin-Walled Structures*

This section presents the dynamic response analysis of various structural models as in Carrera and Varello (2012). The Newmark method was employed for the direct integration of the equations of motion whose matrices were obtained according to the 1D unified formulation.

8.6.2.1 Compact Square Cross-Section

A simply supported square cross-section beam is considered here. The sides of the cross-section are equal to 0.1 m, whereas the span-to-height ratio L/h is equal to 100. A 1D mesh of 10 B4 FEs along the y-axis was employed for the discretization of the structure.

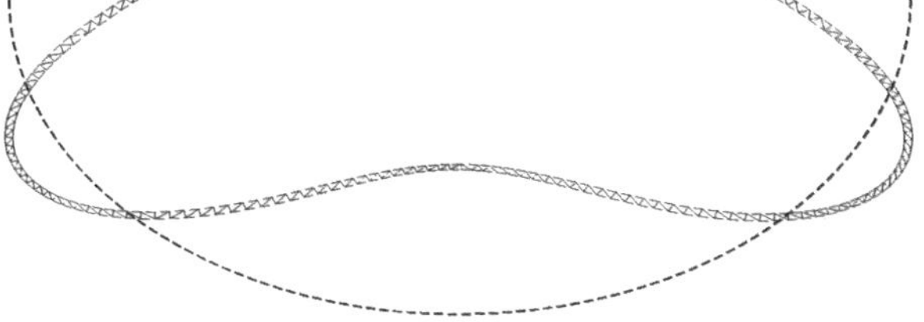

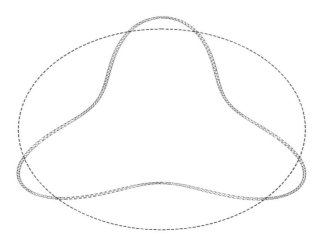

Figure 8.18 A 2D view of three-lobe mode of the thin-walled cylinder obtained with $N = 4$

Table 8.7 First two-lobe frequency of the thin-walled cylinder (Carrera *et al.* 2010b)

Theory	DOFs	f (Hz)
EBBT	155	—
TBT	155	—
$N = 1$	279	—
$N = 2$	558	—
$N = 3$	930	38.755
$N = 4$	1395	25.156
$N = 5$	1953	20.501
$N = 6$	2604	20.450
$N = 7$	3348	17.363
Shell	49 500	17.406
Solid	174 000	18.932

This table presents the frequency values related to the first two-lobe mode of the thin-walled cylinder from different models.

Aluminium was adopted (Young's modulus $E = 69$ GPa, Poisson's ratio $\nu = 0.33$ and density $\rho = 2700\,\text{kg/m}^3$). A harmonic force was applied at the mid-span section of the beam,

$$P_z(t) = P_{z0} \sin(\omega t), \qquad y_L = L/2 \tag{8.95}$$

where $P_{z0} = 1000\,\text{N}$ is the amplitude of the sinusoidal load with angular frequency $\omega = 7\,\text{rad/s}^{-1}$. The analytical undamped dynamic response of an Euler–Bernoulli beam made of isotropic material and loaded with this kind of force was found by Volterra and Zachmanoglou (1965). Let ω_1 be the fundamental angular frequency of the beam corresponding to a bending

Table 8.8 First three-lobe frequency of the thin-walled cylinder (Carrera *et al.* 2010b)

Theory	DOFs	f (Hz)
EBBT	155	—
TBT	155	—
$N = 1$	279	—
$N = 2$	558	—
$N = 3$	930	—
$N = 4$	1395	75.690
$N = 5$	1953	65.186
$N = 6$	2604	52.386
$N = 7$	3348	50.372
$N = 8$	4185	40.102
Shell	49 500	40.427
Solid	174 000	46.444

This table presents the frequency values related to the first three-lobe mode of the thin-walled cylinder from different models.

modal shape. When $\omega < \omega_1$, some reference values for the maximum transverse dynamic and static deflections occurring at the load application point are

$$u^{\text{anlt}}_{z\,\max,\,\text{DYN}} \cong \frac{2\,P_{z0}\,L^3}{\pi^4\,E\,I}\,\frac{1}{1-\omega/\omega_1}, \qquad u^{\text{anlt}}_{z\,\max,\,\text{ST}} = \frac{P_{z0}\,L^3}{48\,E\,I} \cong \frac{2\,P_{z0}\,L^3}{\pi^4\,E\,I} \tag{8.96}$$

$$\frac{u^{\text{analanlt}}_{z\,\max,\,\text{DYN}}}{u^{\text{analanlt}}_{z\,\max,\,\text{ST}}} \cong \frac{1}{1-\omega/\omega_1} \tag{8.97}$$

where I is the moment of inertia of the beam cross-section. The analytical solution based on EBBT was used for comparison purposes.

The numerical dynamic response of the system was investigated over the interval [0, 8] s. A convergence study on ΔT was carried out to evaluate the dependence of the results on the time step. The time history of the transverse displacement u_z at the mid-span section is shown in Figure 8.19. The choice of $\Delta t = 0.08$ s represents a coarse time discretization for this problem, whereas a good agreement with the analytical deflection is achieved for $\Delta t = 0.004$ s. It should be noted that the dynamic response is approximately the sum of two sinusoidal functions with angular frequencies equal to ω and ω_1. The maximum dynamic displacement computed through FEs based on EBBT differs in about 0.03% from the analytical value, as reported in Table 8.9. It occurs for $t = 3.816$ s at the mid-span beam section. The static solution of the system is also evaluated by disabling the inertial contribution of the mass matrix. As expected, it is a time-dependent sinusoid with the same frequency as that of the point force. Unlike the dynamic case, the amplitude is constant and equal to $u^{\text{anlt}}_{z\,\max,\,\text{ST}}$.

The time-response analysis was also conducted through refined beam models. Table 8.9 presents the maximum dynamic and static displacements for the third- and seventh-order models. In this case, the increase in N does not reveal any remarkable difference in comparison with EBBT. As it is known, the use of a compact square section for a slender beam subjected to a bending load restricts the local effects of the beam cross-section, which are eventually detectable by the higher-order terms of the beam displacement field. Also, the value of the fundamental bending frequency of the beam, ω_1, is substantially the same for all the theories involved.

8.6.2.2 Thin-Walled Cylinder

A thin-walled circular cross-section is considered in this section. The outer diameter d is equal to 1.05 m, the thickness t is equal to 0.05 m and the length L equal to 0.5 m. Aluminium was adopted.

The cylinder was first analysed through a free vibration analysis (under simply supported BCs). Results from the 3D elasticity equations were adopted for comparison purposes (from Armenàkas *et al.* (1969) and Soldatos and Hadjigeorgiou (1990)). Results were compared through a frequency parameter defined as

$$\bar{\omega} = \frac{\Omega\,\pi\,L}{t\,\sqrt{2}} = \omega\,L\,\sqrt{\frac{\rho\,(1+\nu)}{E}} \tag{8.98}$$

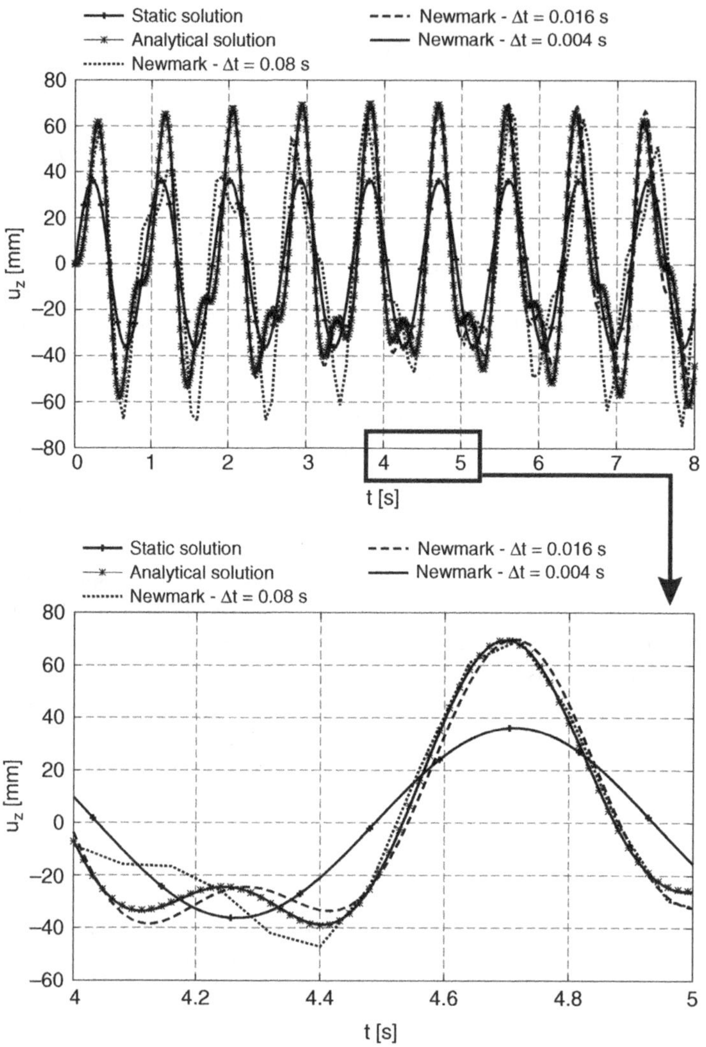

Figure 8.19 Influence of the time step (Δt) on the transverse displacement at the mid-span section of a square cross-section beam modelled through EBBT

where Ω is the frequency parameter used in Armenàkas *et al.* (1969) and ω is the vibrational natural angular frequency. Different values of the circumferential half-wave number n were investigated, whereas the axial half-wave number m was set to 1. In Table 8.10, the two first frequency parameters $\bar{\omega}$ are given. Classical beam theories are completely ineffective in studying this kind of structure, therefore it is necessary to enhance the displacement field with higher-order terms to describe correctly the dynamic behaviour of the cylinder. A clamped–clamped cylinder (see Figure 8.20) was also investigated (the outer diameter d is equal to 0.1 m, the thickness is equal to 0.001 m and the span-to-diameter ratio (L/d) is equal to 10). Four

Table 8.9 Dynamic response analysis of a square cross-section beam with $\Delta t = 0.004$ s (Carrera and Varello 2012)

Model	$u_{z\,\max,\,DYN}$	$u_{z\,\max,\,ST}$	ω_1
Analytical	70.0116	36.2319	14.4030
EBBT	69.9886	36.2318	14.4024
$N = 3$	70.0232	36.2427	14.4006
$N = 7$	70.0233	36.2428	14.4006

This table shows the maximum displacement in the dynamic and static case ($u_{z\,\max,\,DYN}$ and $u_{z\,\max,\,ST}$) and the natural oscillatory frequency (ω_1) given by different models.

Table 8.10 Comparison of frequency parameters $\bar{\omega}$ based on the present 1D CUF and 3D analysis for a simply supported circular cylindrical shell ($m = 1$) (Carrera and Varello 2012)

n		$N = 4$	$N = 7$	$N = 9$	Exact 3D	3D
2	I	1.0804	1.0620	1.0620	1.0623	1.0624
	II	2.3758	2.3745	2.3744	2.3744	2.3745
4	I	0.9937	0.8838	0.8819	0.8823	0.8826
	II	2.9118	2.7160	2.7159	2.7159	2.7159
6	I	1.7500	0.8388	0.8112	0.8093	0.8096
	II	4.1441	3.1562	3.1534	3.1533	3.1533

This table shows the free vibration analysis of a simply supported cylinder through different models; results are compared with those by Armenàkas *et al.* (1969) (Exact 3D) and Soldatos and Hadjigeorgiou (1990) (3D).

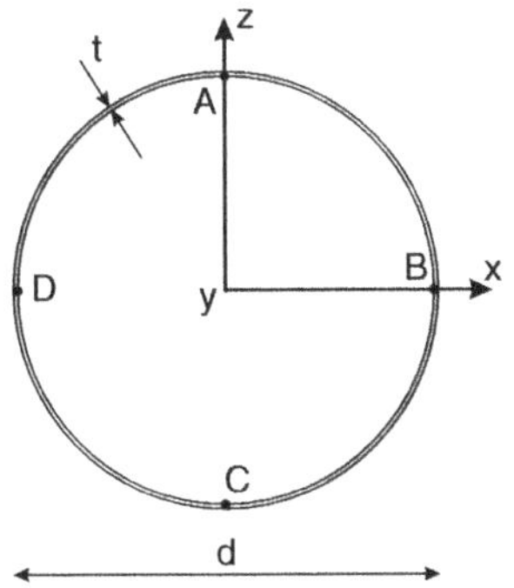

Figure 8.20 Thin-walled circular cross-section

Table 8.11 Displacements (mm) of the loading points A and D for different FE models at $t = 0$ s (Carrera and Varello 2012)

Model	u_{xD}	Error $u_{xD}(\%)$	$u_{zA}(\%)$	Error u_{zA}	DOFs
EBBT	−0.9906	−95.78	−1.7157	−82.64	93
TBT	−1.1453	−95.12	−1.9838	−79.93	155
$N = 1$	−2.0937	−91.08	−1.4362	−85.47	271
$N = 3$	−2.9313	−87.51	−3.5311	−64.27	930
$N = 4$	−5.9690	−74.56	−6.8900	−30.29	1395
$N = 7$	−15.7213	−32.99	−9.3591	−5.31	3348
$N = 10$	−19.7523	−15.81	−9.7314	−1.54	6138
$N = 14$	−21.1939	% − 9.67	−9.8418	−0.43	11 160
Nastran	−23.4628	—	−9.8840	—	250 000

This table shows the displacements of two points of the thin-walled cylinder through different models.

particular points were considered over the mid-span cross-section where four concentrated forces were applied as time-dependent sinusoids with amplitude $P_{z0} = 10\,000$ N and a phase shift,

$$\begin{aligned} P_{zA}(t) &= P_{z0}\sin(\omega t + \phi_A), & \phi_A &= 0 \\ P_{xB}(t) &= P_{z0}\sin(\omega t + \phi_B), & \phi_B &= 30 \\ P_{zC}(t) &= -P_{z0}\sin(\omega t + \phi_C), & \phi_C &= 60 \\ P_{xD}(t) &= -P_{z0}\sin(\omega t + \phi_D), & \phi_D &= 90 \end{aligned} \tag{8.99}$$

where the angular frequency is $\omega = 100\,\text{rad/s}^{-1}$. The dynamic response of the structure was evaluated over the time interval [0, 0.025] s by involving classical as well as refined 1D models.

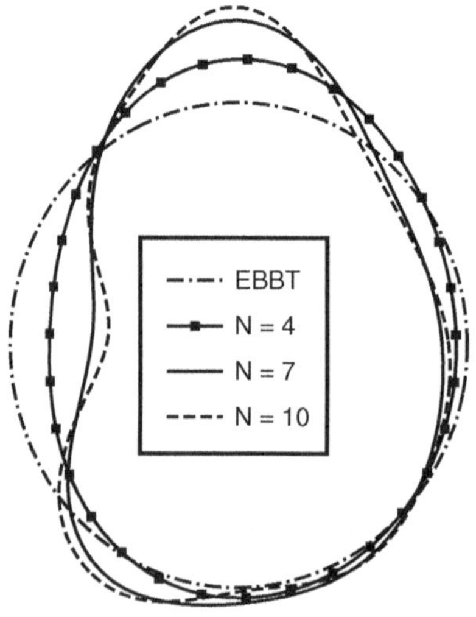

Figure 8.21 Deformation of the mid-span circular cross-section, $t = 0.025$ s

Table 8.11 shows the displacements of two points at $t = 0$ s. The configuration at the final time instant $t = 0.025$ s is shown in Figure 8.21 (mid-span cross-section). With $N = 4$ no local effect was detected near the loading points; it only detects a global deflection of the circular section. On the contrary, with $N = 7$ and $N = 10$ the proposed 1D model perfectly detects the local deformations typical of a shell-like behaviour.

References

Armenàkas AE, Gazis DC and Herrmann G 1969 *Free Vibrations of Circular Cylindrical Shells*. Pergamon Press.

Ballarini R 2003 The da Vinci-Euler–Bernoulli beam theory? *Mechanical Engineering Magazine*. Available Online, http://www.memagazine.org/contents/current/webonly/webex418.html.

Carrera E and Brischetto S 2008a Analysis of thickness locking in classical, refined and mixed multilayered plate theories. *Composite Structures* **82**(4), 549–562.

Carrera E and Brischetto S 2008b Analysis of thickness locking in classical, refined and mixed theories for layered shells. *Composite Structures* **85**(1), 83–90.

Carrera E, Giunta G, Nali P and Petrolo M 2010a Refined beam elements with arbitrary cross-section geometries. *Computers and Structures* **88**(5–6), 283–293. DOI: 10.1016/j.compstruc.2009.11.002.

Carrera E, Giunta G and Petrolo M 2010b A modern and compact way to formulate classical and advanced beam theories. In: *Developments and Applications in Computational Structures Technology*, Ch. 4. Saxe-Coburg Publications.

Carrera E, Giunta G and Petrolo M 2011 *Beam Structures: Classical and Advanced Theories*. John Wiley & Sons, Ltd.

Carrera E, Miglioretti F and Petrolo M 2012a Computations and evaluations of higher-order theories for free vibration analysis of beams. *Journal of Sound and Vibrations* **331**(2), 4269–4284. DOI: 10.1016/j.jsv.2012.04.017.

Carrera E and Petrolo M 2010 Guidelines and recommendations to construct theories for metallic and composite plates. *AIAA Journal* **48**(12), 2852–2866.

Carrera E and Petrolo M 2011 On the effectiveness of higher-order terms in refined beam theories. *Journal of Applied Mechanics* **78**. DOI: 10.1115/1.4002207.

Carrera E, Petrolo M and Varello A 2012b Advanced beam formulations for free vibration analysis of conventional and joined wings. *Journal of Aerospace Engineering* **25**(2), 282–293. DOI: 10.1061/(ASCE)AS.1943-5525.0000130.

Carrera E, Petrolo M and Zappino E 2012c Performance of CUF approach to analyze the structural behavior of slender bodies. *Journal of Structural Engineering* **138**(2), 285–297. DOI: 10.1061/(ASCE)ST.1943-541X.0000402.

Carrera E and Varello A 2012 Dynamic response of thin-walled structures by variable kinematic one-dimensional models. *Journal of Sound and Vibration* **331**(24), 5268–5282. DOI: 10.1016/j.jsv.2012.07.006.

Cinefra M, Carrera E and Nali P 2010 MITC technique extended to variable kinematic multilayered plate elements *Composite Structures* **92**, 1888–1895.

Cowper GR 1966 The shear coefficient in Timoshenko's beam theory. *Journal of Applied Mechanics*, **33**(2), 335–340.

Euler L 1744 De curvis elasticis. In: *Methodus Inveniendi Lineas Curvas Maximi Minimive Proprietate Gaudentes, Sive Solutio Problematis Isoperimetrici Lattissimo Sensu Accept*. Bousquet.

Gruttmann F, Sauer R and Wagner W 1999 Shear stresses in prismatic beams with arbitrary cross-sections. *International Journal for Numerical Methods in Engineering* **45**, 865–889.

Gruttmann F and Wagner W 2001 Shear correction factors in Timoshenko's beam theory for arbitrary shaped cross-sections. *Computational Mechanics* **27**, 199–207.

Kapania K and Raciti S 1989 Recent advances in analysis of laminated beams and plates, Part I: Shear effects and buckling. *AIAA Journal* **27**(7), 923–935.

Pai PF and Schulz MJ 1999 Shear correction factors and an energy consistent beam theory. *International Journal of Solids and Structures* **36**, 1523–1540.

Reddy JN 1997 On locking-free shear deformable beam finite elements. *Computer Methods in Applied Mechanics and Engineering* **149**, 113–132.

Reti L 1974 *The Unknown Leonardo*. McGraw-Hill.

Soldatos KP and Hadjigeorgiou VP 1990 Three-dimensional solution of the free vibration problem of homogeneous isotropic cylindrical shells and panels. *Journal of Sound and Vibration* **137**, 369–384, DOI: 10.1115/1.4002207.

Timoshenko SP 1921 On the corrections for shear of the differential equation for transverse vibrations of prismatic bars. *Philosophical Magazine* **41**, 744–746.

Timoshenko SP 1922 On the transverse vibrations of bars of uniform cross-section. *Philosophical Magazine* **43**, 125–131.

Volterra E and Zachmanoglou EC 1965 *Dynamics of Vibrations*. C.E. Merrill Books.

Washizu K 1968 *Variational Methods in Elasticity and Plasticity*. Pergamon Press.

9

One-Dimensional Models with a Physical Volume/Surface-Based Geometry and Pure Displacement Variables, the Lagrange Expansion Class

Different expansion functions can be implemented through the unified formulation. In a displacement-based approach, the displacement field of the structure cross-section can be defined by different classes of functions, such as polynomials, harmonics and exponentials. In the previous chapter, TE models based on Taylor-like polynomial expansions were described. In this chapter a second class of 1D CUF models based on Lagrange polynomial expansions is described. Table 9.1 shows the FN for 1D models that is considered in this chapter. The Lagrange expansion 1D models – hereafter referred to as LE models – have the following main characteristics:

1. LE model variables and BCs can be located above the physical surfaces of the structure. This feature is particular relevant in a CAD–FEM coupling scenario.
2. The unknown variables of the problem are pure displacement components. No rotations or higher-order variables are exploited to describe the displacement field of an LE model.
3. Locally refined models can be easily built since Lagrange polynomial sets can be arbitrarily spread above the cross-section.

Each of these capabilities will be outlined in the following sections of this chapter and numerical examples will be provided in order to highlight the enhanced capabilities of 1D CUF LE models in terms of 3D-like accuracies and very low computational costs.

Finite Element Analysis of Structures Through Unified Formulation, First Edition.
Erasmo Carrera, Maria Cinefra, Marco Petrolo and Enrico Zappino.

Table 9.1 A schematic description of the CUF and the related FN of the stiffness matrix for 1D models

Equilibrium equations in Strong Form $\longrightarrow$ $\delta L_i = \int_V \delta\boldsymbol{u}\boldsymbol{k}\boldsymbol{u}dV + \int_S \ldots dS$

$$\underbrace{\begin{bmatrix} k_{xx} & k_{xy} & k_{xz} \\ k_{yx} & k_{yy} & k_{yz} \\ k_{zx} & k_{zy} & k_{zz} \end{bmatrix}}_{\boldsymbol{k}} \underbrace{\begin{Bmatrix} u_x \\ u_y \\ u_z \end{Bmatrix}}_{\boldsymbol{u}} = \underbrace{\begin{Bmatrix} p_x \\ p_y \\ p_z \end{Bmatrix}}_{\boldsymbol{p}}$$

$$k_{xx} = -(\lambda + 2G)\,\partial_{xx} - G\,\partial_{zz} - G\,\partial_{yy};$$
$$k_{xy} = -\lambda\,\partial_{xy} - G\,\partial_{yx};$$
$$k_{xz} = \ldots$$

$\lambda=(E\nu)/[(1+\nu)(1-2\nu)]$; $G=E/[2(1+\nu)]$

$\boldsymbol{u} = \boldsymbol{u}(x,y,z)$
$\delta\boldsymbol{u} = \delta\boldsymbol{u}(x,y,z)$

The diagonal (e.g. k_{xx}) and the non-diagonal (e.g. k_{xy}) terms can be obtained through proper index permutations.

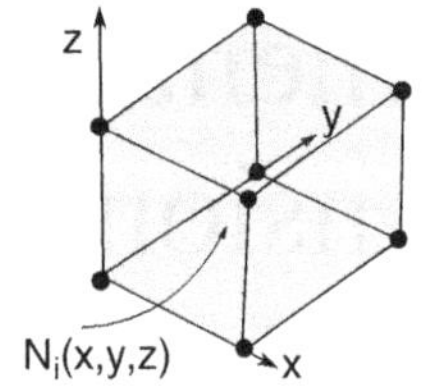

$\boldsymbol{u} = N_i(x,y,z)\boldsymbol{u}_i$
$\delta\boldsymbol{u} = N_j(x,y,z)\delta\boldsymbol{u}_j$

3D FEM Formulation $\longrightarrow$ $\delta L_i = \delta\boldsymbol{u}_j\boldsymbol{k}^{ij}\boldsymbol{u}_i$

$$k^{ij}_{xx} = (\lambda + 2G)\int_V N_{j,x}N_{i,x}dV + G\int_V N_{j,z}N_{i,z}dV + G\int_V N_{j,y}N_{i,y}dV;$$
$$k^{ij}_{xy} = \lambda\int_V N_{j,y}N_{i,x}dV + G\int_V N_{j,x}N_{i,y}dV$$

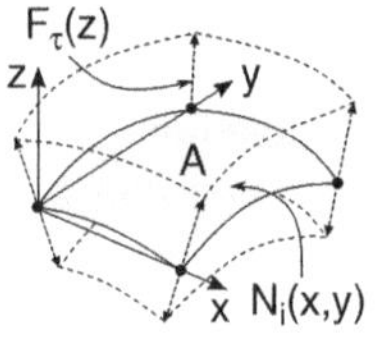

$\boldsymbol{u} = N_i(x,y)F_\tau(z)\boldsymbol{u}_{\tau i}$
$\delta\boldsymbol{u} = N_j(x,y)F_s(z)\delta\boldsymbol{u}_{sj}$

2D FEM Formulation $\longrightarrow$ $\delta L_i = \delta\boldsymbol{u}_{sj}\boldsymbol{k}^{\tau sij}\boldsymbol{u}_{\tau i}$

$$k^{\tau sij}_{xx} = (\lambda + 2G)\int_\Omega N_{i,x}N_{j,x}d\Omega\int_h F_\tau F_s dz$$
$$+ G\int_\Omega N_iN_j d\Omega\int_h F_{\tau,z}F_{s,z}dz + G\int_V N_{i,y}N_{j,y}d\Omega\int_h F_\tau F_s dz;$$
$$k^{\tau sij}_{xy} = \lambda\int_\Omega N_{i,y}N_{j,y}d\Omega\int_h F_\tau F_s dz + G\int_\Omega N_{i,x}N_{j,y}d\Omega\int_h F_\tau F_s dz$$

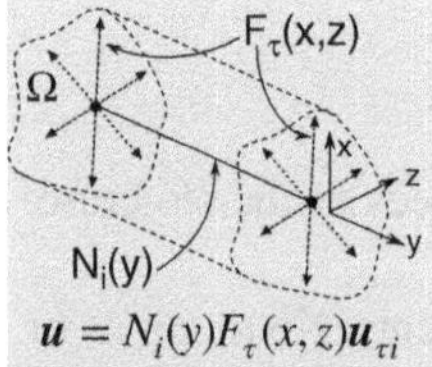

$\boldsymbol{u} = N_i(y)F_\tau(x,z)\boldsymbol{u}_{\tau i}$
$\delta\boldsymbol{u} = N_j(y)F_s(x,z)\delta\boldsymbol{u}_{sj}$

1D FEM Formulation $\longrightarrow$ $\delta L_i = \delta\boldsymbol{u}_{sj}\boldsymbol{k}^{\tau sij}\boldsymbol{u}_{\tau i}$

$$k^{\tau sij}_{xx} = (\lambda + 2G)\int_l N_iN_j dy\int_A F_{\tau,x}F_{s,x}dA$$
$$+ G\int_l N_iN_j dy\int_A F_{\tau,z}F_{s,z}dA + G\int_l N_{i,y}N_{j,y}dy\int_A F_\tau F_s dA;$$
$$k^{\tau sij}_{xy} = \lambda\int_l N_{i,y}N_j dy\int_A F_\tau F_{s,x}dA + G\int_l N_iN_{j,y}dy\int_A F_{\tau,x}F_s dA$$

CUF leads to the automatic implementation of any theory of structures through 4 loops (i.e. 4 indexes):

- τ and s deal with the functions that approximate the displacement field and its virtual variation along the plate/shell thickness ($F_\tau(z), F_s(z)$) or over the beam cross-section ($F_\tau(x,z), F_s(x,z)$);
- i and j deal with the shape functions of the FE model, (3D:$N_i(x,y,z), N_j(x,y,z)$; 2D:$N_i(x,y), N_j(x,y)$; 1D:$N_i(y), N_j(y)$).

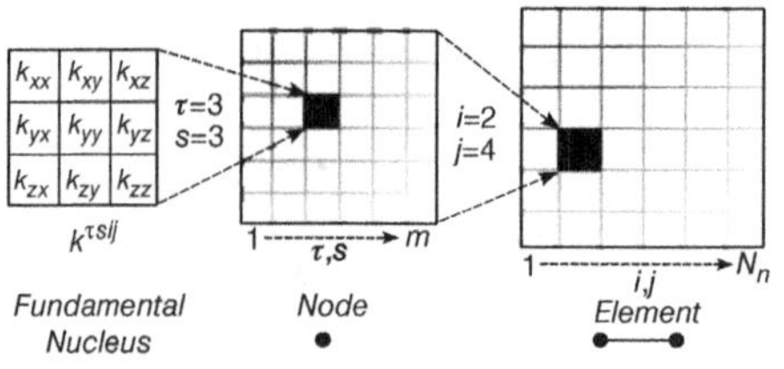

This table shows the essential features of the CUF for 1D models.

9.1 Physical Volume/Surface Approach

Finite element analyses (FEAs) are typically conducted on structural models whose geometries are derived from CAD tools. CAD and FEA representations of the geometry may differ significantly and the proper FE modelling of a CAD-based structure is a critical and lengthy task which can also affect the accuracy of the results.

There are two main aspects that should be carefully considered in a CAD–FEA scenario:

1. The FE discretization is, by definition, a process that leads to a modified geometry of a structure. Mesh refinements or higher-order shape functions are typical remedies to this problem (see also Section 1.2).
2. Many structural elements (e.g. beams, plates and shells) require the definition of reference surfaces or axes where these elements and the problem unknowns lie. The definition of reference surfaces/axes is particularly critical when 3D CAD geometries are provided. This difficulty increases if FEA is, for instance, used to conduct the topological optimization of the geometry of a structure.

The present 1D FE formulation offers significant advantages related to the second point listed above, as LE elements can deal directly with the 3D geometry given by a CAD model.

Figure 9.1 shows the typical steps required to implement a 1D beam element, where both classical (EBBT, TBT) and refined (TE) models are considered. Starting from the 3D

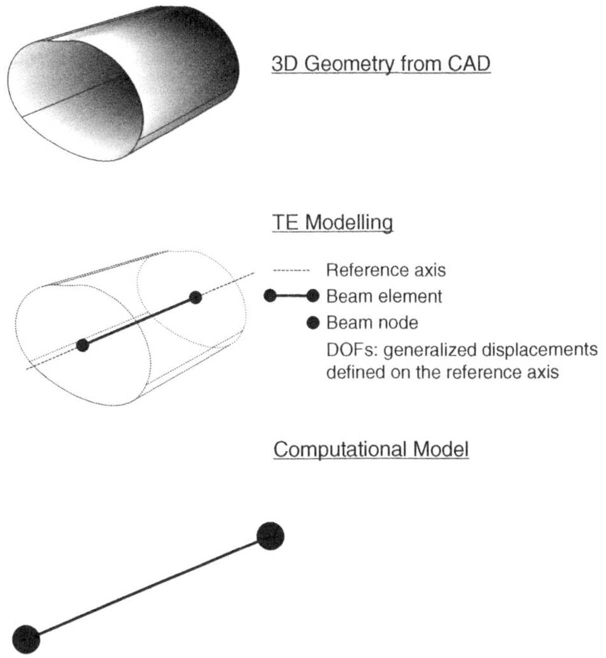

Figure 9.1 Geometrical considerations of the TE modelling approach, where a line is used to model the entire 3D volume

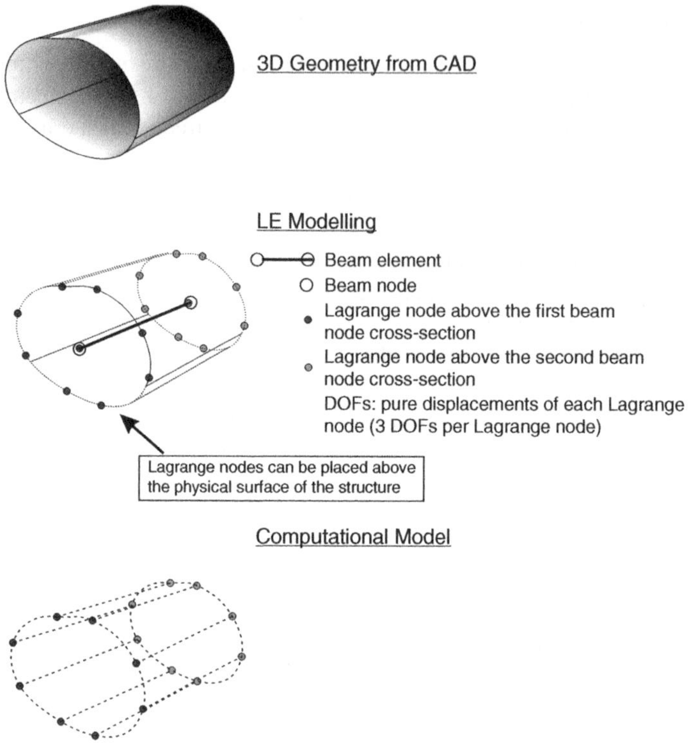

Figure 9.2 Geometrical considerations of the LE modelling approach, where the lines and nodes lie on the external surface of the physical body

geometry of the structure, an axis is defined and used to create the FE discretization. The problem unknowns are defined along this axis. The geometrical characteristics of the cross-section are retained by means of the surface integration of the expansion functions within the FE matrices. This process can be particularly critical when multiple CAD–FEA iterative processes are required (e.g. in an optimization problem) since it can be difficult to redefine a 3D geometry starting from a 1D FE model. Figure 9.2 shows the LE modelling approach. In this case, the cross-section nodes can be directly located along the surface contour of the 3D structure. This implies that the FE unknowns lie above the physical surface of the structure. The definition of reference surfaces or axes is not required and a 3D CAD geometry can be used directly for the FEA. In other words, a 1D FE can be used for a 3D geometrical description. This important feature makes LE models extremely attractive for:

1. easily creating FE models derived from 3D CAD;
2. improving the CAD–FEA coupling capabilities in an iterative design scenario.

Figure 9.3 presents a summary of the main differences between the different modelling approaches described above.

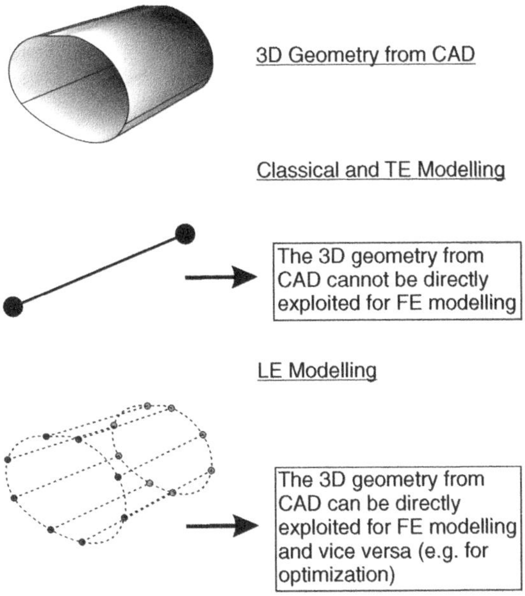

Figure 9.3 Different geometrical modelling approaches, TE vs LE; it is clear that TE formulations introduce fictitious entities

9.2 Lagrange Polynomials and Isoparametric Formulation

The LE 1D models represent the second class of 1D models developed in the framework of the CUF. In an LE model, the F_τ expansion functions coincide with the Lagrange polynomials. The use of LE as F_τ does not imply a reformulation of the problem equations and matrices, which is typical in the CUF environment. LE, on the other hand, uses a different approach with respect to TE since the isoparametric formulation is exploited. This section will first describe LE polynomials and then LE models will be given together with a number of numerical examples. Descriptions provided in the following will focus on the specific issues related to the CUF LE models. Details on Lagrange polynomials and the isoparametric formulation not directly related to CUF LE models are covered in Bathe (1996, Oñate (2009) and Zienkiewicz *et al.* (2005). In these books, this formulation is usually provided for 2D FEs.

9.2.1 Lagrange Polynomials

Lagrange polynomials are usually given in terms of normalized – or natural – coordinates. This choice is not compulsory since LE polynomials can also be implemented in terms of actual coordinates. However, the normalized formulation was preferred since it offers many advantages. Quadrilateral and triangular sets are presented in this book and each set is named according to its number of nodes. Only 2D polynomials are presented hereafter since only these are employed on the cross-section of the model.

The simplest quadrilateral Lagrange polynomial is the four-point (L4) set as shown in Figure 9.4 and the polynomials are given by Equation (9.1) where α and β are the

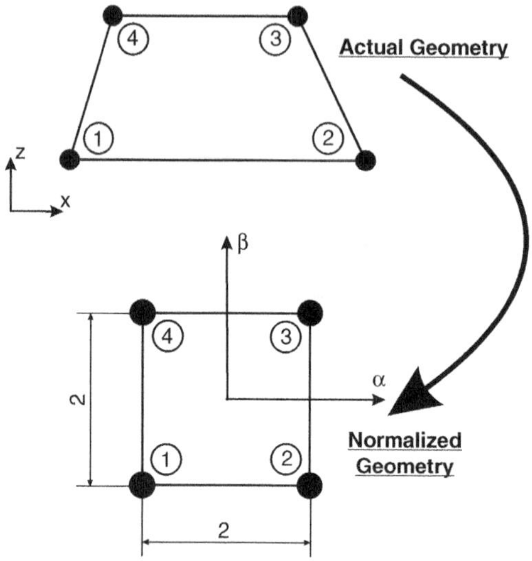

Figure 9.4 Four-node Lagrange element (L4) in actual and normalized geometry

normalized coordinates and α_τ and β_τ are the coordinates of the four nodes given in Table 9.2:

$$F_\tau = \frac{1}{4}(1 + \alpha\alpha_\tau)(1 + \beta\beta_\tau), \qquad \tau = 1, 2, 3, 4 \tag{9.1}$$

L4 can be seen as a linear expansion (terms 1, α and β) plus a bilinear term ($\alpha\beta$). The second L-set is given by L9 in Figure 9.5. L9 polynomials and point coordinates are given by Equation (9.2) and Table 9.3:

$$\begin{aligned} F_\tau &= \frac{1}{4}(\alpha^2 + \alpha\alpha_\tau)(\beta^2 + \beta\beta_\tau), \qquad \tau = 1, 3, 5, 7 \\ F_\tau &= \frac{1}{2}\beta_\tau^2(\beta^2 + \beta\beta_\tau)(1 - \alpha^2) + \frac{1}{2}\alpha_\tau^2(\alpha^2 + \alpha\alpha_\tau)(1 - \beta^2), \\ &\quad \tau = 2, 4, 6, 8 \\ F_\tau &= (1 - \alpha^2)(1 - \beta^2), \qquad \tau = 9 \end{aligned} \tag{9.2}$$

Table 9.2 L4 point normalized coordinates, see Figure 9.4

Point	α_τ	β_τ
1	−1	−1
2	1	−1
3	1	1
4	−1	1

This table presents the normalized coordinates of the four points of an L4 element.

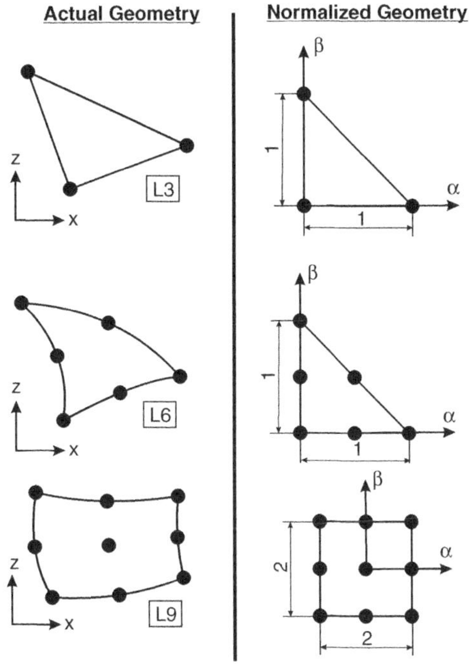

Figure 9.5 Three-, six- and nine-node Lagrange elements (L3, L6 and L9) in actual and normalized geometry

L9 can be seen as a parabolic expansion plus two cubic terms ($\alpha\beta^2$ and $\alpha^2\beta$) and a quartic term ($\alpha^2\beta^2$).

The simplest triangular set is the L3 set shown in Figure 9.5. L3 polynomials are given by Equation (9.3) and the node coordinates are shown in Table 9.4:

$$F_1 = 1 - \alpha - \beta \qquad F_2 = \alpha \qquad F_3 = \beta \tag{9.3}$$

Table 9.3 L9 point normalized coordinates, see Figure 9.5

Point	α_τ	β_τ
1	−1	−1
2	0	−1
3	1	−1
4	1	0
5	1	1
6	0	1
7	−1	1
8	−1	0
9	0	0

This table presents the normalized coordinates of the nine points of an L9 element.

Table 9.4 L3 point normalized coordinates, see Figure 9.5

Point	α_τ	β_τ
1	0	0
2	1	0
3	0	1

This table presents the normalized coordinates of the three points of an L3 element.

L3 has only linear terms. The L6 set is shown in Figure 9.5, and the polynomials and points are shown in Equation (9.4) and Table 9.5 respectively:

$$
\begin{aligned}
F_1 &= 1 - 3(\alpha + \beta) + 2(\alpha^2 + 2\,\alpha\beta + \beta^2), & F_2 &= 4\alpha(1 - \alpha - \beta) \\
F_3 &= \alpha(2\alpha - 1), & F_4 &= 4\alpha\beta \\
F_5 &= \alpha(2\beta - 1), & F_6 &= 4\beta(1 - \alpha - \beta)
\end{aligned}
\tag{9.4}
$$

L6 has only linear and parabolic terms.

9.2.2 *Isoparametric Formulation*

Isoparametric formulations can be 1D, 2D or 3D. Such formulations are exploited in CUF LE FEs to deal with:

1. 1D shape functions along the longitudinal axis of the structure;
2. 2D expansion functions to describe the displacement field of the structure on its cross-section.

The 2D isoparametric formulation is described hereafter, as it is adopted to implement LE models. The same 2D formulation is usually employed for 2D FEs (plate models).

Table 9.5 L6 point normalized coordinates, see Figure 9.5

Point	α_τ	β_τ
1	0	0
2	0.5	0
3	1	0
4	0.5	0.5
5	0	1
6	0	0.5

This table presents the normalized coordinates of the six points of an L6 element.

As shown in the chapter on TE, computation of the FN in a 1D formulation requires an evaluation of the surface integrals in Cartesian coordinates (x and z),

$$\int_A F_{\tau,x} F_{s,z} dx\, dz \tag{9.5}$$

where the 2D integration domain (A) can be of arbitrary shape. These integrals are formally independent of the adopted class of F_τ functions; that is, if Lagrange polynomials are adopted – instead of TE – the surface integrals will not formally change. If normalized coordinates (α and β) are accounted for, the integrals can be computed above a fixed 2D domain, regardless of the actual geometry. For instance, if quadrilateral domains are considered, the following will hold:

$$\int_A F_{\tau,x}(x,z) F_{s,z}(x,z) dx\, dz = \int_{-1}^{+1} \int_{-1}^{+1} F_{\tau,x}(\alpha,\beta) F_{s,z}(\alpha,\beta) |J(\alpha,\beta)| d\alpha\, d\beta \tag{9.6}$$

where $|J|$ is the Jacobian determinant of the transformation. In some cases, the new integral in the normalized coordinates can be computed analytically, but numerical techniques often have to be employed.

Partial derivatives have to be computed, with respect to the normalized coordinates, according to the chain rule:

$$\begin{aligned} F_{\tau,x} &= F_{\tau,\alpha}\, \alpha_{,x} + F_{\tau,\beta}\, \beta_{,x} \\ F_{\tau,z} &= F_{\tau,\alpha}\, \alpha_{,z} + F_{\tau,\beta}\, \beta_{,z} \end{aligned} \tag{9.7}$$

The evaluation of Equation (9.7) requires the following explicit relationships:

$$\alpha = \alpha(x,z), \quad \beta = \beta(x,z) \tag{9.8}$$

These relationships are often difficult to establish and, for this reason, it is preferable to use the chain rule as follows:

$$\begin{aligned} F_{\tau,\alpha} &= F_{\tau,x}\, x_{,\alpha} + F_{\tau,z}\, z_{,\alpha} \\ F_{\tau,\beta} &= F_{\tau,x}\, x_{,\beta} + F_{\tau,z}\, z_{,\beta} \end{aligned} \tag{9.9}$$

Equation 9.9 also holds for F_τ and can be rewritten in matrix form,

$$\begin{Bmatrix} F_{\tau,\alpha} \\ F_{\tau,\beta} \end{Bmatrix} = \underbrace{\begin{bmatrix} x_{,\alpha} & z_{,\alpha} \\ x_{,\beta} & z_{,\beta} \end{bmatrix}}_{\mathbf{J}} \begin{Bmatrix} F_{\tau,x} \\ F_{\tau,z} \end{Bmatrix} \tag{9.10}$$

The inverse relationship is given by

$$\left\{\begin{array}{c} F_{\tau,x} \\ \\ F_{\tau,z} \end{array}\right\} = \frac{1}{|J|}\begin{bmatrix} z_{,\beta} & -z_{,\alpha} \\ \\ -x_{,\alpha} & x_{,\beta} \end{bmatrix}\left\{\begin{array}{c} F_{\tau,\alpha} \\ \\ F_{\tau,\beta} \end{array}\right\} \tag{9.11}$$

The four terms of the Jacobian matrix can be computed if a known relation between the Cartesian and normalized coordinates exists,

$$x = x(\alpha, \beta), \quad z = z(\alpha, \beta) \tag{9.12}$$

The isoparametric formulation is generally used to obtain this relation and, therefore, to compute the Jacobian and to associate the actual geometries to the normalized geometry of the LE. It is important to note that Equation (9.11) requires that the inverse of **J** exists. In order to fulfil this requirement, attention should be paid to the geometry of the the Lagrange element. Major distortions, or folded back elements, can cause a singularity problem.

The term *isoparametric* means that the same functions are adopted to interpolate the displacement field and the geometry of a structural element. In the CUF LE, Lagrange polynomials are employed,

$$\begin{gathered} \mathbf{u} = F_\tau \mathbf{u}_\tau \\ x = F_\tau\, x_\tau, \qquad\qquad z = F_\tau\, z_\tau \end{gathered} \tag{9.13}$$

where x_τ and z_τ are the actual coordinates of the Lagrange nodes. If an L4 element is considered, the geometry will be interpolated as

$$\begin{aligned} x &= F_1\, x_1 + F_2\, x_2 + F_3\, x_3 + F_4\, x_4 \\ z &= F_1\, z_1 + F_2\, z_2 + F_3\, z_3 + F_4\, z_4 \end{aligned} \tag{9.14}$$

Typical derivatives for the Jacobian matrix are given by

$$\begin{aligned} x_{,\alpha} &= F_{1,\alpha}\, x_1 + F_{2,\alpha}\, x_2 + F_{3,\alpha}\, x_3 + F_{4,\alpha}\, x_4 \\ z_{,\beta} &= F_{1,\beta}\, z_1 + F_{2,\beta}\, z_2 + F_{3,\beta}\, z_3 + F_{4,\beta}\, z_4 \end{aligned} \tag{9.15}$$

The integrals in Equation (9.6) can have analytical solutions but, in practice, numerical integrations are required. Gauss quadrature formulae are extensively adopted in FEA for these purposes. In the 1D CUF, these formulae are employed to compute 1D shape function integrals. In the 1D LE CUF, Gauss quadrature is also implemented to compute the F_τ integrals on the cross-section domain

$$\int_{-1}^{+1}\int_{-1}^{+1} F_\tau F_s |J| d\alpha\, d\beta = \sum_{h,k} w_h\, w_k\, F_\tau(\alpha_h, \beta_k)\, F_\tau(\alpha_h, \beta_k)\, |J(\alpha_h, \beta_k)| \tag{9.16}$$

where w_h and w_k are the integration weights and α_h and β_k are the integration points. The weights and points depend on the adopted set of Lagrange polynomials – three-, four-, six- or nine-node elements in the present discussion – which are given in Appendix A.

9.3 LE Displacement Fields and Cross-section Elements

Cross-section elements are introduced in this section. A cross-section element is described by a set of Lagrange polynomials defined on a given number of points. The elements are used to define the cross-section displacement field. As stated above, three-, four-, six- and nine-node elements are described below and are referred to as L3, L4, L6 and L9 according to the Lagrange polynomials they are based on. This means, for instance, that the L4 cross-section element is based on the four-point Lagrange polynomials given in Equation (9.1). The L-elements based on other polynomial sets (e.g. L16) could also be implemented easily. L4 leads to the following displacement field:

$$\begin{aligned} u_x &= F_1\, u_{x_1} + F_2\, u_{x_2} + F_3\, u_{x_3} + F_4\, u_{x_4} \\ u_y &= F_1\, u_{y_1} + F_2\, u_{y_2} + F_3\, u_{y_3} + F_4\, u_{y_4} \\ u_z &= F_1\, u_{z_1} + F_2\, u_{z_2} + F_3\, u_{z_3} + F_4\, u_{z_4} \end{aligned} \tag{9.17}$$

Figure 9.6 shows an L4 element and its nodes. The unknown variables, $u_{x_1}, \ldots, u_{z_4}$, are the three displacements components of each node. This means the following:

1. The problem unknowns are only physical translational displacements.
2. The problem unknowns can be placed on the physical surfaces of the body.

These two fundamental characteristics are valid for each L-element, regardless of the number of nodes. A typical L-element modelling approach is shown in Figure 9.7, where the following modelling steps are pointed out:

1. The 3D body is discretized at the cross-section level by means of L-elements. Their number depends on the geometry of the structure and on the BCs (geometrical or mechanical). For the sake of simplicity, only one L4 was adopted in Figure 9.7 and a two-node (B2) beam element was considered. More complex discretizations are described in the following sections.
2. In the FE LE formulation, the cross-section discretization determines the number of DOFs of each beam node. If an L4 element is used, 12 DOFs per beam node will be exploited.

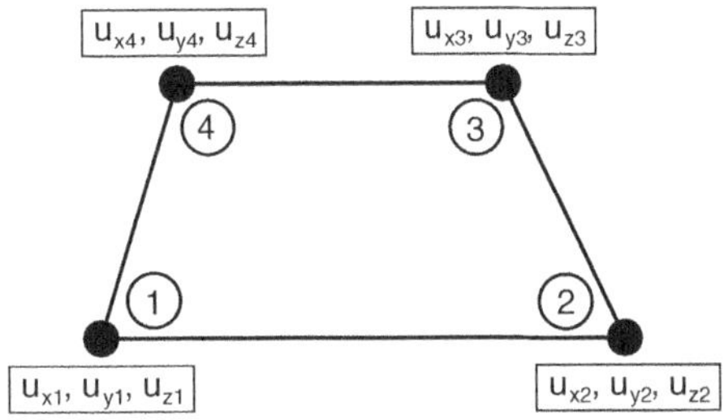

Figure 9.6 L4 element DOFs; only pure displacement unknowns are employed

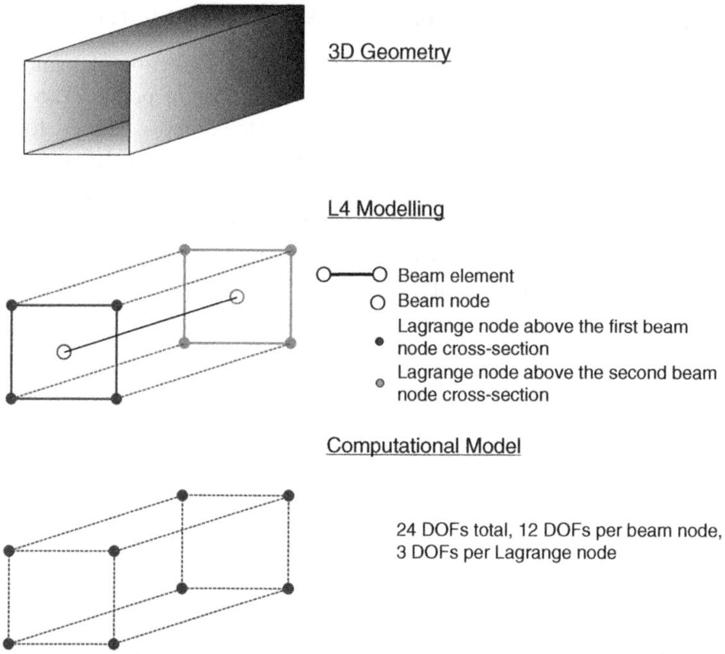

Figure 9.7 An example of LE modelling via an L4 element

3. A classical FE approach, based on beam elements, is employed to build FE matrices. It is important to emphasize that – unlike classical and TE models – the unknown variables of the computational model do not lie on the beam element axis.

The L3 displacement field is given by

$$
\begin{aligned}
u_x &= F_1\, u_{x_1} + F_2\, u_{x_2} + F_3\, u_{x_3} \\
u_y &= F_1\, u_{y_1} + F_2\, u_{y_2} + F_3\, u_{y_3} \\
u_z &= F_1\, u_{z_1} + F_2\, u_{z_2} + F_3\, u_{z_3}
\end{aligned}
\tag{9.18}
$$

Nine unknowns are employed, as shown in Figure 9.8. L6 is based on the following 18 unknown field:

$$
\begin{aligned}
u_x &= F_1\, u_{x_1} + F_2\, u_{x_2} + F_3\, u_{x_3} + F_4\, u_{x_4} + F_5\, u_{x_5} + F_6\, u_{x_6} \\
u_y &= F_1\, u_{y_1} + F_2\, u_{y_2} + F_3\, u_{y_3} + F_4\, u_{y_4} + F_5\, u_{y_5} + F_6\, u_{y_6} \\
u_z &= F_1\, u_{z_1} + F_2\, u_{z_2} + F_3\, u_{z_3} + F_4\, u_{z_4} + F_5\, u_{z_5} + F_6\, u_{z_6}
\end{aligned}
\tag{9.19}
$$

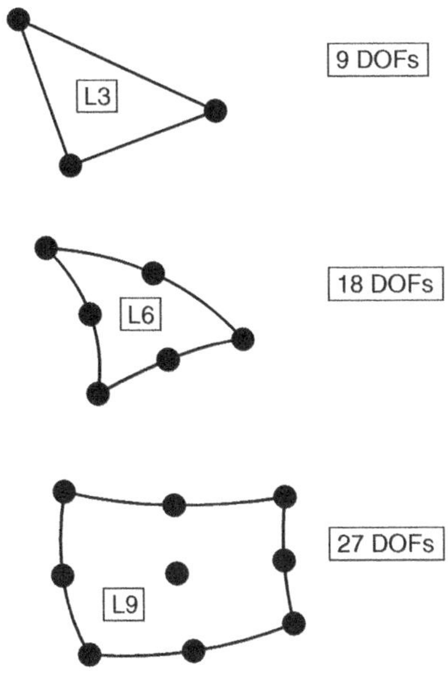

Figure 9.8 L3, L6 and L9 element DOFs

Eventually, the L9 element becomes

$$\begin{aligned} u_x &= F_1\, u_{x_1} + F_2\, u_{x_2} + F_3\, u_{x_3} + F_4\, u_{x_4} + F_5\, u_{x_5} + F_6\, u_{x_6} + F_7\, u_{x_7} + F_8\, u_{x_8} + F_9\, u_{x_9} \\ u_y &= F_1\, u_{y_1} + F_2\, u_{y_2} + F_3\, u_{y_3} + F_4\, u_{y_4} + F_5\, u_{y_5} + F_6\, u_{y_6} + F_7\, u_{y_7} + F_8\, u_{y_8} + F_9\, u_{y_9} \\ u_z &= F_1\, u_{z_1} + F_2\, u_{z_2} + F_3\, u_{z_3} + F_4\, u_{z_4} + F_5\, u_{z_5} + F_6\, u_{z_6} + F_7\, u_{z_7} + F_8\, u_{z_8} + F_9\, u_{z_9} \end{aligned} \tag{9.20}$$

Example 9.3.1 *Let us consider the L4 element in Figure 9.6, where a strain component (ε_{xx}) has to be computed at a generic point (α_k, β_k). The inputs are the coordinates of the four nodes (x_i, z_i, $i = 1, 2, 3, 4$) and their displacements.*

The strain component that has to be computed is given by

$$\varepsilon_{xx} = u_{x,x}$$

where u_x is given by

$$u_x = F_1\, u_{x_1} + F_2\, u_{x_2} + F_3\, u_{x_3} + F_4\, u_{x_4}$$

According to Equation (9.11), the derivative of u_x is

$$u_{x,x} = \frac{1}{|J(\alpha_k, \beta_k)|}(z_{,\beta}\, u_{x,\alpha} - z_{,\alpha}\, u_{x,\beta})_{\alpha=\alpha_k,\beta=\beta_k}$$

where

$$
\begin{aligned}
z_{,\beta} &= -\frac{1}{4}(1-\alpha_k)\,z_1 - \frac{1}{4}(1+\alpha_k)\,z_2 + \frac{1}{4}(1+\alpha_k)\,z_3 + \frac{1}{4}(1-\alpha_k)\,z_4 \\
u_{x,\alpha} &= -\frac{1}{4}(1-\beta_k)\,u_{x1} - \frac{1}{4}(1+\beta_k)\,u_{x2} + \frac{1}{4}(1+\beta_k)\,u_{x3} + \frac{1}{4}(1-\beta_k)\,u_{x4} \\
z_{,\alpha} &= -\frac{1}{4}(1-\beta_k)\,z_1 - \frac{1}{4}(1+\beta_k)\,z_2 + \frac{1}{4}(1+\beta_k)\,z_3 + \frac{1}{4}(1-\beta_k)\,z_4 \\
u_{x,\beta} &= -\frac{1}{4}(1-\alpha_k)\,u_{x1} - \frac{1}{4}(1+\alpha_k)\,u_{x2} + \frac{1}{4}(1+\alpha_k)\,u_{x3} + \frac{1}{4}(1-\alpha_k)\,u_{x4}
\end{aligned}
$$

and $|J(\alpha_k, \beta_k)|$ can be computed according to Equation (9.10).

This procedure is usually carried out, for example, to compute strain components at given points for the result postprocessing, starting from the displacement vector.

9.3.1 FE Formulation and FN

The adoption of LE models does not imply any formal changes in the problem governing equations or FNs. All the equations given in Section 8.4 are valid, regardless of the expansion polynomials (e.g. Taylor or Lagrange) and their order (e.g. $N = 1$, $N = 4$, L9, etc.). The FE formulation has been adopted in this book and its description can be found in the aforementioned section. The formal expressions for the stiffness and mass matrices and the loading vector do not depend on the expansion that is adopted, and this means that the FNs given for TE are also valid for LE. For the sake of clarity, the expression for the FN of the stiffness matrix is recalled below (the nine components of the nucleus have already been introduced in Table 9.1):

$$
\begin{aligned}
k_{xx}^{\tau sij} &= C_{22}\int_A F_{\tau,x}F_{s,x}dx\,dz\int_l N_iN_jdy + C_{66}\int_A F_{\tau,z}F_{s,z}dx\,dz\int_l N_iN_jdy \\
&\quad + C_{44}\int_A F_\tau F_s dx\,dz\int_l N_{i,y}N_{j,y}dy \\
k_{xy}^{\tau sij} &= C_{23}\int_A F_\tau F_{s,x}dx\,dz\int_l N_{i,y}N_jdy + C_{44}\int_A F_{\tau,x}F_s dx\,dz\int_l N_iN_{j,y}dy \\
k_{xz}^{\tau sij} &= C_{12}\int_A F_{\tau,z}F_{s,x}dx\,dz\int_l N_iN_jdy + C_{66}\int_A F_{\tau,x}F_{s,z}dx\,dz\int_l N_iN_jdy \\
k_{yx}^{\tau sij} &= C_{44}\int_A F_\tau F_{s,x}dx\,dz\int_l N_{i,y}N_jdy + C_{23}\int_A F_{\tau,x}F_s dx\,dz\int_l N_iN_{j,y}dy \\
k_{yy}^{\tau sij} &= C_{55}\int_A F_{\tau,z}F_{s,z}dx\,dz\int_l N_iN_jdy + C_{44}\int_A F_{\tau,x}F_{s,x}dx\,dz\int_l N_iN_jdy \\
&\quad + C_{33}\int_A F_\tau F_s dx\,dz\int_l N_{i,y}N_{j,y}dy \\
k_{yz}^{\tau sij} &= C_{55}\int_A F_\tau F_{s,z}dx\,dz\int_l N_{i,y}N_jdy + C_{13}\int_A F_{\tau,z}F_s dx\,dz\int_l N_iN_{j,y}dy
\end{aligned}
$$

$$k_{zx}^{\tau sij} = C_{12}\int_A F_{\tau,x}F_{s,z}dx\,dz \int_l N_iN_jdy + C_{66}\int_A F_{\tau,z}F_{s,x}dx\,dz\int_l N_iN_jdy$$
$$k_{zy}^{\tau sij} = C_{13}\int_A F_{\tau}F_{s,z}dx\,dz \int_l N_{i,y}N_jdy + C_{55}\int_A F_{\tau,z}F_{s}dx\,dz\int_l N_iN_{j,y}dy$$
$$k_{zz}^{\tau sij} = C_{11}\int_A F_{\tau,z}F_{s,z}dx\,dz \int_l N_iN_jdy + C_{66}\int_A F_{\tau,x}F_{s,x}dx\,dz\int_l N_iN_jdy$$
$$+ C_{55}\int_A F_{\tau}F_{s}dx\,dz\int_l N_{i,y}N_{j,y}dy \quad (9.21)$$

This expression is independent of the order type of the displacement expansion.

As for the TE case, the nodal stiffness matrix is built by expanding τ and s. An FN is computed for each τs-set and assembled accordingly. Figure 9.9 shows the nodal stiffness matrix for the L4 case. Each 3×3 block represents an FN. The nodal stiffness matrix is then assembled within the beam element and the global stiffness matrices, as shown in Figures 8.7 and 8.8, respectively.

Example 9.3.2 *Let us consider a rectangular beam element of length equal to L. A B2 element is used and, in order to model the cross-section, an L4 element is adopted:*

$$\tau, s = 1,2,3,4, \qquad F_\tau = \frac{1}{4}(1+\alpha\,\alpha_\tau)(1+\beta\,\beta_\tau)$$
$$i,j = 1,2, \qquad N_1 = 1-\frac{y}{L}, N_2 = \frac{y}{L}$$

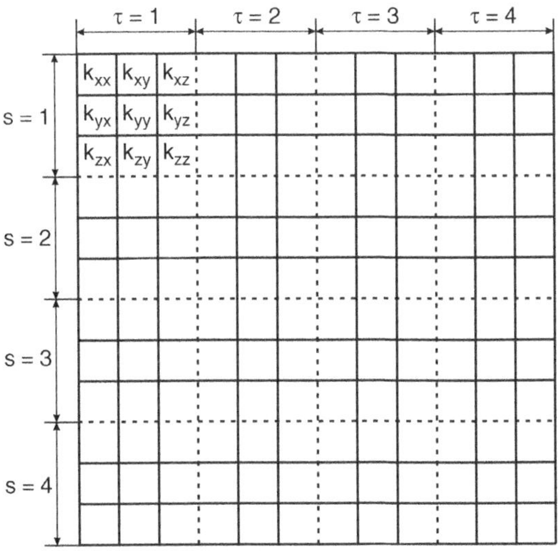

Figure 9.9 L4 nodal stiffness matrix, where an FN is shown for $\tau, s = 1$

The cross-section coordinates range from $-a$ to $+a$ in the x direction and from $-b$ to $+b$ in the z direction. The k_{xx} component has to be computed for $\tau, s = 2$, $i = 1$ and $j = 2$:

$$
\begin{aligned}
k_{xx}^{2212} = {} & C_{22} \int_{-a}^{+a} \int_{-b}^{+b} F_{2,x}\, F_{2,x}\, dx\, dz \int_0^L N_1\, N_2 dy \\
& + C_{66} \int_{-a}^{+a} \int_{-b}^{+b} F_{2,z}\, F_{2,z}\, dx\, dz \int_0^L N_1\, N_2 dy \\
& + C_{44} \int_{-a}^{+a} \int_{-b}^{+b} F_2\, F_2 dx\, dz \int_0^L N_{1,y}\, N_{2,y} dy
\end{aligned}
$$

The surface integrals are computed by means of the normalized coordinates. The Jacobian determinant is given by

$$|J| = \left(z_{,\beta}\, x_{,\alpha} - z,\alpha\, x_{,\beta}\right) = ab$$

since

$$
\begin{aligned}
x_{,\alpha} &= F_{1,\alpha}\, x_1 + F_{2,\alpha}\, x_2 + F_{3,\alpha}\, x_3 + F_{4,\alpha}\, x_4 \\
&= F_{1,\alpha}\, (-a) + F_{2,\alpha}\, a + F_{3,\alpha}\, a + F_{4,\alpha}\, (-a) = a \\
z_{,\alpha} &= 0 \\
z_{,\beta} &= b \\
x_{,\beta} &= 0
\end{aligned}
\tag{9.22}
$$

In this case $|J|$ is constant, although in most cases $|J| = |J(\alpha, \beta|$.

The derivatives of the expansion functions are given by

$$
\begin{aligned}
F_{2,x} &= \frac{1}{|J|}\left(z_{,\beta}\, F_{2,\alpha} - z_{,\alpha}\, F_{2,\beta}\right) = \frac{1}{4a}(1-\beta) \\
F_{2,z} &= \frac{1}{|J|}\left(-x_{,\alpha}\, F_{2,\alpha} + x_{,\beta}\, F_{2,\beta}\right) = -\frac{1}{4b}(1-\beta)
\end{aligned}
$$

By utilizing the previous expressions, the FN component becomes

$$
\begin{aligned}
k_{xx}^{2212} = {} & C_{22} \frac{b}{16a} \int_{-1}^{+1} \int_{-1}^{+1} (1-\beta)^2 d\alpha\, d\beta \int_0^L N_1\, N_2 dy \\
& + C_{66} \frac{a}{16b} \int_{-1}^{+1} \int_{-1}^{+1} (1-\beta)^2 d\alpha\, d\beta \int_0^L N_2\, N_1 dy \\
& C_{44} \frac{ab}{16} \int_{-1}^{+1} \int_{-1}^{+1} (1-\alpha)^2 (1-\beta)^2 d\alpha\, d\beta \int_0^L N_{2,y} N_{1,y} dy \\
= {} & \frac{1}{18} \frac{bL}{a} C_{22} + \frac{1}{18} \frac{aL}{b} C_{66} - \frac{4}{9} \frac{ab}{L}\, C_{44}
\end{aligned}
$$

Figure 9.10 shows the position of k_{xx}^{2212} within the element stiffness matrix.

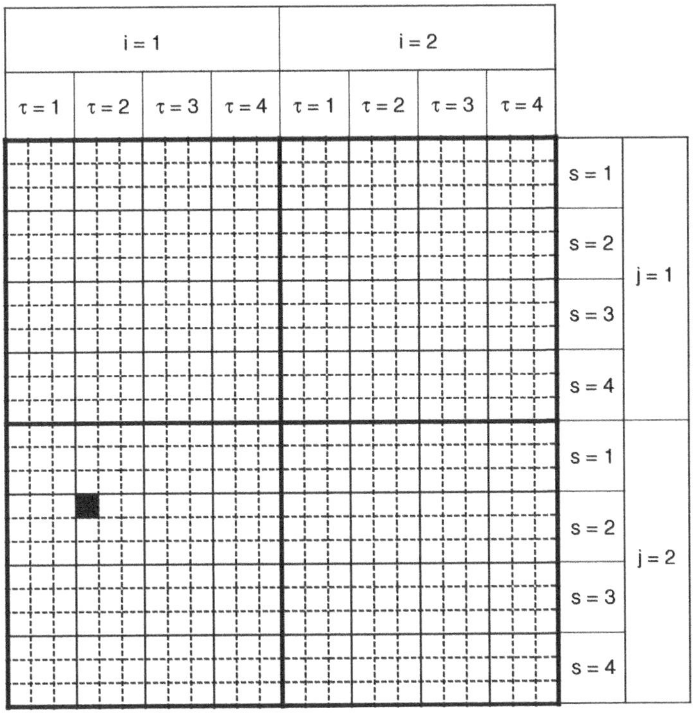

Figure 9.10 Position of k_{xx}^{2212} within the element stiffness matrix

9.4 Cross-section Multi-elements and Locally Refined Models

Cross-sections can be discretized by means of multiple LE elements. This is a fundamental function of LE elements. Multi-elements are generally adopted for three main purposes:

1. To refine the cross-section displacement field without increasing the polynomial expansion order.
2. To impose the geometrical discontinuities above the cross-section.
3. To refine the structural model locally.

Figure 9.11 shows a typical example in which two L4 elements are used to model the cross-section displacement field. A refined model is obtained by combining two piecewise linear elements. The assembly of the FE matrices requires the definition of a local and global connectivity of the cross-section nodes, as shown in Figure 9.12. The definition of the global connectivity should take into account the band matrix assembly, an aspect that has not been analysed in this book. The stiffness matrices of each L4 element are computed and assembled on the basis of the global connectivity, see Figures 9.13, 9.14 and 9.15. This assembly technique is analogous to the one that is commonly used for 2D and 3D FEs. Continuity of the displacements is imposed at the interface nodes. Different L-elements can be assembled simultaneously, e.g. a combination of L4 and L9 elements can be used in a cross-section.

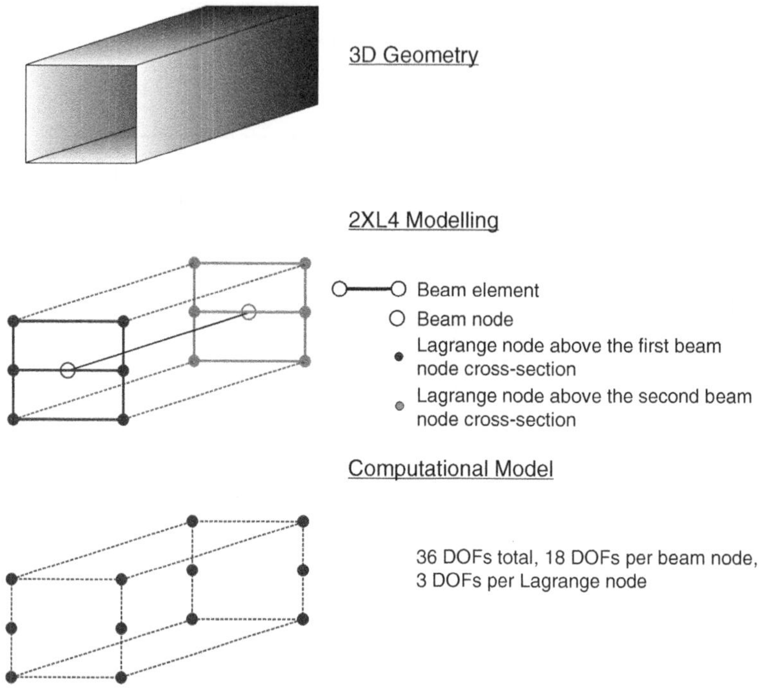

Figure 9.11 An example of LE modelling with two L4 elements

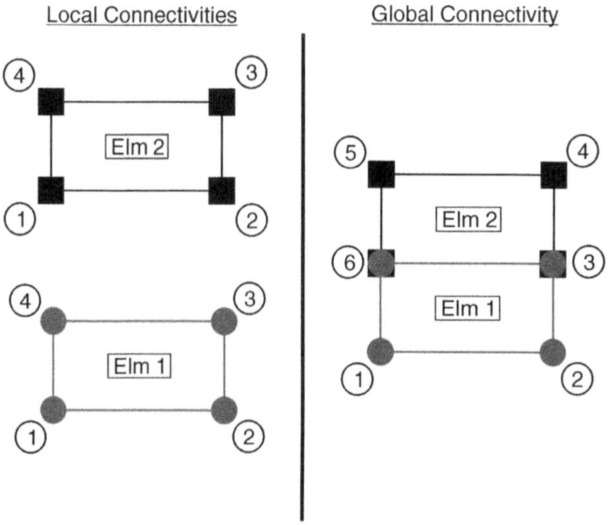

Figure 9.12 An example of two L4 elements assembled within a beam node

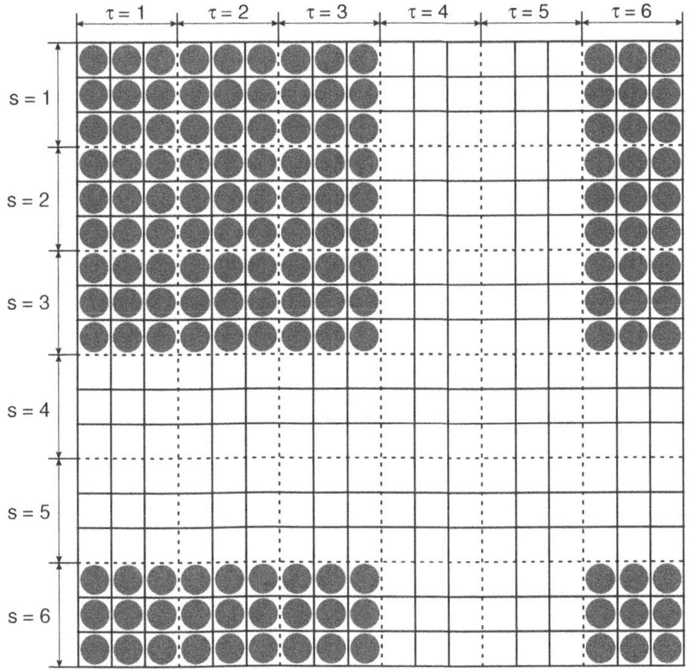

Figure 9.13 Element 1 assembly within a beam node

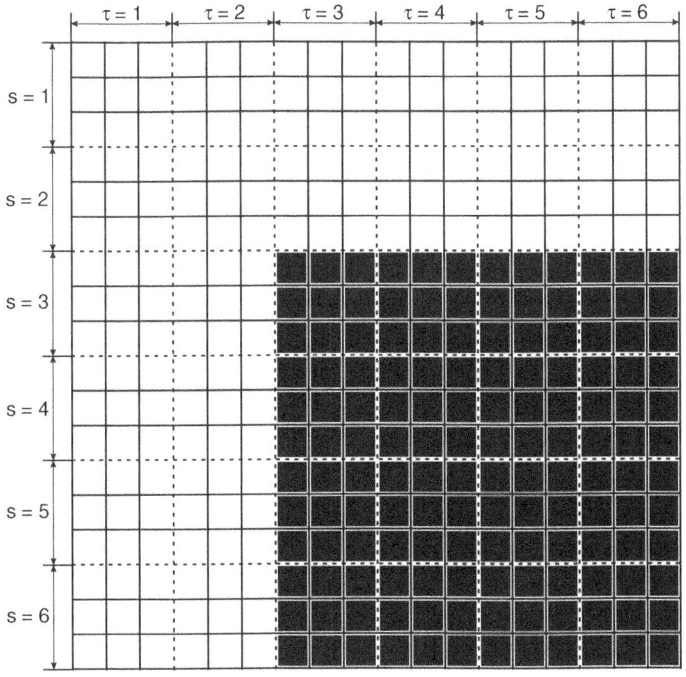

Figure 9.14 Element 2 assembly within a beam node

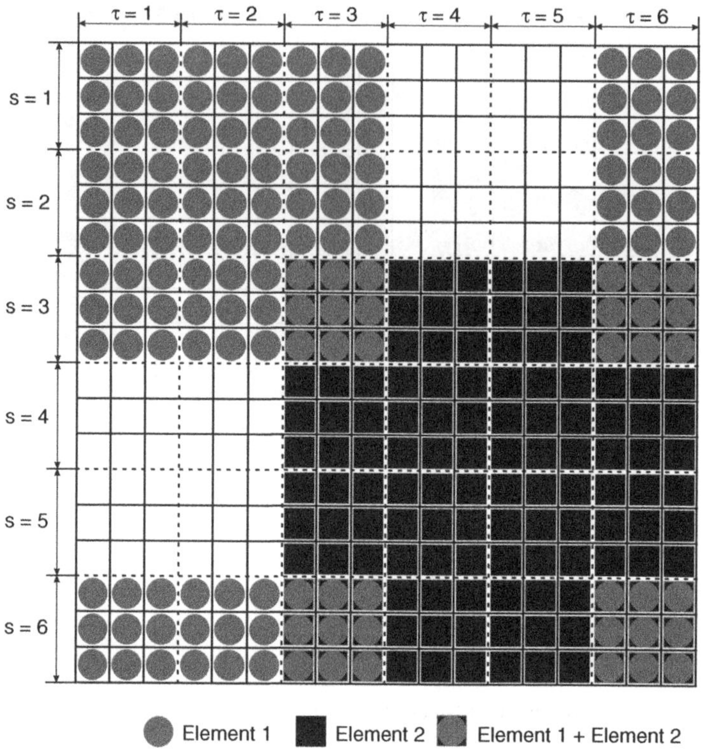

Figure 9.15 Two L4 elements assembls within a beam node

Geometrical discontinuities can be easily modelled. Figure 9.16 shows an example in which nodes 6 and 7 are physically disconnected – although located in the same position – and Figure 9.17 shows the new stiffness matrix.

The overall number of DOFs per cross-section (N_{cs}) (i.e. per beam node) is given by

$$N_{cs} = 3 \times N_{cn} \tag{9.23}$$

where N_{cn} is the number of cross-section nodes.

Locally refined models are of particular interest when local effects play an important role in a structural problem. A typical example is that of a thin-walled structure under point loads. Figure 9.18 shows a thin-walled cross-section and two point loads acting on the top and bottom edges. In this case, the deformed displacement field is affected to a great extent by local effects close to the loading points, while the loads have much less influence on the vertical edges. A proper detection of the in-plane deformed configuration requires higher-order polynomials, since a fairly complex deformed shape has to be modelled. If TE models are adopted, higher-order expansions are compulsory. The drawback of TE models is that a single expansion set can be adopted, which means that both the highly deformed and the barely deformed zones of the cross-section are modelled by means of higher-order models. It would be preferable to tune the refinement locally in order to optimize computational costs. Local refinements

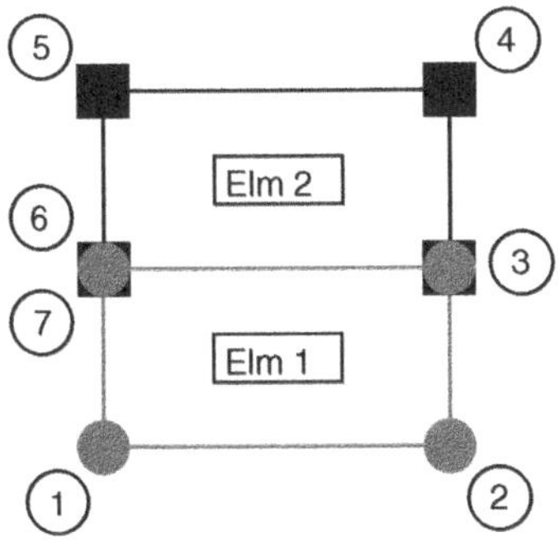

Figure 9.16 An example of two L4 elements assembled within a beam node, where two cross-section nodes are disconnected

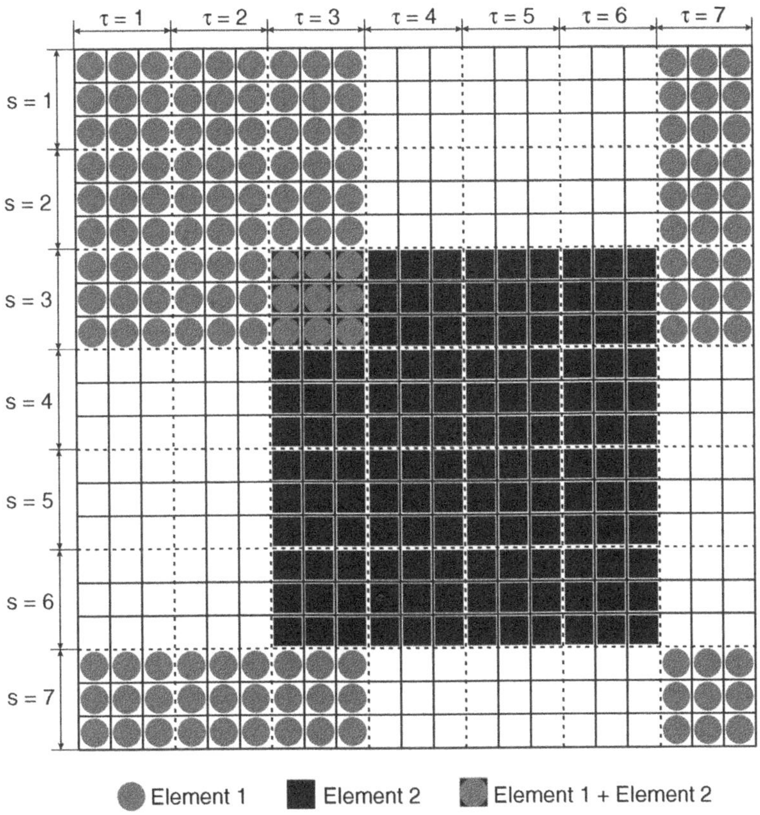

Figure 9.17 Two L4-element assembly, where two cross-section nodes are disconnected

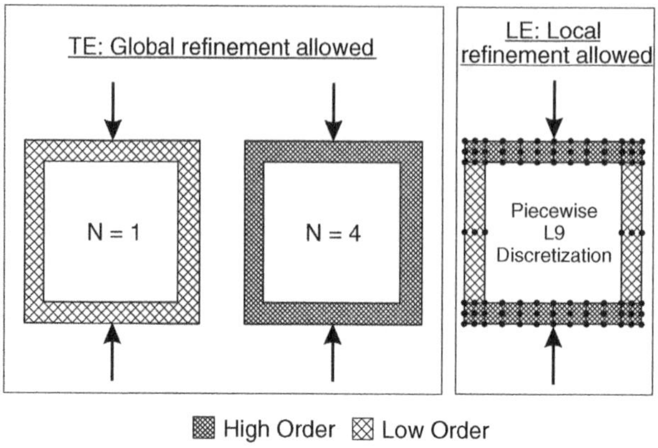

Figure 9.18 Global vs local refinements on the cross-section

can be implemented straightforwardly with LE models, since a finer cross-section mesh (or higher-order sets) can be used when needed. For the sake of completeness, it should be noted that locally refined models could also be obtained with TE models, but this process would require the imposition of compatibility conditions through, for instance, Lagrange multipliers.

9.5 Numerical Examples

9.5.1 Mesh Refinement and Convergence Analysis

Compact rectangular cross-sections are considered in this section in order to evaluate the effect of cross-section discretization in terms of displacements and stresses. The cross-section discretization should, in fact, be tailored through a proper choice of the L-elements (L3, L4, L6 or L9) and their distribution on the cross-section plane. Mechanical and geometrical BCs are usually the main factors that need to be taken into account in a cross-section discretization process.

A cantilever beam is considered here, where the cross-section is square ($h = 0.2$ m) and the length-to-thickness ratio (L/h) is equal to 100. The material is isotropic, with $E = 75$ GPa and $\nu = 0.33$. A point load (P_z) is applied at the centre of the free tip (0, L, 0). The magnitude of P_z is equal to −50 N. The loaded point vertical displacement (u_z) is evaluated. EBBT is used for comparison purposes ($u_{z_b} = P_z L^3/3EI$), where I is the cross-section moment of inertia. Table 9.6 shows the displacement values for different meshes and beam models. Classical theories (EBBT and TBT) are accounted for as well as TE models ($N = 1$ and $N = 2$). Four different L-element sets are adopted: 1 L4, 2 × 1 L4 (two L4 elements in x direction), 1 × 2 L4 (two L4 elements in the z direction), and 1 L9. The following considerations hold:

1. L9 gives results which are equivalent to those of an $N = 2$ model. This means that the cubic and quartic polynomial terms ($\alpha^2\beta$, $\alpha\beta^2$ and $\alpha^2\beta^2$) do not play a very significant role in this problem, since these terms are those retained in L9 and which are missing in $N = 2$.
2. L4 has slower convergence rates than L9. More details about this issue will be given in the following sections. The subdivision of the cross-section into several L4 elements is very effective.

Table 9.6 Effect of the number of L-elements on u_z (Carrera and Petrolo 2011a)

Model	$u_z \times 10^2$ (m)
$u_{z_b} \times 10^2 = -1.333$ m	
TE	
EBBT	−1.333
TBT	−1.333
$N = 1$	−1.333
$N = 2$	−1.331
LE	
1 L4	−1.115
2 × 1 L4	−1.254
1 × 2 L4	−1.262
2 × 2 L4	−1.268
1 L9	−1.331

This table shows the effect of the cross-section discretization on the free-tip transverse displacement.

3. The improvement offered by the cross-section discretization into L4 elements is related to the total number of elements and their distribution above the cross-section. A refinement in the z direction is more effective than one in the x direction, when a P_z load is applied.

A rectangular cantilever beam is now considered ($h = 0.1$ m, $b = h/4$ and $L/h = 6$). A point load (P_z) is applied to the free-tip bottom edge $(0, L, -h/2)$, $P_z = -1$ N. Two cross-section L9 distributions are adopted, 1 L9 and 3 × 3 L9. Table 9.7 presents the vertical displacements and stress values at different points. Comparisons with a solid model are reported and the computational costs of each model are provided in terms of DOFs. These results suggest the following:

1. A general good match is found between the present formulation and the solid model solution. A slight difference can be observed in the vertical displacement, due to the local effects given by the point load.
2. Cross-section discretization refinement is an effective method that leads to a 3D solid solution. Shear stress distributions, in particular, are improved through the adoption of a refined cross-section model.
3. The present formulation has much lower computational costs than a solid model.

9.5.2 *Considerations on PL*

The slow convergence rate of L4 is related to the bilinear term in the displacement field expression ($\alpha\beta$). As seen in the previous chapter, the PL correction can be activated if linear expansions are considered. A bilinear term makes the correction detrimental. In this section,

Table 9.7 Displacement and stress values of the rectangular beam (Carrera and Petrolo 2011a)

Solid 27 000 DOFs	1 L9 4941 DOFs	3 × 3 L9 8967 DOFs	$[x, y, z]$
	$u_z \times 10^7$ m		
4.770	4.652	4.682	$[0, L, -h/2]$
	$\sigma_{yy} \times 10^{-4}$ Pa		
1.292	1.296	1.291	$[0, L/10, +h/2]$
	$\sigma_{yz} \times 10^{-2}$ Pa		
−6.168	−4.277	−6.086	$[b/2, L/10, 0]$

This table shows the effect of the L-element discretization on displacement and stress fields.

the equivalence of a bilinear TE and an L4 model is shown. Then, the detrimental effects of the PL correction are described by means of L3 elements.

The TE bilinear model is given as

$$
\begin{aligned}
u_x &= u_{x_1} + x\,u_{x_2} + z\,u_{x_3} + xz\,u_{x_5} \\
u_y &= u_{y_1} + x\,u_{y_2} + z\,u_{y_3} + xz\,u_{y_5} \\
u_z &= u_{z_1} + x\,u_{z_2} + z\,u_{z_3} + xz\,u_{z_5}
\end{aligned} \tag{9.24}
$$

This model can be obtained through a penalization technique that will be presented in the following chapters and which was first proposed by Carrera and Petrolo (2011b). The same square cross-section problem as previously addressed is considered; Table 9.8 shows the results in terms of transverse displacements. The following can be stated:

1. L4 is equivalent to the bilinear TE term.
2. A PL correction corrupts the effectiveness of both models, because of the presence of a bilinear term.
3. A single bilinear term is not enough to eliminate PL.
4. As a general remark, it can be stated that a beam model, based on a bilinear displacement field, should not be used because of convergence issues. Linear models (e.g. TBT or $N = 1$) with the PL correction, or at least a second-order model, would be preferable to predict the bending behaviour of a compact beam.

Table 9.8 Role of the bilinear term and of PL on u_z (Carrera and Petrolo 2011a)

Correction	TE bilinear	1 L4
	$u_z \times 10^2$ m, $u_{z_b} \times 10^2 = -1.333$ m	
Activated	−1.866	−1.868
Deactivated	−1.115	−1.115

This table shows the effect of PL and its correction on a square bent beam.

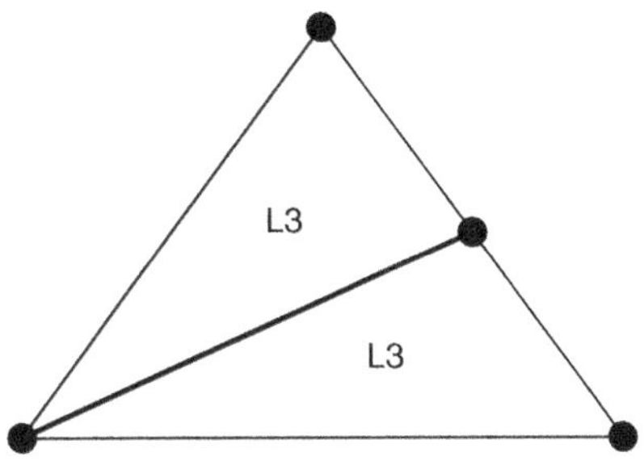

Figure 9.19 Two L3 discretizations of a triangular cross-section

An equilateral triangular cross-section is now considered in order to investigate PL. The edge of the triangle is 1 m and L/b is equal to 20. A vertical force (P_z) is applied at the centre point ($P_z = -30\,\text{N}$). Two cross-section discretizations are used, 1 L3 and 2 L3. The latter is shown in Figure 9.19. Table 9.9 gives the vertical centre point displacement values for both beam models and the effects of the PL correction. The following considerations hold:

1. The PL correction is beneficial in the case of 1 L3, because a linear description of the cross-section displacement field is given. This confirms what was previously stated pertaining to the role of the bilinear term.
2. The correction is detrimental in the case of 2 L3, because the displacement field is stepwise linear, and therefore it is overall higher than the first-order model. However, more than 2 L3 elements are needed to eliminate PL.

9.5.3 *Thin-Walled Structures and Open Cross-Sections*

Thin-walled structures are of interest in many engineering fields. For instance, aerospace and automotive structures are, for instance, based to a great extent on thin-walled components. The typical structural behaviour of thin walls includes in- and out-of-plane warping and local effects. The 2D plate and shell FEs are the most common ways of analysing thin walls. Over the last few decades, 1D beam models for thin-walled structures have been developed. Such models are advantageous as far as computational costs are concerned, but these models are

Table 9.9 Effect of the PL correction on u_z (Carrera and Petrolo 2011a)

Correction	1 L3	2 L3
$u_z \times 10^5$ m, $u_{z_b} \times 10^5 = -5.912$ m		
Activated	−5.917	−7.300
Deactivated	−3.995	−4.567

This table shows the effect of PL and its correction on a triangular bent beam.

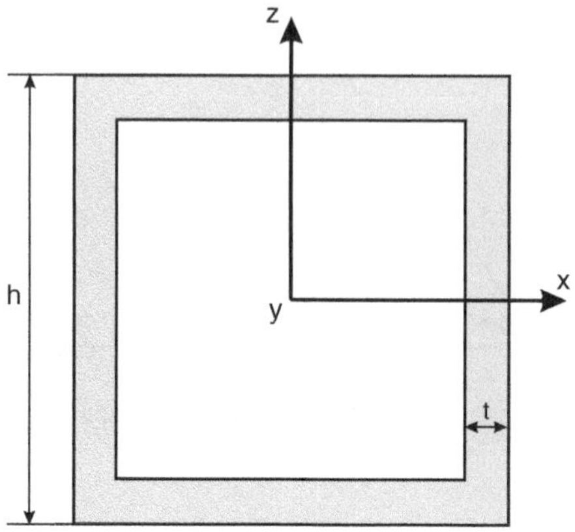

Figure 9.20 Hollow square cross-section

often prone to problems and difficult to exploit for many classes of structural problems. The capabilities of LE models are especially powerful in the analysis of thin-walled structures and, in particular, in the modelling of geometrical discontinuities, such as cuts. The numerical examples in this section have the aim of showing the enhanced capabilities of LE models to deal with such structures, without the need to resort to ad hoc formulations.

A clamped–clamped hollow square cross-section is considered (see Figure 9.20, $L/h = 20$, $h/t = 10$. $h = 1$ m). A point load ($P_z = 1$ N) is applied at (0, $L/2$, $-h/2$). The material is isotropic with $E = 75$ GPa and $\nu = 0.33$. Three cross-section discretizations are used, as shown in Figure 9.21. The 8 L9 mesh is symmetric, whereas 9 *L9* and 11 L9 are refined in the proximity of the loaded point.

Table 9.10 presents the transverse displacement of the loaded point together with the number of DOFs of each model. The first row shows the solid model results obtained in MSC Nastran. Increasing-order TE models are considered in the second and fifth rows. The results of the LE model are shown in the last three rows. The following statements hold:

1. There is an excellent match between the 1D CUF and the solid models.
2. The computational costs for the 1D CUF are much lower than for the solid models.
3. An appropriate distribution of L9 elements on the cross-section is effective in improving the accuracy of the solution. Local refinement of a beam model is also possible and leads to solid-like accuracy.
4. Local refinements make LE models able to detect a more accurate solution than TE ones, with reduced computational costs, since the TE refinements are spread uniformly over the cross-section and there is no distinction between scarcely and severely deformed zones.

The refinement capabilities of LE models are further investigated by means of a second loading case. Two point loads ($P_z = \pm 1$ N) are applied at (0, $L/2$, $\mp h/2$). The L9 distributions are those shown in Figures 9.21a and 9.21c (symmetric and asymmetric distributions are considered). Table 9.11 gives the displacements of the two loaded points, $u_{z_{top}}$ and $u_{z_{bot}}$, respectively.

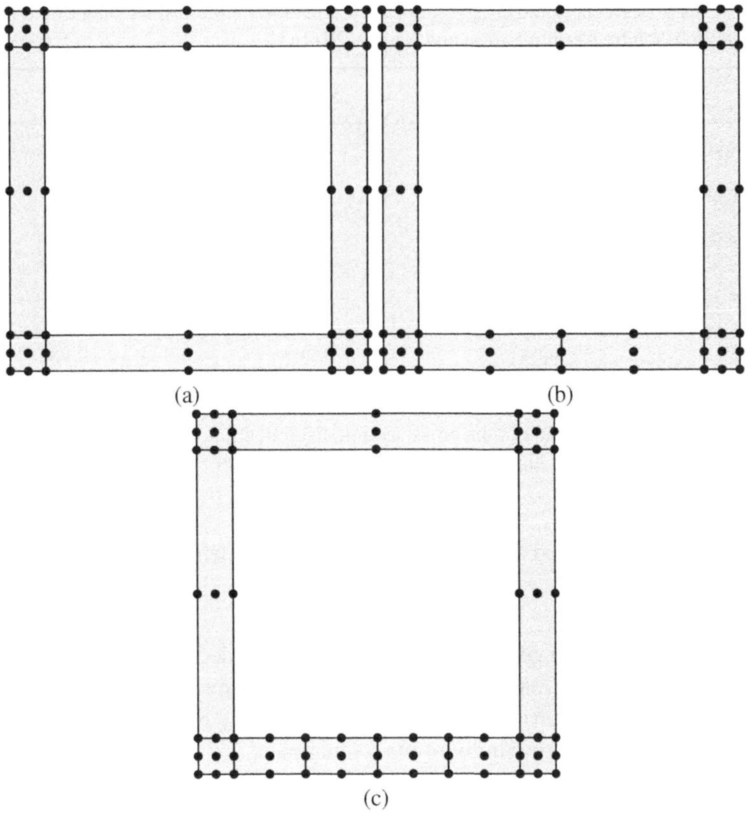

Figure 9.21 Hollow square cross-section discretizations: (a) 8 L9; (b) 9 L9; (c) 11 L9

Table 9.10 Loaded point transverse displacement of the hollow square beam (Carrera and Petrolo 2011a)

	DOFs	$u_z \times 10^8$ (m)
Solid	128 952	1.374
	TE	
EBBT	155	1.129
$N = 4$	1395	1.209
$N = 8$	4185	1.291
$N = 11$	7254	1.309
	LE	
8 L9	4464	1.277
9 L9	5022	1.308
11 L9	6138	1.326

This table presents the transverse displacement of the hollow square beam loading point for different models and their computational costs.

Table 9.11 Effects of the cross-section discretization on the displacement field of the hollow square beam (Carrera and Petrolo 2011a)

	DOFs	$u_{z_{top}} \times 10^9$ (m)	$u_{z_{bot}} \times 10^9$ (m)
Solid	128 952	−1.716	1.716
		TE	
EBBT	155	0.0	0.0
$N = 4$	1395	−0.178	0.178
$N = 8$	4185	−1.046	1.046
$N = 11$	7254	−1.270	1.270
		LE	
8 L9	4464	−0.985	0.985
11 L9	6138	−0.972	1.456

This table shows the effect of the cross-section discretization, where finer meshes in the bottom edge lead to higher flexibility than in the coarser top edge.

Figure 9.22 shows the deformed cross-section ($y = L/2$) for each L9 distribution. The following considerations hold:

1. Due to the symmetry of the geometry and the load, the loaded points should have identical vertical displacements (in magnitude). This result is obtained in all the models considered, except in the asymmetric L9 distribution. The locally refined model leads to higher displacement values in the proximity of the refinement.
2. The solution improvement provided by LE models is greater and computationally cheaper than that provided by TE ones.
3. As expected, classical beam models (e.g. EBBT), based on constant transverse displacements distributions, provide null displacement fields for such loading cases.

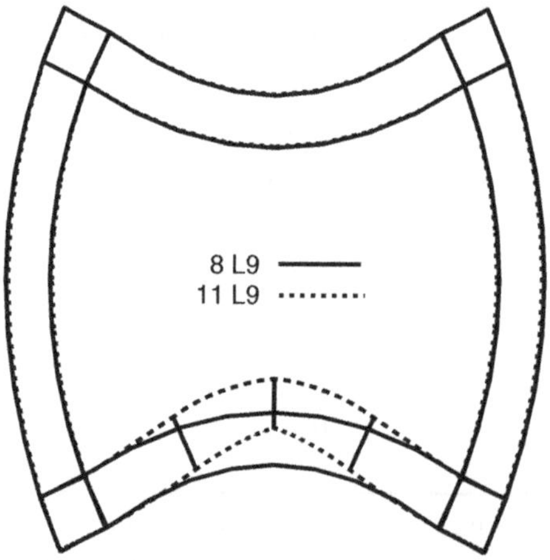

Figure 9.22 Effects of the L9 mesh on the hollow square cross-section

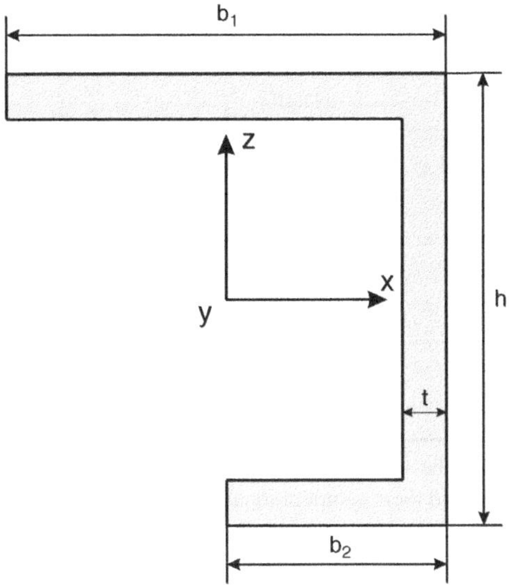

Figure 9.23 C-shaped cross-section

The first open cross-section considered is a C-shaped beam (see Figure 9.23, $L/h = 20$, $h/t = 10$, $h = b_2 = 1$ m, $b_1 = b_2/2$). Two point loads ($P_z = \mp 1$ N) are applied at (0, L, ± 0.4). Two L9 meshes are adopted and are shown in Figure 9.24. The 9 L9 mesh is finer in the proximity of the loading points.

Table 9.12 presents the transverse displacement (u_z) at ($-b_2/2, L, 0.4$). Figure 9.25 shows the free-tip deformed cross-section obtained with the LE and with the solid model. The following considerations hold:

1. The 9 L9 model is able to detect the solid solution perfectly, with a significant reduction in computational costs.

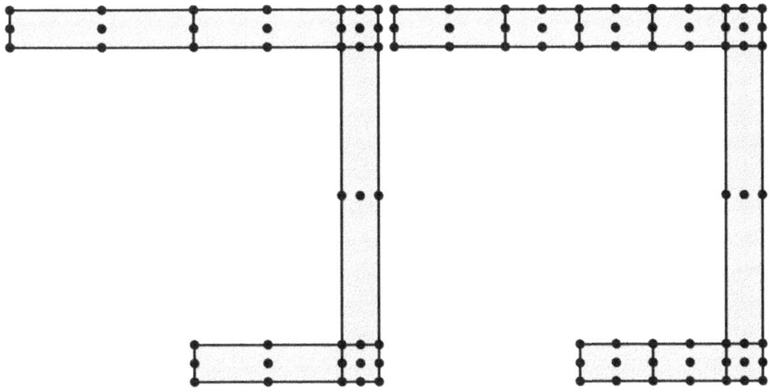

(a) (b)

Figure 9.24 C-shaped cross-section discretizations: (a) 6 L9; (b) 9 L9

Table 9.12 Vertical displacement at $(-b_2/2, L, 0.4)$, C-shaped beam (Carrera and Petrolo 2011a)

	DOFs	$u_z \times 10^8$ (m)
Solid	84 600	−3.067
	TE	
EBBT	155	0.0
$N = 4$	1395	−0.245
$N = 8$	4185	−2.161
$N = 11$	7254	−2.565
	LE	
6 L9	3627	−2.930
9 L9	5301	−2.982

This table shows the transverse displacement of the C-shaped beam in different models and their computational costs

2. TE models require higher than 11th-order expansions to match the solid model solution. Consequently, the difference in computational costs between TE and LE models appears to be higher for open than for closed cross-sections.
3. As seen previously, classical models are totally unable to detect the displacement field of such structural problems.

A flexural–torsional load is considered as a second loading case. A point force ($P_z = -1$ N) is applied at $(b_1, L, -h/2)$. In this case, two L/h values are considered, 20 and 10. The 9 L9 model is adopted. The displacement and stress values, at different locations, are presented in Table 9.13. These results suggest the following:

1. The flexural–torsional behaviour of a moderately short, open cross-section beam can be predicted correctly with the present formulation.
2. Stress distributions obtained from the solid model are clearly detected by LE models.

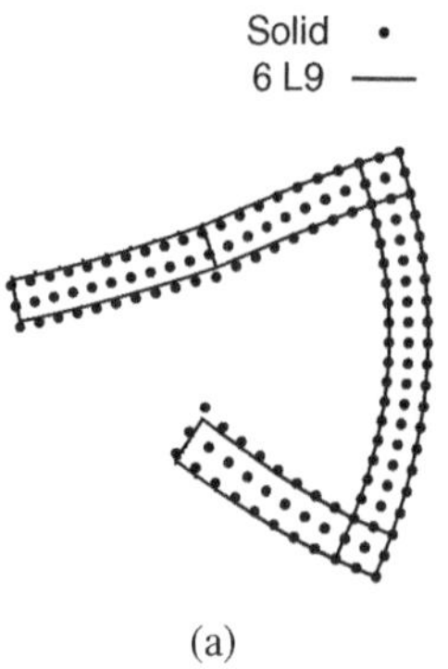

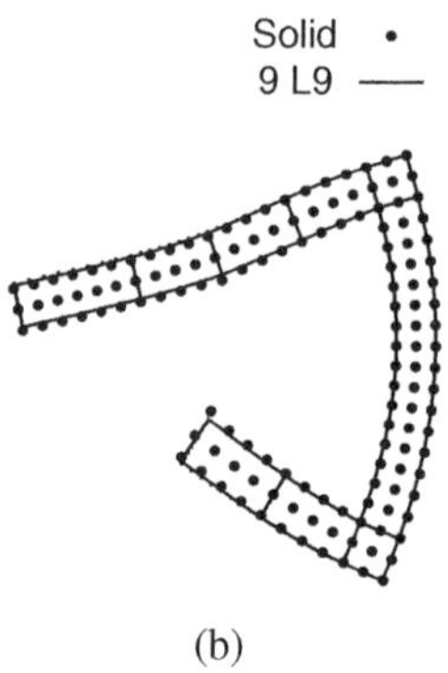

Figure 9.25 C-shaped free-tip deformation: (a) 6 L9; (b) 9 L9

Table 9.13 Displacements and stresses of the C-shaped beam (Carrera and Petrolo 2011a)

	Solid 84 600 DOFs	9 L9 5301 DOFs	(x, y, z)
$L/h = 20$			
	$u_z \times 10^6$ m		
	−1.470	−1.462	$(-b_2/2, L, +h/2)$
	$\sigma_{yy} \times 10^{-2}$ Pa		
	3.880	3.976	$(b_1, L/10, +h/2)$
	$\sigma_{xy} \times 10^{-2}$ Pa		
	−1.636	−1.691	$(b_1, L, -h/2)$
	$\sigma_{yz} \times 10^{-1}$ Pa		
	−2.401	−2.348	$(0.4, L/10, 0)$
$L/h = 10$			
	$u_z \times 10^7$ m		
	−2.280	−2.272	$(-b_2/2, L, +h/2)$
	$\sigma_{yy} \times 10^{-2}$ Pa		
	2.030	2.055	$(b_1, L/10, +h/2)$
	$\sigma_{yz} \times 10^{-2}$ Pa		
	−4.345	−3.837	$(b_1, L, -h/2)$
	$\sigma_{yz} \times 10^{-1}$ Pa		
	−1.930	−1.863	$(0.4, L/10, 0)$

This table shows displacements and stresses of the C-shaped beam at various points for different slenderness ratios.

An open, square cross-section is now considered (see Figure 9.26 for the dimensions and material characteristics, which are the same). Two opposite unit point loads ($\pm P_x$) are applied at $(0, L, -0.45)$. Three L9 distributions are adopted, as can be seen in Figure 9.27, in which the disconnected points are indicated.

Table 9.14 presents the horizontal displacement of the loaded point on the right-hand side. A solid model is used to validate the results. The free-tip deformed cross-section is shown in

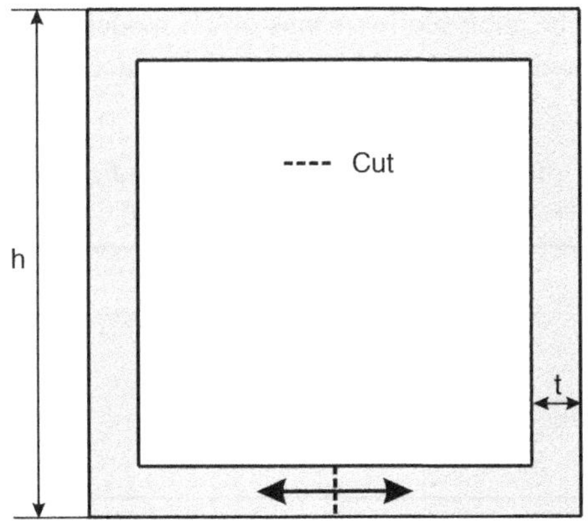

Figure 9.26 Open, hollow square cross-section

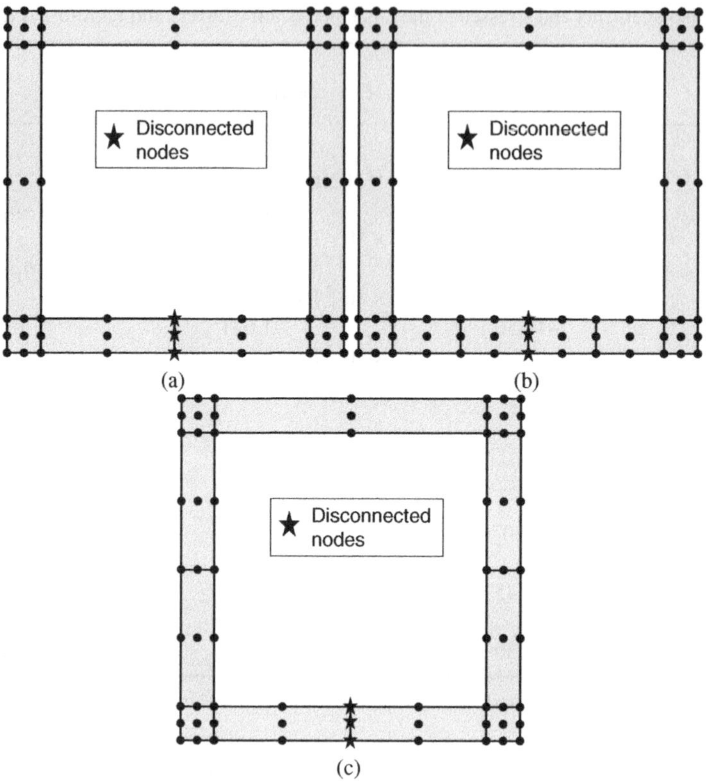

Figure 9.27 Open, hollow square cross-section discretizations: (a) 9 L9; (b) 11 L9; (c) 11 L9*

Figure 9.28. Figure 9.29 shows the 3D deformed configuration. The following considerations hold:

1. LEs are able to deal with cut cross-sections.
2. Such a problem cannot be analysed by means of TE models, since the application of two opposite forces at the same point would imply null displacements.

Table 9.14 Horizontal displacement at $(0, L, -h/2)$, open, hollow square beam (Carrera and Petrolo 2011a)

	DOFs	$u_x \times 10^8$ m
Solid	131 400	5.292
9 L9	5301	4.884
11 L9	6417	4.888
11 L9*	6417	5.116

This table shows the displacements of the open, hollow square beam from different models and their computational costs.

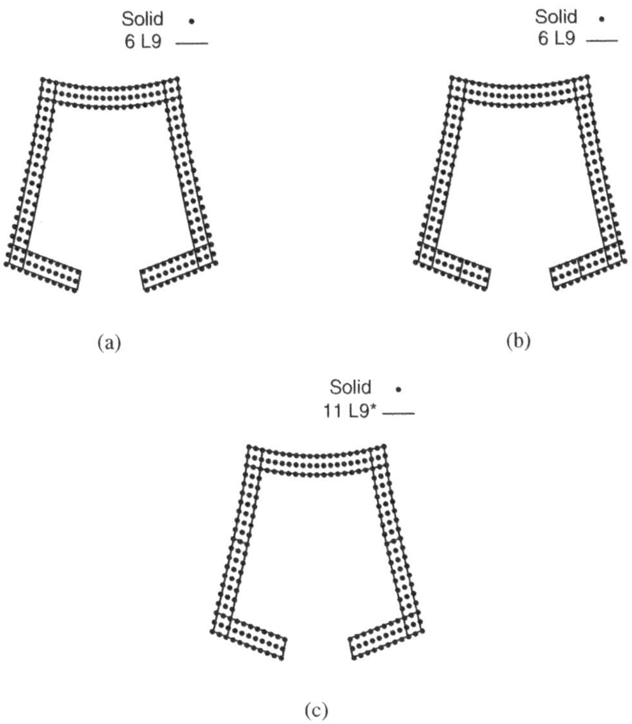

Figure 9.28 Open, hollow square cross-section for free-tip deformation: (a) 9 L9; (b) 11 L9; (c) 11 L9*

3. The most appropriate refined L9 distribution does not necessarily occur in the proximity of load points. In this case, the most effective refinement was the one above the vertical braces of the cross-section which undergo severe bending deformations.

9.5.4 Solid-like Geometrical BCs

Constraints in a beam model can usually be imposed over the entire cross-section, as shown in Figure 9.30a. However, many structural problems require the modelling of localized BCs,

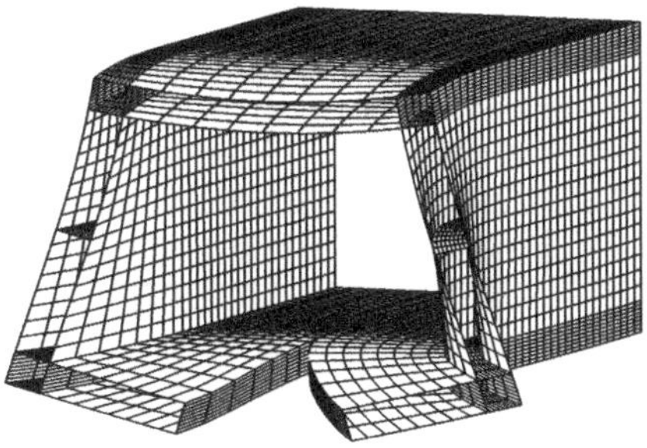

Figure 9.29 The 3D deformed, open, hollow square cross-section

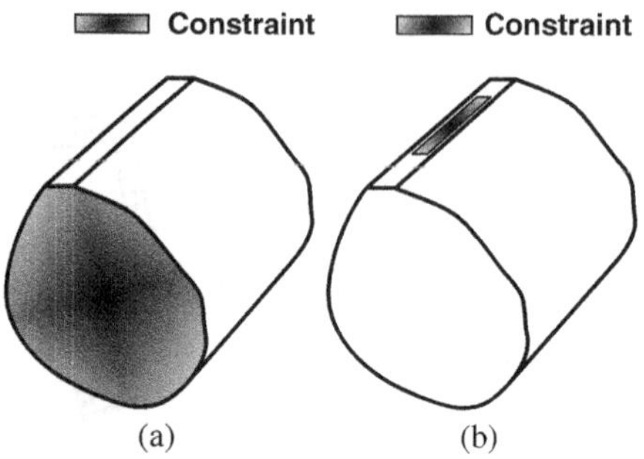

Figure 9.30 Geometrical BCs on a beam: (a) on the cross-section; (b) on a localized portion of the structure

as shown in Figure 9.30b. Localized BCs can be imposed in a beam model through various techniques (e.g. Lagrange multipliers), but supplemental equations and variables are often required. BCs can be directly imposed on each DOF of each cross-section node through LE models. Furthermore, localized BCs can be imposed by acting on the stiffness matrix of the problem, without the need for supplemental equations. Figure 9.31 shows the stiffness matrix

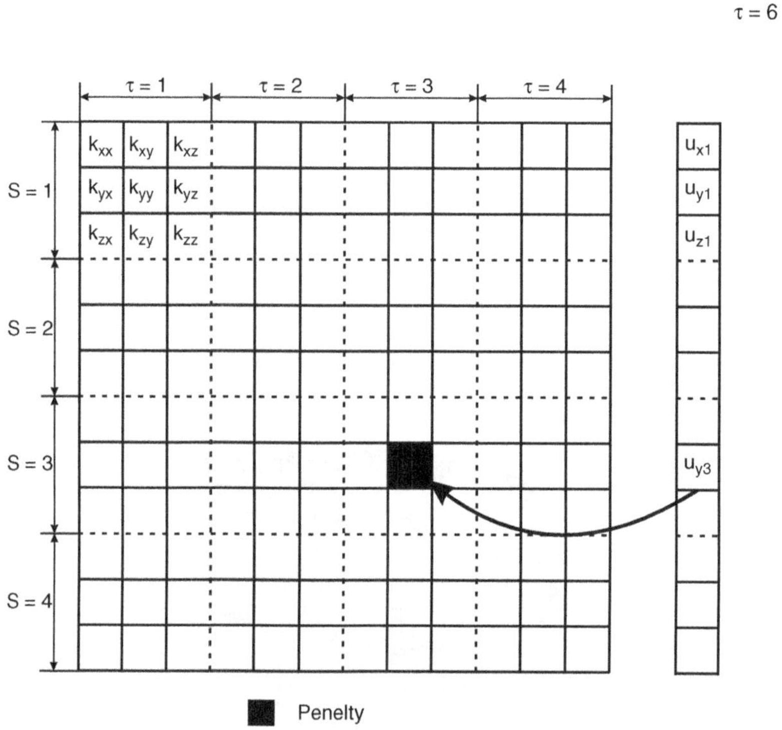

Figure 9.31 Constraining of DOFs in an L4 nodal stiffness matrix

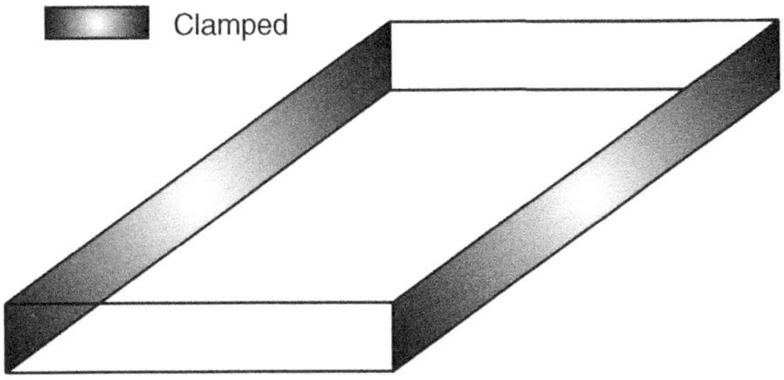

Figure 9.32 Rectangular beam clamped at the lateral edges

and the displacement vector of an L4 element. If a penalty technique is used and, for instance, the u_y component of the third cross-section point has to be set equal to zero, the penalty will be placed on the corresponding main diagonal position of the stiffness matrix. More complex BC sets can be imposed by adopting this method, and BCs can be imposed on the physical surfaces of the structure. A compact rectangular beam clamped at the lateral edges is considered (see Figure 9.32, $L/h = 100$, $b/h = 10$, $h = 0.01$ m). A set of 21 unitary point loads is applied along the mid-span cross-section on the top surface ($z = h/2$) with constant steps in x starting from the edge of the cross-section. Two L9 distributions, 5 L9 and 10 L9, are adopted. Table 9.15 presents the centre-point transverse displacement. The deformed mid-span cross-section is shown in Figure 9.33.

A second numerical assessment is carried out on a circular arch cross-section beam clamped at the lateral edges (see Figure 9.34). The length of the beam (L) is equal to 2 m. The outer (r_1) and inner (r_2) radii are equal to 1 and 0.9 m, respectively. The angle of the arch (θ) is equal to $\pi/4$ rad. Three unitary point loads are applied on the bottom surface at $y = 0$, $y = L/2$, and $y = L$ ($\theta = \theta/2$). Each load acts in the radial direction (from the inner to the outer direction). The L9 cross-section discretization is shown in Figure 9.35. Table 9.16 instead shows the transverse displacement of a point of the mid-span cross-section and Figure 9.36 shows the 3D deformed configuration.

The previously analysed C-section beam is now reconsidered with a new set of BCs (see Figure 9.37). Two unitary point loads (P_z) are applied at (0, 0, 0.4) and (0, L, 0.4). Both forces act in the negative direction. The L9 cross-section distribution is shown in Figure 9.38. The

Table 9.15 Transverse displacement at (0, $L/2$, 0) of the rectangular beam clamped at the lateral edges (Carrera and Petrolo 2011a)

	DOFs	$u_z \times 10^7$ (m)
Solid	17 271	−1.114
5 L9	3069	−0.959
10 L9	5859	−1.110

This table shows the transverse displacement of the rectangular beam clamped at the lateral edges; LE models are compared with solid ones.

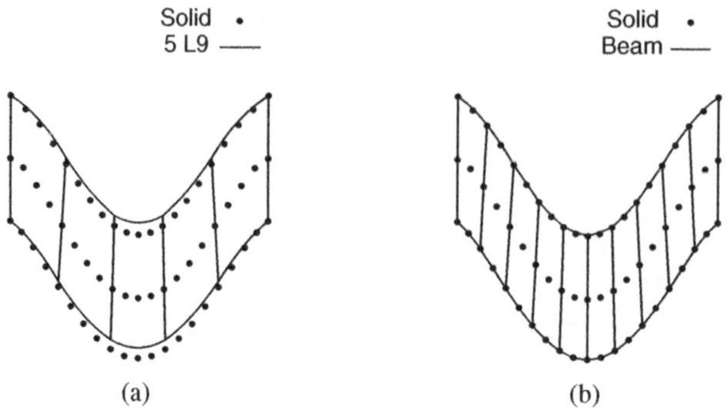

Figure 9.33 Mid-span deformed cross-section of the rectangular beam clamped at the lateral edges: (a) 5 L9; (b) 10 L9

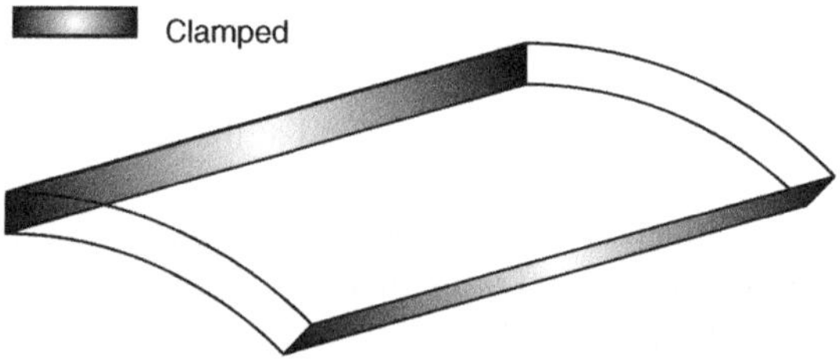

Figure 9.34 Arch beam clamped at the lateral edges

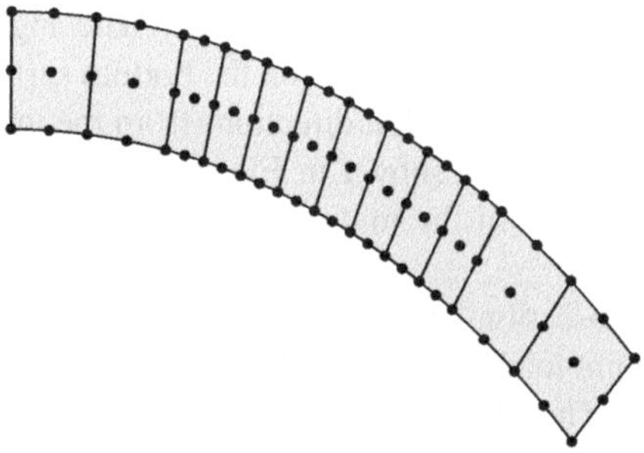

Figure 9.35 L9 mesh for the arch beam clamped at the lateral edges, 12 L9

Table 9.16 Transverse displacement on the external surface of the arch beam ($y = L/2$, $\theta = \theta/2$) (Carrera and Petrolo 2011a)

	DOFs	$u_z \times 10^{10}$ (m)
Solid	43 011	4.797
12 L9	6975	4.809

This table shows the transverse displacement of the arch beam; LE models are compared with solid ones.

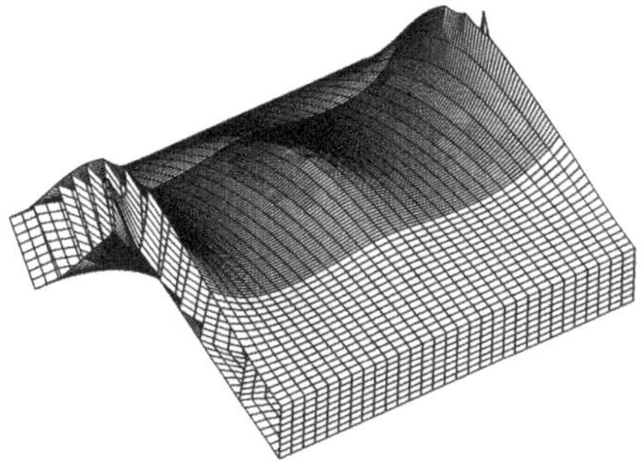

Figure 9.36 Deformed 3D configuration of the arch beam clamped at the lateral edges

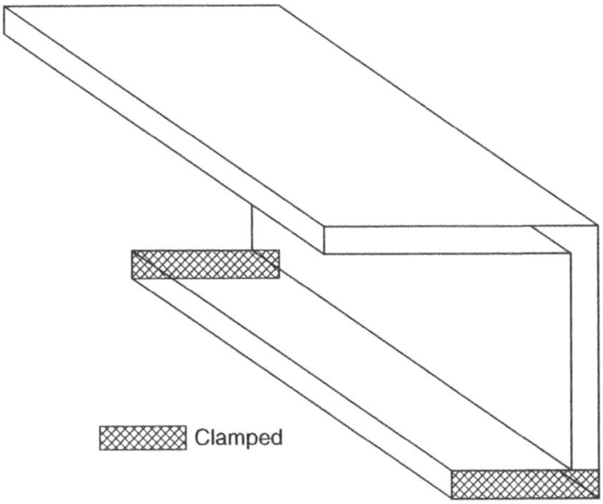

Figure 9.37 C-shaped beam clamped at the bottom flanges

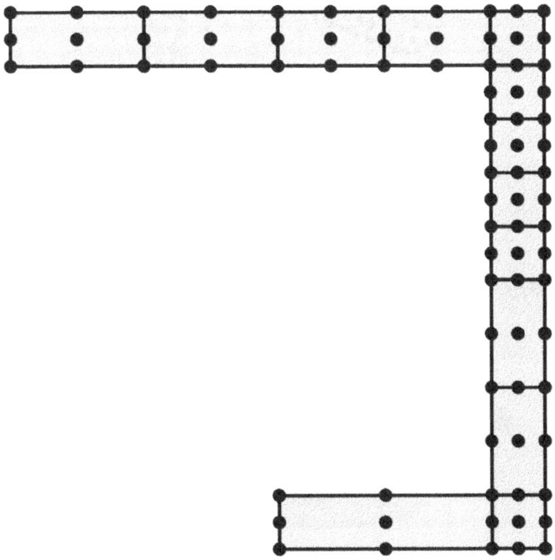

Figure 9.38 L9 mesh for the C-shaped beam clamped at the bottom flanges, 13 L9

Table 9.17 Displacement of the loading point of the C-shaped beam clamped at the bottom flanges (Carrera and Petrolo 2011a)

	DOFs	$u_z \times 10^8$ (m)
Solid	84 600	−3.759
13 L9	7533	−3.662

This table shows the transverse displacement of the C-shaped beam clamped at the bottom flanges; LE models are compared with solid ones.

loading point transverse displacement is presented in Table 9.17. Figures 9.39 and 9.40 show the 2D and 3D deformed configurations. The following considerations hold:

1. The results match perfectly those from solid models in all the cases considered.
2. An analysis of the rectangular cross-section beam highlights the capabilities of LE models in dealing with partially constrained cross-section beams.
3. Constraints can be arbitrarily distributed in the 3D directions, as can be seen from the analysis of the C-shaped beam.
4. The arch beam shows the strength of LE models in dealing with beams that have shell-like characteristics. Furthermore, local effects, due to point loadings, can be detected.

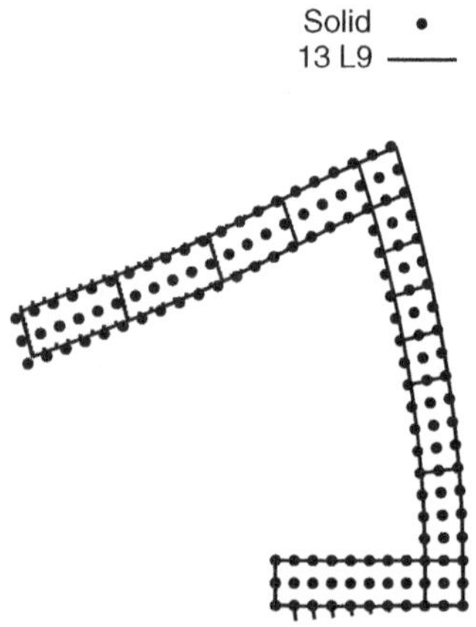

Figure 9.39 Free-tip deformed cross-section of the C-shaped beam clamped at the bottom flanges

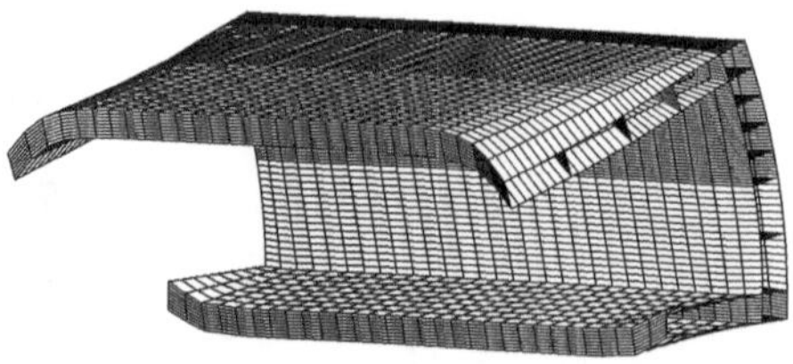

Figure 9.40 Deformed 3D configuration of the C-shaped beam clamped at the bottom flanges

9.6 The Component-Wise Approach for Aerospace and Civil Engineering Applications

The enhanced capabilities of LE models are of particular interest for the analysis of multi-component structures (MCS). Typical examples of MCS are the reinforced shells that are used for aeronautical applications or fibre reinforced composite plates (see Figures 9.41 and 9.42). These structures are composed of multiple components, which can have quite different geometrical characteristics. In a wing structure, for instance, ribs can be modelled as 2D structures, while stiffeners or spars can be modeled as 1D or 3D structures. This implies that an efficient FE modelling of MCS often requires the coupling of different elements – beams/shells/solids – in order to build sufficiently accurate models with a reasonable number of DOFs.

LE models can be exploited to model separately each component of a structure. The resulting approach is denoted as component-wise (CW), since LE models are used to model the unknown variables of each structural component. Figure 9.41 shows the CW approach for a four-stringer wing box, whose components are modelled using LE cross-sectional elements. Each component is considered with its own geometrical and material characteristics. The CW approach does not require coupling techniques, as the FE matrices of each element are formally the same. The approach leads to efficient FE models because the model capabilities can be tuned either by choosing which component requires a more detailed model, or by setting the

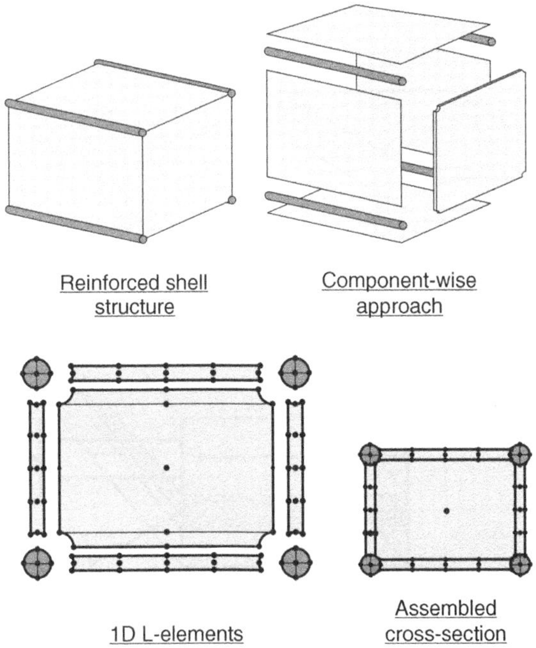

Figure 9.41 Component-wise approach for a reinforced shell structure

Table 9.18 Examples of MCS and modelling strategies

Structure	Components	Typical FEs	Present CW FEs
	Ribs	Shells	1D LEs
Aircraft wing	Spars	Beams/Solids	1D LEs
	Stiffeners	Beams/Solids	1D LEs
	Layers	Shells	1D LEs
Fibre-reinforced composites	Matrix	Solids	1D LEs
	Fibers	Beams/Solids	1D LEs

This table shows typical examples of MCS and their FE modelling.

order of the structural model that has to be used. Table 9.18 presents a brief overview of MCS and their modelling strategies. The CW approach can also be considered as a multiscale approach. Figure 9.42 shows a typical CW strategy for a composite plate; 1D LE models can be simultaneously adopted to model layers (macroscale), matrix and fibres (microscale). Figure 9.43 shows possible CW strategies. This methodology can be very powerful when, for instance, detailed stress fields are required in a specific portion of the structures. Composite structures are not considered in this book, while more details on the use of the CW approach for composites can be found in Carrera *et al.* (2012a) and Carrera *et al.* (2013b). Some of the ideas described below will be reconsidered in Chapter 14.

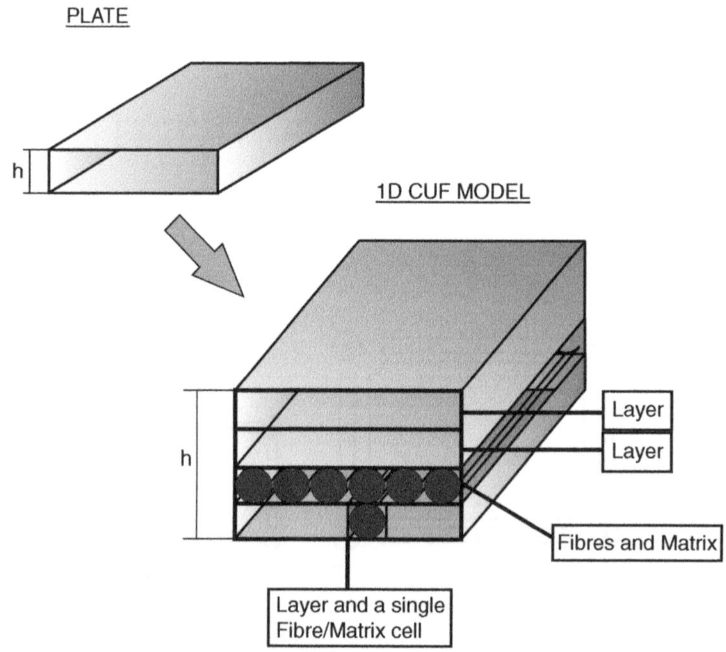

Figure 9.42 Component-wise approach for a fibre-reinforced plate

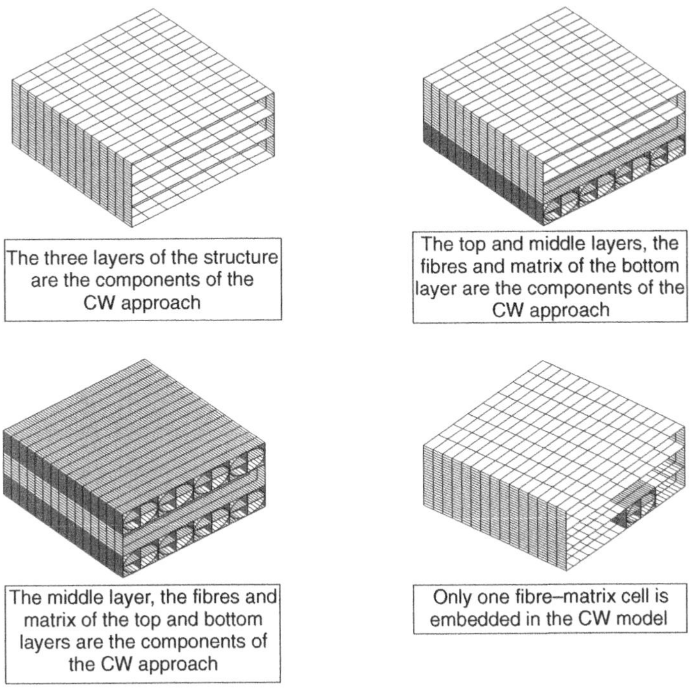
The three layers of the structure are the components of the CW approach
The top and middle layers, the fibres and matrix of the bottom layer are the components of the CW approach
The middle layer, the fibres and matrix of the top and bottom layers are the components of the CW approach
Only one fibre–matrix cell is embedded in the CW model

Figure 9.43 Different CW models for a fibre-reinforced plate

9.6.1 CW Approach for Aeronautical Structures

Primary aircraft structures are mainly composed of reinforced thin shells (Bruhn, 1973), and are also referred to as *semimonocoque* constructions. These reinforced structures are made by assembling three main components: skins (or panels), longitudinal stiffening members (including spar caps) and transversal stiffeners (ribs). The proper detection of stress/strain fields in these structural components is of prime interest to structural analysts; the CW approach can be adopted for this purpose as it offers considerable advantages compared with analytical and classical FE approaches.

The static analysis of a simple spar is considered here (see Figure 9.44). Stringers were taken to be rectangular for the sake of convenience, but their shape does not affect the validity of the proposed analysis. More complex configurations could easily be considered. A CW approach based on LE models is used to analyse the spar and the results are compared with those from 1D TE, classical beam theories and solid elements. Analytical results, based on the simplifying assumptions of semimonocoque models, are provided. According to Bruhn (1973), the internal loads in a statically determinate reinforced-shell structure can be found through the use of static equilibrium equations. Additional equations are necessary in a statically indeterminate structure. These additional equations are obtained from the compatibility conditions, by means of the PVD. This approach is hereafter referred to as the PS (Pure Semimonocoque) model. If EBBT is applied to the idealized semimonocoque assumptions, it is possible to reduce redundancy in statically indeterminate structures. This method is hereafter referred to as the BS (Beam Semimonocoque) model.

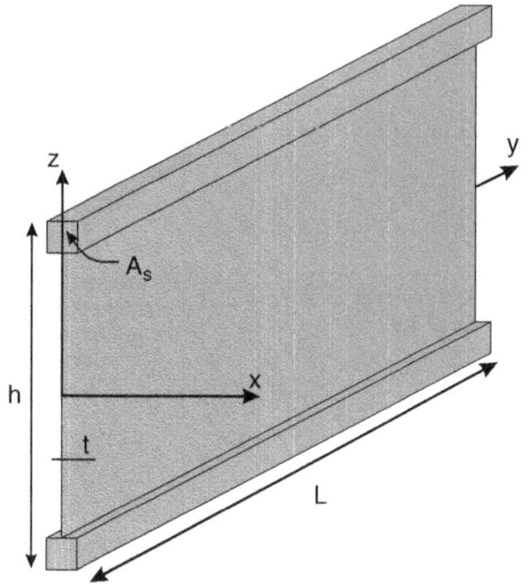

Figure 9.44 Two-stringer spar

The geometrical data are: axial length, $L = 3\,\text{m}$; cross-section height, $h = 1\,\text{m}$; area of the spar caps, $A_s = 0.9 \times 10^{-3}\,\text{m}^2$); web thickness, $t = 1 \times 10^{-3}\,\text{m}$. The whole structure is made of an aluminium alloy material. The material data are: Young's modulus, $E = 75\,\text{GPa}$; Poisson's ratio, $\nu = 0.33$. The beam is clamped at $y = 0$ and a point load, $P_z = -1 \times 10^4$ (N), is applied at $[0, L, 0]$. The transverse displacement (u_z) at the loading point is given in Table 9.19 and the CW results are given in the last two rows. These models were obtained using two different L9

Table 9.19 Transverse displacement (u_z) at the loading point, axial load in the upper stringer (P, at $y = 0$) and mean shear flow on the sheet panel (q, at $y = L/2$), two-stringer spar (Carrera *et al.* 2013c)

	$u_z \times 10^3$ (m)	$P \times 10^{-4}$ (N)	$q \times 10^{-4}$ (N/m)	DOFs
		MSC/Nastran		
Solid	−3.815	2.617	−1.036	76 050
		Analytical methods		
BS	−2.671	3.192	−1.064	—
PS	−3.059	3.192	−1.064	—
		Classical beam theories		
EBBT	−1.827	1.993	−0.274	93
TBT	−2.117	1.993	−0.274	155
		TE		
$N = 3$	−2.514	2.434	−0.665	930
$N = 5$	−2.629	2.350	−0.561	1953
		CW		
4 L9	−3.639	2.833	−1.034	2883
8 L9	−3.639	2.739	−1.035	4743

This table compares results from different models related to a two-stringer spar.

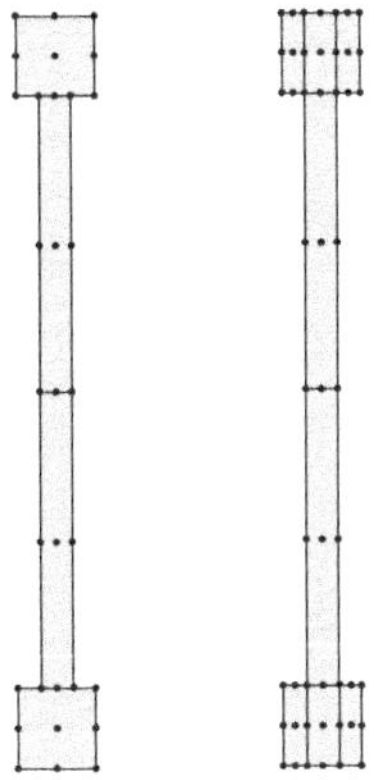

Figure 9.45 L9 distributions of a two-stringer spar

cross-section distributions, as shown in Figure 9.45. The number of DOFs of each model are given in the final column of the table. The analytical results related to BS and PS are evaluated as follows:

$$u_{z_{BS}} = \frac{P_z L^3}{3EI}, \qquad u_{z_{PS}} = \frac{P_z L^3}{3EI} + \frac{P_z L}{AG} \tag{9.25}$$

where I is the cross-section moment of inertia, G is the shear modulus and A is the cross-section area. The stress fields are evaluated in terms of axial loads in the stringers and shear flows in the webs. Table 9.19 also gives the axial load in the upper stringer (P) at $y = 0$, and the mean shear flow in the panel (q) at $y = L/2$. P and q were evaluated for BS and PS models as

$$P = \frac{P_z L}{\bar{h}}, \qquad q = -\frac{P_z}{\bar{h}} \tag{9.26}$$

where $\bar{h}$ is the distance between the centres of the two stringers. The transverse shear stress distribution versus the z-axis is presented in Figure 9.46.

The free vibration analysis of a longeron with three longitudinal stiffeners is considered next. The geometry of the structure is shown in Figure 9.47. The spar is clamped at $y = 0$. The geometrical characteristics are: axial length, $L = 3$ m; cross-sectional height, $h = 1$ m; area of the stringers, $A_s = 1.6 \times 10^{-3}$ m^2; thickness of the panels, $t = 2 \times 10^{-3}$ m; distance between the intermediate stringer and the xy-plane, $b = 0.18$ m. The whole structure is made of the same isotropic material that was adopted previously. The first 15 natural frequencies are given in Table 9.20, together with the number of DOFs for each model. The CW model was obtained by discretizing the cross-section with 5 L9 elements, one for each spar component (stringers and webs). The consistent correspondence between the 4 L9 CW model and the solid model has been further investigated by means of the Modal Assurance Criterion (MAC),

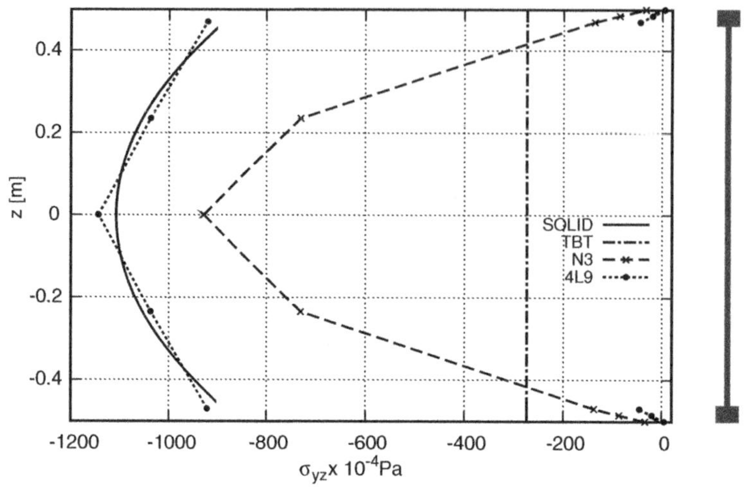

Figure 9.46 Shear stress along the two-stringer spar at $y = L/2, x = 0$

whose graphic representation is shown in Figure 9.48. The MAC is defined as a scalar that represents the degree of consistency (linearity) between two modal vectors (Allemang and Brown 1982):

$$\text{MAC}_{ij} = \frac{|\{\phi_{A_i}\}^T\{\phi_{B_j}\}|^2}{\{\phi_{A_i}\}^T\{\phi_{A_i}\}\{\phi_{B_j}\}\{\phi_{B_j}\}^T} \tag{9.27}$$

where $\{\phi_{A_i}\}$ is the ith eigenvector of model A, while $\{\phi_{B_j}\}$ is the jth eigenvector of model B. The MAC takes on values from zero (no consistent correspondence) to one (representing a consistent correspondence).

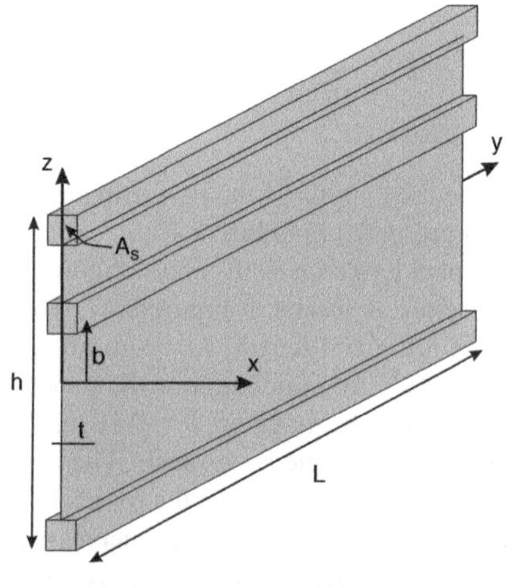

Figure 9.47 Three-stringer spar

Table 9.20 First 15 natural frequencies (Hz) of the three-stringer spar (Carrera *et al.* 2013a)

	EBBT	TBT	$N = 1$	$N = 2$	$N = 3$	$N = 4$	5 L9	Solid
DOFs	93	155	279	558	930	1395	3813	62 580
Mode								
1	3.24^b	3.24^b	3.24^b	3.43^b	3.35^b	3.31^b	3.46^t	3.15^b
2	20.29^b	20.28^b	20.28^b	16.70^t	16.34^t	16.13^t	3.52^b	3.55^t
3	56.81^b	56.74^b	56.74^b	21.39^b	20.97^b	20.75^b	3.76^b	3.82^b
4	111.36^b	108.81^b	108.81^b	55.25^t	52.90^t	51.70^t	14.27^s	13.30^s
5	117.60^b	111.11^b	111.11^b	60.11^b	59.23^b	58.24^b	16.73^s	15.06^s
6	184.30^b	183.57^b	183.57^b	108.19^t	100.81^t	97.87^t	17.67^s	16.33^s
7	275.94^b	274.23^b	269.29^t	109.44^b	105.55^b	102.26^b	21.17^s	19.81^s
8	386.89^b	383.36^b	274.23^b	117.79^b	116.61^b	113.20^b	21.71^t	21.49^t
9	439.21^e	439.20^e	383.36^b	181.03^t	165.23^t	119.39^s	22.95^b	22.81^b
10	517.91^b	455.17^b	439.20^e	194.59^b	183.16^s	161.07^t	25.11^s	24.07^s
11	622.84^b	511.36^b	455.17^b	276.03^t	197.98^b	176.65^s	25.73^s	24.63^s
12	669.05^b	658.20^b	511.36^b	290.25^b	229.97^s	189.01^b	31.21^s	29.69^s
13	830.95^b	817.28^b	658.20^b	325.69^s	248.76^t	243.58^t	37.92^s	36.24^s
14	1104.56^b	972.68^b	807.88^t	393.92^t	290.54^b	258.64^s	45.79^s	43.88^s
15	1317.62^e	1055.78^b	817.28^b	406.78^b	302.06^s	281.59^b	54.86^s	51.64^s

This table shows the first 15 natural frequencies of the three-stringer spar via different models (*b*: bending mode; *t*: torsional mode; *s*: shell-like mode; *e*: extensional mode).

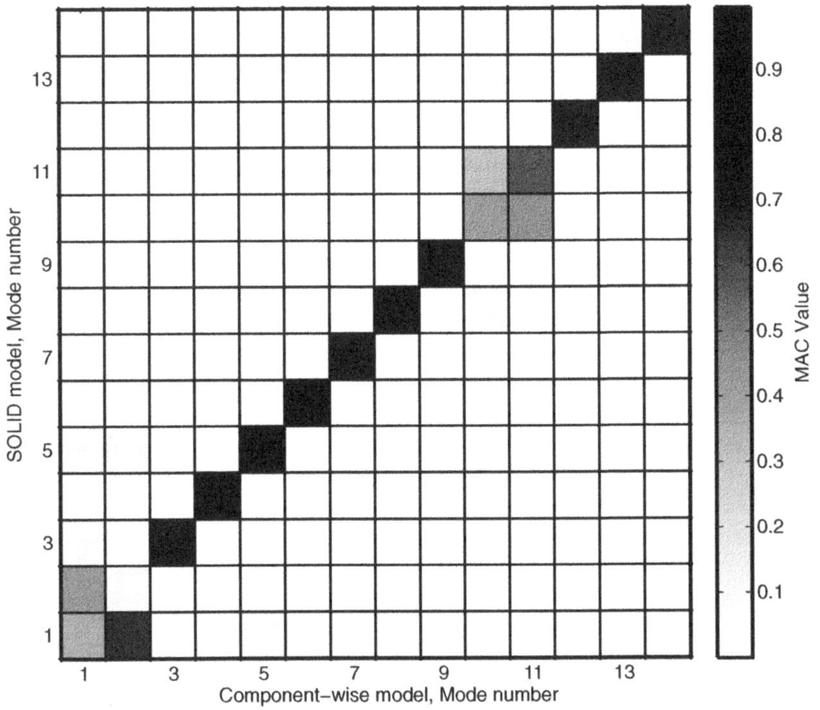

Figure 9.48 Solid vs CW MAC values for the three-stringer spar

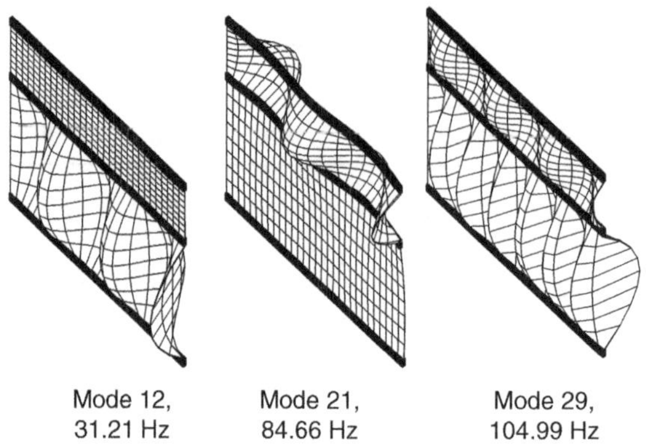

Figure 9.49 Three-stringer spar modal shapes

There is good correspondence between the two models up to the 14th mode. Further refinements of the LE model (i.e. adopting more L-elements to discretize the cross-section of the longeron) would improve this result. Figure 9.49 shows some local modes computed by means of the CW model.

The modal analysis of a complete aircraft wing is now carried out. The cross-section of the wing is shown in Figure 9.50. An NACA 2415 aerofoil was used and two spar webs and four spar caps were added. The aerofoil has a chord (c) equal to 1 m. The span (L) is equal to 6 m. The thickness of the panels is 3×10^{-3} m, whereas the thickness of the spar webs is 5×10^{-3} m. The whole structure is made of the same isotropic material as in the previous cases. The wing is clamped at the root. Two different configurations were considered for the present wing structure. Configuration A has no transverse stiffening members. In Configuration B, the wing is divided into three equal bays, each separated by a rib with a thickness of 6×10^{-3} m. Table 9.21 presents the main modal frequencies of both wing configurations. The results obtained with the CUF models are compared with those from classical beam theories and with those from solid models. The frequencies of the first two shell-like modes are reported in the last two rows.

In order to deal easily with complex structures, such as the one considered in this section, the CW models were inserted into commercial software and postprocessing of the CW model of the wing was performed with MSC/Patran. Two shell-like modes, evaluated by means of the CW model, are shown in Figure 9.51 for Configuration A.

The modal analysis of a simplified aircraft model is considered next. The geometry of the structure is shown in Figure 9.52, where $a = 0.5$ m. The structure has a constant thickness of $0.2 \times a$. The considered material is aluminium. Results from both TE and LE models are

Figure 9.50 Wing cross-section

Table 9.21 Global and local modal frequencies of the complete aircraft wing (Carrera *et al.* 2013a)

	EBBT	TBT	$N = 1$	$N = 2$	$N = 3$	CW	Solid
				Configuration A			
DOFs	93	155	279	558	930	21 312	186 921
				Global modes			
I Bending^x	4.22	4.22	4.22	4.29	4.26	4.23	4.21
I Bending^z	22.10	21.82	21.82	21.95	21.87	21.76	21.69
II Bending^x	26.44	26.36	26.36	26.66	26.25	25.15	24.78
I Torsional	—	—	132.93	50.27	48.46	31.14	29.18
III Bending^x	73.91	73.35	73.35	73.99	71.64	59.26	56.12
II Bending^z	134.66	124.68	124.68	124.99	122.77	118.39	118.00
				Local Modes			
I Shell-like	—	—	—	—	—	86.36	75.13
II Shell-like	—	—	—	—	—	88.94	73.85
				Configuration B			
DOFs	84	140	252	504	840	23 976	171 321
				Global modes			
I Bending^x	4.12	4.12	4.12	4.19	4.17	4.14	4.12
I Bending^z	21.56	21.30	21.30	21.50	21.42	21.28	21.22
II Bending^x	25.71	25.63	25.63	26.00	25.61	25.00	24.92
I Torsional	—	—	131.24	49.57	47.48	39.45	39.22
III Bending^x	71.44	70.90	70.90	71.80	69.49	64.84	63.88
II Bending^z	131.11	121.49	121.49	122.23	120.06	115.76	115.40
						Local modes	
I Shell-like	—	—	—	—	—	85.61	75.01
II Shell-like	—	—	—	—	—	91.54	78.61

This table shows the natural frequencies of the complete wing model via different models (Bending^x: bending mode along the x-axis).

provided. A non-uniform cross-section beam is considered in the TE model of the aircraft, as shown in Figure 9.52. The CW model is obtained by exploiting multiple beam elements, each discretized with an L9 cross-section L-element, as shown in Figure 9.53. The first 10 natural frequencies for fourth- and fifth-order TE models are in the first two columns in Table 9.22. A comparison with the results from the MSC/Nastran commercial code, obtained

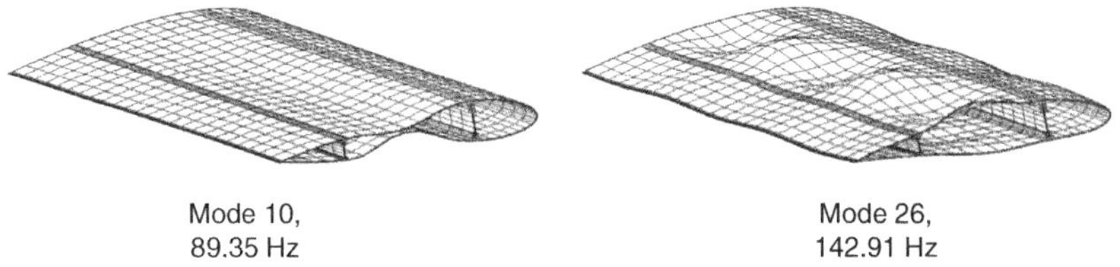

Figure 9.51 Wing model modal shapes

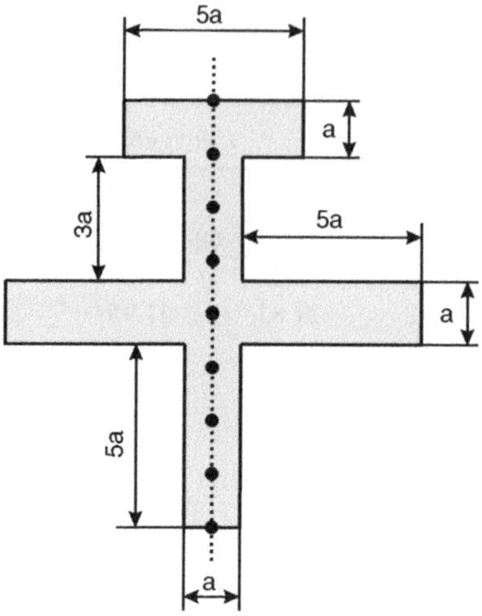

Figure 9.52 Aircraft geometry and TE modelling

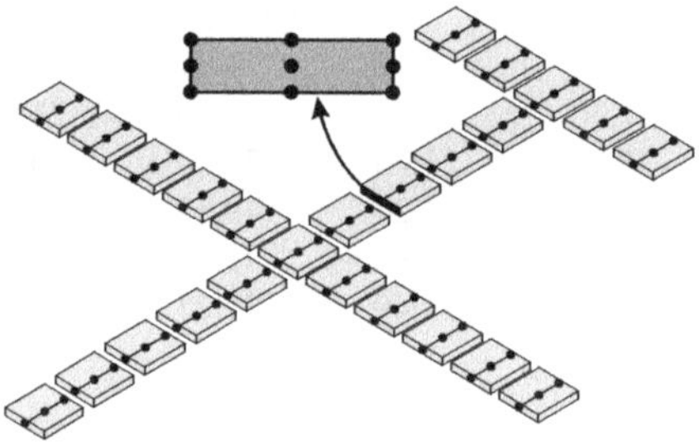

Figure 9.53 Aircraft LE modelling

Table 9.22 First 10 natural frequencies (Hz) of the simplified aircraft model (Carrera *et al.* 2012b)

	$N = 4$	$N = 5$	Shell	CW
DOFs	2880	4032	6120	1845
I	11.427	11.342	10.804	10.944
II	20.824	19.499	17.264	18.201
III	21.591	21.404	20.053	20.208
IV	51.448	51.137	50.372	50.398
V	73.244	52.989	51.162	52.122
VI	62.246	59.476	51.307	52.389
VII	81.915	74.980	65.552	66.962
VIII	74.791	74.612	69.942	70.633
IX	102.908	101.304	75.774	77.844
X	88.310	87.790	87.436	89.572

This table shows the natural frequencies of the complete aircraft model via different models.

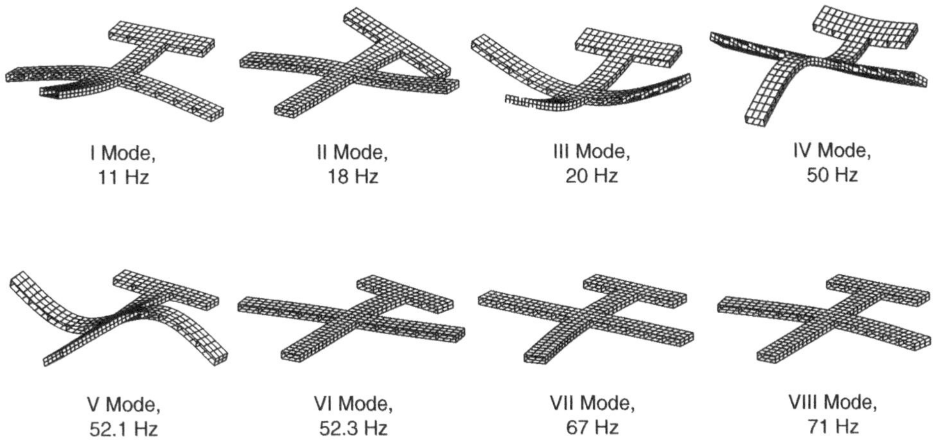

Figure 9.54 Complete aircraft modal shapes from the CW approach

using 2D elements, is given in the third column. The results of the CW model are given in the last column. The first eight natural modes, evaluated by means of the CW model, are shown in Figure 9.54. The results that were obtained suggest the following:

1. Refined beam theories, especially LE ones, can reproduce solid model results.
2. The number of DOFs of the present models is significantly lower than that of solid models.
3. Both MSC/NASTRAN and higher-order CUF models, unlike analytical theories based on idealized stiffened-shell structures and classical 1D models, highlight that the axial stress component (σ_{yy}) is not linear along z and that the shear stress component (σ_{yz}) is not constant along the sheet panel.
4. Classical beam theories and the linear ($N = 1$) TE model are able to detect bending and extensional modes correctly. No torsional modes are detected.
5. A TE model higher than first order is necessary to detect the torsional and shell-like modes. However, very high expansion orders are needed to predict correctly the frequencies of these modal shapes.
6. The CW model matches the solid FE solution and offers a significant reduction in computational costs. It should be noted that CW models can detect typical shell-like modal shapes when the 1D CUF is adopted.
7. The 1D CUF can deal easily with very complex structures, such as wings or aircraft.

9.6.2 CW Approach for Civil Engineering

The free vibration analysis of civil structures, through CW models, is discussed in this section. Two structural configurations are considered, as shown in Figure 9.55. Configuration A is a one-level structure composed of four square columns made of steel (elastic modulus $E = 210$ GPa, density $\rho = 7.5 \times 10^3$ kg/m^3, Poisson ratio $\nu = 0.28$) and the floor is shown in Figure 9.56. The floor is made of a material whose properties are one-fifth of those of the steel alloy. Configuration B is a three-level construction with four columns and three floors. The CW

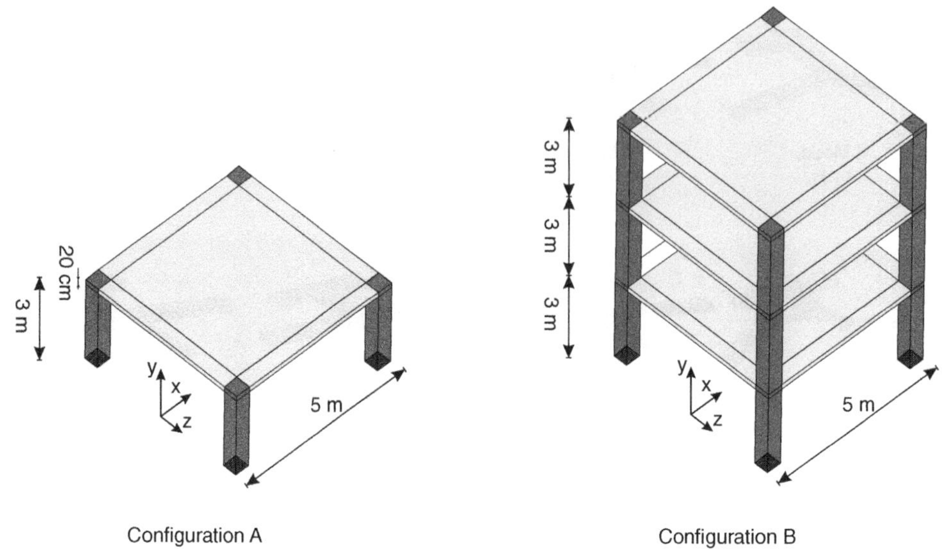

Figure 9.55 Civil structures models

models were obtained from a combination of L9 elements above the cross-sections. The results of the CW analysis have been compared with those from solid analysis by means of MSC/Nastran. Table 9.23 presents the natural frequencies, together with the number of DOFs, for both the LE and solid models. Figures 9.57 and 9.58 show some modal shapes evaluated by means of the CW models. It is important to note that civil structures are obviously multiple component structures and as such can be analysed by means of a 1D CW formulation. In fact, both global and local modes that involve columns and floors are correctly detected by LE models.

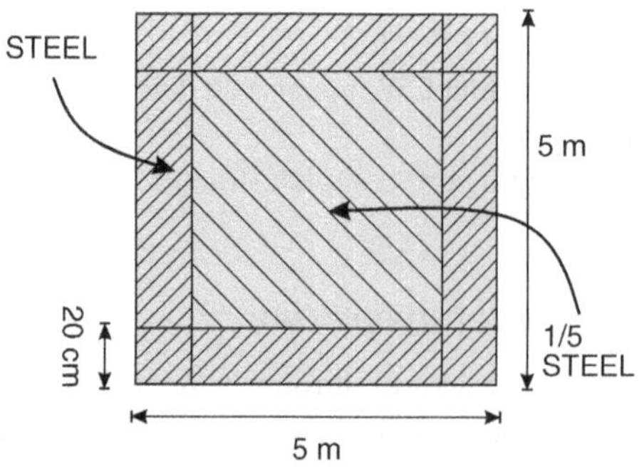

Figure 9.56 Floor model for the civil structures

Table 9.23 Natural frequencies (Hz) of the civil structures (Carrera *et al.* 2012b)

	Configuration A		Configuration B	
	CW	Solid	CW	Solid
DOFs	3396	181 875	6300	78 975
Mode 1	9.43^{b}	8.79^{b}	3.63^{b}	3.07^{b}
Mode 2	9.43^{b}	8.79^{b}	3.63^{b}	3.07^{b}
Mode 3	13.86^{t}	12.44^{t}	5.32^{t}	4.22^{t}
Mode 4	23.07^{f}	21.67^{f}	10.64^{b}	9.69^{b}
Mode 5	36.80^{f}	36.66^{f}	10.64^{b}	9.69^{b}
Mode 6	36.80^{f}	36.66^{f}	15.60^{t}	13.36^{t}
Mode 7	40.59^{f}	37.59^{f}	16.45^{b}	16.42^{b}
Mode 8	72.41^{f}	77.07^{f}	16.45^{b}	16.42^{b}

This table shows the natural frequencies of the civil structures via different models (*b*: bending mode; *t*: torsional mode; *f*: floor mode).

Figure 9.57 Modal shapes of a civil structure (Configuration A)

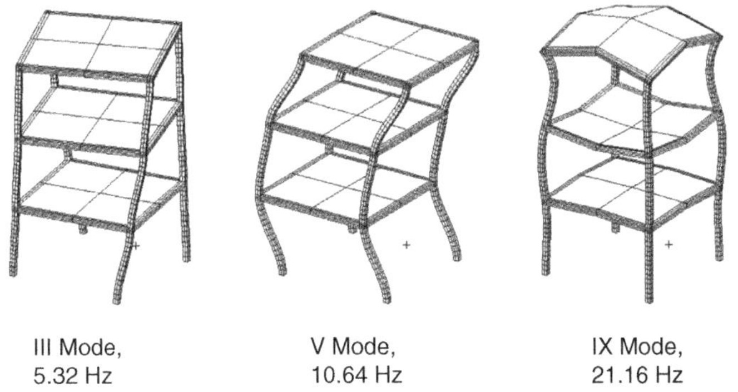

Figure 9.58 Modal shapes of a civil structure (Configuration B)

References

Allemang RJ and Brown DL 1982 A correlation coefficient for modal vector analysis. In: *Proceedings of the International Modal Analysis Conference*, Orlando, Florida, USA, pp. 110–116.

Bathe KJ 1996 *Finite Element Procedures*. Prentice Hall.

Bruhn EF 1973 *Analysis and Design of Flight Vehicle Structures*. Jacobs.

Carrera E, Maiarù M and Petrolo M 2012a Component-wise analysis of laminated anisotropic composites. *International Journal of Solids and Structures* **49**(13). DOI: 10.1016/j.ijsolstr.2012.03.025.

Carrera E, Maiarù M, Petrolo M and Giunta G 2013b A refined 1D element for the structural analysis of single and multiple fiber/matrix cells. *Composite Structures* **96**. DOI: 10.1016/j.compstruct.2012.09.012.

Carrera E, Pagani A and Petrolo M 2013a Component-wise method applied to vibration of wing structures. *Journal of Applied Mechanics*, **80** DOI: 10.1115/1.4007849.

Carrera E, Pagani A and Petrolo M 2013c Classical, refined and component-wise analysis of reinforced-shell wing structures. *AIAA Journal* **51**, 1255–1268.

Carrera E, Pagani A, Petrolo M and Zappino E 2012b A component-wise approach in structural analysis. *Computer Methods for Engineering Sciences*, Ch. 4. Saxe-Coburg.

Carrera E and Petrolo M 2011a Refined beam elements with only displacement variables and plate/shell capabilities. *Meccanica* **47**(3), 537–556.

Carrera E and Petrolo M 2011b On the effectiveness of higher-order terms in refined beam theories. *Journal of Applied Mechanics* **78**. DOI: 10.1115/1.4002207.

Oñate E 2009 *Structural Analysis with the Finite Element Method: Linear Statics*. Volume 1. *Basis and Solids*. Springer.

Zienkiewicz OC, Taylor RL and Zhu JZ 2005 *The Finite Element Method: Its Basis and Fundamentals*. Sixth Edition. Elsevier.

10

Two-Dimensional Plate Models with *N*th-Order Displacement Field, the Taylor Expansion Class

Plates are 2D structures in which one dimension, in general the thickness h, is at least one order of magnitude lower than the in-plane dimensions a and b (Figure 10.1). This permits the reduction of the 3D problem to a 2D one. Such a reduction can be seen as a transformation of the problem defined at each point $Q_V(x, y, z)$ of the 3D continuum plate into a problem defined at each point $Q_\Omega(x, y)$ of the plate surface Ω. The 2D modelling of plates is a classical problem in the TOS. The elimination of the z coordinate can be performed through several methodologies that lead to a significant number of approaches and techniques. For instance, the unknown variables can be axiomatically assumed along z. This means that, for a given point $Q_\Omega(x, y)$ in the plane, the distribution of the unknowns along the thickness will be given by a polynomial expansion in z. The main feature of the unified formulation is the possibility of arbitrarily choosing the kind of expansion and the number of terms.

This chapter presents 2D flat elements based on Taylor expansions of the displacement variables. First of all, classical models (by Kirchhoff and Reissner–Mindlin) will be briefly described together with the more general complete linear expansion case. Higher-order models will then be presented and the unified formulation introduced. The PVD will be employed to derive governing equations and the FE formulation. Table 10.1 shows the FN for 2D models that is introduced in this chapter. The shear locking phenomenon and its correction will be discussed and numerical results will be provided.

10.1 Classical Models and the Complete Linear Expansion

The first mathematical formulations of plates under stretching and bending were provided by Kirchhoff (1850) and Reissner and Mindlin (Reissner, 1945; Mindlin 1951). Their models represent the classical plate theories. They are reference models for analysing 2D flat structures

Finite Element Analysis of Structures Through Unified Formulation, First Edition.
Erasmo Carrera, Maria Cinefra, Marco Petrolo and Enrico Zappino.

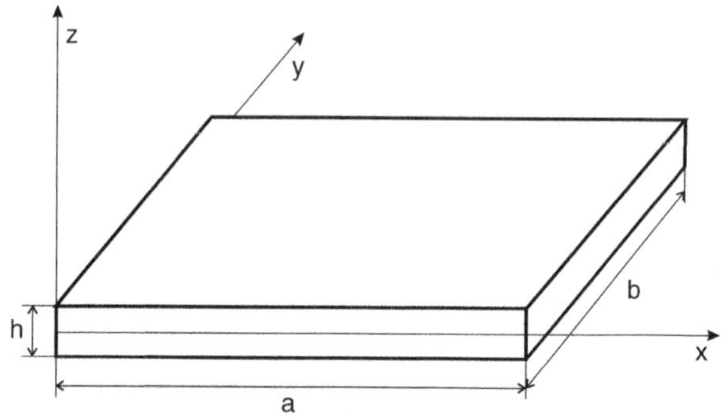

Figure 10.1 Plate geometry and reference system

under stretching and bending loads or for computing natural modes. These theories will be briefly described in this section together with the case of complete linear expansion.

10.1.1 Classical Plate Theory

The Kirchhoff plate model, hereafter referred to as CPT (Classical Plate Theory), was derived from the following a priori assumptions:

1. Straight lines perpendicular to the mid-surface (i.e. transverse normals) before deformation remain straight after deformation.
2. The transverse normals do not experience elongation (i.e. they are inextensible).
3. The transverse normals rotate such that they remain perpendicular to the mid-surface after deformation.

According to the first hypothesis, the in-plane displacements u_x and u_y are linear versus the thickness coordinate z

$$\begin{aligned} u_x(x,y,z) &= u_{x_0}(x,y) + \phi_x(x,y)z \\ u_y(x,y,z) &= u_{y_0}(x,y) + \phi_y(x,y)z \end{aligned} \tag{10.1}$$

where ϕ_x and ϕ_y are the rotations of a transverse normal about the y- and x-axes, respectively. The notation where ϕ_x denotes the rotation of a transverse normal about the y-axis and ϕ_y denotes the rotation about the x-axis may be a little confusing, and they do not follow the right-hand rule. However, the notation has been used extensively in the literature, and we will not depart from it. If (β_x, β_y) denote the rotations about the x- and y-axes that respectively, follow the right-hand rule, then

$$\beta_x = \phi_y, \quad \beta_y = -\phi_x \tag{10.2}$$

On the basis of the second hypothesis, the transverse displacement u_z is independent of the transverse (or thickness) coordinate and the transverse normal strain ε_{zz} is disregarded:

$$u_z(x,y,z) = u_{z_0}(x,y) \quad \Rightarrow \quad \varepsilon_{zz} = \frac{\partial u_z}{\partial z} = 0 \tag{10.3}$$

Table 10.1 A schematic description of the CUF and the related FN of the stiffness matrix for 2D models

---

Equilibrium equations in Strong Form $\longrightarrow$ $\delta L_i = \int_V \delta \boldsymbol{u} \boldsymbol{k} \boldsymbol{u} dV + \int_S \ldots dS$

$$\underbrace{\begin{bmatrix} k_{xx} & k_{xy} & k_{xz} \\ k_{yx} & k_{yy} & k_{yz} \\ k_{zx} & k_{zy} & k_{zz} \end{bmatrix}}_{\boldsymbol{k}} \underbrace{\begin{Bmatrix} u_x \\ u_y \\ u_z \end{Bmatrix}}_{\boldsymbol{u}} = \underbrace{\begin{Bmatrix} p_x \\ p_y \\ p_z \end{Bmatrix}}_{\boldsymbol{p}}$$

$$k_{xx} = -(\lambda + 2G)\, \partial_{xx} - G\, \partial_{zz} - G\, \partial_{yy};$$
$$k_{xy} = -\lambda\, \partial_{xy} - G\, \partial_{yx};$$
$$k_{xz} = \ldots$$
$$\lambda = (E\nu)/[(1+\nu)(1-2\nu)]; \quad G = E/[2(1+\nu)]$$

$\boldsymbol{u} = \boldsymbol{u}(x, y, z)$

$\delta \boldsymbol{u} = \delta \boldsymbol{u}(x, y, z)$

The diagonal (e.g. k_{xx}) and the non-diagonal (e.g. k_{xy}) terms can be obtained through proper index permutations.

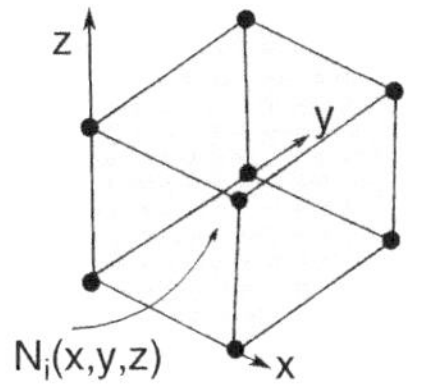

$\boldsymbol{u} = N_i(x, y, z)\boldsymbol{u}_i$

$\delta \boldsymbol{u} = N_j(x, y, z)\delta \boldsymbol{u}_j$

3D FEM Formulation $\longrightarrow$ $\delta L_i = \delta \boldsymbol{u}_j \boldsymbol{k}^{ij} \boldsymbol{u}_i$

$$k^{ij}_{xx} = (\lambda + 2G) \int_V N_{j,x} N_{i,x} dV + G \int_V N_{j,z} N_{i,z} dV + G \int_V N_{j,y} N_{i,y} dV;$$

$$k^{ij}_{xy} = \lambda \int_V N_{j,y} N_{i,x} dV + G \int_V N_{j,x} N_{i,y} dV$$

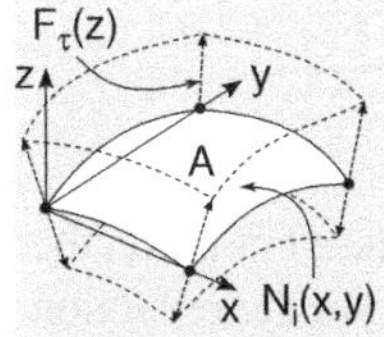

$\boldsymbol{u} = N_i(x, y)F_\tau(z)\boldsymbol{u}_{\tau i}$

$\delta \boldsymbol{u} = N_j(x, y)F_s(z)\delta \boldsymbol{u}_{sj}$

2D FEM Formulation $\longrightarrow$ $\delta L_i = \delta \boldsymbol{u}_{sj} \boldsymbol{k}^{\tau sij} \boldsymbol{u}_{\tau i}$

$$k^{\tau sij}_{xx} = (\lambda + 2G) \int_\Omega N_{i,x} N_{j,x} d\Omega \int_h F_\tau F_s dz$$
$$+ G \int_\Omega N_i N_j d\Omega \int_h F_{\tau,z} F_{s,z} dz + G \int_V N_{i,y} N_{j,y} d\Omega \int_h F_\tau F_s dz;$$

$$k^{\tau sij}_{xy} = \lambda \int_\Omega N_{i,y} N_{j,y} d\Omega \int_h F_\tau F_s dz + G \int_\Omega N_{i,x} N_{j,y} d\Omega \int_h F_\tau F_s dz$$

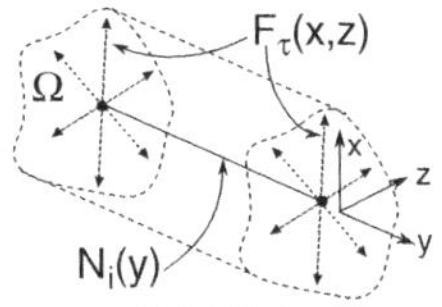

$\boldsymbol{u} = N_i(y)F_\tau(x, z)\boldsymbol{u}_{\tau i}$

$\delta \boldsymbol{u} = N_j(y)F_s(x, z)\delta \boldsymbol{u}_{sj}$

1D FEM Formulation $\longrightarrow$ $\delta L_i = \delta \boldsymbol{u}_{sj} \boldsymbol{k}^{\tau sij} \boldsymbol{u}_{\tau i}$

$$k^{\tau sij}_{xx} = (\lambda + 2G) \int_l N_i N_j dy \int_A F_{\tau,x} F_{s,x} dA$$
$$+ G \int_l N_i N_j dy \int_A F_{\tau,z} F_{s,z} dA + G \int_l N_{i,y} N_{j,y} dy \int_A F_\tau F_s dA;$$

$$k^{\tau sij}_{xy} = \lambda \int_l N_{i,y} N_j dy \int_A F_\tau F_{s,x} dA + G \int_l N_i N_{j,y} dy \int_A F_{\tau,x} F_s dA$$

CUF leads to the automatic implementation of any theory of structures through 4 loops (i.e. 4 indexes):

- $\boldsymbol{\tau}$ and $\boldsymbol{s}$ deal with the functions that approximate the displacement field and its virtual variation along the plate/shell thickness ($F_\tau(z), F_s(z)$) or over the beam cross-section ($F_\tau(x, z), F_s(x, z)$);
- $\boldsymbol{i}$ and $\boldsymbol{j}$ deal with the shape functions of the FE model, (3D:$N_i(x, y, z), N_j(x, y, z)$; 2D:$N_i(x, y), N_j(x, y)$; 1D:$N_i(y), N_j(y)$).

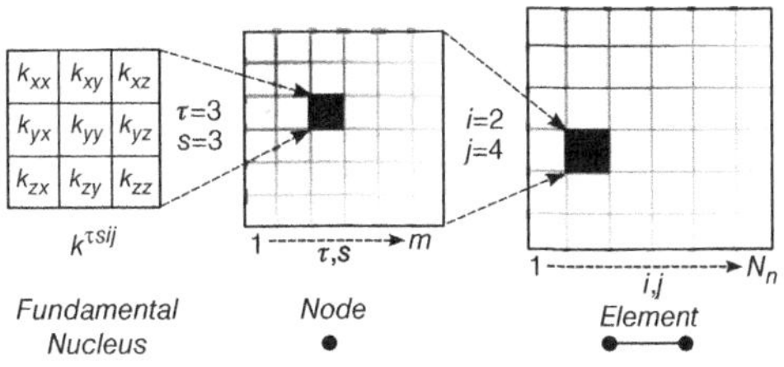

This table shows the essential features of the CUF for 2D models.

On the basis of the third hypothesis and according to the definition of shear strains, shear deformations γ_{xz} and γ_{yz} are disregarded,

$$\gamma_{xz} = \gamma_{yz} = 0 \tag{10.4}$$

Equations (10.1), (10.3) and (10.4) allow the rotation angles to be obtained as functions of the derivatives of the transverse displacement

$$\begin{cases} \gamma_{xz} = \dfrac{\partial u_z}{\partial x} + \dfrac{\partial u_x}{\partial z} = \dfrac{\partial u_{z_0}}{\partial x} + \phi_x = 0 \\ \gamma_{xz} = \dfrac{\partial u_z}{\partial y} + \dfrac{\partial u_y}{\partial z} = \dfrac{\partial u_{z_0}}{\partial y} + \phi_y = 0 \end{cases} \Rightarrow \begin{cases} \phi_x = -\dfrac{\partial u_{z_0}}{\partial x} \\ \phi_y = -\dfrac{\partial u_{z_0}}{\partial y} \end{cases} \tag{10.5}$$

The displacement field of CPT is then

$$\begin{aligned} u_x &= u_{x_0} - \frac{\partial u_{z_0}}{\partial x} z \\ u_y &= u_{y_0} - \frac{\partial u_{z_0}}{\partial y} z \\ u_z &= u_{z_0} \end{aligned} \tag{10.6}$$

where $(u_{x_0}, u_{y_0}, u_{z_0})$ are the displacements along the coordinate lines of a material point on the xy-plane. Note that the form of the displacement field in Equation (10.6) allows the reduction of the 3D problem to one of studying the deformation of the reference plane $z = 0$ (or mid-plane). Once the midplane displacements $(u_{x_0}, u_{y_0}, u_{z_0})$ are known, the displacements of any arbitrary point (u_x, u_y, u_z) in the 3D continuum can be determined using Equation (10.6). From a mathematical point of view, the CPT displacement field can be seen as a Maclaurin-like series expansion in which a zeroth-order approximation is used for the transverse component and an expansion order N equal to one is adopted for the in-plane displacements. CPT presents three unknown variables and the relations among them have been derived from kinematic considerations. Figure 10.2 shows the typical distribution of displacement components according to CPT: linear for u_x and u_y and constant for u_z. Also the physical meaning of the derivatives of transverse displacement, $u_{z,x}$ and $u_{z,y}$, is represented.

According to the kinematic hypotheses, CPT accounts for the in-plane strains only. On the basis of their definition, and of the CPT displacement field, these strains are

$$\begin{aligned} \varepsilon_{xx} &= \frac{\partial u_x}{\partial x} = \frac{\partial u_{x_0}}{\partial x} - \frac{\partial^2 u_{z_0}}{\partial x^2} z = k_x^x + k_{xx}^z z \\ \varepsilon_{yy} &= \frac{\partial u_y}{\partial y} = \frac{\partial u_{y_0}}{\partial y} - \frac{\partial^2 u_{z_0}}{\partial y^2} z = k_y^y + k_{yy}^z z \\ \gamma_{xy} &= \frac{\partial u_x}{\partial y} + \frac{\partial u_y}{\partial x} = \frac{\partial u_{x_0}}{\partial y} + \frac{\partial u_{y_0}}{\partial x} - 2\frac{\partial^2 u_{z_0}}{\partial xy} z = k_y^x + k_x^y + 2k_{xy}^z z \end{aligned} \tag{10.7}$$

k_x^x, k_y^y, k_y^x and k_x^y have the physical meaning of membrane deformation, whereas k_{xx}^z, k_{yy}^z and k_{xy}^z, being the second-order derivatives of the transverse displacement, represent the curvatures in

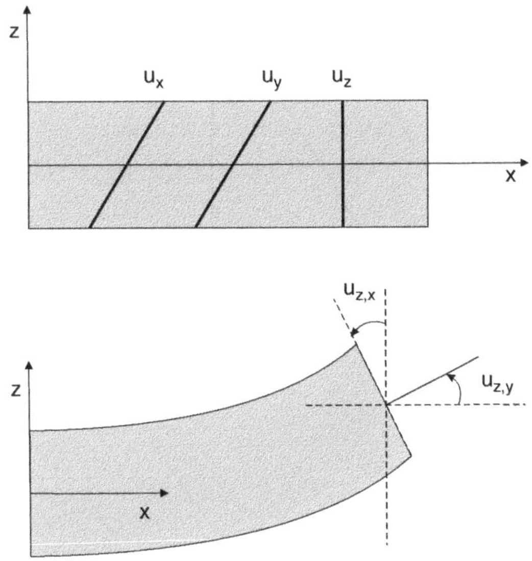

Figure 10.2 Distribution of displacements in Classical Plate Theory

the case of infinitesimal deformations and small rotations. The corresponding in-plane stresses are obtained by means of the reduced constitutive equations

$$\begin{Bmatrix} \sigma_{xx} \\ \sigma_{yy} \\ \tau_{xy} \end{Bmatrix} = \frac{E}{1-\nu^2} \begin{bmatrix} 1 & \nu & 0 \\ \nu & 1 & 0 \\ 0 & 0 & (1-\nu)/2 \end{bmatrix} \begin{Bmatrix} \varepsilon_{xx} \\ \varepsilon_{yy} \\ \gamma_{xy} \end{Bmatrix} \tag{10.8}$$

10.1.2 First-Order Shear Deformation Theory

In the Reissner–Mindlin theory, also called first-order shear deformation theory (FSDT), the third part of Kirchhoff's hypothesis is removed, therefore the transverse normals do not remain perpendicular to the mid-surface after deformation. In this way, transverse shear strains γ_{xz} and γ_{yz} are included in the theory. However, the inextensibility of the transverse normal remains, so displacement u_z is constant in the thickness direction z. The displacement field in the case of FSDT is

$$\begin{aligned} u_x(x,y,z) &= u_{x_0}(x,y) + \phi_x(x,y)z \\ u_y(x,y,z) &= u_{y_0}(x,y) + \phi_y(x,y)z \\ u_z(x,y,z) &= u_{z_0}(x,y) \end{aligned} \tag{10.9}$$

The quantities $(u_{x_0}, u_{y_0}, u_{z_0}, \phi_x, \phi_y)$ will be the unknowns. For thin plates, i.e. when the plate in-plane characteristic dimension-to-thickness ratio is of the order of 50 or more, the rotation functions ϕ_x and ϕ_y should approach the respective slopes of the transverse deflection

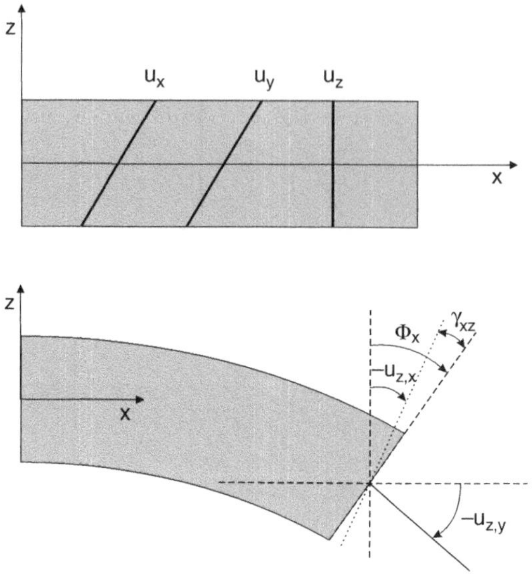

Figure 10.3 Distribution of displacements in FSDT

$-\partial u_{z_0}/\partial x$ and $-\partial u_{z_0}/\partial y$. Figure 10.3 shows the typical distribution of displacement components according to FSDT: linear for u_x and u_y and constant for u_z. Also the physical meaning of the rotations, ϕ_x and ϕ_y, is represented.

The strain components are obtained by substituting the displacement field (Equations (10.9)) in the geometrical relations. Only strain ε_{zz} is zero, therefore the non-null strains are

$$\begin{aligned}
\varepsilon_{xx} &= \frac{\partial u_x}{\partial x} = u_{x_0,x} + \phi_{x,x} z \\
\varepsilon_{yy} &= \frac{\partial u_y}{\partial y} = u_{y_0,y} + \phi_{y,y} z \\
\gamma_{xy} &= \frac{\partial u_x}{\partial y} + \frac{\partial u_y}{\partial x} = u_{x_0,y} + u_{y_0,x} + \phi_{x,y} z + \phi_{y,x} z \\
\gamma_{xz} &= \frac{\partial u_x}{\partial z} + \frac{\partial u_z}{\partial x} = \phi_x + u_{z_0,x} \\
\gamma_{xy} &= \frac{\partial u_x}{\partial y} + \frac{\partial u_y}{\partial x} = \phi_y + u_{z_0,y}
\end{aligned} \tag{10.10}$$

The constitutive relations are used to obtain the in-plane stresses and the shear stress components

$$\begin{Bmatrix} \sigma_{xx} \\ \sigma_{yy} \\ \tau_{xy} \end{Bmatrix} = \frac{E}{1-\nu^2} \begin{bmatrix} 1 & \nu & 0 \\ \nu & 1 & 0 \\ 0 & 0 & (1-\nu)/2 \end{bmatrix} \begin{Bmatrix} \varepsilon_{xx} \\ \varepsilon_{yy} \\ \gamma_{xy} \end{Bmatrix} \tag{10.11}$$

$$\tau_{xz} = \kappa G \gamma_{xz}, \qquad \tau_{yz} = \kappa G \gamma_{yz}$$

where κ is the shear correction factor. Similar to TBT, the shear predicted by FSDT should be corrected since the model yields a constant value along the thickness, whereas it is at least parabolic in order to satisfy the stress-free BCs on the unloaded top and bottom faces of the plate. In the literature, there are many methods for computing κ for the FSDT: see, for instance, Babuska *et al.* (1994) and Rössle (1999). The shear correction factor is not discussed here, but it will be shown that the adoption of higher-order models represents a general approach to avoid shear correction factors.

10.1.3 The Complete Linear Expansion Case

The complete linear expansion model involves a first-order ($N = 1$) Taylor polynomial to describe the through-the-thickness displacement field

$$\begin{array}{rl|l} u_x = & u_{x_0} & + z\, u_{x_1} \\ u_y = & u_{y_0} & + z\, u_{y_1} \\ u_z = & \underbrace{u_{z_0}}_{N=0} & \underbrace{+ z\, u_{z_1}}_{N=1} \end{array} \tag{10.12}$$

The plate model given in Equation (10.12) has six displacement variables: three constant ($N = 0$) and three linear ($N = 1$). Strain components are given by

$$\begin{aligned} \varepsilon_{xx} &= \partial u_x/\partial x = && u_{x_0,x} + z\, u_{x_1,x} \\ \varepsilon_{yy} &= \partial u_y/\partial y = && u_{y_0,y} + z\, u_{y_1,y} \\ \varepsilon_{zz} &= \partial u_z/\partial z = && u_{z_1} \\ \gamma_{xy} &= \partial u_x/\partial y + \partial u_y/\partial x = && u_{x_0,y} + u_{y_0,x} + z\, u_{x_1,y} + z\, u_{y_1,x} \\ \gamma_{xz} &= \partial u_x/\partial z + \partial u_z/\partial x = && u_{x_1} + u_{z_0,x} + z\, u_{z_1,x} \\ \gamma_{yz} &= \partial u_y/\partial z + \partial u_z/\partial y = && u_{y_1} + u_{z_0,y} + z\, u_{z_1,y} \end{aligned} \tag{10.13}$$

The linear model leads to a constant distribution of ε_{zz} along the thickness and a linear distribution of other strain components.

The adoption of the $N = 1$ model is necessary to introduce the through-the-thickness stretching of the plate, given by ε_{zz}. As demonstrated by the results provided below, this thickness stretching cannot be negligible when the plate is very thick.

10.1.4 An FE Based on $N = 1$

The $N = 1$ model has six displacement variables, which implies that, in an FE formulation, each node has six generalized displacement variables. The aim of this section is to provide the

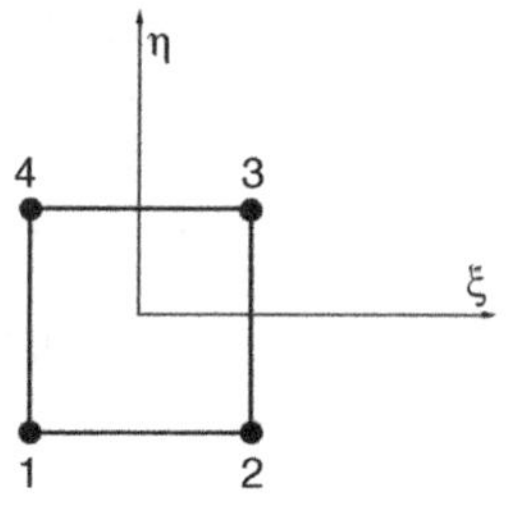

Figure 10.4 Q4 element

FE formulation for the complete linear expansion case. The derivation of the governing FE equations begins with the definition of the nodal displacement vector

$$\mathbf{u}_i = \left\{ u_{x_0} \quad u_{y_0} \quad u_{z_0} \quad u_{x_1} \quad u_{y_1} \quad u_{z_1} \right\}^T \tag{10.14}$$

The displacement variables are interpolated in the plane of the plate by means of the shape functions N_i

$$\mathbf{u} = N_i \mathbf{u}_i \tag{10.15}$$

Plate elements with four nodes, Q4, are considered here (see Figure 10.4), with the following shape functions:

$$\begin{aligned} N_1 &= \frac{1}{4}(1-\xi)(1-\eta) \\ N_2 &= \frac{1}{4}(1+\xi)(1-\eta) \\ N_3 &= \frac{1}{4}(1+\xi)(1+\eta) \\ N_4 &= \frac{1}{4}(1-\xi)(1+\eta) \end{aligned} \tag{10.16}$$

where the natural coordinates ξ and η vary from -1 to $+1$. Considering a rectangular plate, these are defined as follows:

$$\begin{aligned} \xi &= \frac{2(x-x_1)-a}{a} \\ \eta &= \frac{2(y-y_1)-b}{b} \end{aligned} \tag{10.17}$$

x_1, y_1 are the global coordinates of node 1 of the element and a, b are the dimensions of the plate along the x- and y-axis, respectively.

The total number of DOFs of the structural model will be given by

$$\begin{aligned}\text{DOFs} = &\underbrace{3\times 2}_{\text{DOFs per node}} \times [(\underbrace{2}_{\text{nodes per edge}} - 1) \times \underbrace{N_{Ex}}_{\text{elements along } x} + 1] \\ &\times [(\underbrace{2}_{\text{nodes per edge}} - 1) \times \underbrace{N_{Ey}}_{\text{elements along } y} + 1]\end{aligned} \tag{10.18}$$

The PVD is employed to compute the FEM matrices

$$\delta L_{int} = \delta L_{ext} \tag{10.19}$$

where

$$\delta L_{int} = \int_V \left(\delta\boldsymbol{\varepsilon}_p^T\boldsymbol{\sigma}_p + \delta\boldsymbol{\varepsilon}_n^T\boldsymbol{\sigma}_n\right) dV \tag{10.20}$$

L_{int} stands for the strain energy, L_{ext} is the work of the external loadings, and δ stands for the virtual variation. Using Equation (10.16), a compact form of the virtual variation of the internal work can be obtained following the well-known FE procedure

$$\delta L_{int} = \delta\mathbf{u}_j^T\mathbf{k}^{ij}\mathbf{u}_i \tag{10.21}$$

where $\mathbf{k}^{ij}$ is the stiffness matrix. For a given i,j pair, the stiffness matrix has the form

$$\left[\begin{array}{ccc|ccc} k_{xx}^{1,1} & k_{xy}^{1,1} & k_{xz}^{1,1} & k_{xx}^{1,z} & k_{xy}^{1,z} & k_{xz}^{1,z} \\ k_{yx}^{1,1} & k_{yy}^{1,1} & k_{yz}^{1,1} & k_{yx}^{1,z} & k_{yy}^{1,z} & k_{yz}^{1,z} \\ k_{zx}^{1,1} & k_{zy}^{1,1} & k_{zz}^{1,1} & k_{zx}^{1,z} & k_{zy}^{1,z} & k_{zz}^{1,z} \\ \hline k_{xx}^{z,1} & k_{xy}^{z,1} & k_{xz}^{z,1} & k_{xx}^{z,z} & k_{xy}^{z,z} & k_{xz}^{z,z} \\ k_{yx}^{z,1} & k_{yy}^{z,1} & k_{yz}^{z,1} & k_{yx}^{z,z} & k_{yy}^{z,z} & k_{yz}^{z,z} \\ k_{zx}^{z,1} & k_{zy}^{z,1} & k_{zz}^{z,1} & k_{zx}^{z,z} & k_{zy}^{z,z} & k_{zz}^{z,z} \end{array}\right]_{ij} \tag{10.22}$$

where the superscripts indicate the expansion functions that are involved in each component of the stiffness matrix, i.e. 1 and z. For the sake of clarity, the explicit expression of two components is reported here:

$$\begin{aligned} k_{xx}^{1,1} &= \tilde{C}_{11}\int_A 1\cdot 1\,dz\int_\Omega N_{i,x}N_{j,x}d\Omega + C_{66}\int_A 1\cdot 1\,dz\int_\Omega N_{i,y}N_{j,y}d\Omega \\ k_{yz}^{1,z} &= C_{23}\int_A 1\cdot\frac{\partial z}{\partial z}\,dz\int_\Omega N_{i,y}N_j d\Omega = C_{23}\int_A 1\cdot 1\,dz\int_\Omega N_{i,y}N_j d\Omega \end{aligned} \tag{10.23}$$

where Ω indicates the in-plane domain and A the through-the-thickness domain.

10.2 CPT, FSDT and $N = 1$ Model in Unified Form

CPT, FSDT and $N = 1$ models can be obtained in a unified manner via a condensed notation that represents the basic step towards the CUF. In this section, the $N = 1$ model will be reformulated by means of a new notation, while FSDT and CPT will be obtained as particular cases.

10.2.1 Unified Formulation of the $N = 1$ Model

The $N = 1$ stiffness matrix was given in Equation (10.22). That matrix can be considered as being composed of four 3×3 sub-matrices, as in the following:

$$\begin{array}{ccc} & 1 & z \\ 1 & \begin{bmatrix} k_{xx}^{1,1} & k_{xy}^{1,1} & k_{xz}^{1,1} \\ k_{yx}^{1,1} & k_{yy}^{1,1} & k_{yz}^{1,1} \\ k_{zx}^{1,1} & k_{zy}^{1,1} & k_{zz}^{1,1} \end{bmatrix} & \begin{bmatrix} k_{xx}^{1,z} & k_{xy}^{1,z} & k_{xz}^{1,z} \\ k_{yx}^{1,z} & k_{yy}^{1,z} & k_{yz}^{1,z} \\ k_{zx}^{1,z} & k_{zy}^{1,z} & k_{zz}^{1,z} \end{bmatrix} \\ z & \begin{bmatrix} k_{xx}^{z,1} & k_{xy}^{z,1} & k_{xz}^{z,1} \\ k_{yx}^{z,1} & k_{yy}^{z,1} & k_{yz}^{z,1} \\ k_{zx}^{z,1} & k_{zy}^{z,1} & k_{zz}^{z,1} \end{bmatrix} & \begin{bmatrix} k_{xx}^{z,z} & k_{xy}^{z,z} & k_{xz}^{z,z} \\ k_{yx}^{z,z} & k_{yy}^{z,z} & k_{yz}^{z,z} \\ k_{zx}^{z,z} & k_{zy}^{z,z} & k_{zz}^{z,z} \end{bmatrix} \end{array} \tag{10.24}$$

Each sub-matrix has a fixed pair of expansion functions that are used in the explicit computation of the integrals along the thickness, as shown in Equation (10.23). As highlighted for 1D models, the formal expression of each component of the sub-matrices does not depend on the thickness functions. For example, if one considers the following components, they have the same formal expression:

$$\begin{aligned} k_{xx}^{1,1} &= C_{11} \int_A 1 \cdot 1 \, dz \int_\Omega N_{i,x} N_{j,x} d\Omega + C_{66} \int_A 1 \cdot 1 \, dz \int_\Omega N_{i,y} N_{j,y} d\Omega \\ k_{xx}^{1,z} &= C_{11} \int_A 1 \cdot z \, dz \int_\Omega N_{i,x} N_{j,x} d\Omega + C_{66} \int_A 1 \cdot z \, dz \int_\Omega N_{i,y} N_{j,y} d\Omega \end{aligned} \tag{10.25}$$

This implies that the sub-matrix can be considered as a fundamental invariant nucleus which can be used to build the global stiffness matrix. Let us introduce the following notation for the expansion functions, F_τ:

$$\begin{aligned} F_{\tau=0} &= 1 \\ F_{\tau=1} &= z \end{aligned} \tag{10.26}$$

The displacement field in Equation (10.12) becomes

$$\begin{aligned} u_x &= F_0 \, u_{x_0} + F_1 \, u_{x_1} = F_\tau \, u_{x_\tau} \\ u_y &= F_0 \, u_{y_0} + F_1 \, u_{y_1} = F_\tau \, u_{y_\tau} \\ u_z &= F_0 \, u_{z_0} + F_1 \, u_{z_1} = F_\tau \, u_{z_\tau} \end{aligned} \tag{10.27}$$

where the repeated index indicates summation according to the Einstein notation. The displacement vector can be written as

$$\mathbf{u} = F_\tau \mathbf{u}_\tau, \qquad \tau = 0, 1 \tag{10.28}$$

If an FE formulation is introduced and four-node elements are adopted, the nodal unknown vector is given by

$$\mathbf{u}_{\tau i} = \left\{ u_{x_{\tau i}} \quad u_{y_{\tau i}} \quad u_{z_{\tau i}} \right\}^T, \qquad \tau = 0, 1, \qquad i = 1, \ldots, 4 \tag{10.29}$$

The compact form of the internal work in Equation (10.21) becomes

$$\delta L_{int} = \delta \mathbf{u}_{sj}^T \mathbf{k}^{\tau sij} \mathbf{u}_{\tau i} \tag{10.30}$$

where

- τ and s are the thickness function indexes;
- i and j are the shape function indexes.

Coherently with the notation introduced, the matrix in Equation (10.24) can be expressed as

$$\begin{array}{c|c|c} & 0 & 1 \\ 0 & \begin{bmatrix} k_{xx}^{0,0} & k_{xy}^{0,0} & k_{xz}^{0,0} \\ k_{yx}^{0,0} & k_{yy}^{0,0} & k_{yz}^{0,0} \\ k_{zx}^{0,0} & k_{zy}^{0,0} & k_{zz}^{0,0} \end{bmatrix} & \begin{bmatrix} k_{xx}^{0,1} & k_{xy}^{0,1} & k_{xz}^{0,1} \\ k_{yx}^{0,1} & k_{yy}^{0,1} & k_{yz}^{0,1} \\ k_{zx}^{0,1} & k_{zy}^{0,1} & k_{zz}^{0,1} \end{bmatrix} \\ 1 & \begin{bmatrix} k_{xx}^{1,0} & k_{xy}^{1,0} & k_{xz}^{1,0} \\ k_{yx}^{1,0} & k_{yy}^{1,0} & k_{yz}^{1,0} \\ k_{zx}^{1,0} & k_{zy}^{1,0} & k_{zz}^{1,0} \end{bmatrix} & \begin{bmatrix} k_{xx}^{1,1} & k_{xy}^{1,1} & k_{xz}^{1,1} \\ k_{yx}^{1,1} & k_{yy}^{1,1} & k_{yz}^{1,1} \\ k_{zx}^{1,1} & k_{zy}^{1,1} & k_{zz}^{1,1} \end{bmatrix} \end{array} \tag{10.31}$$

Each 3×3 block is the FN of the stiffness matrix. A component of the nucleus is given for different combinations of thickness functions. For example, the components of Equation (10.23) become

$$\begin{aligned} k_{xx}^{1,1} &= C_{11} \int_A F_0 \cdot F_0 \, dz \int_\Omega N_{i,x} N_{j,x} d\Omega + C_{66} \int_A F_0 \cdot F_0 \, dz \int_\Omega N_{i,y} N_{j,y} d\Omega \\ k_{yz}^{1,z} &= C_{23} \int_A F_0 \cdot \frac{\partial F_1}{\partial z} \, dz \int_\Omega N_{i,y} N_j d\Omega \end{aligned} \tag{10.32}$$

10.2.2 CPT and FSDT as Particular Cases of $N = 1$

CPT and FSDT are particular cases of $N = 1$, and they can be obtained by acting on the full linear expansion. As far as FSDT is concerned, the displacement field is given by

$$\begin{aligned} u_x &= u_{x_0} + z\, u_{x_1} \\ u_y &= u_{y_0} + z\, u_{y_1} \\ u_z &= u_{z_0} \end{aligned} \tag{10.33}$$

That is, a linear distribution of in-plane displacements and a constant transverse displacement distribution are accounted for. Starting from the $N = 1$ case, two possible techniques can be used to obtain FSDT: (1) the rearranging of rows and columns of the stiffness matrix; (2) penalization of the stiffness terms related to u_{z_1} (the latter is preferred here in the numerical applications). The main diagonal terms have to be considered, i.e, $i = j$ and $\tau = s$; moreover, only the component with $\tau, s = 1$ has to be penalized. Therefore,

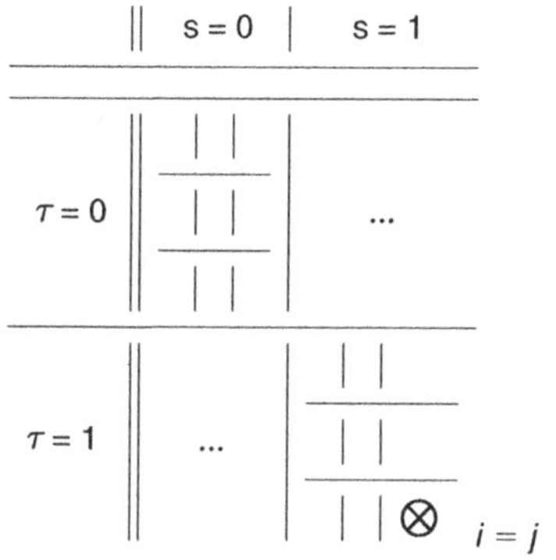

(10.34)

CPT can be obtained through the penalization of γ_{xz} and γ_{yz}. The condition can be imposed using a penalty value χ in the constitutive equations

$$\begin{aligned} \tau_{xy} &= \chi C_{55} \gamma_{xy} \\ \tau_{zy} &= \chi C_{44} \gamma_{zy} \end{aligned} \tag{10.35}$$

10.3 CUF of *N*th Order

CPT and FSDT are the simplest theories that permit the kinematic behaviour of most thin plates to be described adequately. Refined theories can represent the kinematics better, may not require shear correction factors, and can yield more accurate results in the case of thick plates. However, they involve considerably more computational effort. Therefore, such theories should be chosen case by case, depending on the structural problem analysed, in order to consider only those higher-order terms that are necessary.

This section presents a unified approach to deal with plate models of any order. In the present unified formulation, the displacement field is obtained through a formal expression regardless of the order of the theory (N), which is considered as an input of the analysis.

Table 10.2 Taylor polynomials

N	M	F_τ
0	1	$F_0 = 1$
1	2	$F_1 = z$
2	3	$F_2 = z^2$
3	4	$F_3 = z^3$
$\vdots$	$\vdots$	$\vdots$
N	$N+1$	$F_N = z^N$

This table presents the compact form of the Taylor polynomials.

The unified formulation of the through-the-thickness displacement field is described by an expansion of generic functions (F_τ)

$$\mathbf{u}(x, y, z) = F_\tau(z)\mathbf{u}_\tau(x, y), \qquad \tau = 0, 1, \ldots, N \tag{10.36}$$

where F_τ are functions of the thickness coordinate z, $\mathbf{u}_\tau$ is the displacement vector depending on the in-plane coordinates x, y, and N is the order of expansion of the model. According to the Einstein notation, the repeated subscript τ indicates summation. Note that τ assumes values from 0 to N (differently from the 1D CUF) because, in 2D modelling, it coincides with the polynomial order of the corresponding thickness function. The choice of F_τ and N is arbitrary; different base functions of any order can be taken into account to model the displacement field of a structure along the thickness. One possible choice is the adoption of Taylor polynomials consisting of the base z^i, where i is a positive integer. Table 10.2 presents F_τ, as functions of the order of expansion N, and M, which is the number of terms in the expansion. Each row shows the expansion terms of an Nth order theory. The second-order model ($N = 2$) exploits a parabolic expansion of the Taylor polynomials

$$\begin{aligned}
u_x &= \underbrace{u_{x_0}}_{N=0} \;\Big\|\; + \underbrace{z\, u_{x_1}}_{N=1} \;\Big\|\; + \underbrace{z^2\, u_{x_2}}_{N=2} \\
u_y &= u_{y_0} \;\Big\|\; + z\, u_{y_1} \;\Big\|\; + z^2\, u_{y_2} \\
u_z &= u_{z_0} \;\Big\|\; + z\, u_{z_1} \;\Big\|\; + z^2\, u_{z_2}
\end{aligned} \tag{10.37}$$

The 2D model given by Equation (10.37) has nine displacement variables. The strain components can be derived through partial derivatives, for instance the shear and normal components are

$$\begin{aligned}
\gamma_{xz} &= u_{x,z} + u_{z,x} = u_{x_1} + 2\, z\, u_{x_2} + u_{z_0,x} + z\, u_{z_1,x} + z^2\, u_{z_2,x} \\
\varepsilon_{zz} &= u_{z,z} = u_{z_1} + 2\, z\, u_{z_2}
\end{aligned} \tag{10.38}$$

The CUF can also deal with *reduced* models by exploiting a lower number of variables, see Carrera *et al.* (2011). An example of a reduced model is given by

$$\begin{aligned} u_x &= u_{x_0} + z\, u_{x_1} \\ u_y &= u_{y_0} + z\, u_{y_1} \\ u_z &= u_{z_0} + z\, u_{z_1} + z^2\, u_{z_2} \end{aligned} \tag{10.39}$$

The 2D model given in Equation (10.39) has seven displacement variables: three constants, three linear and one parabolic. More details about reduced models will be provided in the following chapters.

10.3.1 $N = 3$ and $N = 4$

The third-order model ($N = 3$) exploits a cubic expansion of the Taylor polynomials

$$\begin{aligned} u_x &= \quad \dots \quad \Big\| + z^3\, u_{x_3} \\ u_y &= \quad \dots \quad \Big\| + z^3\, u_{y_3} \\ u_z &= \underbrace{\quad \dots \quad}_{N \leq 2} \Big\| + \underbrace{z^3\, u_{z_3}}_{N=3} \end{aligned} \tag{10.40}$$

The 2D model given by Equation (10.40) has 12 displacement variables. The shear and normal strain components are

$$\begin{aligned} \gamma_{xz} &= \quad u_{x,z} + u_{z,x} = \quad u_{x_1} + 2\, z\, u_{x_2} + 3\, z^2\, u_{x_3} + u_{z_0,x} + z\, u_{z_1,x} + z^2\, u_{z_2,x} + z^3\, u_{z_3,x} \\ \varepsilon_{zz} &= \quad u_{z,z} = \quad u_{z_1} + 2\, z\, u_{z_2} + 3\, z^2\, u_{z_3} \end{aligned} \tag{10.41}$$

Similarly, the fourth-order model ($N = 4$) exploits a quartic expansion of the Taylor polynomials

$$\begin{aligned} u_x &= \quad \dots \quad \Big\| + z^4\, u_{x_4} \\ u_y &= \quad \dots \quad \Big\| + z^4\, u_{y_4} \\ u_z &= \underbrace{\quad \dots \quad}_{N \leq 3} \Big\| + \underbrace{z^4\, u_{z_4}}_{N=4} \end{aligned} \tag{10.42}$$

The plate model given by Equation (10.42) has 15 displacement variables. Usually in 2D modelling, a fourth-order model is sufficient for reaching the convergence solution. Figure 10.5 shows the typical distribution of the displacements for different orders of expansions.

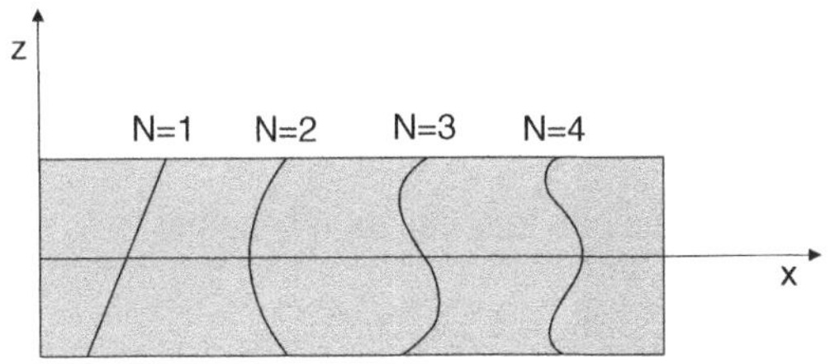

Figure 10.5 Distribution of displacements according to linear and higher-order theories in the unified formulation

10.4 Governing Equations, the FE Formulation and the FN

The governing equations are derived by means of the PVD. Starting from the unified formulation of the displacement field in Equation (10.36), stiffness, mass and loading arrays are obtained in terms of FNs whose form is independent of the order of the model. The weak form of the governing equations is obtained here by means of the FEM, which can easily handle arbitrary geometries, loading and boundary conditions. Closed-form solutions were obtained by Brischetto *et al.* (2012).

10.4.1 Governing Equations

According to the PVD, the following equation holds:

$$\delta L_{int} = \delta L_{ext} - \delta L_{ine} \tag{10.43}$$

L_{int} represents the strain energy, L_{ext} is the work of the external loads and L_{ine} is the work of the inertial loads; δ stands for a virtual variation. Stress (σ) and strain (ε) components are grouped as follows:

$$\begin{aligned} \boldsymbol{\sigma}_p &= \left\{ \sigma_{xx} \quad \sigma_{yy} \quad \tau_{xy} \right\}^T, \ \boldsymbol{\varepsilon}_p = \left\{ \varepsilon_{xx} \quad \varepsilon_{yy} \quad \gamma_{xy} \right\}^T \\ \boldsymbol{\sigma}_n &= \left\{ \tau_{xz} \quad \tau_{yz} \quad \sigma_{zz} \right\}^T, \ \boldsymbol{\varepsilon}_n = \left\{ \gamma_{xz} \quad \gamma_{yz} \quad \varepsilon_{zz} \right\}^T \end{aligned} \tag{10.44}$$

Linear strain–displacement relations can be rewritten as

$$\begin{aligned} \boldsymbol{\varepsilon}_p &= \boldsymbol{b}_p \, \mathbf{u} \\ \boldsymbol{\varepsilon}_n &= (\boldsymbol{b}_{np} + \boldsymbol{b}_{nz}) \mathbf{u} \end{aligned} \tag{10.45}$$

where

$$\boldsymbol{b}_p = \begin{bmatrix} \partial/\partial x & 0 & 0 \\ 0 & \partial/\partial y & 0 \\ \partial/\partial y & \partial/\partial x & 0 \end{bmatrix}, \quad \boldsymbol{b}_{np} = \begin{bmatrix} 0 & 0 & \partial/\partial x \\ 0 & 0 & \partial/\partial y \\ 0 & 0 & 0 \end{bmatrix}, \quad \boldsymbol{b}_{nz} = \begin{bmatrix} \partial/\partial z & 0 & 0 \\ 0 & \partial/\partial z & 0 \\ 0 & 0 & \partial/\partial z \end{bmatrix} \tag{10.46}$$

From Hooke's law,

$$\boldsymbol{\sigma} = \boldsymbol{C}\boldsymbol{\varepsilon} \tag{10.47}$$

According to Equation (10.44), the previous equation becomes

$$\begin{aligned}\boldsymbol{\sigma}_p &= \boldsymbol{C}_{pp}\boldsymbol{\varepsilon}_p + \boldsymbol{C}_{pn}\boldsymbol{\varepsilon}_n \\ \boldsymbol{\sigma}_n &= \boldsymbol{C}_{np}\boldsymbol{\varepsilon}_p + \boldsymbol{C}_{nn}\boldsymbol{\varepsilon}_n\end{aligned} \tag{10.48}$$

In the case of isotropic material the matrices $\boldsymbol{C}_{pp}$, $\boldsymbol{C}_{nn}$, $\boldsymbol{C}_{pn}$ and $\boldsymbol{C}_{np}$ are

$$\boldsymbol{C}_{pp} = \begin{bmatrix} C_{11} & C_{12} & 0 \\ C_{12} & C_{22} & 0 \\ 0 & 0 & C_{66} \end{bmatrix}, \quad \boldsymbol{C}_{nn} = \begin{bmatrix} C_{55} & 0 & 0 \\ 0 & C_{44} & 0 \\ 0 & 0 & C_{33} \end{bmatrix}, \quad \boldsymbol{C}_{pn} = \boldsymbol{C}_{np}^T = \begin{bmatrix} 0 & 0 & C_{13} \\ 0 & 0 & C_{23} \\ 0 & 0 & 0 \end{bmatrix} \tag{10.49}$$

According to the grouping of the stress and strain components in Equation (10.44), the virtual variation of the strain energy is considered as the sum of two contributions,

$$\delta L_{int} = \int_A \int_\Omega \delta\varepsilon_n^T \sigma_n \, d\Omega \, dz + \int_A \int_\Omega \delta\varepsilon_p^T \sigma_p \, d\Omega \, dz \tag{10.50}$$

The virtual variation of the work of the inertial loadings is

$$\delta L_{ine} = \int_V \rho \ddot{\boldsymbol{u}} \delta \boldsymbol{u}^T dV \tag{10.51}$$

where ρ stands for the density of the material and $\ddot{\boldsymbol{u}}$ is the acceleration vector.

The application of surface and point loads is now discussed to derive the variation of the external work. A generic surface load acting on a horizontal surface of the plate is first considered ($p_\alpha(x, y)$), where α can be equal to x, y or z. The subscript (α) indicates the direction of the load. The virtual variation of the external work due to p_α is given by

$$\delta L_{ext}^{p_\alpha} = \int_\Omega \delta u_\alpha \, p_\alpha \, d\Omega \tag{10.52}$$

The loading vector in the case of a generic concentrated load $\mathbf{P}$ is

$$\mathbf{P} = \left\{ P_{u_x} \quad P_{u_y} \quad P_{u_z} \right\}^T \tag{10.53}$$

and the work due to $\mathbf{P}$ is

$$\delta L_{ext} = \mathbf{P} \delta \mathbf{u}^T \tag{10.54}$$

10.4.2 FE Formulation

The nodal displacement vector is introduced,

$$\mathbf{u}_{\tau i} = \left\{ u_{x_{\tau i}} \quad u_{y_{\tau i}} \quad u_{z_{\tau i}} \right\}^T, \qquad \tau = 0, 1, \ldots, N, \quad i = 1, 2, \ldots, N_{EN} \tag{10.55}$$

where the subscript i indicates the element node and N_{EN} stands for the number of nodes per element. If a linear model is considered ($N = 1$), and a four-node element is adopted, the element unknowns will be

$$\mathbf{u}_{\tau i} = \begin{Bmatrix} u_{x_{01}} & u_{y_{01}} & u_{z_{01}} & u_{x_{11}} & u_{y_{11}} & u_{z_{11}} \\ u_{x_{02}} & u_{y_{02}} & u_{z_{02}} & u_{x_{12}} & u_{y_{12}} & u_{z_{12}} \\ u_{x_{03}} & u_{y_{03}} & u_{z_{03}} & u_{x_{13}} & u_{y_{13}} & u_{z_{13}} \\ u_{x_{04}} & u_{y_{04}} & u_{z_{04}} & u_{x_{14}} & u_{y_{14}} & u_{z_{14}} \end{Bmatrix}^T \tag{10.56}$$

The displacement variables are interpolated in the domain of the plate by means of the shape functions (N_i),

$$\mathbf{u} = N_i F_\tau \mathbf{u}_{\tau i} \tag{10.57}$$

Plate elements with four nodes (Q4) are considered, whose shape functions have been given in Equation (10.16). However, one can consider plate elements with different numbers of nodes: eight nodes (Q8) and nine nodes (Q9).

The plate model order is given by the expansion along the thickness. The number of nodes for each element is related to the approximation in the domain (x, y). An Nth order plate model is therefore a theory that exploits an Nth order Taylor expansion to describe the displacement field through the thickness. The choice of the plate model, the plate element and the mesh (i.e. the number of plate elements) determines the total number of DOFs of the structural model,

$$\text{DOFs} = \underbrace{3 \times M}_{\text{number of DOFs per node}} \times \underbrace{N_N}_{\text{total number of nodes}} \tag{10.58}$$

10.4.3 Stiffness Matrix

In the CUF, FE matrices are formulated in terms of the FNs. A compact form of the stiffness matrix can be obtained through Equations (10.45), (10.48) and (10.57),

$$\delta L_{int} = \delta \mathbf{u}_{sj}^T \mathbf{k}^{\tau sij} \mathbf{u}_{\tau i} \tag{10.59}$$

where $\mathbf{k}^{ij\tau s}$ is the stiffness matrix written in the form of the FN. The FN is a 3×3 array which is formally independent of the order of the structural model. Its explicit expression is

$$
\begin{aligned}
k_{xx}^{\tau sij} &= Z_{pp11}^{\tau s} \int_{\Omega} N_{i,x} N_{j,x}\, d\Omega + Z_{pp66}^{\tau s} \int_{\Omega} N_{i,y} N_{j,y}\, d\Omega + Z_{nn55}^{\tau_{,z} s_{,z}} \int_{\Omega} N_i N_j\, d\Omega \\
k_{xy}^{\tau sij} &= Z_{pp12}^{\tau s} \int_{\Omega} N_{i,y} N_{j,x}\, d\Omega + Z_{pp66}^{\tau s} \int_{\Omega} N_{i,x} N_{j,y}\, d\Omega \\
k_{xz}^{\tau sij} &= Z_{pn13}^{\tau_{,z} s} \int_{\Omega} N_i N_{j,x}\, d\Omega + Z_{nn55}^{\tau s_{,z}} \int_{\Omega} N_{i,x} N_j\, d\Omega \\
k_{yx}^{\tau sij} &= Z_{pp12}^{\tau s} \int_{\Omega} N_{i,x} N_{j,y}\, d\Omega + Z_{pp66}^{\tau s} \int_{\Omega} N_{i,y} N_{j,x}\, d\Omega \\
k_{yy}^{\tau sij} &= Z_{pp22}^{\tau s} \int_{\Omega} N_{i,y} N_{j,y}\, d\Omega + Z_{pp66}^{\tau s} \int_{\Omega} N_{i,x} N_{j,x}\, d\Omega + Z_{nn44}^{\tau_{,z} s_{,z}} \int_{\Omega} N_i N_j\, d\Omega \\
k_{yz}^{\tau sij} &= Z_{pn23}^{\tau_{,z} s} \int_{\Omega} N_i N_{j,y}\, d\Omega + Z_{nn44}^{\tau s_{,z}} \int_{\Omega} N_{i,y} N_j\, d\Omega \\
k_{zx}^{\tau sij} &= Z_{nn55}^{\tau_{,z} s} \int_{\Omega} N_i N_{j,x}\, d\Omega + Z_{np13}^{\tau s_{,z}} \int_{\Omega} N_{i,x} N_j\, d\Omega \\
k_{zy}^{\tau sij} &= Z_{nn44}^{\tau_{,z} s} \int_{\Omega} N_i N_{j,y}\, d\Omega + Z_{np23}^{\tau s_{,z}} \int_{\Omega} N_{i,y} N_j\, d\Omega \\
k_{zz}^{\tau sij} &= Z_{nn55}^{\tau s} \int_{\Omega} N_{i,x} N_{j,x}\, d\Omega + Z_{nn44}^{\tau s} \int_{\Omega} N_{i,y} N_{j,y}\, d\Omega + Z_{nn33}^{\tau_{,z} s_{,z}} \int_{\Omega} N_i N_j\, d\Omega
\end{aligned}
\tag{10.60}
$$

where the following notation is introduced to indicate the line integrals along the thickness direction:

$$
\begin{aligned}
&\left(\mathbf{Z}_{pp}^{\tau s}, \mathbf{Z}_{pn}^{\tau s}, \mathbf{Z}_{np}^{\tau s}, \mathbf{Z}_{nn}^{\tau s}\right) = \left(\mathbf{C}_{pp}, \mathbf{C}_{pn}, \mathbf{C}_{np}, \mathbf{C}_{nn}\right) E_{\tau s} \\
&\left(\mathbf{Z}_{pn}^{\tau s_{,z}}, \mathbf{Z}_{nn}^{\tau s_{,z}}, \mathbf{Z}_{np}^{\tau_{,z} s}, \mathbf{Z}_{nn}^{\tau_{,z} s}, \mathbf{Z}_{nn}^{\tau_{,z} s_{,z}}\right) = \left(\mathbf{C}_{pn} E_{\tau s_{,z}}, \mathbf{C}_{nn} E_{\tau s_{,z}}, \mathbf{C}_{np} E_{\tau_{,z} s}, \mathbf{C}_{nn} E_{\tau_{,z} s}, \mathbf{C}_{nn} E_{\tau_{,z} s_{,z}}\right) \\
&\left(E_{\tau s}, E_{\tau s_{,z}}, E_{\tau_{,z} s}, E_{\tau_{,z} s_{,z}}\right) = \int_A \left(F_\tau F_s, F_\tau F_{s,z}, F_{\tau,z} F_s, F_{\tau,z} F_{s,z}\right) dz
\end{aligned}
\tag{10.61}
$$

Some of these integrals are also written in explicit form in Table 10.1, at the beginning of this chapter. It should be noted that no assumptions were made about the approximation order. It is therefore possible to obtain refined plate models without changing the formal expression for the FN. The assembly procedure for the global stiffness matrix starting from the FN is the same as for beam models, discussed in Section 8.4.3.

10.4.4 Mass Matrix

The assembly technique for the FN can be utilized for every FE matrix or vector. In order to obtain the mass matrix, Equation (10.51) is rewritten using Equations (10.57):

$$\delta L_{ine} = \int_{\Omega} \delta \mathbf{u}_{sj}^T N_j \left[\int_A (\rho \, \mathbf{I}) F_\tau F_s dz \right] N_i \ddot{\boldsymbol{u}}_{\tau i} d\Omega \tag{10.62}$$

where $\ddot{q}$ is the nodal acceleration vector. This equation can be rewritten in the following compact manner:

$$\delta L_{ine} = \delta \mathbf{u}_{sj}^T \mathbf{m}^{\tau sij} \ddot{\boldsymbol{u}}_{\tau i} \tag{10.63}$$

where $\mathbf{m}^{ij\tau s}$ is the mass matrix in the form of the FN. Its components are

$$\begin{aligned} m_{xx}^{\tau sij} &= m_{yy}^{\tau sij} = m_{zz}^{\tau sij} = \rho \int_A F_\tau F_s dz \int_\Omega N_i N_j d\Omega \\ m_{xy}^{\tau sij} &= m_{xz}^{\tau sij} = m_{yx}^{\tau sij} = m_{yz}^{\tau sij} = m_{zx}^{\tau sij} = m_{zy}^{\tau sij} = 0 \end{aligned} \tag{10.64}$$

The undamped dynamic problem can be written as follows:

$$\boldsymbol{M}\ddot{\boldsymbol{U}} + \boldsymbol{K}\boldsymbol{U} = \boldsymbol{P} \tag{10.65}$$

where $\boldsymbol{u}$ is the vector of the nodal unknowns and $\boldsymbol{P}$ is the loading vector. Introducing harmonic solutions, it is possible to compute the natural frequencies (ω_i) by solving an eigenvalue problem:

$$(-\omega_i^2 \boldsymbol{M} + \boldsymbol{K})\boldsymbol{U}_i = 0 \tag{10.66}$$

where $\boldsymbol{U}_i$ is the ith eigenvector.

10.4.5 Loading Vector

Surface loads are addressed first. The F_τ expansions and the nodal displacements were introduced in Equation (10.52),

$$\delta L_{ext}^{p_\alpha} = \int_\Omega F_s(z_p)\, N_j \, \delta u_{\alpha_{sj}} \, p_\alpha \, d\Omega \tag{10.67}$$

where z_p stands for the loading application coordinate.

Equation (10.54) is used for the point load case and becomes

$$\delta L_{ext} = F_s N_j \mathbf{P} \delta \mathbf{u}_{sj}^T \tag{10.68}$$

This equation is used to assemble the loading vector by detecting the displacement variables that have to be loaded.

10.4.6 Numerical Integration

The computation of the FNs requires the evaluation of surface integrals in Cartesian coordinates (x and y), such as

$$\int_{\Omega} N_{i,x} N_{j,y} dxdy \tag{10.69}$$

where Ω is the plate element domain. If normalized coordinates ξ and η are accounted for (see Figure 10.4), integrals can be computed independently of the actual geometry of the element. For instance, if quadrilateral elements are considered, the following will arise:

$$\int_{\Omega} N_{i,x} N_{i,y} dx\, dy = \int_{-1}^{+1} \int_{-1}^{+1} N_{i,\xi} N_{j,\eta} |J| d\xi\, d\eta \tag{10.70}$$

where $|J|$ is the Jacobian determinant of the transformation. In some cases, the new integral in the normalized coordinates can be computed analytically, but more often numerical techniques have to be employed.

Partial derivatives have to be computed with respect to the normalized coordinates according to the chain rule,

$$\begin{aligned} \frac{\partial N_i}{\partial x} &= \frac{\partial N_i}{\partial \xi}\frac{\partial \xi}{\partial x} + \frac{\partial N_i}{\partial \eta}\frac{\partial \eta}{\partial x} \\ \frac{\partial N_i}{\partial y} &= \frac{\partial N_i}{\partial \xi}\frac{\partial \xi}{\partial y} + \frac{\partial N_i}{\partial \eta}\frac{\partial \eta}{\partial y} \end{aligned} \tag{10.71}$$

The evaluation of Equation (10.71) requires the following explicit relationships:

$$\xi = \xi(x, y), \quad \eta = \eta(x, y) \tag{10.72}$$

as in Equations (10.17). These explicit relationships are often difficult to establish, and for this reason it is preferable to use the chain rule as in the following:

$$\begin{aligned} \frac{\partial N_i}{\partial \xi} &= \frac{\partial N_i}{\partial x}\frac{\partial x}{\partial \xi} + \frac{\partial N_i}{\partial y}\frac{\partial y}{\partial \xi} \\ \frac{\partial N_i}{\partial \eta} &= \frac{\partial N_i}{\partial x}\frac{\partial x}{\partial \eta} + \frac{\partial N_i}{\partial y}\frac{\partial y}{\partial \eta} \end{aligned} \tag{10.73}$$

Equation (10.73) also holds for N_j and can be rewritten in matrix form,

$$\begin{Bmatrix} \dfrac{\partial N_i}{\partial \xi} \\ \dfrac{\partial N_i}{\partial \eta} \end{Bmatrix} = \underbrace{\begin{bmatrix} \dfrac{\partial x}{\partial \xi} & \dfrac{\partial y}{\partial \xi} \\ \dfrac{\partial x}{\partial \eta} & \dfrac{\partial y}{\partial \eta} \end{bmatrix}}_{\mathbf{J}} \begin{Bmatrix} \dfrac{\partial N_i}{\partial x} \\ \dfrac{\partial N_i}{\partial y} \end{Bmatrix} \tag{10.74}$$

The inverse relationship is given by

$$\begin{Bmatrix} \dfrac{\partial N_i}{\partial x} \\ \dfrac{\partial N_i}{\partial y} \end{Bmatrix} = \frac{1}{|J|} \begin{bmatrix} \dfrac{\partial y}{\partial \eta} & -\dfrac{\partial y}{\partial \xi} \\ -\dfrac{\partial x}{\partial \xi} & \dfrac{\partial x}{\partial \eta} \end{bmatrix} \begin{Bmatrix} \dfrac{\partial N_i}{\partial \xi} \\ \dfrac{\partial N_i}{\partial \eta} \end{Bmatrix} \tag{10.75}$$

It is important to note that Equation (10.75) requires that the inverse of $\mathbf{J}$ exists. In order to fulfil this requirement, attention should be paid to the geometry of the element. Major distortions or folded back elements can cause such a singularity problem.

Similarly, the integrals of thickness functions F_τ along the thickness coordinate z are performed considering the normalized coordinate $\zeta = 2z/h$. For example,

$$\int_A F_{\tau,z} F_{s,z} dz = \int_{-1}^{+1} F_{\tau,z} F_{s,z} \frac{h}{2}\, d\zeta \tag{10.76}$$

where $h/2$ is the Jacobian of this transformation.

The integrals in Equations (10.70) and (10.76) can have analytical solutions but, in practice, numerical integrations are required. Gauss quadrature formulae are massively adopted in FE formulations. In the unified formulation framework, the Gauss rule is employed also to solve the integrals along the thickness. In this manner, the integrals can be calculated independently of the thickness functions F_τ, F_s.

For example, the integrals in Equations (10.70) and (10.76) are solved as follows:

$$\begin{aligned} &\int_{-1}^{+1}\int_{-1}^{+1} N_{i,\xi} N_{j,\eta} |J| d\xi\, d\eta = \sum_{h,k} w_h\, w_k\, N_{i,\xi}(\xi_h, \eta_k)\, N_{j,\eta}(\xi_h, \eta_k)\, |J(\xi_h, \eta_k)| \\ &\int_{-1}^{+1} F_{\tau,z} F_{s,z} \frac{h}{2}\, d\zeta = \sum_l w_l\, F_{\tau,z}(\zeta_l)\, F_{s,z}(\zeta_l)\, \frac{h}{2} \end{aligned} \tag{10.77}$$

where w_h, w_k and w_l are the integration weights, and ξ_h, η_k and ζ_l are the integration points, in the domain and along the thickness respectively. Weights and points depend on the set of shape functions and thickness functions adopted and are provided in Appendix A.

10.5 Locking Phenomena

Two types of locking phenomena that affect plate FEs are described in this section, namely thickness and shear locking. The former is due to deformation coupling that arises in CPT, FSDT and $N = 1$ models. The latter is a numerical phenomenon related to the FE formulation. The effects of both lockings can be extremely detrimental; however, a number of correction techniques have been proposed as will be shown in the following section.

10.5.1 Poisson Locking and its Correction

The thickness locking mechanism, also known as the Poisson locking phenomenon, is caused by the use of simplified kinematic assumptions in the plate analysis (Carrera and Brischetto

2008). The 2D structures can be analysed as a particular case of the 3D continuum by eliminating, via a priori integration, the thickness coordinate z. The introduction of 2D approximations can produce some undesired effects which are not in the 3D solution. One of these is Poisson locking, which is related to the use of plane-strain/plane-stress hypotheses in thin-plate theory. The analysis of thin problems is, in fact, often associated with plane-stress assumptions (thin surface problem) while the plane-strain hypothesis usually refers to beam theory. Discussion of plane strain, plane stress and/or plane elasto-static problems can also be found in the paper by Carrera and Brischetto (2008). However, in most 2D theories, contradictory assumptions are made on strain fields. Plane-strain assumptions are

$$\gamma_{xz} = \gamma_{yz} = \varepsilon_{zz} = 0 \tag{10.78}$$

and they are used in place of the more natural plane-stress assumptions

$$\tau_{xz} = \tau_{yz} = \sigma_{zz} = 0 \tag{10.79}$$

This contradiction introduces a 'locking mechanism' that makes the plate model inapplicable in some cases. Thickness locking (TL, also known as Poisson locking, PL) is the name assigned to this mechanism: TL does not permit 2D analysis with constant or linear transverse displacement u_z through the thickness (which means transverse strain ε_{zz} is zero or constant) to lead to the 3D solution in thin-plate problems. A well-known technique to contrast TL consists of modifying the elastic stiffness coefficients by forcing the 'contradictory' condition, known as the transverse normal stress zero condition:

$$\sigma_{zz} = 0 \tag{10.80}$$

PL appears if a plate theory shows a constant distribution of transverse normal strain ε_{zz}. In order to avoid TL, the plate theories require at least a parabolic distribution of the transverse displacement component u_z. For these theories, Hooke's law given in Equation (10.48) is suitable. For further details, see the complete discussion reported in the books by Washizu (1968) and Librescu (1975).

By imposing the condition σ_{zz} in Equation (10.48), the modified stiffness coefficients (reduced stiffness coefficients) can be obtained. The latter are indicated as C'_{ij}, and they are used in 2D theories with transverse displacement u_z constant or linear in the thickness direction,

$$\begin{aligned} &C'_{11} = C'_{22} = \frac{E}{1-\nu^2}, \quad C'_{12} = \frac{\nu E}{1-\nu^2} \\ &C'_{13} = C_{13}, \quad C'_{23} = C_{23}, \quad C'_{33} = C_{33} \\ &C'_{44} = C_{44}, \quad C'_{55} = C_{55}, \quad C'_{66} = C_{66} \end{aligned} \tag{10.81}$$

Equation (10.81) means that, to avoid the PL phenomenon, Equation (10.48) must be used, but by considering the reduced stiffness coefficients C'_{11}, C'_{12} and C'_{22}.

10.5.2 Shear Locking and its Correction

The shear locking phenomenon in Reissner–Mindlin-type FEs remains a milestone in FEA. Shear locking consists of the numerical difficulty (in the convergence rate sense) of reducing Reissner–Mindlin plate elements to exact or Kirchhoff solutions in the case of thin-plate analysis. In fact, the inclusion of both bending and shear stiffness in a unique rotational DOF makes it difficult to obtain zero transverse shear energy in thin-plate structures, as in physics. The mesh convergence rate becomes very small and a huge number of elements would be necessary to fulfil the constraint of zero transverse shear energy. A pioneering remedy, proposed in the late 1960s and early 1970s, is based on a reduced and/or selective numerical integration technique of transverse shear stiffness contributions. The reduced integration permits a significant increase in the convergence rate through the evaluation of the transverse shear strain energy at a plate integration point in which the corresponding shear strains are very small. The problem has been clearly explained and dealt with in detail in many FE books, such as those by Zienkiewicz *et al.* (2005), Bathe (1996) and Hughes (1987), among others.

An alternative technique to reduced integration was introduced in the 1980s by Bathe and Dvorkin (1985) and McNeal (1982). In Bathe and Dvorkin (1985), this technique was denominated mixed interpolation of tensorial components (MITC) and was proposed for linear four-node Q4 plate elements. The MITC formulation permits the transverse shear locking phenomenon to be eliminated by introducing an independent FE approximation into the element domain for the transverse shear strains. Further studies on the MITC method for Reissner–Mindlin problems were carried out by Brezzi *et al.* (1990) and Della Croce and Scapolla (1994).

This section is devoted to the description of both selective-reduced integration and the mixed interpolation techniques in the unified formulation framework.

10.5.2.1 Selective-Reduced Integration

The internal energy can be conveniently written by splitting the in-plane, shear and normal contributions

$$\delta L_{int} = \int_A \int_\Omega \delta \epsilon_p^T \sigma_p \, d\Omega \, dz + \int_A \int_\Omega \left(\delta \epsilon_s^T \sigma_s + \delta \epsilon_{zz}^T \sigma_{zz} \right) \, d\Omega \, dz \tag{10.82}$$

where $\epsilon_s = \{\gamma_{xz}, \gamma_{yz}\}$ and $\sigma_s = \{\tau_{xz}, \tau_{yz}\}$ determine the shear energy. Due to the locking phenomenon, this energy is overestimated, the plate appears stiffer than it really is (the displacements are underestimated) and the solution exhibits a loss of convergence.

According to the separation made in the PVD, the stiffness matrix also can be derived by isolating the shear terms. Referring to the governing equations (Equation (10.59)), $\boldsymbol{K}$ can be written by separating the shear terms from the others, i.e.

$$\boldsymbol{K} = \boldsymbol{K}_{pn} + \boldsymbol{K}_s \tag{10.83}$$

where $\boldsymbol{K}_{pn}$ includes both in-plane and normal contributions.

An expedient to contrast the shear locking is to integrate the shear terms of the stiffness matrix, $\boldsymbol{K}_s$, using a number of Gauss points that are lower than necessary to calculate the integrals exactly. The terms contained in $\boldsymbol{K}_{pn}$ are integrated on 2×2 Gauss points, while only one central Gauss point is used for $\boldsymbol{K}_s$. In this way, the shear energy is underestimated and the effect of the locking is balanced.

Unfortunately the reduced integration techniques, since they introduce, 'by definition', an additional error into the evaluation of the stiffness matrices, could lead to an increase in spurious and/or hour glass-type 'non-physical' deformation modes which could destroy the related FE solutions. Such a drawback cannot be tolerated, especially in the case of nonlinear analysis, in which previous load-step solutions (which can be greatly affected by spurious modes) are used to build the current equilibrium. Non-convergent solutions can be obtained in practical applications.

10.5.2.2 MITC4 Plate Element

A four-node element based on the MITC method was proposed by Bathe and Dvorkin (1985) in order to contrast the shear locking. According to the MITC4 technique, the transverse shear strains γ_{xz} and γ_{yz} are calculated in a different manner from the other tensorial components. The shear strains are interpolated in sample points (M, N, P, Q) of the domain Ω, as indicated in Figure 10.6.

Therefore, the transverse strain array ε_n can be rewritten as

$$\varepsilon_n = \begin{Bmatrix} \gamma_{xz} \\ \gamma_{yz} \\ \varepsilon_{zz} \end{Bmatrix} = \begin{Bmatrix} \frac{1}{2}(1+\xi)\gamma_{xz}^N + \frac{1}{2}(1-\xi)\gamma_{xz}^Q \\ \frac{1}{2}(1+\eta)\gamma_{yz}^P + \frac{1}{2}(1-\eta)\gamma_{yz}^M \\ \varepsilon_{zz} \end{Bmatrix} \tag{10.84}$$

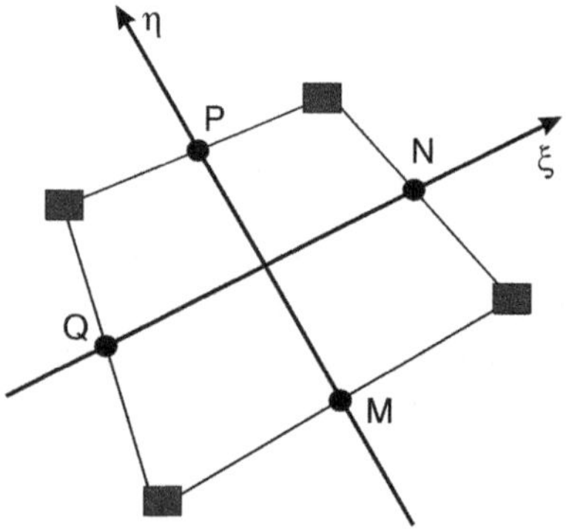

Figure 10.6 Sample points (M, N, P, Q) to approximate the shear contribution in the MITC4 element

where $\gamma_{xz}^N, \gamma_{xz}^Q, \gamma_{yz}^P, \gamma_{yz}^M$ are the strains calculated at the sample points. In matrix form, this can be expressed as

$$\varepsilon_n = \begin{bmatrix} \frac{1}{2}(1+\xi) & \frac{1}{2}(1-\xi) & 0 & 0 & 0 \\ 0 & 0 & \frac{1}{2}(1+\eta) & \frac{1}{2}(1-\eta) & 0 \\ 0 & 0 & 0 & 0 & 1 \end{bmatrix} \begin{Bmatrix} \gamma_{xz}^N \\ \gamma_{xz}^Q \\ \gamma_{yz}^P \\ \gamma_{yz}^M \\ \varepsilon_{zz} \end{Bmatrix} \tag{10.85}$$

The strains in the sample points can be calculated using the geometrical relations. Substituting the displacements (Equation (10.57)) in the geometrical relations (Equations (10.45)), the strains in the sample points are calculated as follows:

$$\begin{Bmatrix} \gamma_{xz}^N \\ \gamma_{xz}^Q \\ \gamma_{yz}^P \\ \gamma_{yz}^M \\ \varepsilon_{zz} \end{Bmatrix} = \begin{bmatrix} 0 & 0 & F_s N_{j,x}^N \\ 0 & 0 & F_s N_{j,x}^Q \\ 0 & 0 & F_s N_{j,y}^P \\ 0 & 0 & F_s N_{j,y}^M \\ 0 & 0 & 0 \end{bmatrix} u_{sj} + \begin{bmatrix} F_{s,z} N_j^N & 0 & 0 \\ F_{s,z} N_j^Q & 0 & 0 \\ 0 & F_{s,z} N_j^P & 0 \\ 0 & F_{s,z} N_j^M & 0 \\ 0 & 0 & F_{s,z} N_j \end{bmatrix} u_{sj} \tag{10.86}$$

Finally, recalling the interpolating functions used for the MITC method,

$$N_a = \frac{1}{2}(1+\xi), \quad N_b = \frac{1}{2}(1-\xi), \quad N_c = \frac{1}{2}(1+\eta), \quad N_d = \frac{1}{2}(1-\eta) \tag{10.87}$$

one can rewrite ε_n as

$$\varepsilon_{n_{sj}} = F_s B_{1j} u_{sj} + F_{s,z} B_{2j} u_{sj} \tag{10.88}$$

where the introduced matrices are

$$B_{1j} = \begin{bmatrix} 0 & 0 & N_a N_{j,x}^N + N_b N_{j,x}^Q \\ 0 & 0 & N_c N_{j,y}^P + N_d N_{j,y}^M \\ 0 & 0 & 0 \end{bmatrix} \quad B_{2j} = \begin{bmatrix} N_a N_j^N + N_b N_j^Q & 0 & 0 \\ 0 & N_c N_j^P + N_d N_j^M & 0 \\ 0 & 0 & N_j \end{bmatrix} \tag{10.89}$$

The governing equations are derived from the PVD, as shown in Section 10.4.1. But, in the case of the MITC4 plate element, the FN of the stiffness matrix is different. The components of $\boldsymbol{k}^{\tau sij}$ are

$$
\begin{aligned}
k_{xx}^{\tau sij} &= C_{55} N_i^N N_j^N \int_\Omega N_a N_a \, d\Omega \int_A F_{s,z} F_{\tau,z} \, dz + C_{55} N_i^Q N_j^N \int_\Omega N_b N_a \, d\Omega \int_A F_{s,z} F_{\tau,z} \, dz \\
&+ C_{55} N_i^N N_j^Q \int_\Omega N_a N_b \, d\Omega \int_A F_{s,z} F_{\tau,z} \, dz + C_{55} N_i^Q N_j^Q \int_\Omega N_b N_b \, d\Omega \int_A F_{s,z} F_{\tau,z} \, dz \\
&+ C_{11} \int_\Omega N_{i,x} N_{j,x} \, d\Omega \int_A F_s F_\tau \, dz + C_{66} \int_\Omega N_{i,y} N_{j,y} \, d\Omega \int_A F_s F_\tau \, dz \\
k_{yx}^{\tau sij} &= C_{12} \int_\Omega N_{i,y} N_{j,x} \, d\Omega \int_A F_s F_\tau \, dz + C_{66} \int_\Omega N_{i,x} N_{j,y} \, d\Omega \int_A F_s F_\tau \, dz \\
k_{zx}^{\tau sij} &= C_{55} N_{i,x}^N N_j^N \int_\Omega N_a N_a \, d\Omega \int_A F_{s,z} F_\tau \, dz + C_{55} N_{i,x}^N N_j^Q \int_\Omega N_a N_b \, d\Omega \int_A F_{s,z} F_\tau \, dz \\
&+ C_{55} N_{i,x}^Q N_j^N \int_\Omega N_b N_a \, d\Omega \int_A F_{s,z} F_\tau \, dz + C_{55} N_{i,x}^Q N_j^Q \int_\Omega N_b N_b \, d\Omega \int_A F_{s,z} F_\tau \, dz \\
&+ C_{13} \int_\Omega N_i N_{j,x} \, d\Omega \int_A F_s F_{\tau,z} \, dz \\
k_{xy}^{\tau sij} &= C_{66} \int_\Omega N_{i,y} N_{j,x} \, d\Omega \int_A F_s F_\tau \, dz + C_{12} \int_\Omega N_{i,x} N_{j,y} \, d\Omega \int_A F_s F_\tau \, dz \\
k_{yy}^{\tau sij} &= C_{44} N_i^M N_j^M \int_\Omega N_d N_d \, d\Omega \int_A F_{s,z} F_{\tau,z} \, dz + C_{44} N_i^P N_j^M \int_\Omega N_c N_d \, d\Omega \int_A F_{s,z} F_{\tau,z} \, dz \\
&+ C_{44} N_i^M N_j^P \int_\Omega N_d N_c \, d\Omega \int_A F_{s,z} F_{\tau,z} \, dz + C_{44} N_i^P N_j^P \int_\Omega N_c N_c \, d\Omega \int_A F_{s,z} F_{\tau,z} \, dz \\
&+ C_{66} \int_\Omega N_{i,x} N_{j,x} \, d\Omega \int_A F_s F_\tau \, dz + C_{22} \int_\Omega N_{i,y} N_{j,y} \, d\Omega \int_A F_s F_\tau \, dz
\end{aligned}
\tag{10.90}
$$

$$
\begin{aligned}
k_{zy}^{\tau sij} &= C_{44} N_{i,y}^M N_j^M \int_\Omega N_d N_d \, d\Omega \int_A F_{s,z} F_\tau \, dz + C_{44} N_{i,y}^M N_j^P \int_\Omega N_d N_c \, d\Omega \int_A F_{s,z} F_\tau \, dz \\
&+ C_{44} N_{i,y}^P N_j^M \int_\Omega N_c N_d \, d\Omega \int_A F_{s,z} F_\tau \, dz + C_{44} N_{i,y}^P N_j^P \int_\Omega N_c N_c \, d\Omega \int_A F_{s,z} F_\tau \, dz \\
&+ C_{45} N_{i,x}^Q N_j^P \int_\Omega N_b N_c \, d\Omega \int_A F_{s,z} F_\tau \, dz + C_{23} \int_\Omega N_i N_{j,y} \, d\Omega F_s F_{\tau,z} \\
k_{xz}^{\tau sij} &= C_{55} N_i^N N_{j,x}^N \int_\Omega N_a N_a \, d\Omega \int_A F_s F_{\tau,z} \, dz + C_{55} N_i^Q N_{j,x}^N \int_\Omega N_b N_a \, d\Omega \int_A F_s F_{\tau,z} \, dz \\
&+ C_{55} N_i^N N_{j,x}^Q \int_\Omega N_a N_b \, d\Omega \int_A F_s F_{\tau,z} \, dz + C_{55} N_i^Q N_{j,x}^Q \int_\Omega N_b N_b \, d\Omega \int_A F_s F_{\tau,z} \, dz \\
&+ C_{13} \int_\Omega N_{i,x} N_j \, d\Omega \int_A F_{s,z} F_\tau \, dz
\end{aligned}
$$

$$
\begin{aligned}
k_{yz}^{\tau sij} &= C_{44}N_i^M N_{j,y}^M \int_\Omega N_d N_d \, d\Omega \int_A F_s F_{\tau,z} \, dz + C_{44}N_i^P N_{j,y}^M \int_\Omega N_c N_d \, d\Omega \int_A F_s F_{\tau,z} \, dz \\
&+ C_{44}N_i^M N_{j,y}^P \int_\Omega N_d N_c \, d\Omega \int_A F_s F_{\tau,z} \, dz + C_{44}N_i^P N_{j,y}^P \int_\Omega N_c N_c \, d\Omega \int_A F_s F_{\tau,z} \, dz \\
&+ C_{23} \int_\Omega N_{i,y} N_j \, d\Omega \int_A F_{s,z} F_\tau \, dz \\
k_{zz}^{\tau sij} &= C_{44}N_{i,y}^M N_{j,y}^M \int_\Omega N_d N_d \, d\Omega \int_A F_s F_\tau \, dz + C_{44}N_{i,y}^P N_{j,y}^M \int_\Omega N_c N_d \, d\Omega \int_A F_s F_\tau \, dz \\
&+ C_{55}N_{i,x}^N N_{j,x}^N \int_\Omega N_a N_a \, d\Omega \int_A F_s F_\tau \, dz + C_{55}N_{i,x}^Q N_{j,x}^N \int_\Omega N_b N_a \, d\Omega \int_A F_s F_\tau \, dz \\
&+ C_{44}N_{i,y}^M N_{j,y}^P \int_\Omega N_d N_c \, d\Omega \int_A F_s F_\tau \, dz + C_{44}N_{i,y}^P N_{j,y}^P \int_\Omega N_c N_c \, d\Omega \int_A F_s F_\tau \, dz \\
&+ C_{55}N_{i,x}^N N_{j,x}^Q \int_\Omega N_a N_b \, d\Omega \int_A F_s F_\tau \, dz + C_{55}N_{i,x}^Q N_{j,x}^Q \int_\Omega N_b N_b \, d\Omega \int_A F_s F_\tau \, dz \\
&+ C_{33} \int_\Omega N_i N_j \, d\Omega \int_A F_{s,z} F_{\tau,z} \, dz
\end{aligned}
\tag{10.91}
$$

The mass matrix and the loading vector are the same.

10.6 Numerical Applications

A numerical investigation has been performed to assess the effectiveness of the MITC method applied to plate elements based on the CUF.

The analytical solution (A), obtained using the closed-form solution proposed in Carrera (1998), is used for comparison purposes. The MITC solutions are compared with various numerical integration schemes: normal (N) (2×2 Gauss points), reduced (R) (1×1 Gauss point) and selective (S) (2×2 Gauss points for bending stiffness and 1×1 for transverse shear stiffness contributions).

The material data are given in Table 10.3. The results are restricted to non-dimensional transverse displacement $\bar{u}_z = u_z \cdot 100Eh^3/(p_z^+ a^4)$, calculated at the centre of the simply supported plate. The plate is subjected to a bisinusoidal load applied at the top surface of the plate in the z direction:

$$
p_z^+ = \hat{p}_z^+ \sin\left(\frac{m\pi x}{a}\right) \sin\left(\frac{n\pi y}{b}\right) \tag{10.92}
$$

where m and n are the wave numbers and considered to be equal to one. $\hat{p}_z^+ = 10^3$ Pa is the amplitude of the load. A 15×15 mesh and a square plate geometry, with plate dimensions $a = b = 1$ m, were adopted.

Various integration schemes are compared with the analytical solution and MITC technique. The shear locking mechanism is evident for the thin plate in the normal integration case. The MITC4 element prevents shear locking without any sub-integration scheme.

Table 10.3 Material data: aluminium plate

	Aluminium
E	72.0 GPa
G	27.1 GPa
ν	0.33
ρ	2810 kg/m^3

This table shows the physical properties of the aluminium.

Table 10.4 Transverse displacement $\bar{u}_z$ at the centre of the plate. Theory FSDT (Cinefra *et al.* 2010)

$\bar{u}_z(z=0)$	$a/h=10$	$a/h=50$	$a/h=100$	$a/h=500$	$a/h=1000$
A	2.8780	2.7487	2.7446	2.7434	2.7432
N	2.6718	0.9571	0.3236	0.0146	0.0037
R	2.8683	2.7386	2.7346	2.7333	2.7331
S	2.8616	2.7320	2.7279	2.7266	2.7266
MITC	2.8611	2.7316	2.7279	2.7266	2.7266

This table compares different methods for solving the shear locking phenomenon by considering various thickness ratios.

Table 10.5 Transversal displacement $\bar{u}_z$ at the centre of the plate. Theory $N=2$ (Cinefra *et al.* 2010)

$\bar{u}_z(z=0)$	$a/h=10$	$a/h=50$	$a/h=100$	$a/h=500$	$a/h=1000$
A	2.8657	2.7482	2.7445	2.7434	2.7432
N	2.6591	0.9567	0.3236	0.0146	0.0037
R	2.8561	2.7382	2.7345	2.7333	2.7331
S	2.8479	2.7299	2.7262	2.7250	2.7252
MITC	2.8474	2.7298	2.7261	2.7250	2.7252

This table compares different methods for solving the shear locking phenomenon by considering various thickness ratios.

The FSDT results are considered in Table 10.4 and higher-order results, related to a parabolic distribution of the displacement field, are given in Table 10.5. It appears clear from this table that the MITC method provides good results also in the case of FEs formulated on the basis of higher-order theories. In other words, the shear locking phenomenon does not depend on the order of the displacement field in the plate thickness. Table 10.6 confirms this for the case of the fourth-order displacement field. Higher-order theories obviously lead to more deformable structures in the case of thick plates; this, however, does not depend on shear locking: it is instead due to the higher transverse shear deformability of thick plates. A compendium of the analyses performed, in the case of very thin plates, is given in Table 10.7. The capability of the MITC technique to remain efficient in the higher-order modelling of plates is evident. The results presented here are taken from the work by Cinefra *et al.* (2010).

Table 10.6 Transverse displacement $\bar{u}_z$ at the centre of the plate. Theory $N = 4$ (Cinefra *et al.* 2010)

$\bar{u}_z(z = 0)$	$a/h = 10$	$a/h = 50$	$a/h = 100$	$a/h = 500$	$a/h = 1000$
A	2.8845	2.7490	2.7447	2.7434	2.7432
N	2.6754	0.9568	0.3236	0.0146	0.0037
R	2.8749	2.7389	2.7346	2.7333	2.7331
S	2.8667	2.7306	2.7263	2.7250	2.7252
MITC	2.8661	2.7306	2.7263	2.7250	2.7252

This table compares different methods for solving the shear locking phenomenon by considering various thickness ratios.

Table 10.7 Transverse displacement $\bar{u}_z$ at the centre of the plate. Thickness ratio $a/h = 1000$ (Cinefra *et al.* 2010)

$\bar{u}_z(z = 0)$	A	N	R	S	MITC
FSDT	2.7432	0.0037	2.7331	2.7266	2.7266
$N = 1$	2.7432	0.0037	2.7331	2.7266	2.7266
$N = 2$	2.7432	0.0037	2.7331	2.7252	2.7252
$N = 3$	2.7432	0.0037	2.7331	2.7252	2.7252
$N = 4$	2.7432	0.0037	2.7331	2.7252	2.7252

This table compares different methods for solving the shear locking phenomenon by considering various theories.

References

Babuska I, dHarcourt JM and Schwab C 1994 Optimal shear correction factors for hierarchical plate models. *Mathematical Modelling and Scientific Computing* **I**, 1–30.

Bathe KJ 1996 *Finite Element Procedures*. Prentice Hall.

Bathe KJ and Dvorkin EN 1985 A four node plate bending element based on Mindlin-Reissner plate theory and mixed interpolation. *International Journal for Numerical Methods in Engineering* **21**, 367–383.

Brezzi F, Bathe KJ and Fortin M 1990 Mixed interpolated elements for Reissner-Mindlin plates. *International Journal for Numerical Methods in Engineering* **28**, 1787–1801.

Brischetto S, Polit O and Carrera E 2012 Refined shell model for the linear analysis of isotropic and composite elastic structures. *European Journal of Mechanics. A, Solids* **34**, 102–119.

Carrera E 1998 Evaluation of layer-wise mixed theories for laminated plates analysis. *AIAA Journal* **26**, 830–839.

Carrera E and Brischetto S 2008 Analysis of thickness locking in classical, refined and mixed multilayered plate theories. *Composite Structures* **82**(4), 549–562.

Carrera E, Miglioretti F and Petrolo M 2011 Guidelines and recommendations on the use of higher-order finite elements for bending analysis of plates. *International Journal for Computational Methods in Engineering Science and Mechanics* **12**, 303–324.

Cinefra M, Carrera E and Nali P 2010 MITC technique extended to variable kinematic multilayered plate elements. *Composite Structures* **92**, 1888–1895.

Della Croce L and Scapolla T 1994 Combining hierarchic high-order and mixed-interpolated finite elements for Reissner-Mindlin plate problems. *Computer Methods in Applied Mechanics and Engineering* **116**, 185–192.

Hughes TJR 1987 *The Finite Element Method: Linear Static and Dynamic Finite Element Analysis*. Prentice Hall.

Kirchhoff G 1850 Über das Gleichgewicht und die Bewegung einer elastishen Scheibe. *Journal für die reine und angewandte Mathematik* **40**, 51–88.

Librescu L 1975 *Elasto-static and Kinetics of Anisotropic and Heterogeneous Shell-type Structures*. Nordhoff International.

McNeal RH 1982 Derivation of element stiffness matrices by assumed strain distribution. *Nuclear Engineering and Design* **70**, 3–12.

Mindlin RD 1951 Influence of rotatory inertia and shear on flexural motions of isotropic, elastic plates, *ASME Journal of Applied Mechanics* **18**, 31–38.

Reissner E 1945 The effect of transverse shear deformation on the bending of elastic plates. *Journal of Applied Mechanics* **12**(2), 69–77.

Rössle A 1999 On the derivation of an asymptotically correct shear correction factor for the Reissner-Mindlin plate model. *Comptes Rendus de l'Academie des Sciences - Series I - Mathematics* **328**(3), 269–274.

Washizu K 1968 *Variational Methods in Elasticity and Plasticity*. Pergamon Press.

Zienkiewicz OC, Taylor RL and Zhu JZ 2005 *The Finite Element Method*. Elsevier Butterworth Heinemann.

11

Two-Dimensional Shell Models with *N*th-Order Displacement Field, the TE Class

A thin shell is defined as a 3D body bounded by two closely spaced curved surfaces, where the distance between the two surfaces should be small compared with the other dimensions. The middle surface of the shell is the locus of the points that lie midway between these surfaces. The distance between the surfaces, measured along the normal to the middle surface, is the thickness of the shell at that point. Shells may be considered as generalizations of a flat plate; conversely, a flat plate can be considered as a special case of a shell with no curvature.

This chapter presents shell FEs based on Taylor expansions of the displacement variables. The exact geometrical description of cylindrical shells is provided first in order to obtain strain–displacement relations. Classical shell models and higher-order models in the unified formulation are then described briefly, as extensions of the plate models. The governing equations are derived from the PVD for the FEA of shell structures. Table 11.1 shows the FN for 2D models that is introduced in this chapter. The nucleus components that are shown in this table are valid in an orthogonal Cartesian reference system. Different expressions are obtained by adopting curvilinear coordinates. Membrane and shear locking phenomena and their correction are discussed and numerical results are given.

In this chapter, the classical tensorial notation is used in which superscripts indicate the covariant components whereas the subscripts indicate the contravariant ones.

11.1 Geometrical Description

The shell can be considered as a solid medium geometrically defined by a mid-surface, given by the coordinates α, β, immersed in physical space and with a parameter representing the thickness z of the medium around this surface.

Finite Element Analysis of Structures Through Unified Formulation, First Edition.
Erasmo Carrera, Maria Cinefra, Marco Petrolo and Enrico Zappino.

Table 11.1 A schematic description of the CUF and the related FN of the stiffness matrix for 2D models

Equilibrium equations in Strong Form $\rightarrow$ $\delta L_i = \int_V \delta \boldsymbol{u} \boldsymbol{k} \boldsymbol{u} dV + \int_S \ldots dS$

$$\underbrace{\begin{bmatrix} k_{xx} & k_{xy} & k_{xz} \\ k_{yx} & k_{yy} & k_{yz} \\ k_{zx} & k_{zy} & k_{zz} \end{bmatrix}}_{\boldsymbol{k}} \underbrace{\begin{Bmatrix} u_x \\ u_y \\ u_z \end{Bmatrix}}_{\boldsymbol{u}} = \underbrace{\begin{Bmatrix} p_x \\ p_y \\ p_z \end{Bmatrix}}_{\boldsymbol{p}}$$

$$k_{xx} = -(\lambda + 2G)\, \partial_{xx} - G\, \partial_{zz} - G\, \partial_{yy};$$
$$k_{xy} = -\lambda\, \partial_{xy} - G\, \partial_{yx};$$
$$k_{xz} = \ldots$$

$\lambda = (E\nu)/[(1+\nu)(1-2\nu)]$; $G = E/[2(1+\nu)]$

$\boldsymbol{u} = \boldsymbol{u}(x, y, z)$

$\delta \boldsymbol{u} = \delta \boldsymbol{u}(x, y, z)$

The diagonal (e.g. k_{xx}) and the non-diagonal (e.g. k_{xy}) terms can be obtained through proper index permutations.

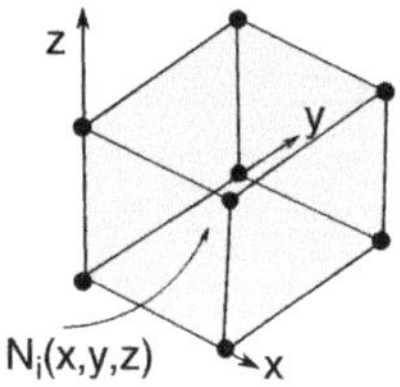

$\boldsymbol{u} = N_i(x, y, z)\boldsymbol{u}_i$

$\delta \boldsymbol{u} = N_j(x, y, z)\delta \boldsymbol{u}_j$

3D FEM Formulation $\rightarrow$ $\delta L_i = \delta \boldsymbol{u}_j \boldsymbol{k}^{ij} \boldsymbol{u}_i$

$$k^{ij}_{xx} = (\lambda + 2G) \int_V N_{j,x} N_{i,x} dV + G \int_V N_{j,z} N_{i,z} dV + G \int_V N_{j,y} N_{i,y} dV;$$

$$k^{ij}_{xy} = \lambda \int_V N_{j,y} N_{i,x} dV + G \int_V N_{j,x} N_{i,y} dV$$

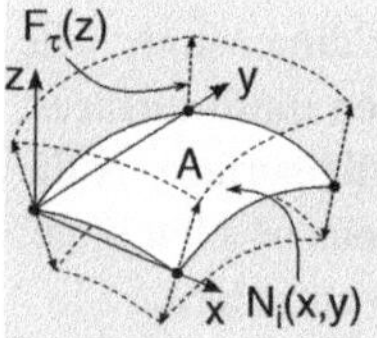

$\boldsymbol{u} = N_i(x, y) F_\tau(z) \boldsymbol{u}_{\tau i}$

$\delta \boldsymbol{u} = N_j(x, y) F_s(z) \delta \boldsymbol{u}_{sj}$

2D FEM Formulation $\rightarrow$ $\delta L_i = \delta \boldsymbol{u}_{sj} \boldsymbol{k}^{\tau sij} \boldsymbol{u}_{\tau i}$

$$k^{\tau sij}_{xx} = (\lambda + 2G) \int_\Omega N_{i,x} N_{j,x} d\Omega \int_h F_\tau F_s dz$$
$$+ G \int_\Omega N_i N_j d\Omega \int_h F_{\tau,z} F_{s,z} dz + G \int_V N_{i,y} N_{j,y} d\Omega \int_h F_\tau F_s dz;$$

$$k^{\tau sij}_{xy} = \lambda \int_\Omega N_{i,y} N_{j,y} d\Omega \int_h F_\tau F_s dz + G \int_\Omega N_{i,x} N_{j,y} d\Omega \int_h F_\tau F_s dz$$

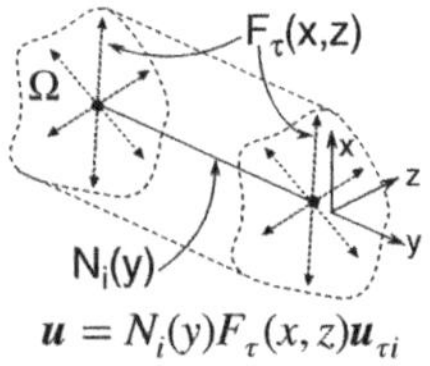

$\boldsymbol{u} = N_i(y) F_\tau(x, z) \boldsymbol{u}_{\tau i}$

$\delta \boldsymbol{u} = N_j(y) F_s(x, z) \delta \boldsymbol{u}_{sj}$

1D FEM Formulation $\rightarrow$ $\delta L_i = \delta \boldsymbol{u}_{sj} \boldsymbol{k}^{\tau sij} \boldsymbol{u}_{\tau i}$

$$k^{\tau sij}_{xx} = (\lambda + 2G) \int_l N_i N_j dy \int_A F_{\tau,x} F_{s,x} dA$$
$$+ G \int_l N_i N_j dy \int_A F_{\tau,z} F_{s,z} dA + G \int_l N_{i,y} N_{j,y} dy \int_A F_\tau F_s dA;$$

$$k^{\tau sij}_{xy} = \lambda \int_l N_{i,y} N_j dy \int_A F_\tau F_{s,x} dA + G \int_l N_i N_{j,y} dy \int_A F_{\tau,x} F_s dA$$

CUF leads to the automatic implementation of any theory of structures through 4 loops (i.e. 4 indexes):

- τ and s deal with the functions that approximate the displacement field and its virtual variation along the plate/shell thickness ($F_\tau(z), F_s(z)$) or over the beam cross-section ($F_\tau(x, z), F_s(x, z)$);
- i and j deal with the shape functions of the FE model, (3D:$N_i(x, y, z), N_j(x, y, z)$; 2D:$N_i(x, y), N_j(x, y)$; 1D:$N_i(y), N_j(y)$).

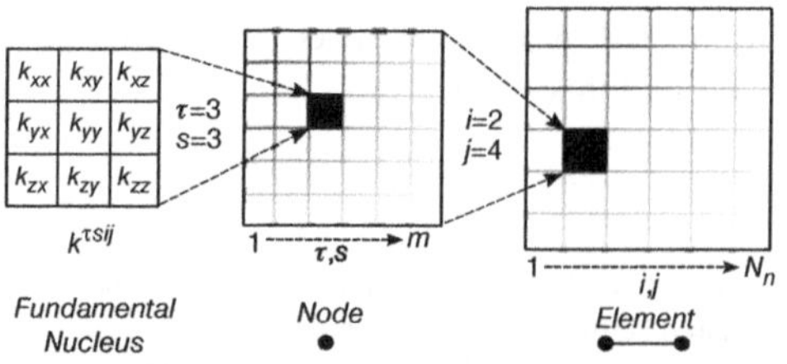

This table shows the essential features of the CUF for 2D models.

The Cartesian coordinates (x, y, z) can be expressed as a function of the curvilinear coordinates α, β, z by means of the so-called 3D chart $\boldsymbol{\Phi}$, i.e.

$$\boldsymbol{\Phi}(\alpha, \beta, z) = \boldsymbol{\phi}(\alpha, \beta) + z\boldsymbol{a}_z(\alpha, \beta) \tag{11.1}$$

Therefore, the 3D chart is defined by means of the 2D chart $\boldsymbol{\phi}$ and a unit vector $\boldsymbol{a}_z$, introduced below.

Starting from $\boldsymbol{\Phi}$, one can calculate the 3D *covariant basis* $(\boldsymbol{g}_\alpha, \boldsymbol{g}_\beta, \boldsymbol{g}_z)$ as follows:

$$\boldsymbol{g}_m = \frac{\partial \boldsymbol{\Phi}(\alpha, \beta, z)}{\partial m}, \quad m = \alpha, \beta, z \tag{11.2}$$

$\boldsymbol{g}_m$ is a local basis and it is defined at each point of the shell volume. The 3D *contravariant basis* $(\boldsymbol{g}^\alpha, \boldsymbol{g}^\beta, \boldsymbol{g}^z)$ can be inferred from the 3D covariant basis by the relations

$$\boldsymbol{g}_m \cdot \boldsymbol{g}^n = \delta_m^n, \quad m, n = \alpha, \beta, z \tag{11.3}$$

where δ denotes Kronecker's delta ($\delta_m^n = 1$ if $m = n$ and 0 otherwise).

On the other hand, the vectors $\boldsymbol{a}_l$ $(l = \alpha, \beta)$ and $\boldsymbol{a}_z$, which form the covariant basis of the plane tangent to the mid-surface at each point, are calculated from the 2D chart as follows:

$$\boldsymbol{a}_l = \frac{\partial \boldsymbol{\phi}(\alpha, \beta)}{\partial l}, \quad l = \alpha, \beta, \quad \boldsymbol{a}_z = \frac{\boldsymbol{a}_\alpha \wedge \boldsymbol{a}_\beta}{\|\boldsymbol{a}_\alpha \wedge \boldsymbol{a}_\beta\|} \tag{11.4}$$

Similar to the 3D case, a contravariant basis of the tangent plane $(\boldsymbol{a}^\alpha, \boldsymbol{a}^\beta)$ can be defined by the relations

$$\boldsymbol{a}_l \cdot \boldsymbol{a}^r = \delta_l^r, \quad l, r = \alpha, \beta \tag{11.5}$$

The 3D base vectors can be defined by means of the basis of the tangent plane, using the following relations:

$$\begin{aligned} \boldsymbol{g}_l &= \left(\delta_l^r - z b_l^r\right)\boldsymbol{a}_r = \mu_l^r \boldsymbol{a}_r, \quad l, r = \alpha, \beta \\ \boldsymbol{g}_z &= \boldsymbol{a}_z \end{aligned} \tag{11.6}$$

where the tensors b_l^r and μ_l^r have been introduced; b_l^r takes into account the curvature of the shell and is

$$b_l^r = \boldsymbol{a}_{,l}^r \cdot \boldsymbol{a}_z$$

In the case of the contravariant basis, one has

$$\boldsymbol{g}^l = m_r^l \boldsymbol{a}^r, \quad l, r = \alpha, \beta \tag{11.7}$$

where m_r^l is the inverse of the tensor μ_l^r introduced in the previous relations (Equations (11.6)),

$$m_r^l = (\mu^{-1})_r^l$$

One can conclude that, knowing the 2D chart $\boldsymbol{\phi}$, it is possible to define each point in the volume of the shell and all the quantities (3D base vectors, 2D base vectors and so on) presented here. The last are used to write the strain–displacement relations of the shell.

11.2 Classical Models and Unified Formulation

The most common mathematical models used to describe shell structures can be classified into two classes, on the basis of different physical assumptions. The Koiter model is based on the Kirchhoff–Love hypothesis and indicated hereafter as the Classical Shell Theory (CST). The Naghdi model is based on the Reissner–Mindlin assumptions that take into account the transverse shear deformation, and it is indicated hereafter as the first-order shear deformation theory (FSDT). In the first model, normals to the reference surface remain normal in the deformed states and do not change in length. This means that transverse shear and transverse normal strains are negligible compared with the other strains. In the second model, one or more CST postulates are removed and, for example, the effects of transverse shear stresses can be taken into account. For more details on the assumptions of the CST and FSDT models, the reader can refer to the works by Koiter (1970) and Naghdi (1972), respectively. The formulation of these models is very similar to that of the CPT and FSDT presented for the plates in Section 10.1. The difference is that, in the case of shells, the displacement components are expressed in curvilinear coordinates.

With (u_α, u_β, u_z) being the displacement components along the coordinates (α, β, z), respectively, the displacement field for the CST model is

$$\begin{aligned} u_\alpha(\alpha, \beta, z) &= u_{\alpha_0}(\alpha, \beta) - u_{z_0,\alpha}(\alpha, \beta)z \\ u_\beta(\alpha, \beta, z) &= u_{\beta_0}(\alpha, \beta) - u_{z_0,\beta}(\alpha, \beta)z \\ u_z(\alpha, \beta, z) &= u_{z_0}(\alpha, \beta) \end{aligned} \tag{11.8}$$

where $(\,.\,)_{,i}$ indicates the partial derivative with respect to the coordinate i.

On the other hand, the displacement field of the FSDT model is

$$\begin{aligned} u_\alpha(\alpha, \beta, z) &= u_{\alpha_0}(\alpha, \beta) + \phi_\alpha(\alpha, \beta)z \\ u_\beta(\alpha, \beta, z) &= u_{\beta_0}(\alpha, \beta) + \phi_\beta(\alpha, \beta)z \\ u_z(\alpha, \beta, z) &= u_{z_0}(\alpha, \beta) \end{aligned} \tag{11.9}$$

where ϕ_α and ϕ_β are the rotations of a transverse normal about the β- and α-axes, respectively.

As in the case of plates, the CST and FSDT models can be seen as particular cases of the $N = 1$ model. By acting on the full linear expansion, they can be obtained by applying penalty techniques as shown in Section 10.2.2. Generalizing the displacement field according to the unified formulation, the displacement field is obtained through a formal expression regardless of the order of the theory (N), which is considered as an input of the analysis.

As in the plates, the unified formulation of the through-the-thickness displacement field $\mathbf{u} = (u_\alpha, u_\beta, u_z)$ is described by an expansion of generic functions (F_τ)

$$\mathbf{u}(\alpha, \beta, z) = F_\tau(z)\mathbf{u}_\tau(\alpha, \beta), \qquad \tau = 0, 1, \ldots, N \tag{11.10}$$

but, for shells, the thickness functions F_τ depend on the thickness coordinate z and the displacement components $\mathbf{u}_\tau$ depend on the coordinates α, β. N is the order of expansion of the model, and the Taylor polynomials, consisting of the base z^i, with $i = 0, 1, \ldots, N$, are chosen as thickness functions. Therefore, the complete expansion of an Nth-order model is

$$
\begin{array}{lll}
u_\alpha = u_{\alpha_0} + z\,u_{\alpha_1} + & \ldots & \| + z^N\, u_{\alpha_N} \\
u_\beta = u_{\beta_0} + z\,u_{\beta_1} + & \ldots & \| + z^N\, u_{\beta_N} \\
u_z = u_{z_0} + z\,u_{z_1} + & \ldots & \| + z^N\, u_{z_N}
\end{array}
\tag{11.11}
$$

Usually, orders of expansion up to $N = 4$ are sufficient to reproduce the 3D solution of shell problems.

11.3 Geometrical Relations for Cylindrical Shells

The geometrical relations for shells are derived by considering the linear part of the 3D Green–Lagrange strain tensor (Chapelle and Bathe 2003), as is expressed in the following formula:

$$\varepsilon'_{ij} = (\boldsymbol{g}_i \mathbf{u}_{,j} + \boldsymbol{g}_j \mathbf{u}_{,i}), \quad i, j = \alpha, \beta, z \tag{11.12}$$

where the subscript comma indicates the partial derivative of the displacements with respect to the curvilinear coordinates (α, β, z) and $\boldsymbol{g}_i$ are the 3D base vectors of the curvilinear reference system.

The derivatives of the displacements are calculated in the following way:

$$\mathbf{u}_{,m} = F_\tau \mathbf{u}_{\tau,m}, \quad m = \alpha, \beta, z \tag{11.13}$$

where

$$\mathbf{u}_{\tau,m} = \left(u_{l_\tau} \boldsymbol{a}^l + u_{z_\tau} \boldsymbol{a}^z\right)_{,m}, \quad l = \alpha, \beta \tag{11.14}$$

Then, one has

$$\left(u_{l_\tau} \boldsymbol{a}^l\right)_{,m} = u_{l_{\tau|m}} \boldsymbol{a}^l + b^l_m u_{l_\tau} \boldsymbol{a}_z, \quad m, l = \alpha, \beta \tag{11.15}$$

where | indicates the covariant derivative, defined as follows:

$$u_{l_{\tau|m}} = u_{l_{\tau,m}} - \Gamma^r_{lm} u_{r_\tau}, \quad r = \alpha, \beta \tag{11.16}$$

The Γ^r_{lm} is the surface *Christoffel symbol* and reads

$$\Gamma^r_{lm} = \boldsymbol{a}_{l,m} \cdot \boldsymbol{a}^r$$

For more details about the mathematical definitions, the reader can refer to the book by Chapelle and Bathe (2003).

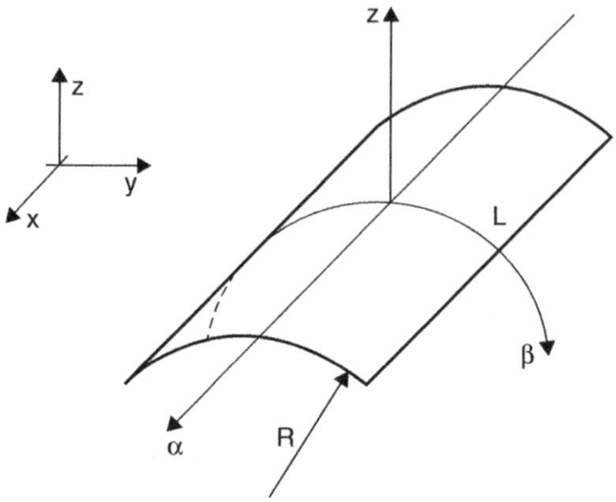

Figure 11.1 Cylindrical shell

If one considers the cylindrical geometry, the 2D chart is as follows:

$$\{x, y, z\}^T = \phi(\alpha, \beta) = \begin{bmatrix} \alpha \\ R\sin(\beta/R) \\ R\cos(\beta/R) \end{bmatrix} \tag{11.17}$$

where R is the curvature radius of the cylinder, as indicated in Figure 11.1. Using the 2D chart, one can calculate the 3D base vectors (Equations (11.6)) and the derivatives of the displacements (Equation (11.13)), following the procedure explained in the previous section. Substituting in the 3D Green–Lagrange strain tensor (Equation (11.12)), one obtains the following strain–displacement relations, valid for the cylinder:

$$\begin{aligned}
\varepsilon'_{\alpha\alpha} &= F_s u_{\alpha_{s,\alpha}} \\
\varepsilon'_{\beta\beta} &= F_s \left[\left(1 + \frac{z}{R}\right)\frac{u_{z_s}}{R} + \left(1 + \frac{z}{R}\right) u_{\beta_{s,\beta}}\right] \\
\varepsilon'_{\alpha\beta} &= F_s \left[u_{\alpha_{s,\beta}} + \left(1 + \frac{z}{R}\right) u_{\beta_{s,\alpha}}\right] = \varepsilon_{\beta\alpha} \\
\varepsilon'_{\alpha z} &= u_{z_{s,\alpha}} F_s + u_{\alpha_s} F_{s,z} \\
\varepsilon'_{\beta z} &= F_s \left[u_{z_{s,\beta}} - \frac{u_{\beta_s}}{R}\right] + F_{s,z}\left[\left(1 + \frac{z}{R}\right) u_{\beta_s}\right] \\
\varepsilon'_{zz} &= u_{z_s} F_{s,z}
\end{aligned} \tag{11.18}$$

The geometrical relations in matrix form are

$$\begin{aligned}
\boldsymbol{\varepsilon}_p &= (\boldsymbol{b}_p + \boldsymbol{a}_p)\mathbf{u} \\
\boldsymbol{\varepsilon}_n &= (\boldsymbol{b}_{np} + \boldsymbol{b}_{nz} - \boldsymbol{a}_n)\mathbf{u}
\end{aligned} \tag{11.19}$$

where the subscript p indicates the in-plane strain components $(\varepsilon'_{\alpha\alpha}, \varepsilon'_{\beta\beta}, \varepsilon'_{\alpha\beta})$ and n the transverse components $(\varepsilon'_{\alpha z}, \varepsilon'_{\beta z}, \varepsilon'_{zz})$. The differential operators used above are defined as follows:

$$\boldsymbol{b}_p = \begin{bmatrix} \partial_\alpha & 0 & 0 \\ 0 & H\partial_\beta & 0 \\ \partial_\beta & H\partial_\alpha & 0 \end{bmatrix}, \quad \boldsymbol{b}_{np} = \begin{bmatrix} 0 & 0 & \partial_\alpha \\ 0 & 0 & \partial_\beta \\ 0 & 0 & 0 \end{bmatrix}, \quad \boldsymbol{b}_{nz} = \partial_z \cdot \boldsymbol{a}_{nz} = \partial_z \cdot \begin{bmatrix} 1 & 0 & 0 \\ 0 & H & 0 \\ 0 & 0 & 1 \end{bmatrix} \tag{11.20}$$

$$\boldsymbol{a}_p = \begin{bmatrix} 0 & 0 & 0 \\ 0 & 0 & H/R \\ 0 & 0 & 0 \end{bmatrix}, \quad \boldsymbol{a}_n = \begin{bmatrix} 0 & 0 & 0 \\ 0 & 1/R & 0 \\ 0 & 0 & 0 \end{bmatrix} \tag{11.21}$$

and $H = (1 + z/R)$.

The strain components ε'_{ij} are expressed in the 3D contravariant basis $(\boldsymbol{g}^\alpha, \boldsymbol{g}^\beta, \boldsymbol{g}^z)$. In order to derive the governing equations, it is necessary to refer all the quantities (displacements, strains and stresses) to the basis $(\boldsymbol{a}^\alpha, \boldsymbol{a}^\beta, \boldsymbol{a}^z)$. Therefore, the strains ε'_{ij} must be transformed in ε_{ij} according to the following relations:

$$\begin{aligned} \varepsilon_{ij} &= m_i^l m_j^r \varepsilon'_{lr} \\ \varepsilon_{lz} &= m_l^r \varepsilon'_{rz} \\ \varepsilon_{zz} &= \varepsilon'_{zz}, \qquad\qquad i,j,l,r = \alpha,\beta \end{aligned} \tag{11.22}$$

where m is the tensor introduced in Equation (11.7) and, in the case of cylindrical geometry, its components are

$$m_\alpha^\alpha = 1, \quad m_\beta^\alpha = m_\alpha^\beta = 0, \quad m_\beta^\beta = \left(1 + \frac{z}{R}\right)^{-1} = H^{-1} \tag{11.23}$$

In this case, the matrices of differential operators in Equations (11.19) are

$$\boldsymbol{b}_p = \begin{bmatrix} \partial_\alpha & 0 & 0 \\ 0 & \partial_\beta/H & 0 \\ \partial_\beta/H & \partial_\alpha & 0 \end{bmatrix}, \quad \boldsymbol{b}_{np} = \begin{bmatrix} 0 & 0 & \partial_\alpha \\ 0 & 0 & \partial_\beta/H \\ 0 & 0 & 0 \end{bmatrix}, \quad \boldsymbol{b}_{nz} = \partial_z \cdot \boldsymbol{a}_{nz} = \partial_z \cdot \begin{bmatrix} 1 & 0 & 0 \\ 0 & 1 & 0 \\ 0 & 0 & 1 \end{bmatrix} \tag{11.24}$$

$$\boldsymbol{a}_p = \begin{bmatrix} 0 & 0 & 0 \\ 0 & 0 & 1/HR \\ 0 & 0 & 0 \end{bmatrix}, \quad \boldsymbol{a}_n = \begin{bmatrix} 0 & 0 & 0 \\ 0 & 1/HR & 0 \\ 0 & 0 & 0 \end{bmatrix} \tag{11.25}$$

If the shell is very thin, the basis $(\boldsymbol{g}^\alpha, \boldsymbol{g}^\beta, \boldsymbol{g}^z)$ can be considered coincident with $(\boldsymbol{a}^\alpha, \boldsymbol{a}^\beta, \boldsymbol{a}^z)$ and this transformation can be neglected, $\varepsilon'_{ij} = \varepsilon_{ij}$. The geometrical relations (Equations (11.19)) remain valid for the plate, in which the radius of curvature R is infinite.

11.4 Governing Equations, FE Formulation and the FN

The derivation of the governing equations is the same as that presented for the plates. However, the expression for stiffness matrix components changes with respect to the plate because it takes into account the curvature terms deriving from the shell geometry.

11.4.1 Governing Equations

In the case of cylindrical shell structures, the PVD is written as

$$\int_{\Omega}\int_{A}\left(\delta\epsilon_p^T\boldsymbol{\sigma}_p + \delta\epsilon_n^T\boldsymbol{\sigma}_n\right) H\, d\Omega\, dz = \int_{\Omega}\int_{A} \delta\boldsymbol{u}\,\boldsymbol{p}\, H\, d\Omega\, dz - \int_{\Omega}\int_{A} \rho\ddot{\boldsymbol{u}}\delta\boldsymbol{u}^T\, H\, d\Omega\, dz \quad (11.26)$$

where Ω is the in-plane integration domain given by the mid-surface of the shell and A is the thickness domain. Moreover, the factor H has been introduced in order to refer the integration domain to the mid-surface of the shell.

The constitutive equations remain equal to those for the plate,

$$\begin{aligned}\boldsymbol{\sigma}_p &= \boldsymbol{C}_{pp}\boldsymbol{\varepsilon}_p + \boldsymbol{C}_{pn}\boldsymbol{\varepsilon}_n \\ \boldsymbol{\sigma}_n &= \boldsymbol{C}_{np}\boldsymbol{\varepsilon}_p + \boldsymbol{C}_{nn}\boldsymbol{\varepsilon}_n\end{aligned} \quad (11.27)$$

where the stress (σ) and strain (ε) components are

$$\begin{aligned}\boldsymbol{\sigma}_p &= \left\{\sigma_{\alpha\alpha} \quad \sigma_{\beta\beta} \quad \sigma_{\alpha\beta}\right\}^T, \quad \boldsymbol{\varepsilon}_p = \left\{\varepsilon_{\alpha\alpha} \quad \varepsilon_{\beta\beta} \quad \varepsilon_{\alpha\beta}\right\}^T \\ \boldsymbol{\sigma}_n &= \left\{\sigma_{\alpha z} \quad \sigma_{\beta z} \quad \sigma_{zz}\right\}^T, \quad \boldsymbol{\varepsilon}_n = \left\{\varepsilon_{\alpha z} \quad \varepsilon_{\beta z} \quad \varepsilon_{zz}\right\}^T\end{aligned} \quad (11.28)$$

Substituting the constitutive equations (Equations (11.27)) and the geometrical relations (Equations (11.19)) in the PVD (Equation (11.26)), one obtains the governing equations.

11.4.2 FE Formulation

According to the FEM, the displacement field (Equation (11.10)) is interpolated by means of the shape functions N_i,

$$\mathbf{u} = F_\tau N_i \mathbf{u}_{\tau i} \quad (11.29)$$

with $\tau = 0, 1, \ldots, N$ and $i = 1, \ldots, N_n$, where N_n is the number of nodes of the element and $\mathbf{u}^k_{\tau i}$ are the nodal unknown variables.

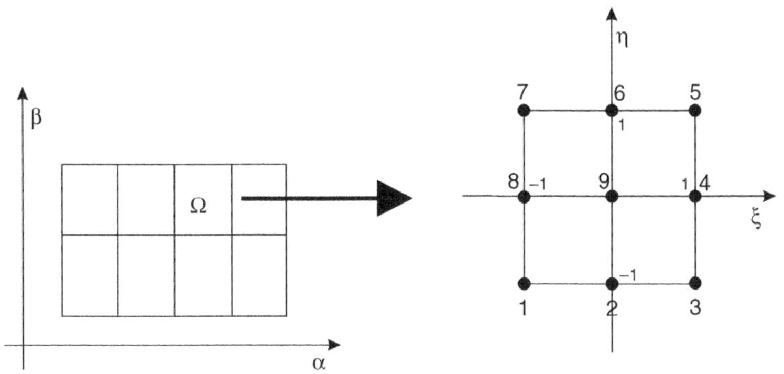

Figure 11.2 Local reference system of the element

The shell elements considered here have nine nodes and the Lagrangian shape functions are used to interpolate the displacements. The shape functions, expressed in the local reference system of the element (ξ, η) (see Figure 11.2), are as follows:

$$
\begin{aligned}
N_1 &= \frac{1}{4}(\xi^2 - \xi)(\eta^2 - \eta) \\
N_2 &= \frac{1}{2}(1 - \xi^2)(\eta^2 - \eta) \\
N_3 &= \frac{1}{4}(\xi^2 + \xi)(\eta^2 - \eta) \\
N_4 &= \frac{1}{2}(\xi^2 + \xi)(1 - \eta^2) \\
N_5 &= \frac{1}{4}(\xi^2 + \xi)(\eta^2 + \eta) \\
N_6 &= \frac{1}{2}(1 - \xi^2)(\eta^2 + \eta) \\
N_7 &= \frac{1}{4}(\xi^2 - \xi)(\eta^2 + \eta) \\
N_8 &= \frac{1}{2}(\xi^2 - \xi)(1 - \eta^2) \\
N_9 &= (1 - \xi^2)(1 - \eta^2)
\end{aligned}
\tag{11.30}
$$

11.5 Membrane and Shear Locking Phenomenon

It is known that when an FEM is used to discretize a physical model, the numerical locking phenomenon may arise from hidden constraints that are not well represented in the FE approximation. In the Naghdi model, both transverse shear and membrane constraints appear as the shell thickness diminishes, and locking may arise. One locking phenomenon that is already clearly understood is the *shear locking* of plates (see Section 10.5.2). Shear locking is also present in shells, but another and more severe type of locking, referred to as *membrane locking*, also appears (Pitkäranta, 1992). The most common approach proposed to overcome the locking phenomenon is the use of standard displacement formulations with higher-order elements (see Hakula *et al.* (1996) and Chinosi *et al.* (1998)). The numerical results of higher-order elements show that they are able to contrast locking in the shell problem in its displacement formulation. However, in the case of very small thickness and when the element does not have a high degree, as needed, the numerical solution exhibits a loss in the convergence rate. Another remedy for locking is the use of several techniques of reduced-selective integration (see Zienkiewicz *et al.* (1971) and Stolarski and Belytschko (1981)), but they fail because the additional DOFs at the centre of the element produce spurious modes. Finally, in the scientific literature, it is possible to find many examples of modified variational forms that can be used to overcome the locking problem which provide good results: see Arnold and Brezzi (1997); Bathe and Dvorkin (1986); Bucalem and Bathe (1993); Chang *et al.* (1989); Chinosi and Della Croce (1998); Huang (1987); and Rhiu and Lee (1987).

Based on the work of Bathe and colleagues, this section presents the mixed interpolation of tensorial components (MITC) technique, extended to variable kinematic models of the unified formulation, for the case of shell elements with nine nodes. Derivation of the stiffness matrix is also performed.

11.5.1 MITC9 Shell Element

Considering the components of the strain tensor in the local coordinate system (ξ, η), the MITC shell elements are formulated by using – instead of the strain components directly computed from the displacements – an interpolation of these strain components within each element, following a specific interpolation strategy for each component. The interpolation points corresponding to each strain component, called the *tying points*, are shown in Figure 11.3 for a nine-node shell element.

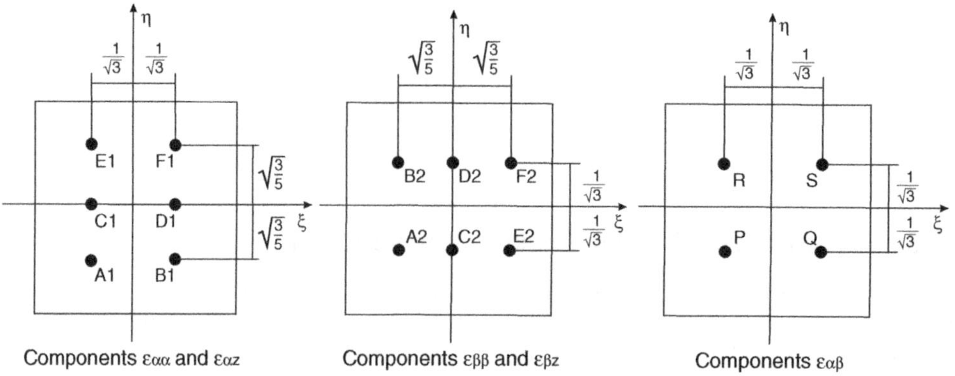

Figure 11.3 Tying points for MITC9 shell FEs

The strain components $\varepsilon_{\alpha\alpha}$ and $\varepsilon_{\alpha z}$ are interpolated on the points $A1, B1, C1, D1, E1, F1$ by means of the following functions:

$$
\begin{aligned}
N_{A1} &= \frac{5\sqrt{3}}{12}\eta\left(\eta - \sqrt{\frac{3}{5}}\right)\left(\frac{1}{\sqrt{3}} - \xi\right)\\
N_{B1} &= \frac{5\sqrt{3}}{12}\eta\left(\eta - \sqrt{\frac{3}{5}}\right)\left(\frac{1}{\sqrt{3}} + \xi\right)\\
N_{C1} &= -\frac{5\sqrt{3}}{6}\left(\eta + \sqrt{\frac{3}{5}}\right)\left(\eta - \sqrt{\frac{3}{5}}\right)\left(\frac{1}{\sqrt{3}} - \xi\right)\\
N_{D1} &= -\frac{5\sqrt{3}}{6}\left(\eta + \sqrt{\frac{3}{5}}\right)\left(\eta - \sqrt{\frac{3}{5}}\right)\left(\frac{1}{\sqrt{3}} + \xi\right)\\
N_{E1} &= \frac{5\sqrt{3}}{12}\left(\eta + \sqrt{\frac{3}{5}}\right)\eta\left(\frac{1}{\sqrt{3}} - \xi\right)\\
N_{F1} &= \frac{5\sqrt{3}}{12}\left(\eta + \sqrt{\frac{3}{5}}\right)\eta\left(\frac{1}{\sqrt{3}} + \xi\right)
\end{aligned}
\tag{11.31}
$$

For the components $\varepsilon_{\beta\beta}$ and $\varepsilon_{\beta z}$, the interpolating functions on the points $(A2, B2, C2, D2, E2, F2)$ are

$$
\begin{aligned}
N_{A2} &= \frac{5\sqrt{3}}{12}\xi\left(\xi - \sqrt{\frac{3}{5}}\right)\left(\frac{1}{\sqrt{3}} - \eta\right)\\
N_{B2} &= \frac{5\sqrt{3}}{12}\xi\left(\xi - \sqrt{\frac{3}{5}}\right)\left(\frac{1}{\sqrt{3}} + \eta\right)\\
N_{C2} &= -\frac{5\sqrt{3}}{6}\left(\xi + \sqrt{\frac{3}{5}}\right)\left(\xi - \sqrt{\frac{3}{5}}\right)\left(\frac{1}{\sqrt{3}} - \eta\right)\\
N_{D2} &= -\frac{5\sqrt{3}}{6}\left(\xi + \sqrt{\frac{3}{5}}\right)\left(\xi - \sqrt{\frac{3}{5}}\right)\left(\frac{1}{\sqrt{3}} + \eta\right)\\
N_{E2} &= \frac{5\sqrt{3}}{12}\left(\xi + \sqrt{\frac{3}{5}}\right)\xi\left(\frac{1}{\sqrt{3}} - \eta\right)\\
N_{F2} &= \frac{5\sqrt{3}}{12}\left(\xi + \sqrt{\frac{3}{5}}\right)\xi\left(\frac{1}{\sqrt{3}} + \eta\right)
\end{aligned}
\tag{11.32}
$$

Finally, the interpolating functions on the points P, Q, R, S for the strain $\varepsilon_{\alpha\beta}$ are

$$
\begin{aligned}
N_P &= \frac{3}{4}\left(\frac{1}{\sqrt{3}} - \xi\right)\left(\frac{1}{\sqrt{3}} - \eta\right) \\
N_Q &= \frac{3}{4}\left(\frac{1}{\sqrt{3}} + \xi\right)\left(\frac{1}{\sqrt{3}} - \eta\right) \\
N_R &= \frac{3}{4}\left(\frac{1}{\sqrt{3}} - \xi\right)\left(\frac{1}{\sqrt{3}} + \eta\right) \\
N_S &= \frac{3}{4}\left(\frac{1}{\sqrt{3}} + \xi\right)\left(\frac{1}{\sqrt{3}} + \eta\right)
\end{aligned}
\tag{11.33}
$$

The strain component ε_{zz} is directly calculated from the displacements.

For convenience, the interpolating functions can be arranged in the following vectors:

$$
\begin{aligned}
\boldsymbol{N}_{m1} &= [N_{A1}, N_{B1}, N_{C1}, N_{D1}, N_{E1}, N_{F1}] \\
\boldsymbol{N}_{m2} &= [N_{A2}, N_{B2}, N_{C2}, N_{D2}, N_{E2}, N_{F2}] \\
\boldsymbol{N}_{m3} &= [N_P, N_Q, N_R, N_S]
\end{aligned}
\tag{11.34}
$$

Considering the MITC method, the strain components are written as follows:

$$
\begin{aligned}
\boldsymbol{\varepsilon}_p &= \begin{Bmatrix} \varepsilon_{\alpha\alpha} \\ \varepsilon_{\beta\beta} \\ \varepsilon_{\alpha\beta} \end{Bmatrix} = \begin{bmatrix} \boldsymbol{N}_{m1} & 0 & 0 \\ 0 & \boldsymbol{N}_{m2} & 0 \\ 0 & 0 & \boldsymbol{N}_{m3} \end{bmatrix} \begin{Bmatrix} \boldsymbol{\varepsilon}_{\alpha\alpha_{m1}} \\ \boldsymbol{\varepsilon}_{\beta\beta_{m2}} \\ \boldsymbol{\varepsilon}_{\alpha\beta_{m3}} \end{Bmatrix} = \boldsymbol{N}_1 \begin{Bmatrix} \boldsymbol{\varepsilon}_{\alpha\alpha_{m1}} \\ \boldsymbol{\varepsilon}_{\beta\beta_{m2}} \\ \boldsymbol{\varepsilon}_{\alpha\beta_{m3}} \end{Bmatrix} \\
\boldsymbol{\varepsilon}_n &= \begin{Bmatrix} \varepsilon_{\alpha z} \\ \varepsilon_{\beta z} \\ \varepsilon_{zz} \end{Bmatrix} = \begin{bmatrix} \boldsymbol{N}_{m1} & 0 & 0 \\ 0 & \boldsymbol{N}_{m2} & 0 \\ 0 & 0 & 1 \end{bmatrix} \begin{Bmatrix} \boldsymbol{\varepsilon}_{\alpha z_{m1}} \\ \boldsymbol{\varepsilon}_{\beta z_{m2}} \\ \varepsilon_{zz} \end{Bmatrix} = \boldsymbol{N}_2 \begin{Bmatrix} \boldsymbol{\varepsilon}_{\alpha z_{m1}} \\ \boldsymbol{\varepsilon}_{\beta z_{m2}} \\ \varepsilon_{zz} \end{Bmatrix}
\end{aligned}
\tag{11.35}
$$

where the matrices $\boldsymbol{N}_1$ and $\boldsymbol{N}_2$ have been introduced and the vectors of the strains are

$$
\begin{aligned}
&\boldsymbol{\varepsilon}_{\alpha\alpha_{m1}} = \begin{Bmatrix} \varepsilon_{\alpha\alpha}(A1) \\ \varepsilon_{\alpha\alpha}(B1) \\ \varepsilon_{\alpha\alpha}(C1) \\ \varepsilon_{\alpha\alpha}(D1) \\ \varepsilon_{\alpha\alpha}(E1) \\ \varepsilon_{\alpha\alpha}(F1) \end{Bmatrix}, \quad \boldsymbol{\varepsilon}_{\beta\beta_{m2}} = \begin{Bmatrix} \varepsilon_{\beta\beta}(A2) \\ \varepsilon_{\beta\beta}(B2) \\ \varepsilon_{\beta\beta}(C2) \\ \varepsilon_{\beta\beta}(D2) \\ \varepsilon_{\beta\beta}(E2) \\ \varepsilon_{\beta\beta}(F2) \end{Bmatrix}, \quad \boldsymbol{\varepsilon}_{\alpha\beta_{m3}} = \begin{Bmatrix} \varepsilon_{\alpha\beta}(P) \\ \varepsilon_{\alpha\beta}(Q) \\ \varepsilon_{\alpha\beta}(R) \\ \varepsilon_{\alpha\beta}(S) \end{Bmatrix} \\
&\boldsymbol{\varepsilon}_{\alpha z_{m1}} = \begin{Bmatrix} \varepsilon_{\alpha z}(A1) \\ \varepsilon_{\alpha z}(B1) \\ \varepsilon_{\alpha z}(C1) \\ \varepsilon_{\alpha z}(D1) \\ \varepsilon_{\alpha z}(E1) \\ \varepsilon_{\alpha z}(F1) \end{Bmatrix}, \quad \boldsymbol{\varepsilon}_{\beta z_{m2}} = \begin{Bmatrix} \varepsilon_{\beta z}(A2) \\ \varepsilon_{\beta z}(B2) \\ \varepsilon_{\beta z}(C2) \\ \varepsilon_{\beta z}(D2) \\ \varepsilon_{\beta z}(E2) \\ \varepsilon_{\beta z}(F2) \end{Bmatrix}
\end{aligned}
\tag{11.36}
$$

The notation (m) indicates that the strain is calculated at the tying point m using the geometrical relations (Equations (11.19)). For example, if one considers the strain $\varepsilon_{\alpha\alpha}$ calculated at the point $A1$, one has

$$\boldsymbol{\varepsilon}_{\alpha\alpha}(A1) = F_s(\boldsymbol{b}_p + \boldsymbol{a}_p)_{(1,:)} N_j(\xi_{A1}, \eta_{A1})\mathbf{u}_{sj} \tag{11.37}$$

where $(1, :)$ means that the first row of the matrix $(\boldsymbol{b}_p + \boldsymbol{a}_p)$ is considered and ξ_{A1}, η_{A1} are the coordinates of the point $A1$.

According to this interpolation, the geometrical relations can be rewritten as follows:

$$\begin{aligned} \boldsymbol{\varepsilon}^s_{p_{jm}} &= F_s \boldsymbol{B}_{3_{jm}} \mathbf{u}_{sj} \\ \boldsymbol{\varepsilon}^s_{n_{jm}} &= F_s \boldsymbol{B}_{1_{jm}} \mathbf{u}_{sj} + F_{s_{,z}} \boldsymbol{B}_{2_{jm}} \mathbf{u}_{sj} \end{aligned} \tag{11.38}$$

where m indicates a loop on the tying points and the matrices introduced are

$$\begin{aligned} \boldsymbol{B}_{1_{jm}} &= \boldsymbol{N}_2 \begin{bmatrix} (\boldsymbol{b}_{n\Omega} - \boldsymbol{a}_n)_{(1,:)}(N_j\boldsymbol{I})_{m1} \\ (\boldsymbol{b}_{n\Omega} - \boldsymbol{a}_n)_{(2,:)}(N_j\boldsymbol{I})_{m2} \\ (\boldsymbol{b}_{n\Omega} - \boldsymbol{a}_n)_{(3,:)}(N_j\boldsymbol{I}) \end{bmatrix}, \quad \boldsymbol{B}_{2_{jm}} = \boldsymbol{N}_2 \begin{bmatrix} \boldsymbol{a}_{nz(1,:)}(N_j\boldsymbol{I})_{m1} \\ \boldsymbol{a}_{nz(2,:)}(N_j\boldsymbol{I})_{m2} \\ \boldsymbol{a}_{nz(3,:)}(N_j\boldsymbol{I}) \end{bmatrix}, \\ \boldsymbol{B}_{3_{jm}} &= \boldsymbol{N}_1 \begin{bmatrix} (\boldsymbol{b}_p + \boldsymbol{a}_p)_{(1,:)}(N_j\boldsymbol{I})_{m1} \\ (\boldsymbol{b}_p + \boldsymbol{a}_p)_{(2,:)}(N_j\boldsymbol{I})_{m2} \\ (\boldsymbol{b}_p + \boldsymbol{a}_p)_{(3,:)}(N_j\boldsymbol{I})_{m3} \end{bmatrix} \end{aligned} \tag{11.39}$$

Also, the constitutive equations are rewritten considering the loop on the tying points by means of the index n

$$\begin{aligned} \boldsymbol{\sigma}^\tau_{p_{in}} &= \boldsymbol{C}_{pp}\boldsymbol{\varepsilon}^\tau_{p_{in}} + \boldsymbol{C}_{pn}\boldsymbol{\varepsilon}^\tau_{n_{in}} \\ \boldsymbol{\sigma}^\tau_{n_{in}} &= \boldsymbol{C}_{np}\boldsymbol{\varepsilon}^\tau_{p_{in}} + \boldsymbol{C}_{nn}\boldsymbol{\varepsilon}^\tau_{n_{in}} \end{aligned} \tag{11.40}$$

By substituting the constitutive equations (Equations (11.40)), the geometrical relations (Equations (11.38)) and the displacement field (Equation (11.29)) in the PVD (Equation (11.26)), one obtains

$$\begin{aligned} &\delta\mathbf{u}^T_{sj}\left\{\int_A F_s\left(\int_\Omega [\boldsymbol{B}^T_{3_{jm}}(\boldsymbol{C}_{pp}\boldsymbol{B}_{3_{in}} + \boldsymbol{C}_{pn}\boldsymbol{B}_{1_{in}}) + \boldsymbol{B}^T_{1_{jm}}(\boldsymbol{C}_{np}\boldsymbol{B}_{3_{in}} + \boldsymbol{C}_{nn}\boldsymbol{B}_{1_{in}})]\, d\Omega\right) F_\tau H\, dz\right\}\mathbf{u}_{\tau i} \\ &+ \delta\mathbf{u}^T_{sj}\left\{\int_A F_s\left(\int_\Omega [(\boldsymbol{B}^T_{3_{jm}}\boldsymbol{C}_{pn} + \boldsymbol{B}^T_{1_{jm}}\boldsymbol{C}_{nn})\boldsymbol{B}_{2_{in}}]\, d\Omega\right) F_{\tau,z} H\, dz\right\}\mathbf{u}_{\tau i} \\ &+ \delta\mathbf{u}^T_{sj}\left\{\int_A F_{s,z}\left(\int_\Omega [\boldsymbol{B}^T_{2_{jm}}(\boldsymbol{C}_{np}\boldsymbol{B}_{3_{in}} + \boldsymbol{C}_{nn}\boldsymbol{B}_{1_{in}})]\, d\Omega\right) F_\tau H\, dz\right\}\mathbf{u}_{\tau i} \\ &+ \delta\mathbf{u}^T_{sj}\left\{\int_A F_{s,z}\left(\int_\Omega [\boldsymbol{B}^T_{2_{jm}}\boldsymbol{C}_{nn}\boldsymbol{B}_{2_{in}}]\, d\Omega\right) F_{\tau,z} H\, dz\right\}\mathbf{u}_{\tau i} \\ &= \delta\mathbf{u}^T_{sj}\left\{\int_A F_s\left(\int_\Omega N_j p\, d\Omega\right) H\, dz\right\} \end{aligned} \tag{11.41}$$

Therefore, the following governing equation system can be written:

$$\boldsymbol{K}\mathbf{u} = \boldsymbol{P} \tag{11.42}$$

where the vector of unknowns is given by the nodal displacements $\mathbf{u} = \mathbf{u}_{s_j}$ and the FN is dependent on the indexes i, j.

11.5.2 Stiffness Matrix

In order to write the explicit expression for the FN of the stiffness matrix, the following matrices are introduced:

$$N_m = \begin{bmatrix} N_{A1} & N_{A2} & N_P \\ N_{B1} & N_{B2} & N_Q \\ N_{C1} & N_{C2} & N_R \\ N_{D1} & N_{D2} & N_S \\ N_{E1} & N_{E2} & 0 \\ N_{F1} & N_{F2} & 0 \end{bmatrix} \tag{11.43}$$

where it is assumed that $N_m(:,1) = N_{m1}$, $N_m(:,2) = N_{m2}$ and $N_m(:,3) = N_{m3}$. Similarly, the shape functions and the corresponding derivatives with respect to α and β are calculated at the tying points and arranged in the following matrices:

$$N_{i_m} = \begin{bmatrix} N_{i_{A1}} & N_{i_{A2}} & N_{i_P} \\ N_{i_{B1}} & N_{i_{B2}} & N_{i_Q} \\ N_{i_{C1}} & N_{i_{C2}} & N_{i_R} \\ N_{i_{D1}} & N_{i_{D2}} & N_{i_S} \\ N_{i_{E1}} & N_{i_{E2}} & 0 \\ N_{i_{F1}} & N_{i_{F2}} & 0 \end{bmatrix}, \quad N_{i,1_m} = \begin{bmatrix} N_{i,\alpha_{A1}} & N_{i,\alpha_{A2}} & N_{i,\alpha_P} \\ N_{i,\alpha_{B1}} & N_{i,\alpha_{B2}} & N_{i,\alpha_Q} \\ N_{i,\alpha_{C1}} & N_{i,\alpha_{C2}} & N_{i,\alpha_R} \\ N_{i,\alpha_{D1}} & N_{i,\alpha_{D2}} & N_{i,\alpha_S} \\ N_{i,\alpha_{E1}} & N_{i,\alpha_{E2}} & 0 \\ N_{i,\alpha_{F1}} & N_{i,\alpha_{F2}} & 0 \end{bmatrix},$$

$$N_{i,2_m} = \begin{bmatrix} N_{i,\beta_{A1}} & N_{i,\beta_{A2}} & N_{i,\beta_P} \\ N_{i,\beta_{B1}} & N_{i,\beta_{B2}} & N_{i,\beta_Q} \\ N_{i,\beta_{C1}} & N_{i,\beta_{C2}} & N_{i,\beta_R} \\ N_{i,\beta_{D1}} & N_{i,\beta_{D2}} & N_{i,\beta_S} \\ N_{i,\beta_{E1}} & N_{i,\beta_{E2}} & 0 \\ N_{i,\beta_{F1}} & N_{i,\beta_{F2}} & 0 \end{bmatrix} \tag{11.44}$$

and it is assumed that

$$\begin{aligned} &N_{i_m}(:,1) = N_{i_{m1}}, \quad N_{i_m}(:,2) = N_{i_{m2}}, \quad N_{i_m}(:,3) = N_{i_{m3}} \\ &N_{i,\alpha_m}(:,1) = N_{i,\alpha_{m1}}, \quad N_{i,\alpha_m}(:,2) = N_{i,\alpha_{m2}}, \quad N_{i,\alpha_m}(:,3) = N_{i,\alpha_{m3}} \\ &N_{i,\beta_m}(:,1) = N_{i,\beta_{m1}}, \quad N_{i,\beta_m}(:,2) = N_{i,\beta_{m2}}, \quad N_{i,\beta_m}(:,3) = N_{i,\beta_{m3}} \end{aligned} \tag{11.45}$$

Therefore, the components of the FN $\mathbf{k}^{\tau sij}$, obtained using the differential matrices in Equations (11.20) and (11.21), are

$$\begin{aligned}
k_{\alpha\alpha}^{\tau sij} &= C_{55} N_{i_{m1}} \int_\Omega N_{m1} N_{n1}\, d\Omega\, N_{j_{n1}} \int_A H F_{\tau,z} F_{s,z}\, dz \\
&+ C_{11} N_{i,\alpha_{m1}} \int_\Omega N_{m1} N_{n1}\, d\Omega\, N_{j,\alpha_{n1}} \int_A H F_\tau F_s\, dz \\
&+ C_{66} N_{i,\beta_{m3}} \int_\Omega N_{m3} N_{n3}\, d\Omega\, N_{j,\beta_{n3}} \int_A H F_\tau F_s\, dz \\
k_{\alpha\beta}^{\tau sij} &= C_{12} N_{i,\alpha_{m1}} \int_\Omega N_{m1} N_{n2}\, d\Omega\, N_{j,\beta_{n2}} \int_A H^2 F_\tau F_s\, dz \\
&+ C_{66} N_{i,\beta_{m3}} \int_\Omega N_{m3} N_{n3}\, d\Omega\, N_{j,\alpha_{n3}} \int_A H^2 F_\tau F_s\, dz \\
k_{\alpha z}^{\tau sij} &= C_{13} N_{i,\alpha_{m1}} \int_\Omega N_{m1} N_j\, d\Omega \int_A H F_\tau F_{s,z}\, dz \\
&+ C_{12} \frac{1}{R} N_{i,\alpha_{m1}} \int_\Omega N_{m1} N_{n2}\, d\Omega\, N_{j_{n2}} \int_A H^2 F_\tau F_s\, dz \\
&+ C_{55} N_{i_{m1}} \int_\Omega N_{m1} N_{n1}\, d\Omega\, N_{j,\alpha_{n1}} \int_A H F_{\tau,z} F_s\, dz
\end{aligned} \tag{11.46}$$

$$\begin{aligned}
k_{\beta\alpha}^{\tau sij} &= C_{12} N_{i,\beta_{m2}} \int_\Omega N_{m2} N_{n1}\, d\Omega\, N_{j,\alpha_{n1}} \int_A H^2 F_\tau F_s\, dz \\
&+ C_{66} N_{i,\alpha_{m3}} \int_\Omega N_{m3} N_{n3}\, d\Omega\, N_{j,\beta_{n3}} \int_A H^2 F_\tau F_s\, dz \\
k_{\beta\beta}^{\tau sij} &= C_{22} N_{i,\beta_{m2}} \int_\Omega N_{m2} N_{n2}\, d\Omega\, N_{j,\beta_{n2}} \int_A H^3 F_\tau F_s\, dz \\
&+ C_{66} N_{i,\alpha_{m3}} \int_\Omega N_{m3} N_{n3}\, d\Omega\, N_{j,\alpha_{n3}} \int_A H^3 F_\tau F_s\, dz \\
&+ C_{44} \frac{1}{R^2} N_{i_{m2}} \int_\Omega N_{m2} N_{n2}\, d\Omega\, N_{j_{n2}} \int_A H F_\tau F_s\, dz \\
&- C_{44} \frac{1}{R} N_{i_{m2}} \int_\Omega N_{m2} N_{n2}\, d\Omega\, N_{j_{n2}} \int_A H^2 F_\tau F_{s,z}\, dz \\
&- C_{44} \frac{1}{R} N_{i_{m2}} \int_\Omega N_{m2} N_{n2}\, d\Omega\, N_{j_{n2}} \int_A H^2 F_{\tau,z} F_s\, dz \\
&+ C_{44} N_{i_{m2}} \int_\Omega N_{m2} N_{n2}\, d\Omega\, N_{j_{n2}} \int_A H^3 F_{\tau,z} F_{s,z}\, dz
\end{aligned} \tag{11.47}$$

$$
\begin{aligned}
k_{\beta z}^{\tau sij} &= C_{22}\frac{1}{R}N_{i,\beta_{m2}}\int_{\Omega} N_{m2}N_{n2}\,d\Omega\, N_{j_{n2}}\int_{A} H^3 F_{\tau} F_{s}\,dz \\
&+ C_{23}N_{i,\beta_{m2}}\int_{\Omega} N_{m2}N_{j}\,d\Omega \int_{A} H^2 F_{\tau} F_{s,z}\,dz \\
&- C_{44}\frac{1}{R}N_{i_{m2}}\int_{\Omega} N_{m2}N_{n2}\,d\Omega\, N_{j,\beta_{n2}}\int_{A} H F_{\tau} F_{s}\,dz \\
&+ C_{44}N_{i_{m2}}\int_{\Omega} N_{m2}N_{n2}\,d\Omega\, N_{j,\beta_{n2}}\int_{A} H^2 F_{\tau,z} F_{s}\,dz \\
k_{z\alpha}^{\tau sij} &= C_{55}N_{i,\alpha_{m1}}\int_{\Omega} N_{m1}N_{n1}\,d\Omega\, N_{j_{n1}}\int_{A} H F_{\tau} F_{s,z}\,dz \\
&+ C_{12}\frac{1}{R}N_{i_{m2}}\int_{\Omega} N_{m2}N_{n1}\,d\Omega\, N_{j,\alpha_{n1}}\int_{A} H^2 F_{\tau} F_{s}\,dz \\
&+ C_{13}\int_{\Omega} N_{i}N_{n1}\,d\Omega\, N_{j,\alpha_{n1}}\int_{A} H F_{\tau,z} F_{s}\,dz \\
k_{z\beta}^{\tau sij} &= C_{22}\frac{1}{R}N_{i_{m2}}\int_{\Omega} N_{m2}N_{n2}\,d\Omega\, N_{j,\beta_{n2}}\int_{A} H^3 F_{\tau} F_{s}\,dz \\
&+ C_{23}\int_{\Omega} N_{i}N_{n2}\,d\Omega\, N_{j,\beta_{n2}}\int_{A} H^2 F_{\tau,z} F_{s}\,dz \\
&- C_{44}\frac{1}{R}N_{i,\beta_{m2}}\int_{\Omega} N_{m2}N_{n2}\,d\Omega\, N_{j_{n2}}\int_{A} H F_{\tau} F_{s}\,dz \\
&+ C_{44}N_{i,\beta_{m2}}\int_{\Omega} N_{m2}N_{n2}\,d\Omega\, N_{j_{n2}}\int_{A} H^2 F_{\tau} F_{s}\,dz \qquad (11.48) \\
k_{zz}^{\tau sij} &= C_{22}\frac{1}{R^2}N_{i_{m2}}\int_{\Omega} N_{m2}N_{n2}\,d\Omega\, N_{j_{n2}}\int_{A} H^3 F_{\tau} F_{s}\,dz \\
&+ C_{23}\frac{1}{R}N_{i_{m2}}\int_{\Omega} N_{m2}N_{j}\,d\Omega \int_{A} H^2 F_{\tau} F_{s,z}\,dz \\
&+ C_{23}\frac{1}{R}\int_{\Omega} N_{i}N_{n2}\,d\Omega\, N_{j_{n2}}\int_{A} H^2 F_{\tau,z} F_{s}\,dz \\
&+ C_{33}\int_{\Omega} N_{i}N_{j}\,d\Omega \int_{A} H F_{\tau,z} F_{s,z}\,dz \\
&+ C_{55}N_{i,\alpha_{m1}}\int_{\Omega} N_{m1}N_{n1}\,d\Omega\, N_{j,\alpha_{n1}}\int_{A} H F_{\tau} F_{s}\,dz \\
&+ C_{44}N_{i,\beta_{m2}}\int_{\Omega} N_{m2}N_{n2}\,d\Omega\, N_{j,\beta_{n2}}\int_{A} H F_{\tau} F_{s}\,dz
\end{aligned}
$$

where $m, n = 1, \ldots, 6$ are sum indexes that refer to tying points. This FN is expanded on the indexes $i, j = 1, \ldots, 9$ and $\tau, s = 0, \ldots, N$ in order to build the stiffness matrix of each element.

Then, the stiffness matrices are assembled at element level by imposing the compatibility conditions. Note that, in the case of shells, the integrals in the domain Ω are calculated using the Jacobian of the transformation from the normalized coordinates (ξ, η) to the curvilinear coordinates (α, β). The formal expression of the Jacobian matrix is the same as that for plates in Equation (10.74).

11.6 Numerical Applications

Some discriminating problems taken from the literature have been considered in order to test the efficiency and the robustness of the MITC9 shell element based on the CUF. The first test is called the *pinched shell* test. The essential shape of this structure is shown in Figure 11.4. The pinched shell is simply supported at each end by a rigid diaphragm and singularly loaded by two opposing forces acting at the midpoint of the shell. Due to the symmetry of the structure, the computations have been performed, using a uniform decomposition, on an octave of the shell ($\overline{ABCD}$). The physical data given in Table 11.2 have been assumed.

The following symmetry conditions are applied:

$$\begin{aligned} u_{\beta_\tau}(\alpha, 0) &= 0 \\ u_{\alpha_\tau}(0, \beta) &= 0 \\ u_{\beta_\tau}(\alpha, R\pi/2) &= 0 \end{aligned} \tag{11.49}$$

and the following simply supported conditions are prescribed on the boundary:

$$u_{\beta_\tau}(L/2, \beta) = u_{z_\tau}(L/2, \beta) = 0 \tag{11.50}$$

with $\tau = 0, 1, \ldots, N$.

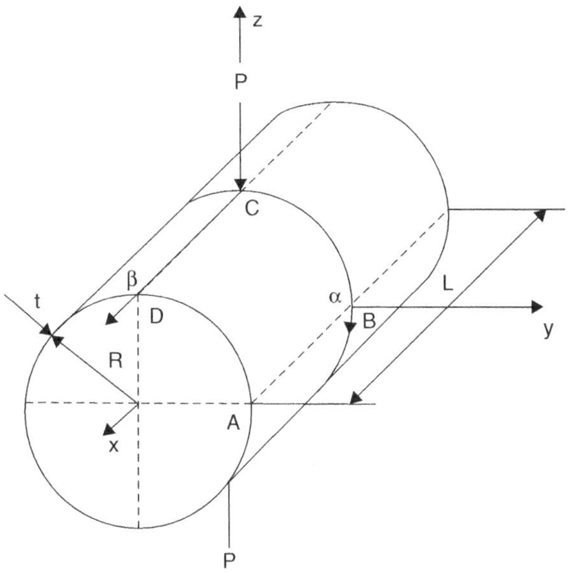

Figure 11.4 Pinched shell

Table 11.2 Physical data for pinched shell

Pinched shell		
Young's modulus	E	3×10^6 psi
Poisson's ratio	ν	0.3
Load	P	1 lb
Length	L	600 in
Radius	R	300 in
Thickness	t	3 in

This table shows the mechanical and geometrical properties of the pinched shell.

In Table 11.3 the transverse displacement at the loaded point C is presented for several decompositions $n \times n$ and different theories. The exact solution is provided by Flügge (1960) and it is 1.8248×10^{-5} in. The results show that the element has good properties of convergence and robustness by increasing the mesh. The results obtained with higher-order theories are greater than the reference value because Flügge refers to a classical shell theory. In this case, higher-order theories are able to capture nonlinear effects, such as the the local stretching of the shell along the thickness due to the concentrated load. Therefore, the solution calculated with the CST model is very close to the exact solution for mesh 13×13, while the FSDT model that takes into account the shear energy gives a higher value, as can be expected. The theory with linear expansion ($N = 1$) produces such a higher value because the correction of Poisson locking has been applied. Note that in the case of cylindrical shell structures this correction gives some problems for $N = 1$. The remaining theories provide almost the same results and they converge to the same value (1.842×10^{-5} in) by increasing the order of expansion and the number of elements used.

The second test deals with a cylindrical shell known in the literature as a *barrel vault*. The shell is described in Figure 11.5. This typical shell is used in civil engineering using conventional processes by Scordelis and Lo (1964). The shell is simply supported on diaphragms and is free on the other sides. It is loaded by its own weight p. The physical data given in Table 11.4 have been assumed. The covariant components of the vertical load are: $p_\alpha = 0$, $p_\beta = -p\sin(\beta/R)$, $p_z = p\cos(\beta/R)$. The barrel vault has a symmetric structure.

Table 11.3 Pinched shell. Transverse displacement u_z (in) $\times 10^5$ at the loaded point C of the mid-surface. Exact solution: 1.8248×10^{-5} in (Cinefra *et al.* 2013)

Mesh	4×4	10×10	13×13
CST	1.7891	1.8230	1.8251
FSDT	1.7984	1.8363	1.8396
$N = 1$	1.9212	1.9582	1.9615
$N = 2$	1.7805	1.8359	1.8406
$N = 3$	1.7818	1.8379	1.8427
$N = 4$	1.7818	1.8379	1.8427

This table compares different theories by increasing the mesh.

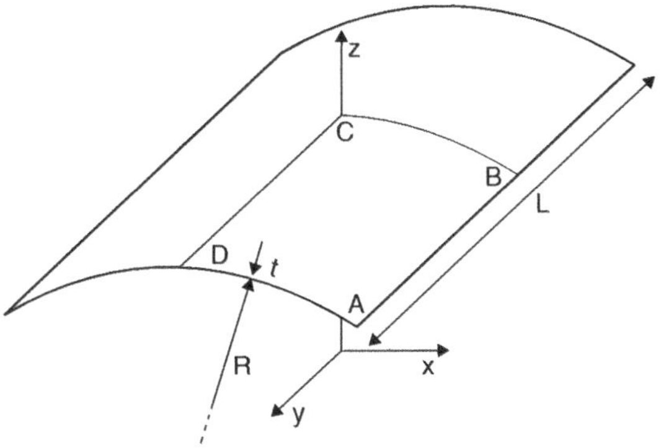

Figure 11.5 Scordelis-Lo roof

Thus, the computations have been performed only on a quarter of the shell, using a uniform decomposition. The following symmetry conditions have been assumed:

$$\begin{aligned} u_{\beta_\tau}(\alpha, 0) = 0 \\ u_{\alpha_\tau}(0, \beta) = 0 \end{aligned} \tag{11.51}$$

and the following BCs are prescribed:

$$u_{\beta_\tau}(L/2, \beta) = u_{z_\tau}(L/2, \beta) = 0 \tag{11.52}$$

with $\tau = 0, 1, \dots, N$.

The exact solution for the present problem is given by McNeal and Harder (1985) in terms of transverse displacement at the point B and it is 0.3024 ft. In Table 11.5 this quantity is calculated for several decompositions $n \times n$ and different theories. The table confirms the considerations made for the pinched shell: the results converge to the exact solution by increasing the order of expansion and the number of elements. One difference with respect to the pinched shell is that the higher-order theories and the classical theories provide almost the same results. Indeed,

Table 11.4 Physical data for barrel vault

Barrel vault		
Young's modulus	E	4.32×10^8 lb/ft^2
Poisson's ratio	ν	0.0
Load	p	90 lb/ft^2
Length	L	50 ft
Radius	R	25 ft
Thickness	t	0.25 ft
Angle	θ_0	$2\pi/9$ rad

This table shows the mechanical and geometrical properties of the barrel vault.

Table 11.5 Barrel vault. Transverse displacement u_z (ft) at the point B of the mid-surface S. Exact solution: 0.3024 ft (Cinefra *et al.* 2013)

Mesh	13×13	16×16	20×20
FSDT	0.300 91	0.300 97	0.301 04
$N = 1$	0.300 91	0.300 97	0.301 04
$N = 2$	0.300 91	0.300 97	0.301 04
$N = 3$	0.300 91	0.300 97	0.301 04
$N = 4$	0.300 91	0.300 97	0.301 04

This table compares different theories by increasing the mesh.

the barrel vault does not present local stretching effects in the thickness direction due to the concentrated load. In Figure 11.6, the $N = 4$ solution, in which the correction of both shear and membrane locking has been applied ($m+$), is compared with the solution in which only the shear locking has been corrected (s). One can note that the membrane locking phenomenon is remarkable in the barrel vault and the shell element presented shows high performance in terms of convergence, by increasing the number of elements. One can conclude that the MITC9 shell element based on the CUF is completely locking-free. For more details, the reader can refer to the work by Cinefra *et al.* (2013).

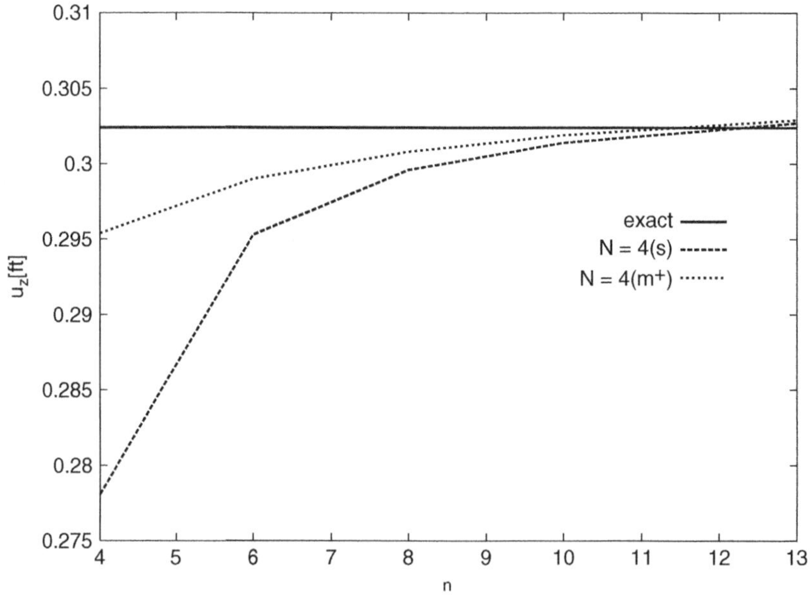

Figure 11.6 Barrel vault. Transverse displacement u_z (ft) at the point B of the mid-surface S by varying the mesh $n \times n$

References

Arnold DN and Brezzi F 1997 Locking free finite element methods for shells. *Mathematics of Computation* **66**, 1–14.

Bathe KJ and Dvorkin E 1986 A formulation of general shell elements – the use of mixed interpolation of tensorial components. *International Journal for Numerical Methods in Engineering* **22**, 697–722.

Bucalem ML and Bathe KJ 1993 High-order MITC general shell elements. *International Journal for Numerical Methods in Engineering* **36**, 3729–3754.

Chang TY, Saleeb AF and Graf W 1989 On the mixed formulation of a nine-node Lagrange shell element. *Computer Methods in Applied Mechanics and Engineering* **73**, 259–281.

Chapelle D and Bathe KJ 2003 *The Finite Element Analysis of Shells: Fundamentals*. Springer.

Chinosi C and Della Croce L 1998 Mixed-interpolated elements for thin shell. *Communications in Numerical Methods in Engineering* **14**, 1155–1170.

Chinosi C, Della Croce L and Scapolla T 1998 Hierarchic finite elements for thin Naghdi shell model. *International Journal of Solids and Structures* **35**, 1863–1880.

Cinefra M, Chinosi C and Della Croce L 2013 MITC9 shell elements based on refined theories for the analysis of isotropic cylindrical structures. *Mechanics of Advanced Materials and Structures* **20**, 91–100.

Flügge W 1960 *Stresses in Shells*. Springer.

Hakula H, Leino Y and Pitkäranta J 1996 Scale resolution, locking, and higher-order finite element modelling of shells. *Computer Methods in Applied Mechanics and Engineering* **133**, 155–182.

Huang HC 1987 Membrane locking and assumed strain shell elements. *Computers and Structures* **27**(5), 671–677.

Koiter WT 1970 On the foundations of the linear theory of thin elastic shell. *Proceedings of the Koninklijke Nederlandse Akademie* **73**, 169–195.

McNeal RH and Harder RL 1985 A proposed standard set of problems to test finite element accuracy. *Finite Elements in Analysis and Design* **1**, 3–20.

Naghdi WT 1972 The theory of shells and plates. *Handbuch der Physik* **6**, 425–640.

Pitkäranta J 1992 The problem of membrane locking in finite element analysis of cylindrical shells. *Numerische Mathematik* **61**, 523–542.

Rhiu JJ and Lee SW 1987 A new efficient mixed formulation for thin shell finite element models. *International Journal for Numerical Methods in Engineering* **24**, 581–604.

Scordelis A and Lo KS 1964 Computer analysis in cylindrical shells. *Journal of American Concrete Institute* **61**, 561–593.

Stolarski H and Belytschko T 1981 Reduced integration for shallow-shell facet elements. In: *New Concepts in Finite Element Analysis. ASME*, pp. 179–194.

Zienkiewicz OC, Taylor RL and Too JM 1971 Reduced integration techniques in general analysis of plates and shells. *International Journal for Numerical Methods in Engineering* **3**, 275–290.

12

Two-Dimensional Models with Physical Volume/Surface-Based Geometry and Pure Displacement Variables, the LE Class

The unified formulation permits different models to be implemented for the analysis of bi-dimensional structures (plates and shells) by varying the class of the thickness functions and the order of expansion along the thickness. Models based on a Taylor-like polynomial expansion were presented in the previous chapters. The second class of CUF models for plates and shells is based on Lagrange polynomials. Lagrange expansion 2D models – hereafter referred to as LE models – are described in this chapter. As mentioned in Chapter 9, LE models have the following main characteristics:

1. The variables and BCs can be located above the physical surfaces of the structure. This feature is particularly relevant for a CAD–FEM coupling scenario.
2. The unknown variables of the problem are the pure displacement components. No rotations or higher-order variables are exploited to describe the displacement field.

These features are described in the following sections of this chapter and numerical examples are provided in order to highlight the enhanced capabilities of 2D CUF LE models, in terms of 3D-like accuracies and very low computational costs.

Table 12.1 shows the FN for 2D models that is described in this chapter.

Finite Element Analysis of Structures Through Unified Formulation, First Edition.
Erasmo Carrera, Maria Cinefra, Marco Petrolo and Enrico Zappino.

Table 12.1 A schematic description of the CUF and the related FN of the stiffness matrix for 2D models

---

Equilibrium equations in Strong Form $\longrightarrow \ \delta L_i = \int_V \delta \boldsymbol{u} \boldsymbol{k} \boldsymbol{u} dV + \int_S \dots dS$

$$\underbrace{\begin{bmatrix} k_{xx} & k_{xy} & k_{xz} \\ k_{yx} & k_{yy} & k_{yz} \\ k_{zx} & k_{zy} & k_{zz} \end{bmatrix}}_{\boldsymbol{k}} \underbrace{\begin{Bmatrix} u_x \\ u_y \\ u_z \end{Bmatrix}}_{\boldsymbol{u}} = \underbrace{\begin{Bmatrix} p_x \\ p_y \\ p_z \end{Bmatrix}}_{\boldsymbol{p}}$$

$$k_{xx} = -(\lambda + 2G)\, \partial_{xx} - G\, \partial_{zz} - G\, \partial_{yy};$$
$$k_{xy} = -\lambda\, \partial_{xy} - G\, \partial_{yx};$$
$$k_{xz} = \dots$$

$\lambda = (Ev)/[(1+v)(1-2v)]; \quad G = E/[2(1+v)]$

$\boldsymbol{u} = \boldsymbol{u}(x, y, z)$

$\delta \boldsymbol{u} = \delta \boldsymbol{u}(x, y, z)$

The diagonal (e.g. k_{xx}) and the non-diagonal (e.g. k_{xy}) terms can be obtained through proper index permutations.

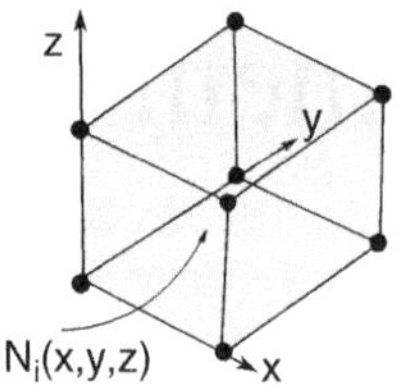

$\boldsymbol{u} = N_i(x, y, z)\boldsymbol{u}_i$

$\delta \boldsymbol{u} = N_j(x, y, z)\delta \boldsymbol{u}_j$

3D FEM Formulation $\longrightarrow \quad \delta L_i = \delta \boldsymbol{u}_j \boldsymbol{k}^{ij} \boldsymbol{u}_i$

$$k^{ij}_{xx} = (\lambda + 2G) \int_V N_{j,x} N_{i,x} dV + G \int_V N_{j,z} N_{i,z} dV + G \int_V N_{j,y} N_{i,y} dV;$$

$$k^{ij}_{xy} = \lambda \int_V N_{j,y} N_{i,x} dV + G \int_V N_{j,x} N_{i,y} dV$$

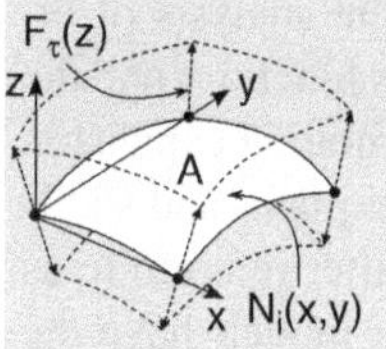

$\boldsymbol{u} = N_i(x, y) F_\tau(z) \boldsymbol{u}_{\tau i}$

$\delta \boldsymbol{u} = N_j(x, y) F_s(z) \delta \boldsymbol{u}_{sj}$

2D FEM Formulation $\longrightarrow \quad \delta L_i = \delta \boldsymbol{u}_{sj} \boldsymbol{k}^{\tau sij} \boldsymbol{u}_{\tau i}$

$$k^{\tau sij}_{xx} = (\lambda + 2G) \int_\Omega N_{i,x} N_{j,x} d\Omega \int_h F_\tau F_s dz$$
$$+ G \int_\Omega N_i N_j d\Omega \int_h F_{\tau,z} F_{s,z} dz + G \int_V N_{i,y} N_{j,y} d\Omega \int_h F_\tau F_s dz;$$

$$k^{\tau sij}_{xy} = \lambda \int_\Omega N_{i,y} N_{j,y} d\Omega \int_h F_\tau F_s dz + G \int_\Omega N_{i,x} N_{j,y} d\Omega \int_h F_\tau F_s dz$$

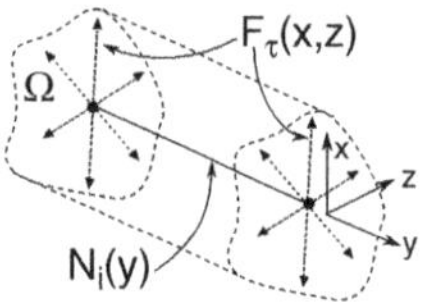

$\boldsymbol{u} = N_i(y) F_\tau(x, z) \boldsymbol{u}_{\tau i}$

$\delta \boldsymbol{u} = N_j(y) F_s(x, z) \delta \boldsymbol{u}_{sj}$

1D FEM Formulation $\longrightarrow \quad \delta L_i = \delta \boldsymbol{u}_{sj} \boldsymbol{k}^{\tau sij} \boldsymbol{u}_{\tau i}$

$$k^{\tau sij}_{xx} = (\lambda + 2G) \int_l N_i N_j dy \int_A F_{\tau,x} F_{s,x} dA$$
$$+ G \int_l N_i N_j dy \int_A F_{\tau,z} F_{s,z} dA + G \int_l N_{i,y} N_{j,y} dy \int_A F_\tau F_s dA;$$

$$k^{\tau sij}_{xy} = \lambda \int_l N_{i,y} N_j dy \int_A F_\tau F_{s,x} dA + G \int_l N_i N_{j,y} dy \int_A F_{\tau,x} F_s dA$$

CUF leads to the automatic implementation of any theory of structures through 4 loops (i.e. 4 indexes):

- τ and s deal with the functions that approximate the displacement field and its virtual variation along the plate/shell thickness ($F_\tau(z), F_s(z)$) or over the beam cross-section ($F_\tau(x, z), F_s(x, z)$);
- i and j deal with the shape functions of the FE model, (3D:$N_i(x, y, z), N_j(x, y, z)$; 2D:$N_i(x, y), N_j(x, y)$; 1D:$N_i(y), N_j(y)$).

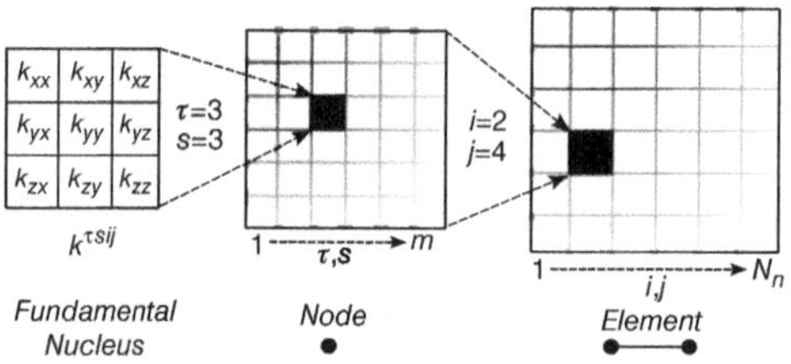

This table shows the essential features of the CUF for 2D models.

12.1 Physical Volume/Surface Approach

In the frame of FEA and CAD, it should be considered that many structural elements require the definition of the reference surfaces on which these elements – and the problem unknowns – lie. Such a definition of the reference surfaces is particularly critical when 3D CAD geometries are considered. The present 2D FE formulation, based on a physical volume/surface approach, offers significant advantages as far as this point is concerned because LE elements can deal directly with the 3D geometry given by a CAD model.

Figure 12.1 shows the FE models of a shell structure based on both Taylor and Lagrange polynomial expansions. For the case of the TE model and starting from the 3D geometry of the structure, the FE discretization requires that a reference mid-surface of the shell is defined. The problem unknowns are defined above this surface. Instead, in the LE modelling approach, the nodes can be located directly on the top and bottom surfaces of the 3D structure. This implies that the FE unknowns lie above the physical surface of the structure. It is not necessary to define reference surfaces and a 3D CAD geometry can be exploited directly for the FEA. In other words, a 2D FE can be used for a 3D geometrical description. Moreover, the LE

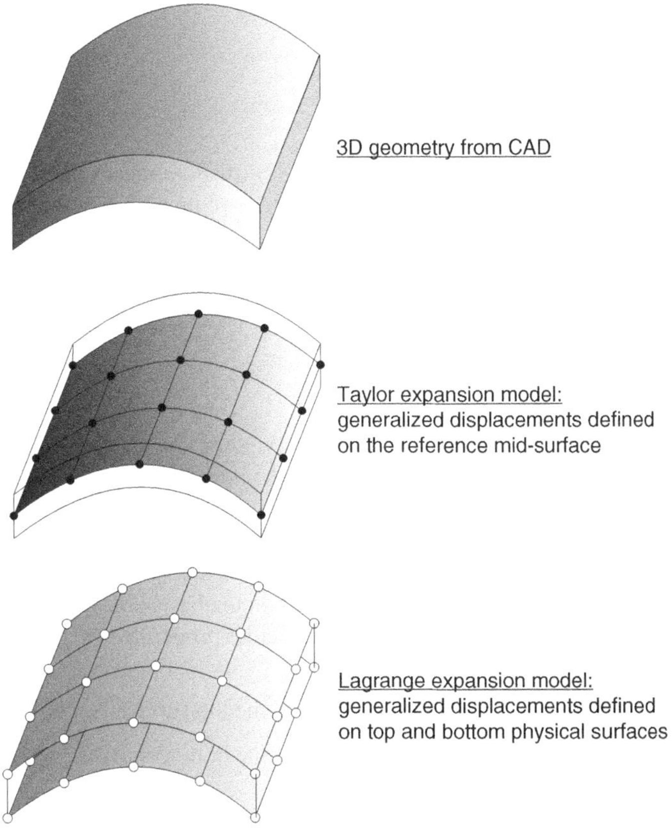

Figure 12.1 Different geometrical modelling approaches, TE model vs LE model

model can be particularly advantageous because it permits BCs to be applied directly to pure displacement components. For example, it is easy to manage problems in which mechanical loads act on the top/bottom surface of the structure. The LE model also become very attractive in view of the analysis of multilayered structures. A layer-wise description of a multilayer presupposes the use of LE models that allow compatibility conditions to be easily imposed between the different layers. For more details on the layer-wise description of multilayered structures, the reader can refer to the work by Carrera (1999).

12.2 LE Model

The LE models represent the second class of 2D models developed in the framework of the CUF. Descriptions provided in this section are tuned on the specific issues related to the CUF. Details of Lagrange polynomials not directly related to the CUF can be found in many excellent books (Bathe 1996; Oñate 2009; Zienkiewicz *et al.* 2005).

In an LE model, the F_τ thickness functions coincide with the Lagrange polynomials. Therefore, the through-the-thickness displacement field is written as

$$
\begin{aligned}
u_\alpha &= F_t\, u_{\alpha_t} + F_b\, u_{\alpha_b} \\
u_\beta &= F_t\, u_{\beta_t} + F_b\, u_{\beta_b} \\
u_z &= F_t\, u_{z_t} + F_b\, u_{z_b}
\end{aligned}
\tag{12.1}
$$

where $(u_{\alpha_t}, u_{\beta_t}, u_{z_t})$ and $(u_{\alpha_b}, u_{\beta_b}, u_{z_b})$ are the values of displacement components at the top and bottom surfaces of the shell, respectively. F_t and F_b are linear Lagrange polynomials and read

$$
\begin{aligned}
F_t &= \frac{1+\zeta}{2} \\
F_b &= \frac{1-\zeta}{2}
\end{aligned}
\tag{12.2}
$$

Lagrange polynomials are usually given in terms of the normalized – also known as natural – thickness coordinate $-1 < \zeta < 1$. This choice is not compulsory since LEs could also be implemented in terms of actual coordinate z; however, the normalized formulation was preferred since it offers many advantages. Only linear expansions are considered in this book, even though higher-order theories can be formulated on the basis of Lagrange polynomials.

These thickness functions have the following particular properties:

$$
\begin{aligned}
&\zeta = 1: F_t = 1 \quad \text{and} \quad F_b = 0 \quad \text{at the top surface} \\
&\zeta = 1: F_b = 1 \quad \text{and} \quad F_t = 0 \quad \text{at the bottom surface}
\end{aligned}
\tag{12.3}
$$

as shown in Figure 12.2. Therefore, the boundary surface values of the displacements are considered as variable unknowns. This fact permits BCs to be applied directly to pure

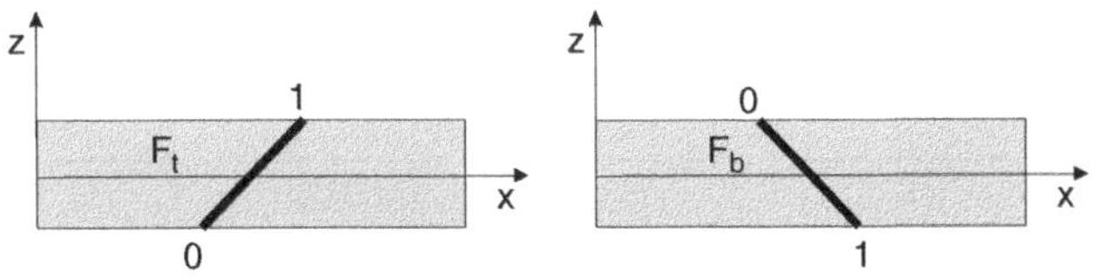

Figure 12.2 Lagrange thickness functions

displacement components and, in the case of multilayered structures, one can easily impose the following compatibility conditions:

$$\mathbf{u}_t^k = \mathbf{u}_b^{k+1} \tag{12.4}$$

where k is the layer index.

In the unified formulation framework, the displacement field preserves the formal expression of Equation (11.10)

$$\mathbf{u} = F_\tau \mathbf{u}_\tau, \qquad \tau = 0, 1, \ldots, N \tag{12.5}$$

The use of LE models as F_τ does not imply a reformulation of the problem equations and matrices, as is typical in the CUF environment. The governing equations, the FE formulation and the FNs are exactly the same as in Sections 10.4 and 11.4, for plates and shells respectively. Note that the LE models presented here refer to shell displacement components, but they remain valid for plates that are shells with $R = \infty$.

12.3 Numerical Examples

In order to show the performances of LE models, the free vibration analysis of a plate with beam-type geometry is considered. The plate is made of aluminium, whose properties were given in Table 10.3. Its dimensions in metres are $a = 1$, $b = 0.1$ and $h = 0.1$, and it is simply supported along the y-direction edges and free along x.

As discussed in Section 10.5.2, the most common approach proposed to overcome the locking phenomenon is the use of reduced-selective integration techniques, but they produce spurious 'non-physical modes' in the dynamic analysis. The MITC technique has been developed to reduce this effect. The static analysis of plate structures, performed in Section 10.6, has shown no differences between the results obtained by reduced-selective integration (S) and those by the MITC technique (*MITC*). The differences become evident when the higher vibration response is checked.

Regarding the plate with beam-type geometry, spurious modes are discussed in Table 12.2. Q4 plate elements are used for the analysis and the adopted meshes are: (30×1), (200×1) and (300×2). The first 40 frequencies are given in hertz. Spurious modes are highlighted by an asterisk. The table confirms that the shear locking phenomenon is reduced by increasing the mesh: spurious vibration modes move to higher-order frequencies in both the S and *MITC* cases. However, the MITC method, compared with selective integration, has two main advantages: it greatly reduces spurious modes; and it moves the spurious modes (if any) to higher-order

Table 12.2 Natural frequencies (Hz) of an aluminium plate, LE model (Cinefra *et al.* 2010)

	Mesh (30×1)		Mesh (200×1)		Mesh (300×2)	
n	(*S*)	(*MITC*)	(*S*)	(*MITC*)	(*S*)	(*MITC*)
2	24.3483	24.3483	20.3704*	24.3122	24.2939	24.2939
3	38.6542*	53.2660	24.3122	26.0570	24.3117	24.3117
4	53.2657	97.7909	26.0570	97.2120	97.1329	97.1330
5	97.7909	213.5543	81.4421*	104.1822	97.2037	97.2045
6	154.9560*	221.5323	97.2120	218.5894	218.3887	218.3895
7	213.5497	397.6177	104.1812	234.2383	218.5473	218.5512
8	221.5323	482.3379	183.0969*	388.2620	387.8490	387.8516
9	349.9235*	628.9913	218.5894	415.9980	388.1291	388.1415
10	397.6178	862.0786	234.2334	605.9767	605.2194	605.2257
11	482.3148	919.5792	325.1388*	649.1461	605.6532	605.6833
12	625.2562*	1274.4017	388.2620	871.4115	870.1253	870.1383
13	628.9913	1356.2359	415.9827	933.2822	870.7433	870.8055
14	862.0057	1553.4594*	507.2958*	1184.1786	1182.1159	1182.1399
15	919.5793	1699.7146	605.9767	1267.9249	1182.9464	1183.0607
16	983.3359*	1969.2746	649.1088	1543.8272	1437.5753*	1540.7081
17	1274.4017	2203.1852	729.2222*	1552.7658*	1540.6676	1541.9296
18	1356.0586	2532.6474	871.4115	1652.5152	1541.7360	1552.7569*
19	1427.2268*	2706.6672	933.2053	1949.8471	1945.1880	1945.2525
20	1699.7146	2794.1037	990.5014*	2086.4225	1946.5168	1946.8243
21	1960.6679*	3111.1787*	1184.1786	2401.6727	2395.0213	2395.1189
22	1968.9087	3483.6325	1267.7831	2530.9717	2396.6294	2397.0938
23	2203.1852	3574.8850	1290.6488*	2568.9487	2530.9470	2530.9470
24	2532.6474	4285.0908	1543.8272	2898.6871	2875.2005*	2889.5946
25	2588.0549*	4581.3664	1629.1165*	3099.3350	2889.4529	2892.0283
26	2705.9940	4677.4287*	1652.2750	3105.6274*	2891.3552	3105.5564*
27	2794.1037	5075.5161	1949.8471	3440.2267	3105.3929*	3427.9134
28	3314.4023*	5214.2676	2005.2968*	3676.7673	3427.7146	3430.8646
29	3483.6326	5734.4497	2086.0404	4025.5857	3429.9218	4009.2618
30	3573.7473	6256.5008*	2328.9033	4300.3824	4008.9906	4012.7916
31	4145.2761*	6289.7471	2401.6727	4654.0210	4011.5084	4632.7837
32	4285.0908	7043.2504	2418.5268*	4658.6805*	4312.9253*	4636.9559
33	4579.5665	7533.2118	2530.9716	4969.2746	4632.4228	4658.4410*
34	5075.5161	7638.9288	2568.3711	5062.0898	4635.2518	5061.8929
35	5086.6807*	7852.7127	2868.0935*	5324.7568	5061.8929	5297.5866
36	5214.2676	8517.4587	2898.6871	5682.5015	5297.1163	5302.4671
37	5731.7504	8969.6656	3098.4969	6036.9894	5300.2514	6002.7471
38	6144.8816*	9470.4077	3353.2385*	6212.0210*	5750.7997*	6008.4033
39	6289.7471	10 167.0191	3440.2267	6439.0906	6002.1455	6211.4534*
40	7039.3813	10 233.4095	3675.5916	6789.8919	6005.5756	6747.3161

This table compares different methods for solving the shear locking phenomenon by considering the first 40 natural frequencies. The asterisk indicates spurious modes.

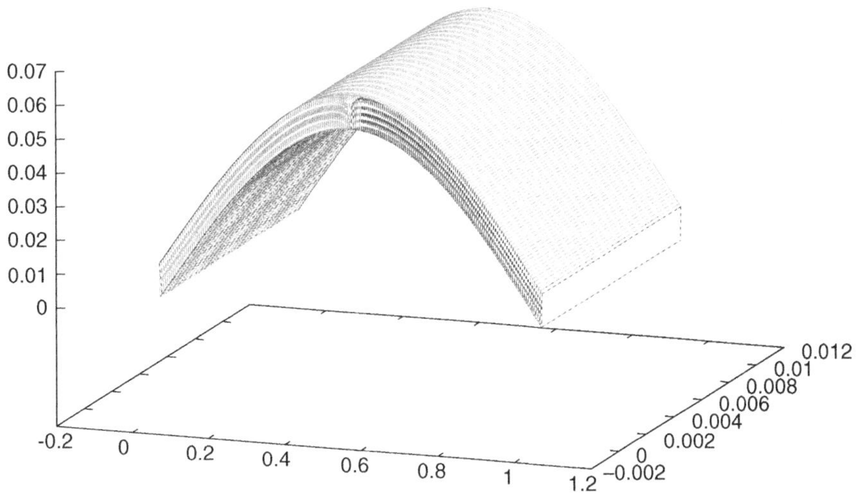

Figure 12.3 Vibrational mode 2 for the (200 × 1) mesh

frequencies. Note the difficulties of the FE plate model in detecting beam vibration modes transverse to the reference surface Ω; the beam has a square section and a refined mesh must be used to find double modes. However, that point does not affect the discussion on spurious modes.

Figures 12.3 and 12.4 show the vibrational modes 2 and 4 for the (200 × 1) mesh. One can see that the LE model easily permits the representation of the 3D deformation of the plate starting from the pure displacements ($\mathbf{u}_t$) and ($\mathbf{u}_b$) calculated with the FEA. The values of the displacements are opportunely scaled for representation reasons.

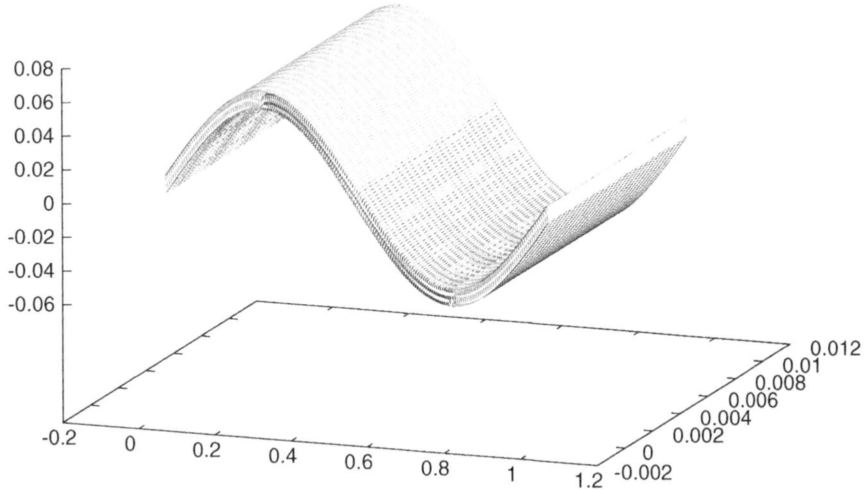

Figure 12.4 Vibrational mode 4 for the (200 × 1) mesh

References

Bathe KJ 1996 *Finite Element Procedures*. Prentice–Hall.

Carrera E 1999 Multilayered shell theories accounting for layerwise mixed description, Part 1: Governing equations. *AIAA Journal* **37**(9), 1107–1116.

Cinefra M, Carrera E and Nali P 2010 MITC technique extended to variable kinematic multilayered plate elements. *Composite Structures* **92**, 1888–1895.

Oñate E 2009 *Structural Analysis with the Finite Element Method: Linear Statics*. Volume 1. *Basis and Solids*. Springer.

Zienkiewicz OC, Taylor RL and Zhu JZ 2005 *The Finite Element Method: Its Basis and Fundamentals*. Sixth Edition. Elsevier.

13

Discussion on Possible Best Beam, Plate and Shell Diagrams

The advanced structural models introduced through the unified formulation are generally based on full expansions of a given order; this means that all the displacement variables of a given order theory are employed. The contribution of each term of a theory varies depending on the structural problem, and some variables can be more important than others in predicting the mechanical behaviour of a structure. Moreover, some terms might not have any influence, since their absence does not affect the accuracy of the solution.

The unified formulation can be used to investigate the influence of each variable of a refined theory on the solution. This capability is hereafter referred to as the mixed axiomatic–asymptotic approach (MAAA). The MAAA can be considered a powerful tool to build:

- reduced refined theories which have the same accuracy as full expansion models but fewer unknown variables;
- the best theory diagram (BTD), where the accuracy of a model can be evaluated against the number of variables.

This chapter describes the MAAA, gives general guidelines to determine the most adequate model for a given problem and introduces the BTD. Numerical examples are presented to evaluate reduced models and to highlight the influence of characteristic parameters, such as the slenderness ratio, the thickness and the loading conditions.

13.1 The MAAA

TOSs can be built on the basis of different approaches, in which the common target of each approach is to evaluate the minimum number of unknown variables in order to solve a given problem against a given accuracy. Axiomatic and asymptotic methods are the main tools for building structural models, as explained in Chapter 3. The MAAA has recently been developed by Carrera and Petrolo (2010, 2011). The choice of its name is due to the fact

Finite Element Analysis of Structures Through Unified Formulation, First Edition.
Erasmo Carrera, Maria Cinefra, Marco Petrolo and Enrico Zappino.

that the MAAA is capable of obtaining asymptotic-like results, starting from axiomatic-like hypotheses. The influence of a variable can in fact be investigated, against the variation of various parameters (e.g. thickness, BCs, etc.), using the MAAA and this analysis is conducted starting from axiomatic-like hypotheses. The smallest number of variables required to fulfil a given accuracy requirement can thus be determined. The basic procedure is described below. This procedure is exactly the same as for 1D and 2D models.

1. The problem data are fixed $\Rightarrow$ *Loadings, BCs, materials*
2. A set of output variables is chosen $\Rightarrow$ σ, ε, u
3. The CUF is used to generate the governing equations for the considered theories $\Rightarrow$ $u = F_\tau u_\tau$
4. A theory is fixed and used to establish the accuracy $\Rightarrow$

$N=0$	$N=1$		$N=2$		
u_{x_1}	u_{x_2}	u_{x_3}	u_{x_4}	u_{x_5}	u_{x_6}
u_{y_1}	u_{y_2}	u_{y_3}	u_{y_4}	u_{y_5}	u_{y_6}
u_{z_1}	u_{z_2}	u_{z_3}	u_{z_4}	u_{z_5}	u_{z_6}

5. Each term is deactivated in trun $\Rightarrow$

▲	▲	▲	▲	▲	▲
▲	▲	△	▲	▲	▲
▲	▲	▲	▲	▲	▲

6. Does the absence of the term affect the solution? $\Rightarrow$

Active term	Inactive term
Yes,▲	No,△

7. The displacement model with the smallest number of terms is then detected for a given structural layout $\Rightarrow$

△	▲	△	▲	△	▲
▲	▲	△	△	▲	△
△	▲	▲	△	▲	△

$\Rightarrow$

$$
\begin{aligned}
u_x &= x\, u_{x_2} + x^2\, u_{x_4} + z^2\, u_{x_6}\\
u_y &= u_{y_1} + x\, u_{y_2} + xz\, u_{y_5}\\
u_z &= x\, u_{z_2} + z\, u_{z_3} + xz\, u_{z_5}
\end{aligned}
$$

A graphic notation is introduced to improve the readability of the results. Table 13.1 shows the locations of each second-order 1D model term within the tabular layout. The first column presents the constant terms ($N = 0$), the second and third columns the linear terms ($N = 1$), and the last three columns show the parabolic terms ($N = 2$). Each term can be activated (black) or deactivated (white), as shown in Table 13.2. On the basis of the adopted notation, the 1D model given in Table 13.2 refers to the following cross-section displacement field:

$$
\begin{aligned}
u_x &= u_{x_1} + x\, u_{x_2} + z\, u_{x_3} + x^2\, u_{x_4} + xz\, u_{x_5} + z^2\, u_{x_6}\\
u_y &= u_{y_1} + x\, u_{y_2} + \qquad\quad + x^2\, u_{y_4} + xz\, u_{y_5} + z^2\, u_{y_6}\\
u_z &= u_{z_1} + x\, u_{z_2} + z\, u_{z_3} + x^2\, u_{z_4} + xz\, u_{z_5} + z^2\, u_{z_6}
\end{aligned}
\tag{13.1}
$$

Table 13.1 Positions of the displacement variables within the table layout

$N=0$	$N=1$		$N=2$		
u_{x_1}	u_{x_2}	u_{x_3}	u_{x_4}	u_{x_5}	u_{x_6}
u_{y_1}	u_{y_2}	u_{y_3}	u_{y_4}	u_{y_5}	u_{y_6}
u_{z_1}	u_{z_2}	u_{z_3}	u_{z_4}	u_{z_5}	u_{z_6}

This table shows the layout notation that defines the position of each displacement variable in the case of a 1D second-order model, $N = 2$.

Table 13.2 Symbolic representation of the reduced kinematic model with u_{y_3} deactivated

▲	▲	▲	▲	▲	▲
▲	▲	△	▲	▲	▲
▲	▲	▲	▲	▲	▲

This table shows the graphic representation of a second-order 1D model that has u_{y_3} deactivated.

The 2D model notation is given in Tables 13.3 and 13.4. The 2D model given in Table 13.4 refers to the following displacement field:

$$\begin{aligned} u_x &= u_{x_1} + z\,u_{x_2} + z^2\,u_{x_3} + z^3\,u_{x_4} + z^4\,u_{x_5} \\ u_y &= u_{y_1} + z\,u_{y_2} + \qquad\quad + z^3\,u_{y_4} + z^4\,u_{y_5} \\ u_z &= u_{z_1} + z\,u_{z_2} + z^2\,u_{z_3} + z^3\,u_{z_4} + z^4\,u_{z_5} \end{aligned} \tag{13.2}$$

The reduced models can be obtained by opportunely rearranging the rows and columns of the FE matrices or through penalty techniques. The latter is the approach that has been adopted in this book.

Example 13.1.1 *Let us evaluate the influence of the u_{z_3} term of a fourth-order 1D model on the torsional analysis of the thin-walled cylinder shown in Figure 8.11 with L/d equal to 10. The torsion is investigated through the application of two concentrated forces (P_z) at $[\pm\, d/2,\ L,\ 0]$. Table 13.5 shows the reduced 1D model considered. The influence of the term is evaluated by means of percentage variations*

$$\delta_u = u/u_{N=4} \times 100, \quad \delta_\sigma = \sigma/\sigma_{N=4} \times 100 \tag{13.3}$$

computed with respect to the full fourth-order model ($N = 4$). Table 13.6 shows the results referring to the three displacement variables computed at $[d/2,\ L,\ 0]$. It can be seen that u_{z_3} influences u_x and u_z, but does not affect u_y.

Example illustrates the procedure used to investigate the effect of a single term on different output variables. This procedure is used to determine all the inactive terms of a refined model for a given structural problem.

Table 13.3 Locations of the displacement variables within the table layout for 2D models

$N=0$	$N=1$	$N=2$	$N=3$	$N=4$
u_{x_1}	u_{x_2}	u_{x_3}	u_{x_4}	u_{x_5}
u_{y_1}	u_{y_2}	u_{y_3}	u_{y_4}	u_{y_5}
u_{z_1}	u_{z_2}	u_{z_3}	u_{z_4}	u_{z_5}

This table shows the layout notation that defines the position of each displacement variable in the case of a fourth-order model for plates, $N = 4$.

Table 13.4 Symbolic representation of the reduced kinematic model with u_{y_3} deactivated (2D models)

▲	▲	▲	▲	▲
▲	▲	△	▲	▲
▲	▲	▲	▲	▲

This table shows the graphic representation of a fourth-order 2D model that has u_{y_3} deactivated.

Table 13.5 Reduced fourth-order model for the torsional analysis of a thin-walled cylinder

●	●	●	●	●	●	●	●	●	●	●	●	●	●	●
●	●	●	●	●	●	●	●	●	●	●	●	●	●	●
●	●	○	●	●	●	●	●	●	●	●	●	●	●	●

This table presents a reduced fourth-order 1D model in which the u_{z_3} variable is deactivated.

Table 13.6 Effect of the absence of u_{z_3} on different output variables

δ_{u_x} (%)	δ_{u_y} (%)	δ_{u_z} (%)
96.9	100.0	118.8

This table shows the effects caused by the absence of u_{z_3} on different output variables.

13.2 Static Analysis of Beams

In this section, the MAAA is used to determine reduced 1D models for the static analysis of cantilever beams. For the sake of clarity, different symbols, which are shown in Table 13.7, are adopted to distinguish the loading cases.

13.2.1 Influence of the Loading Conditions

Bending, torsion and traction load cases are considered for a compact square cantilever beam. A 1D $N = 4$ model is considered as a reference solution in order to evaluate the effectiveness of each displacement variable in detecting the displacement components.

Table 13.7 Symbols that indicate the loading cases and the presence of a displacement variable

Loading case	Active term	Inactive term
Bending	▲	△
Torsion	●	○
Axial	▼	▽

This table shows the symbols that are used to indicate the presence of a displacement variable and the loading case considered in the static analysis of beams.

Table 13.8 Set of active displacement variables for the bending analysis of a square cross-section beam

$M_{eff}/M = 11/45$														
△	△	△	△	▲	△	△	△	△	△	△	▲	△	▲	△
△	△	▲	△	△	△	△	▲	△	▲	△	△	△	△	△
▲	△	△	△	△	▲	△	△	△	△	▲	△	▲	△	▲

This table presents the set of active terms to detect the bending behaviour of a square cross-section beam; 11 out of 45 terms are needed.

Bending is considered as the first loading case. Table 13.8 shows the set of terms that are needed to detect fourth-order accuracy. It can be seen that 11 out of 45 terms are needed and the explicit expression for the 1D model is

$$\begin{aligned} u_x &= xz\, u_{x_5} + x^3z\, u_{x_{12}} + xz^3\, u_{x_{14}} \\ u_y &= z\, u_{y_3} + x^2z\, u_{y_8} + z^3\, u_{y_{10}} \\ u_z &= u_{z_1} + z^2\, u_{z_6} + x^4\, u_{z_{11}} + x^2z^2\, u_{z_{13}} + z^4\, u_{z_{15}} \end{aligned} \tag{13.4}$$

The second loading deals with torsion. The related reduced model is presented in Table 13.9. In this case, 9 out of 45 terms are needed and the 1D model is given by

$$\begin{aligned} u_x &= z\, u_{x_3} + x^2z\, u_{x_8} + z^3\, u_{x_{10}} \\ u_y &= xz\, u_{y_5} + x^3z\, u_{y_{11}} + xz^3\, u_{y_{14}} \\ u_z &= x\, u_{z_2} + x^3\, u_{z_7} + xz^2\, u_{z_9} \end{aligned} \tag{13.5}$$

Traction is considered as a third loading case. The results are given in Table 13.10 and the reduced model is

$$\begin{aligned} u_x &= x\, u_{x_2} + x^3\, u_{x_7} + xz^2\, u_{x_9} \\ u_y &= u_{y_1} + x^2\, u_{y_4} + z^2\, u_{y_6} + x^4\, u_{y_{11}} + x^2z^2\, u_{y_{13}} + z^4\, u_{y_{15}} \\ u_z &= z\, u_{z_3} + x^2z\, u_{z_8} + z^3\, u_{z_{10}} \end{aligned} \tag{13.6}$$

It is important to emphasize that the 1D models in Equations (13.4), (13.5) and (13.6) are substantially different from each other. This means that each loading case needs its own

Table 13.9 Set of active displacement variables for the torsional analysis of a square cross-section beam

$M_{eff}/M = 9/45$														
○	○	●	○	○	○	○	●	○	●	○	○	○	○	○
○	○	○	○	●	○	○	○	○	○	○	●	○	●	○
○	●	○	○	○	○	●	○	●	○	○	○	○	○	○

This table presents the set of active terms to detect the torsional behaviour of a square cross-section beam; 9 out of 45 terms are needed.

Table 13.10 Set of active displacement variables for the traction analysis of a square cross-section beam

$M_{eff}/M = 12/45$														
▽	▼	▽	▽	▽	▽	▼	▽	▼	▽	▽	▽	▽	▽	▽
▼	▽	▽	▼	▽	▼	▽	▽	▽	▽	▼	▽	▼	▽	▼
▽	▽	▼	▽	▽	▽	▽	▼	▽	▼	▽	▽	▽	▽	▽

This table presents the set of active terms to detect the traction behaviour of a square cross-section beam; 12 out of 45 terms are needed.

Table 13.11 Combined set of active displacement variables for the bending, torsional and traction analysis of a square cross-section beam

$M_{eff}/M = 32/45$														
	▼	●		▲		▼	●	▼	●		▲		▲	
▼		▲	▼	●	▼		▲ ●		▲	▼	●	▼	●	▼
▲	●	▼			▲	●	▼	●	▼	▲		▲		▲

This table presents the combined set of active terms to detect the bending, torsional and traction behaviour of a square cross-section beam; 32 out of 45 terms are needed.

reduced beam model. The combined reduced 1D model necessary to detect the fourth-order solution for the bending, torsion and traction loads is presented in Table 13.11.

13.2.2 *Influence of the Cross-section Geometry*

The cross-section geometry is another important parameter that is necessary to determine a refined 1D model. General guidelines state that compact beams need fewer variables than thin-walled beams. Other key aspects concern the symmetry/asymmetry and the presence of closed/open sections. Two cross-section geometries are investigated here: annular and aerofoil-shaped. In both cases, a torsional load is applied to the free tip. Table 13.12 shows the reduced 1D model for the annular cross-section that is equivalent to a full fourth-order model, i.e.

$$
\begin{aligned}
u_x &= x\, u_{x_2} + z\, u_{x_3} + x^3\, u_{x_7} + x^2 z\, u_{x_8} + x z^2\, u_{x_9} + z^3\, u_{x_{10}} \\
u_y &= u_{y_1} + x^2\, u_{y_4} + xz\, u_{y_5} + z^2\, u_{y_6} + x^4\, u_{y_{11}} + x^3 z\, u_{y_{12}} \\
&\quad + x^2 z^2\, u_{y_{13}} + x z^3\, u_{y_{14}} + z^4\, u_{y_{15}} \\
u_z &= x\, u_{z_2} + z\, u_{z_3} + x^3\, u_{z_7} + x^2 z\, u_{z_8} + x z^2\, u_{z_9} + z^3\, u_{z_{10}}
\end{aligned}
\tag{13.7}
$$

Table 13.12 Set of active displacement variables for the torsional analysis of a thin-walled annular beam

$M_{eff}/M = 21/45$														
▽	▼	▼	▽	▽	▽	▼	▼	▼	▼	▽	▽	▽	▽	▽
▼	▽	▽	▼	▼	▼	▽	▽	▽	▽	▼	▼	▼	▼	▼
▽	▼	▼	▽	▽	▽	▼	▼	▼	▼	▽	▽	▽	▽	▽

This table presents the set of active terms to detect the torsional behaviour of a thin-walled annular beam; 21 out of 45 terms are needed.

Table 13.13 Set of active displacement variables for the torsional analysis of an aerofoil-shaped beam

$M_{eff}/M = 45/45$														
●	●	●	●	●	●	●	●	●	●	●	●	●	●	●
●	●	●	●	●	●	●	●	●	●	●	●	●	●	●
●	●	●	●	●	●	●	●	●	●	●	●	●	●	●

This table presents the set of active terms to detect the torsional behaviour of an aerofoil-shaped beam; 45 out of 45 terms are needed.

Table 13.13 shows the equivalent result for an aerofoil-shaped cantilever beam. In this case, all the 45 displacement variables are needed; that is, each term of the fourth-order expansion plays a role in detecting the mechanical behaviour of the structure.

13.2.3 Reduced Models vs Accuracy

Reduced models that are equivalent to full fourth-order expansions were considered in the previous sections. Another important option offered by the present formulation is the possibility to choose the range of accuracy of the refined model. In other words, a reduced model that offers a given accuracy can be determined. The analysis is conducted on an aerofoil-shaped cantilever beam under torsional and bending loads. The input parameter is the error with respect to the fourth-order solution:

$$\overline{\delta_u} = \left\| u - u_{N=4}/u_{N=4} \right\| \times 100 \tag{13.8}$$

For instance, $\overline{\delta_u} = 0$ implies that a full $N = 4$ solution is necessary. Table 13.14 shows the torsion-related results, while the bending case is addressed in Table 13.15. In both cases,

Table 13.14 Set of active displacement variables that offers a given accuracy for an aerofoil-shaped beam under torsion

$\overline{\delta_u} = 0\%$, $M_{eff}/M = 45/45$														
●	●	●	●	●	●	●	●	●	●	●	●	●	●	●
●	●	●	●	●	●	●	●	●	●	●	●	●	●	●
●	●	●	●	●	●	●	●	●	●	●	●	●	●	●
$\overline{\delta_u} \leq 15\%$, $M_{eff}/M = 42/45$														
●	●	●	●	●	●	●	●	●	●	●	●	○	○	●
●	●	●	●	○	●	●	●	●	●	●	●	●	●	●
●	●	●	●	●	●	●	●	●	●	●	●	●	●	●
$\overline{\delta_u} \leq 35\%$, $M_{eff}/M = 25/45$														
●	●	●	○	●	○	○	●	○	●	○	●	○	○	○
●	●	○	●	○	●	●	○	○	○	●	○	○	○	○
●	●	●	●	●	○	●	●	●	●	●	●	○	●	○

This table presents the set of active terms to detect the torsional behaviour of an aerofoil-shaped beam with a given accuracy.

Table 13.15 Set of active displacement variables that offers a given accuracy for an aerofoil-shaped beam under bending

$\overline{\delta_u} = 0\%, M_{eff}/M = 45/45$														
▲	▲	▲	▲	▲	▲	▲	▲	▲	▲	▲	▲	▲	▲	▲
▲	▲	▲	▲	▲	▲	▲	▲	▲	▲	▲	▲	▲	▲	▲
▲	▲	▲	▲	▲	▲	▲	▲	▲	▲	▲	▲	▲	▲	▲
$\overline{\delta_u} \leq 15\%, M_{eff}/M = 23/45$														
▲	▲	▲	▲	▲	△	△	▲	△	△	▲	▲	△	△	△
▲	▲	△	▲	▲	△	△	△	△	△	▲	△	△	△	△
▲	▲	▲	▲	▲	△	▲	△	▲	▲	△	△	▲	▲	△
$\overline{\delta_u} \leq 35\%, M_{eff}/M = 9/45$														
▲	△	▲	△	▲	△	△	△	△	△	△	△	△	△	△
▲	▲	△	△	△	△	△	△	△	△	△	△	△	△	△
△	▲	▲	△	▲	△	△	△	△	▲	△	△	△	△	△

This table presents the set of active terms to detect the bending behaviour of an aerofoil-shaped beam with a given accuracy.

significant reductions in the total number of variables can be observed for the totally different reduced models requested for the torsion and the bending loading cases. These results confirm that the development of reduced higher-order models is decidedly problem dependent. The possibility of dealing with full arbitrary-order models offered by the present unified formulation is fundamental in the analysis of structures of engineering interest in which different loads, geometries and BCs are usually present simultaneously. It should be noted that the present analysis is only for isotropic materials. The case of composite materials, other important parameters such as the orthotropic ratio, and the stacking sequence, can all be expected to play the same role in determining the reduced models such as those seen above. This aspect has been shown by Carrera and Petrolo (2010) and Carrera *et al.* (2011c) in the framework of refined plate models.

13.3 Modal Analysis of Beams

In this section, the MAAA is applied to the modal analysis of beams in order to build reduced 1D models that are able to detect the natural modes and frequencies of beams. The MAAA was recently proposed for modal analysis by Carrera *et al.* (2012).

The accuracy of a reduced model is evaluated on the natural frequency through E_f defined as

$$E_f = \left\| \frac{f - f_{\text{ref}}}{f_{\text{ref}}} \right\| \times 100 \tag{13.9}$$

where f_{ref} denotes the frequency computed by means of the reference model ($N = 4$).

13.3.1 Influence of the Cross-section Geometry

Reduced 1D models for the modal analysis of beams with various cross-sections are dealt with in this section using the MAAA. Four geometries are considered: rectangular, rectangular thin-walled, C-shaped and annular. Three natural modes are considered to evaluate the influence of the generalized variables on the solution:

1. The frequency related to the first bending mode in the z direction, hereafter referred to as *bending z*.
2. The frequency related to the first bending mode in the x direction, hereafter referred to as *bending x*.
3. The frequency related to the first torsional mode, hereafter referred to as *torsional*.

Table 13.16 shows the reduced models for the rectangular beam. These models were obtained by retaining the active terms for a given mode. The explicit expression for the beam model needed to detect the first bending mode along z is

$$\begin{aligned} u_x &= xz\, u_{x_5} + x^3z\, u_{x_{12}} \\ u_y &= z\, u_{y_3} + x\, u_{y_2} + xz\, u_{y_5} \\ u_z &= u_{z_1} + x^2\, u_{z_4} + z^2\, u_{z_6} + x^4\, u_{z_{11}} \end{aligned} \tag{13.10}$$

This model requires only 7 terms out of 45, i.e. 7 DOFs per node. It should be noted that the reduced model related to the bending along x requires fewer terms than that in Equation (13.10) because of the higher flexibility of the rectangular cross-section along z. The torsional mode is the most cumbersome for this cross-section configuration,

$$\begin{aligned} u_x &= z\, u_{x_3} + x^2z\, u_{x_8} + z^3\, u_{x_{10}} \\ u_y &= xz\, u_{y_5} + x^3z\, u_{y_{12}} + xz^3\, u_{y_{14}} \\ u_z &= x\, u_{z_2} + x^3\, u_{z_7} + xz^2\, u_{z_9} \end{aligned} \tag{13.11}$$

Table 13.16 Reduced models for the modal analysis of a cantilever rectangular cross-section beam

First bending z, $M_{eff}/M = 7/45$														
▽	▽	▽	▽	▼	▽	▽	▽	▽	▽	▽	▼	▽	▽	▽
▽	▽	▼	▽	▽	▽	▽	▽	▽	▽	▽	▽	▽	▽	▽
▼	▽	▽	▼	▽	▼	▽	▽	▽	▽	▼	▽	▽	▽	▽
First bending x, $M_{eff}/M = 5/45$														
▼	▽	▽	▼	▽	▽	▽	▽	▽	▽	▽	▽	▽	▽	▽
▽	▼	▽	▽	▽	▽	▼	▽	▽	▽	▽	▽	▽	▽	▽
▽	▽	▽	▽	▼	▽	▽	▽	▽	▽	▽	▽	▽	▽	▽
First torsional, $M_{eff}/M = 9/45$														
▽	▽	▼	▽	▽	▽	▽	▼	▽	▼	▽	▽	▽	▽	▽
▽	▽	▽	▽	▼	▽	▽	▽	▽	▽	▽	▼	▽	▼	▽
▽	▼	▽	▽	▽	▽	▼	▽	▼	▽	▽	▽	▽	▽	▽

This table presents the reduced models required to detect the first bending and torsional modes of a cantilever rectangular cross-section beam.

Table 13.17 Reduced models for the modal analysis of a cantilever rectangular thin-walled cross-section beam

First bending z, $M_{eff}/M = 10/45$														
○	○	○	○	●	○	○	○	○	○	○	●	○	●	○
○	○	●	○	○	○	○	●	○	●	○	○	○	○	○
●	○	○	●	○	●	○	○	○	○	●	○	○	○	○
First bending x, $M_{eff}/M = 6/45$														
●	○	○	●	○	○	○	○	○	○	○	○	○	○	●
○	●	○	○	○	○	●	○	○	○	○	○	○	○	○
○	○	○	○	●	○	○	○	○	○	○	○	○	○	○
First torsional, $M_{eff}/M = 8/45$														
○	○	●	○	○	○	○	●	○	●	○	○	○	○	○
○	○	○	○	●	○	○	○	○	○	○	●	○	●	○
○	●	○	○	○	○	○	○	●	○	○	○	○	○	○

This table presents the reduced models required to detect the first bending and torsional modes of a cantilever rectangular, thin-walled cross-section beam.

The beam model that is necessary to detect all three modes is

$$
\begin{aligned}
u_x &= u_{x_1} + z\, u_{x_3} + x^2\, u_{x_4} + xz\, u_{x_5} + x^2 z\, u_{x_8} + z^3\, u_{x_{10}} + x^3 z\, u_{x_{12}} \\
u_y &= x\, u_{y_2} + z\, u_{y_3} + xz\, u_{y_5} + x^3\, u_{y_7} + x^3 z\, u_{y_{12}} + xz^3\, u_{y_{14}} \\
u_z &= u_{z_1} + x\, u_{z_2} + x^2\, u_{z_4} + xz\, u_{z_5} + z^2\, u_{z_6} + x^3\, u_{z_7} + xz^2\, u_{z_9} + x^4\, u_{z_{11}}
\end{aligned}
\tag{13.12}
$$

Tables 13.17, 13.18 and 13.19 present the reduced models for the remaining three geometries. The symmetry of the annular cross-section makes the reduced beam models for bending symmetric. Table 13.20 shows the reduced models needed to detect all the considered modes.

Table 13.18 Reduced models for the modal analysis of a cantilever C-shaped cross-section beam

First bending z, $M_{eff}/M = 20/45$														
△	△	▲	△	▲	△	△	▲	△	▲	△	▲	△	▲	△
△	△	▲	△	▲	△	△	▲	△	▲	△	▲	△	▲	△
▲	▲	△	▲	△	▲	▲	△	▲	△	▲	△	▲	△	△
First bending x, $M_{eff}/M = 17/45$														
▲	▲	△	▲	△	▲	△	△	△	△	△	△	△	△	▲
▲	▲	△	▲	△	▲	▲	△	△	△	▲	△	▲	△	▲
△	△	▲	△	▲	△	△	△	△	▲	△	△	△	▲	△
First torsional, $M_{eff}/M = 19/45$														
△	△	▲	△	▲	△	△	▲	△	▲	△	▲	△	▲	△
△	△	▲	△	▲	△	△	▲	△	▲	△	▲	△	▲	△
▲	▲	△	▲	△	▲	▲	△	▲	△	▲	△	△	△	△

This table presents the reduced models required to detect the first bending and torsional modes of a cantilever C-shaped cross-section beam.

Table 13.19 Reduced models for the modal analysis of a cantilever annular cross-section beam

First bending z, $M_{eff}/M = 9/45$														
□	□	□	□	■	□	□	□	□	□	□	■	□	■	□
□	□	■	□	□	□	□	■	□	■	□	□	□	□	□
■	□	□	■	□	■	□	□	□	□	□	□	□	□	□
First bending x, $M_{eff}/M = 9/45$														
■	□	□	■	□	■	□	□	□	□	□	□	□	□	□
□	■	□	□	□	□	■	□	■	□	□	□	□	□	□
□	□	□	□	■	□	□	□	□	□	□	■	□	■	□
Torsional, $M_{eff}/M = 3/45$														
□	□	■	□	□	□	□	□	□	□	□	□	□	□	□
□	□	□	□	□	□	□	□	□	□	□	□	□	□	□
■	■	□	□	□	□	□	□	□	□	□	□	□	□	□

This table presents the reduced models required to detect the first bending and torsional modes of a cantilever annular cross-section beam.

For instance, the beam model needed to detect the first bending and torsional modes of an annular cross-section beam is

$$
\begin{aligned}
u_x &= u_{x_1} + z\, u_{x_3} + x^2\, u_{x_4} + xz\, u_{x_5} + z^2\, u_{x_6} + x^3 z\, u_{x_{12}} + xz^3\, u_{x_{14}} \\
u_y &= x\, u_{y_2} + z\, u_{y_3} + x^3\, u_{y_7} + x^2 z\, u_{y_8} + xz^2\, u_{y_9} + z^3\, u_{y_{10}} \\
u_z &= u_{z_1} + x\, u_{z_2} + x^2\, u_{z_4} + xz\, u_{z_5} + z^2\, u_{z_6} + x^3 z\, u_{z_{12}} + xz^3\, u_{z_{14}}
\end{aligned}
\tag{13.13}
$$

Table 13.20 Combined reduced models for the modal analysis of cantilever beams with various cross-sections

Rectangular, $M_{eff}/M = 21/45$														
▼	▽	▼	▼	▼	▽	▽	▼	▽	▼	▽	▼	▽	▽	▽
▽	▼	▼	▽	▼	▽	▼	▽	▽	▽	▽	▼	▽	▼	▽
▼	▼	▽	▼	▼	▼	▼	▽	▼	▽	▼	▽	▽	▽	▽
Rectangular thin-walled, $M_{eff}/M = 24/45$														
●	○	●	●	●	○	○	●	○	●	○	●	○	●	●
○	●	●	○	●	○	●	●	○	●	○	●	○	●	○
●	●	○	●	●	●	○	○	●	○	●	○	○	○	○
C-shaped, $M_{eff}/M = 37/45$														
▲	▲	▲	▲	▲	▲	△	▲	△	▲	△	▲	△	▲	▲
▲	▲	▲	▲	▲	▲	▲	▲	△	▲	▲	▲	▲	▲	▲
▲	▲	▲	▲	▲	▲	▲	△	▲	▲	▲	△	▲	▲	△
Annular, $M_{eff}/M = 20/45$														
■	□	■	■	■	■	□	□	□	□	□	■	□	■	□
□	■	■	□	□	□	■	■	■	■	□	□	□	□	□
■	■	□	■	■	■	□	□	□	□	□	■	□	■	□

This table presents the reduced models required to detect the first bending and torsional modes of cantilever beams with various cross-sections.

These results suggest:

1. Different sets of displacement variables are needed to detect different modes, as stated previously.
2. Thin walls and the asymmetry of the cross-section play a fundamental role in determining the number of terms needed to detect a given mode. The asymmetry, in particular, seems to be of primary importance.
3. As a general guideline, it can be stated that if asymmetric cross-sections are considered, full models should be adopted, whereas symmetric geometries should be analysed by means of reduced models, since great reductions in computational costs are possible.

13.3.2 Influence of BCs

The influence of BCs is investigated in this section through the MAAA. A rectangular beam is considered and three different BCs are considered: clamped at both ends, hinged at both ends and simply supported.

Tables 13.21, 13.22 and 13.23 show the reduced models for the three modes considered for all the BCs under investigation. Table 13.24 presents the combined models for all the modes. The following comments can be made:

1. The BCs play a significant role in the construction of reduced models. Their effect is equivalent to that observed for the cross-section geometry.
2. Clamped–clamped and hinged–hinged conditions instead require similar models and the simply supported condition needs the most cumbersome 1D model.
3. In general, significant reductions in the number of generalized variables can be obtained for all the BCs considered.

Table 13.21 Reduced models for the modal analysis of a clamped–clamped rectangular cross-section beam

First bending z, $M_{eff}/M = 6/45$														
○	○	○	○	○	●	○	○	○	○	○	○	○	○	○
○	○	●	○	○	○	○	○	○	●	○	○	○	○	○
●	○	○	●	○	●	○	○	○	○	○	○	○	○	○
First bending x, $M_{eff}/M = 5/45$														
●	○	○	●	○	●	○	○	○	○	○	○	○	○	○
○	●	○	○	○	○	○	○	○	○	○	○	○	○	○
○	○	○	○	●	○	○	○	○	○	○	○	○	○	○
First torsional, $M_{eff}/M = 8/45$														
●	○	●	○	○	○	○	○	○	○	○	○	○	○	○
○	●	●	○	○	○	○	○	○	○	○	●	○	●	○
●	●	○	○	○	○	○	○	○	○	○	○	○	○	○

This table presents the reduced models required to detect the first bending and torsional modes of a clamped–clamped rectangular cross-section beam.

Table 13.22 Reduced models for the modal analysis of a hinged–hinged rectangular cross-section beam

First bending z, $M_{eff}/M = 5/45$														
△	△	△	△	△	▲	△	△	△	△	△	△	△	△	△
△	△	▲	△	△	△	△	△	△	△	△	△	△	△	△
▲	△	△	▲	△	▲	△	△	△	△	△	△	△	△	△
First bending x, $M_{eff}/M = 4/45$														
▲	△	△	▲	△	△	△	△	△	△	△	△	△	△	△
△	▲	△	△	△	△	△	△	△	△	△	△	△	△	△
△	△	△	△	▲	△	△	△	△	△	△	△	△	△	△
First torsional, $M_{eff}/M = 6/45$														
△	△	▲	△	△	△	△	△	△	△	△	△	△	△	△
△	△	△	△	▲	△	△	△	△	△	△	▲	△	▲	△
△	▲	△	△	△	△	△	△	▲	△	△	△	△	△	△

This table presents the reduced models required to detect the first bending and torsional modes of a hinged–hinged rectangular cross-section beam.

Table 13.23 Reduced models for the modal analysis of a simply supported rectangular cross-section beam

First bending z, $M_{eff}/M = 13/45$														
■	■	□	■	□	■	□	□	□	□	□	□	□	□	□
■	■	□	■	□	■	■	□	■	□	□	□	■	□	□
□	□	■	□	■	□	□	□	□	□	□	□	□	□	□
First bending x, $M_{eff}/M = 8/45$														
□	□	■	□	■	□	□	□	□	□	□	□	□	□	□
□	□	■	□	□	□	□	□	□	■	□	□	□	□	□
■	■	□	■	□	■	□	□	□	□	□	□	□	□	□
First torsional, $M_{eff}/M = 9/45$														
□	□	■	□	□	□	□	□	□	□	□	□	□	□	□
□	□	■	□	■	□	□	■	□	■	□	■	□	■	□
■	■	□	□	□	□	□	□	□	□	□	□	□	□	□

This table presents the reduced models required to detect the first bending and torsional modes of a simply supported rectangular cross-section beam.

13.4 Static Analysis of Plates and Shells

A square plate is analysed in this section through the MAAA; (see also Carrera *et al.* 2011b). The following displacement and stress variables have been considered to build reduced 2D models: u_z, σ_{xx}, σ_{yy} and σ_{zz} at $[a/2, b/2, 0]$, σ_{xz} at $[0, b/2, h/2]$ and σ_{yz} at $[a/2, 0, h/2]$. Four-node plate elements have been used with a uniform mesh of 15×15 elements.

The accuracy of the reduced models obtained through the MAAA is compared with results from other theories that can be obtained as particular cases of the present unified formulation. The previously mentioned classical plate theory (CPT, by Kirchhoff) and the first-order shear deformation theory (FSDT, by Reissner and Mindlin) are considered, together with three other

Table 13.24 Combined reduced models for the modal analysis of cantilever beams with various BCs

Clamped–clamped, $M_{eff}/M = 14/45$														
●	○	●	●	○	●	○	○	○	○	○	○	○	○	○
○	●	●	○	○	○	○	○	○	●	○	●	○	●	○
●	●	○	●	●	●	○	○	○	○	○	○	○	○	○
Hinged–hinged, $M_{eff}/M = 15/45$														
▲	△	▲	▲	△	▲	△	△	△	△	△	△	△	△	△
△	▲	▲	△	▲	△	△	△	△	△	△	▲	△	▲	△
▲	▲	△	▲	▲	▲	△	△	▲	△	△	△	△	△	△
Simply supported, $M_{eff}/M = 25/45$														
■	■	■	■	■	■	□	□	□	□	□	□	□	□	□
■	■	■	■	■	■	■	■	■	■	□	■	■	■	□
■	■	■	■	■	■	□	□	□	□	□	□	□	□	□

This table presents the reduced models required to detect the first bending and torsional modes of cantilever beams with various BCs.

refined models. The first one, hereafter referred to as Pandya, was developed by Pandya and Kant (1988),

$$\begin{aligned} u_x &= u_{x1} + z\, u_{x2} + z^2\, u_{x3} + z^3\, u_{x4} \\ u_y &= u_{y1} + z\, u_{y2} + z^2\, u_{y3} + z^3\, u_{y4} \\ u_z &= u_{z1} \end{aligned} \tag{13.14}$$

The second one, hereafter referred to as Kant-1, was developed by Kant and Manjunatha (1988),

$$\begin{aligned} u_x &= u_{x1} + z\, u_{x2} + z^2\, u_{x3} + z^3\, u_{x4} \\ u_y &= u_{y1} + z\, u_{y2} + z^2\, u_{y3} + z^3\, u_{y4} \\ u_z &= u_{z1} + z\, u_{z2} + z^2\, u_{z3} + z^3\, u_{z4} \end{aligned} \tag{13.15}$$

The third one, hereafter referred to as Kant-2, was developed by Kant (1982),

$$\begin{aligned} u_x &= z\, u_{x2} + z^3\, u_{x4} \\ u_y &= z\, u_{y2} + z^3\, u_{y4} \\ u_z &= u_{z1} + z^2\, u_{z3} \end{aligned} \tag{13.16}$$

13.4.1 Influence of BCs

Reduced 2D models for various BCs are given in this section. The following BCs are considered:

1. *ssss*, four simply supported edges.
2. *cfcf*, two clamped and two free edges.
3. *cccc*, four clamped edges.

Table 13.25 Symbols adopted to distinguish various plate BCs along the four edges

Active term	Inactive term	
▲	△	*ssss*
■	□	*cfcf*
▼	▽	*cccc*

c indicates clamped, *s* simply supported and *f* free.

Table 13.25 shows the symbols that have been adopted to refer to each set of BCs. A bisinusoidal transverse distributed load, $p_z = \bar{p}_z \sin(x/a)\cos(y/b)$, is applied at the top surface. An $N = 4$ solution is used to check the effectiveness of each term. $N = 4$ can, in fact, provide 3D-like solutions for this class of problems.

Table 13.26 presents the reduced 2D models required to detect the 3D-like solution for a moderately thick plate ($a/h = 10$). Each row refers to a displacement or stress variable. Each column considers a different set of BCs. M_{eff} indicates the number of terms (i.e. the number of DOFs per node) of the models that are equivalent to the fourth-order one. The last row shows the reduced models needed to detect all the considered outputs. The latter combined models are used to build Table 13.27, which shows a comparison with the accuracies obtained with the CPT, FSDT and Kant-2 plate models.

It is possible to determine other theories that provide solutions with a certain degree of error, compared with the full fourth-order one. Figure 13.1 shows the number of terms needed to detect σ_{xz} with a given error. The corresponding plate models are also indicated. The analyses that have been carried out suggest:

1. Reduced plate models that are equivalent to a fourth-order theory vary significantly when different output variables are considered.
2. The influence of the BCs is not very significant.
3. Classical models are unable to deal with shear stresses.
4. A significant reduction in computational cost is only obtained when a limited number of output variables have to be detected.
5. The diagram for the number of terms vs the error shows that all the theories derived from the present approach are able to satisfy a given error requirement, with lower computational costs than in the open literature models that have been considered. This shows the strength of the present technique in detecting the best possible theories for a given structural problem.

13.4.2 Influence of the Loading Conditions

Reduced 2D models for various loading conditions are considered in this section for a simply supported plate ($a/h = 10$). Three different loading conditions are taken into account (as shown in Figure 13.2):

1. A bisinusoidal distributed load.
2. A point load.
3. Four point loads.

Table 13.26 Reduced 2D models for a moderately thick plate with various BCs

ssss					*cfcf*					*cccc*				
u_z														
$M_{eff}/M = 6/15$					$M_{eff}/M = 7/15$					$M_{eff}/M = 7/15$				
△	▲	△	▲	△	□	■	□	■	□	▽	▼	▽	▼	▽
△	▲	△	▲	△	□	■	□	■	□	▽	▼	▽	▼	▽
▲	△	▲	△	△	■	■	■	□	□	▼	▼	▼	▽	▽
σ_{xx}														
$M_{eff}/M = 10/15$					$M_{eff}/M = 10/15$					$M_{eff}/M = 10/15$				
▲	▲	△	▲	△	■	■	□	■	□	▼	▼	▽	▼	▽
▲	▲	△	▲	△	■	■	□	■	□	▼	▼	▽	▼	▽
▲	▲	▲	△	▲	■	■	■	□	■	▼	▼	▼	▽	▼
σ_{yy}														
$M_{eff}/M = 10/15$					$M_{eff}/M = 10/15$					$M_{eff}/M = 10/15$				
▲	▲	△	▲	△	■	■	□	■	□	▼	▼	▽	▼	▽
▲	▲	△	▲	△	■	■	□	■	□	▼	▼	▽	▼	▽
▲	▲	▲	△	▲	■	■	■	□	■	▼	▼	▼	▽	▼
σ_{xz}														
$M_{eff}/M = 6/15$					$M_{eff}/M = 6/15$					$M_{eff}/M = 7/15$				
△	▲	△	▲	△	□	■	□	■	□	▽	▼	▽	▼	▽
△	▲	△	▲	△	□	■	□	□	□	▽	▼	▽	▼	▽
▲	△	▲	△	△	■	□	■	□	■	▼	▽	▼	▽	▼
σ_{yz}														
$M_{eff}/M = 6/15$					$M_{eff}/M = 7/15$					$M_{eff}/M = 7/15$				
△	▲	△	▲	△	□	■	□	■	□	▽	▼	▽	▼	▽
△	▲	△	▲	△	□	■	□	■	□	▽	▼	▽	▼	▽
▲	△	▲	△	△	■	□	■	□	■	▼	▽	▼	▽	▼
σ_{zz}														
$M_{eff}/M = 11/15$					$M_{eff}/M = 12/15$					$M_{eff}/M = 13/15$				
▲	▲	▲	△	△	■	■	■	■	□	▼	▼	▼	▼	▽
▲	▲	▲	△	△	■	■	■	□	□	▼	▼	▼	▼	▽
▲	▲	▲	▲	▲	■	■	■	■	■	▼	▼	▼	▼	▼
Combined														
$M_{eff}/M = 13/15$					$M_{eff}/M = 13/15$					$M_{eff}/M = 13/15$				
▲	▲	▲	▲	△	■	■	■	■	□	▼	▼	▼	▼	▽
▲	▲	▲	▲	△	■	■	■	■	□	▼	▼	▼	▼	▽
▲	▲	▲	▲	▲	■	■	■	■	■	▼	▼	▼	▼	▼

This table presents the reduced models required to detect displacement and stress variables for a plate under various BCs.

Table 13.27 Accuracy of different models in detecting displacements and stresses, for a moderately thick plate

	δ_{u_z} (%)	$\delta_{\sigma_{xx}}$ (%)	$\delta_{\sigma_{yy}}$ (%)	$\delta_{\sigma_{xz}}$ (%)	$\delta_{\sigma_{yz}}$ (%)	$\delta_{\sigma_{zz}}$ (%)
			ssss			
Combined	100.0	100.0	100.0	100.0	100.0	100.0
CPT	96.1	98.4	98.4	66.8	66.8	1974.2
FSDT	100.9	98.4	98.4	66.8	66.8	1974.2
Kant-2	99.9	100.2	100.2	100.0	100.0	79.4
			cfcf			
Combined	100.0	100.0	100.0	100.0	100.0	100.0
CPT	91.2	98.3	94.9	239.5	−67.86	1171.3
FSDT	103.1	98.7	97.2	83.2	65.4	1183.7
Kant-2	99.9	100.0	101.0	98.8	99.7	80.4
			cccc			
Combined	100.0	100.0	100.0	100.0	100.0	100.0
CPT	88.4	96.2	96.2	87.5	87.5	1074.0
FSDT	102.8	97.1	97.1	81.4	81.4	1084.8
Kant-2	99.8	99.9	99.9	98.9	98.9	80.0

This table compares different plate models in terms of accuracies for given output variables; the model referred to as ‘Combined’ can be found in Table 13.26.

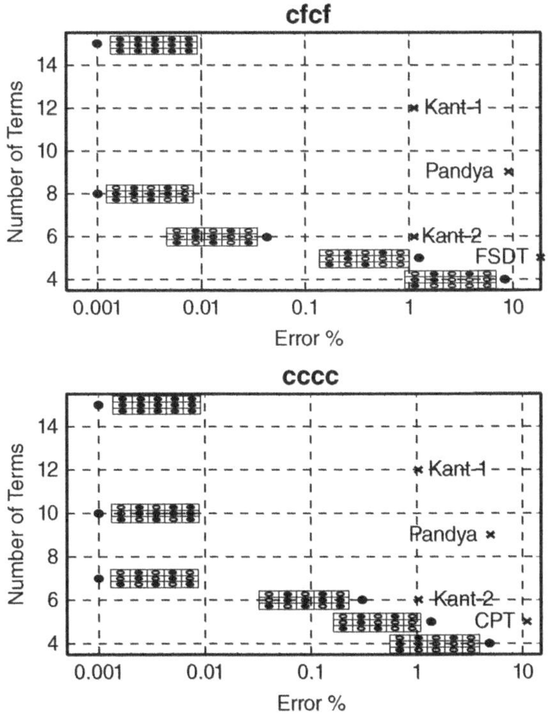

Figure 13.1 Number of terms vs error for different BCs, σ_{xz}, for a moderately thick plate

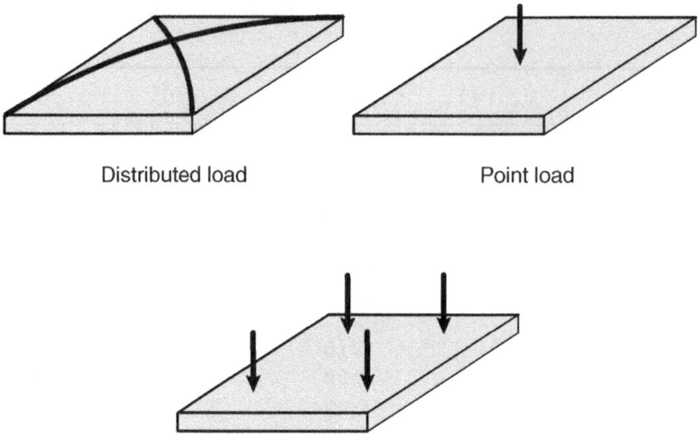

Figure 13.2 Loading conditions for the MAAA analysis of plates

Table 13.28 shows the symbols adopted to distinguish each loading condition. The reduced models needed for different displacement and stress components are shown in Table 13.29. Each row refers to a different output variable, while each column considers a different loading condition. M_{eff} indicates the number of terms of the models that are equivalent to the fourth-order one. The last row shows the plate models that are needed to detect all the considered outputs. The latter combined models have been used to build Table 13.30, which shows a comparison of the accuracies given by the CPT, FSDT and Kant-2 plate models. Figure 13.3 shows the number of terms needed to compute σ_{xz} with a given error. The corresponding plate models are also indicated together with models retrieved from the open literature. The following remarks can be made on the basis of the analyses:

1. The sets of effective displacement variables vary when distributed or concentrated loads are considered, whereas no differences arise if one or multiple point loads are introduced.
2. The effect of the considered output variable is significant in the case of a distributed load, but it is almost negligible if point loads are considered.

Table 13.28 Symbols adopted to distinguish various plate loading conditions

Active term	Inactive term	
▲	△	Distributed load
■	□	Point load
▼	▽	Four point loads

This table shows the graphic notation adopted to indicate the different loading conditions for the MAAA analysis of a plate.

Table 13.29 Reduced 2D models for a simply supported plate under various loading conditions

Distributed load					Point load					Four point loads				
					u_z									
$M_{eff}/M = 6/15$					$M_{eff}/M = 6/15$					$M_{eff}/M = 6/15$				
△	▲	△	▲	△	□	■	□	■	□	▽	▼	▽	▼	▽
△	▲	△	▲	△	□	■	□	■	□	▽	▼	▽	▼	▽
▲	△	▲	△	△	■	□	■	□	□	▼	▽	▼	▽	▽
					σ_{xx}									
$M_{eff}/M = 10/15$					$M_{eff}/M = 7/15$					$M_{eff}/M = 7/15$				
▲	▲	△	▲	△	□	■	□	■	□	▽	▼	▽	▼	▽
▲	▲	△	▲	△	□	■	□	■	□	▽	▼	▽	▼	▽
▲	▲	▲	△	▲	■	□	■	□	■	▼	▽	▼	▽	▼
					σ_{yy}									
$M_{eff}/M = 10/15$					$M_{eff}/M = 7/15$					$M_{eff}/M = 7/15$				
▲	▲	△	▲	△	□	■	□	■	□	▽	▼	▽	▼	▽
▲	▲	△	▲	△	□	■	□	■	□	▽	▼	▽	▼	▽
▲	▲	▲	△	▲	■	□	■	□	■	▼	▽	▼	▽	▼
					σ_{xz}									
$M_{eff}/M = 6/15$					$M_{eff}/M = 7/15$					$M_{eff}/M = 7/15$				
△	▲	△	▲	△	□	■	□	■	□	▽	▼	▽	▼	▽
△	▲	△	▲	△	□	■	□	■	□	▽	▼	▽	▼	▽
▲	△	▲	△	△	■	□	■	□	■	▼	▽	▼	▽	▼
					σ_{yz}									
$M_{eff}/M = 6/15$					$M_{eff}/M = 7/15$					$M_{eff}/M = 7/15$				
△	▲	△	▲	△	□	■	□	■	□	▽	▼	▽	▼	▽
△	▲	△	▲	△	□	■	□	■	□	▽	▼	▽	▼	▽
▲	△	▲	△	△	■	□	■	□	■	▼	▽	▼	▽	▼
					σ_{zz}									
$M_{eff}/M = 11/15$					$M_{eff}/M = 7/15$					$M_{eff}/M = 7/15$				
▲	▲	▲	△	△	□	■	□	■	□	▽	▼	▽	▼	▽
▲	▲	▲	△	△	□	■	□	■	□	▽	▼	▽	▼	▽
▲	▲	▲	▲	▲	■	□	■	□	■	▼	▽	▼	▽	▼
					Combined									
$M_{eff}/M = 13/15$					$M_{eff}/M = 7/15$					$M_{eff}/M = 7/15$				
▲	▲	▲	▲	△	□	■	□	■	□	▽	▼	▽	▼	▽
▲	▲	▲	▲	△	□	■	□	■	□	▽	▼	▽	▼	▽
▲	▲	▲	▲	▲	■	□	■	□	■	▼	▽	▼	▽	▼

This table presents the reduced models required to detect displacement and stress variables for a plate under various loading conditions.

Table 13.30 Accuracy of different models in detecting displacements and stresses for various loading conditions

	δ_{u_z} (%)	$\delta_{\sigma_{xx}}$ (%)	$\delta_{\sigma_{yy}}$ (%)	$\delta_{\sigma_{xz}}$ (%)	$\delta_{\sigma_{xz}}$ (%)	$\delta_{\sigma_{zz}}$ (%)
			Distributed			
Combined	100.0	100.0	100.0	100.0	100.0	100.0
CPT	96.1	98.4	98.4	66.8	66.8	1974.2
FSDT	100.9	98.4	98.4	66.8	66.8	1974.2
Kant-2	99.9	100.3	100.3	100.0	100.0	79.4
			Point load			
Combined	100.0	100.0	100.0	100.0	100.0	100.0
CPT	91.4	83.7	83.7	67.0	67.0	540.1
FSDT	99.9	83.7	83.7	67.0	67.0	540.1
Kant-2	100.0	96.5	96.5	99.9	99.9	39.5
			Four point loads			
Combined	100.0	100.0	100.0	100.0	100.0	100.0
CPT	91.4	83.7	83.7	67.0	67.0	540.1
FSDT	99.9	83.7	83.7	67.0	67.0	540.1
Kant-2	100.0	96.5	96.5	99.9	99.9	39.5

This table compares different plate models in terms of accuracies for given output variables; the model referred to as 'Combined' can be found in Table 13.29.

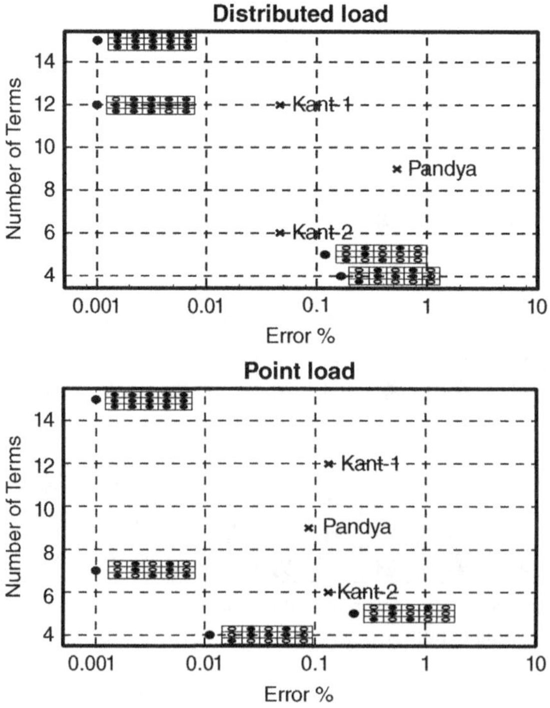

Figure 13.3 Number of terms vs error for different loading conditions, σ_{xz}, for a moderately thick plate

Table 13.31 Symbols adopted to distinguish various displacement and stress components

	Active term	Inactive term
▲	△	u_z
■	□	σ_{xx}
▼	▽	σ_{xz}
◆	◇	σ_{zz}

This table shows the graphic notation adopted to indicate various displacement and stress components for the MAAA analysis of a plate.

3. The validity of the diagram for the number of terms vs the error in providing guidelines for the construction of plate theories to achieve a given accuracy has been confirmed. As the error is fixed, the derived theories generally lie below classical and other refined models; that is, the proposed models are able to fulfil a certain accuracy demand but with lower computational costs.

13.4.3 Influence of the Loading and Thickness

The combined effect of loading and boundary conditions is considered in this section. The influence of the length-to-thickness ratio (a/h) is also investigated. The loading sets considered are the same as those in the previous analyses. Each output variable is associated with a symbol, as shown in Table 13.31.

First, a thick plate is considered assuming $a/h = 5$. Table 13.32 shows the sets of displacement variables that are needed to detect various output variables. Each row refers to a different loading condition; the plate is considered to be simply supported. Table 13.33 shows the plate models that detect all the considered outputs for all the loading and boundary conditions, and for different thicknesses. Comparisons with other theories are given in Table 13.34. The analyses that have been carried out suggest the following comments:

1. The effective displacement variable sets depend on all of the three parameters: loadings, BCs and thickness.
2. A proper analysis of thick plates requires more sophisticated models, since the total number of expansion terms increases as a/h decreases. This result is analogous to that shown by Carrera and Petrolo (2010).
3. It has been confirmed that different output variables require different plate models in order to be detected properly.

13.4.4 Influence of the Thickness Ratio on Shells

The MAAA is now applied to a shell structure, as in Figure 13.4, under cylindrical bending; see also Carrera *et al.* (2011a). The shell is simply supported and a sinusoidal distribution of

Table 13.32 Reduced 2D models for a simply supported thick plate for various displacement and stress components

Distributed load					Point load					Four point loads				
$M_{eff}/M = 9/15$					$M_{eff}/M = 7/15$					$M_{eff}/M = 7/15$				
▲	▲	△	▲	△	△	▲	△	▲	△	△	▲	△	▲	△
▲	▲	△	▲	△	△	▲	△	▲	△	△	▲	△	▲	△
▲	▲	▲	△	△	▲	△	▲	△	▲	▲	△	▲	△	▲
$M_{eff}/M = 11/15$					$M_{eff}/M = 7/15$					$M_{eff}/M = 7/15$				
■	■	□	■	□	□	■	□	■	□	□	■	□	■	□
■	■	□	■	□	□	■	□	■	□	□	■	□	■	□
■	■	■	■	■	■	□	■	□	■	■	□	■	□	■
$M_{eff}/M = 7/15$					$M_{eff}/M = 7/15$					$M_{eff}/M = 6/15$				
▽	▼	▽	▼	▽	▽	▼	▽	▼	▽	▽	▼	▽	▼	▽
▽	▼	▽	▼	▽	▽	▼	▽	▼	▽	▽	▼	▽	▼	▽
▼	▽	▼	▽	▼	▼	▽	▼	▽	▼	▼	▽	▼	▽	▽
$M_{eff}/M = 11/15$					$M_{eff}/M = 7/15$					$M_{eff}/M = 7/15$				
◆	◆	◆	◇	◇	◇	◆	◇	◆	◇	◇	◆	◇	◆	◇
◆	◆	◆	◇	◇	◇	◆	◇	◆	◇	◇	◆	◇	◆	◇
◆	◆	◆	◆	◆	◆	◇	◆	◇	◆	◆	◇	◆	◇	◆

This table presents the reduced models required to detect displacement and stress variables for a simply supported thick plate.

Table 13.33 Sets of effective terms for various thicknesses, boundary and loading conditions

$a/h = 100$				
△□▽◇	▲■▼◆	△□▽◇	△■▼◆	△□▽◇
△□▽◇	▲■▼◆	△□▽◇	△■▼◆	△□▽◇
▲■▼◆	△□▽◆	▲■▼◆	△□▽◇	△□▼◆
$a/h = 10$				
△■▽◆	▲■▼◆	△□▽◆	▲■▼◆	△□▽◇
△■▽◆	▲■▼◆	△□▽◆	▲■▼◆	△□▽◇
▲■▼◆	▲■▽◆	▲■▼◆	△□▽◆	▲■▼◆
$a/h = 5$				
▲■▽◆	▲■▼◆	△□▽◆	▲■▼◆	△□▽◇
▲■▽◆	▲■▼◆	△□▽◆	▲■▼◆	△□▽◇
▲■▼◆	▲■▽◆	▲■▼◆	△■▽◆	▲■▼◆

This table shows the reduced 2D models needed to detect different displacement and stress components for different boundary and loading conditions.

Table 13.34 Accuracy of different models in detecting displacements and stresses for various thicknesses

	δ_{u_z} (%)	$\delta_{\sigma_{xx}}$ (%)	$\delta_{\sigma_{yy}}$ (%)	$\delta_{\sigma_{xz}}$ (%)	$\delta_{\sigma_{xz}}$ (%)	$\delta_{\sigma_{zz}}$ (%)
			$a/h = 100$			
Combined	100.0	100.0	100.0	100.0	100.0	100.0
CPT	102.7	98.5	98.0	158.6	710.2	1337.4
FSDT	102.9	98.5	98.0	80.0	114.3	1337.6
Kant-2	100.0	100.0	100.0	99.9	117.9	99.0
			$a/h = 10$			
Combined	100.0	100.0	100.0	100.0	100.0	100.0
CPT	86.2	82.5	79.2	202.9	−106.014	449.824
FSDT	101.6	82.7	80.24	83.6	72.7	453.2
Kant-2	100.0	96.2	95.7	98.7	97.7	39.8
			$a/h = 5$			
Combined	100.0	100.0	100.0	100.0	100.0	100.0
CPT	60.7	70.6	65.7	228.9	−90.6	189.9
FSDT	102.8	70.9	68.1	88.7	71.6	193.5
Kant-2	100.4	89.7	88.7	96.0	99.1	−23.1

This table compares different plate models in terms of accuracies for various output variables, BCs and thicknesses; the model referred to as 'Combined' can be found in Table 13.33.

transverse pressure is applied at the top surface,

$$p_z = \bar{p}_z \sin(\pi\beta/b) \tag{13.17}$$

where β is the curvilinear coordinate. The displacement u_z and the stresses $\sigma_{\beta\beta}$ and σ_{zz} are computed at $[a/2, b/2, h/2]$, while $\sigma_{\beta z}$ is computed at $[a/2, 0, 0]$.

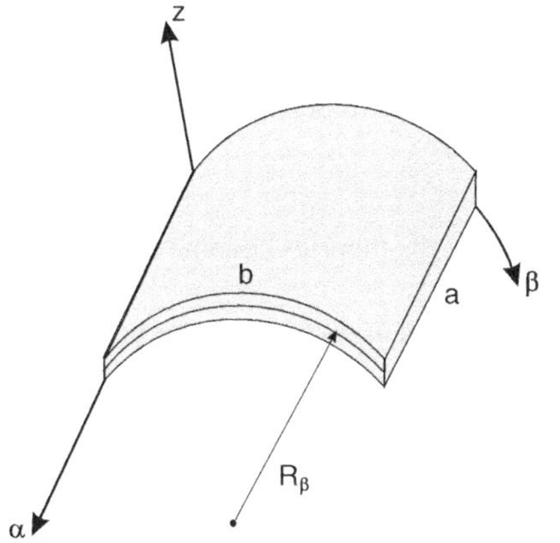

Figure 13.4 Geometry of shell for the MAAA analysis

Table 13.35 Reduced 2D models for the cylindrical bending of isotropic shells with different thickness ratios

$R_\beta/h = 100$					$R_\beta/h = 10$					$R_\beta/h = 4$				
					u_z									
$M_{eff}/M = 4/15$					$M_{eff}/M = 6/15$					$M_{eff}/M = 6/15$				
△	△	△	△	△	△	△	△	△	△	△	△	△	△	△
▲	▲	△	△	△	▲	▲	△	▲	△	▲	▲	△	▲	△
▲	△	▲	△	△	▲	△	▲	▲	△	▲	△	▲	▲	△
					$\sigma_{\beta\beta}$									
$M_{eff}/M = 6/15$					$M_{eff}/M = 8/15$					$M_{eff}/M = 8/15$				
△	△	△	△	△	△	△	△	△	△	△	△	△	△	△
▲	▲	△	△	△	▲	▲	△	▲	△	▲	▲	△	▲	△
▲	▲	▲	▲	△	▲	▲	▲	▲	▲	▲	▲	▲	▲	▲
					$\sigma_{\beta z}$									
$M_{eff}/M = 5/15$					$M_{eff}/M = 5/15$					$M_{eff}/M = 8/15$				
△	△	△	△	△	△	△	△	△	△	△	△	△	△	△
▲	▲	△	▲	△	▲	▲	△	▲	△	▲	▲	△	▲	▲
▲	△	▲	△	△	▲	△	▲	△	△	▲	△	▲	▲	▲
					σ_{zz}									
$M_{eff}/M = 9/15$					$M_{eff}/M = 10/15$					$M_{eff}/M = 10/15$				
△	△	△	△	△	△	△	△	△	△	△	△	△	△	△
▲	▲	▲	▲	△	▲	▲	▲	▲	▲	▲	▲	▲	▲	▲
▲	▲	▲	▲	▲	▲	▲	▲	▲	▲	▲	▲	▲	▲	▲
					Combined									
$M_{eff}/M = 9/15$					$M_{eff}/M = 10/15$					$M_{eff}/M = 10/15$				
△	△	△	△	△	△	△	△	△	△	△	△	△	△	△
▲	▲	▲	▲	△	▲	▲	▲	▲	▲	▲	▲	▲	▲	▲
▲	▲	▲	▲	▲	▲	▲	▲	▲	▲	▲	▲	▲	▲	▲

This table shows different 2D models for detecting displacement and stress components of thin and thick shells.

Table 13.35 shows the reduced 2D models for different thickness ratios. The following remarks can be made:

1. As the thickness ratio decreases, the theories become more computationally expensive (M_{eff} increases).
2. $\sigma_{\beta\beta}$ and $\sigma_{\beta z}$ and all the terms in the u_z expansion are necessary for the exact evaluation of σ_{zz}.
3. In this particular case, the terms in the u_α expansion are not influential, because a cylindrical bending problem has been considered.
4. The constant term of the in-plane displacement u_β is more important in a shell than in a plate because a shell undergoes a membranal deformation, even when it is very thin. This is due to curvature effects.

13.5 The BTD

The construction of reduced models through the MAAA allows one to obtain a diagram, for a given problem, which in terms of accuracy (input) answers the following fundamental questions (see also Carrera *et al.*, 2011b):

- What is the 'minimum' number of terms (N_{min}) that needs to be used in an FE model?
- What terms need to be retained, i.e. what generalized displacement variables need to be used as FE DOFs?

In other words, the MAAA is able to create plots, like the one shown in Figure 13.5, which give the number of terms vs the error. This plot can be defined as the BTD (Best Theory Diagram) since it allows one to edit an arbitrary given theory in order to get a lower number of terms for a given error (vertical shift, Δ_N), or to increase the accuracy while keeping the computational costs constant (horizontal shift, Δ_{error}). The plot appears generally as a hyperbola. The CUF makes the computation of such a plot possible. It should be noted that the diagram has the following properties:

- It changes as the problem changes (a/h, loadings, BCs, etc.).
- It changes as the output variable changes (displacement/stress components, or a combination of these).

The validity of the BTD has been tested by computing the accuracy of all the plate models obtainable as a combination of the 15 terms of the fourth-order theory. The results are shown in Figure 13.6 for the case of a simply supported plate loaded with a distributed load where u_z is considered as the output variable. The BTD perfectly matches the lower boundaries of the region in which all the models lie. This confirms that the BTD can be considered the best theory (i.e. the least cumbersome) for a given error. The BTD can be considered a tool to evaluate any other structural theory.

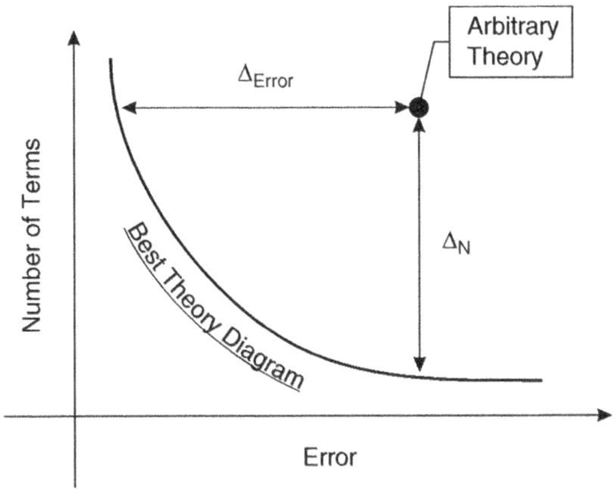

Figure 13.5 The BTD

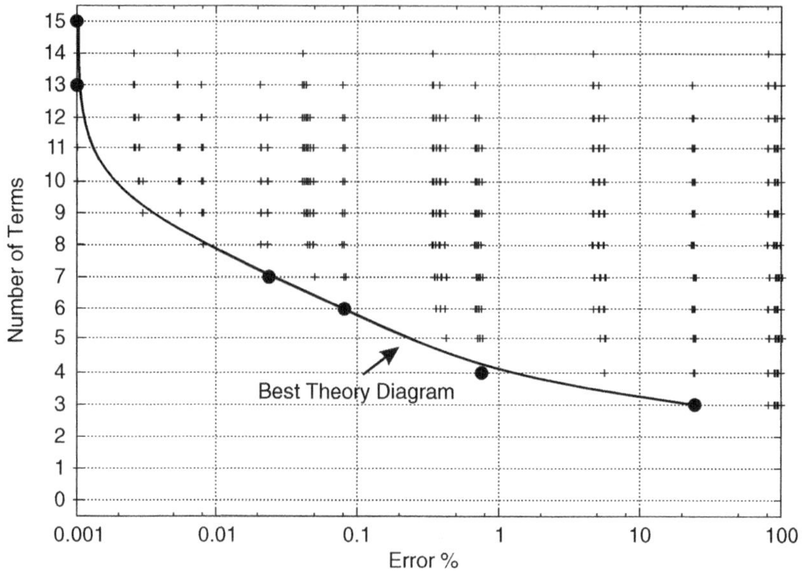

Figure 13.6 Accuracy of all the possible combinations of plate models in computing u_z for the simply supported plate under a distributed load (each '+' indicates a plate model)

References

Carrera E, Cinefra M and Petrolo M 2011a A best theory diagram for metallic and laminated shells. In: *Shell-like Structures: Non-classical Theories and Applications*. Springer, pp. 681–698.

Carrera E, Miglioretti F and Petrolo M 2011b Guidelines and recommendations on the use of higher-order finite elements for bending analysis of plates. *International Journal for Computational Methods in Engineering Science and Mechanics* **12**(6), 303–324.

Carrera E, Miglioretti F and Petrolo M 2011c Accuracy of refined finite elements for laminated plate analysis. *Composite Structures* **93**(5), 1311–1327.

Carrera E, Miglioretti F and Petrolo M 2012 Computations and evaluations of higher-order theories for free vibration analysis of beams. *Journal of Sound and Vibration* **331**, 4269–4284.

Carrera E and Petrolo M 2010 Guidelines and recommendations to construct theories for metallic and composite plates. *AIAA Journal* **48**(12), 2852–2866.

Carrera E and Petrolo M 2011 On the effectiveness of higher-order terms in refined beam theories. *Journal of Applied Mechanics* **78**. DOI: 10.1115/1.4002207.

Kant T 1982 Numerical analysis of thick plates. *Computer Methods in Applied Mechanics and Engineering* **31**, 1–18.

Kant T and Manjunatha B 1988 An unsymmetric FRC laminate C^0 finite element model with 12 degrees of freedom per node. *Engineering Computations* **5**(3), 292–308.

Pandya B and Kant T 1988 Finite element analysis of laminated composite plates using high-order displacement model. *Composites Science and Technology* **32**, 137–155.

14

Mixing Variable Kinematic Models

Many different FE models have been presented in the previous chapters of this book and 1D, 2D and 3D models were described. The solution of a real problem usually requires a mathematical model which includes a combination of 3D, 2D and 1D FEs to be built.

Commercial codes can easily handle different models because they are based above all on the classical assumption of 6 degrees of freedom (three displacements and three rotations) for each node (solids only have three displacements). This approach allows 1D, 2D and 3D models to be connected very easily by imposing the compatibility of displacements and rotations on the shared nodes.

The models presented in this book are not based on the variable kinematic assumption of 6 degrees of freedom per node, and the problem of combining different models therefore becomes more complex. Higher-order theories can be based on different kinematics, and the assembly of different elements therefore requires appropriate techniques.

The CUF-based FEs presented in this book can be divided into two families. The first family includes FEs that only have displacements as unknowns: the 3D model (see Chapter 7), the plate/shell model based on a layer-wise formulation (see Chapter 12) and the beam model based on the Lagrange expansion (see Chapter 9). The second family includes models in which the unknowns are not just displacements, such as the plate model based on an equivalent single layer formulation (see Chapter 10) or the 1D model based on the Taylor expansion (see Chapter 8).

The models in the first family can be assembled by using the classical approach, imposing compatibility of the displacements at the nodes, as in commercial codes. Whichever model is used, 1D or 2D or 3D, two nodes can easily be connected by simply adding the related stiffness to the assembly procedure. In the following sections, this approach is referred to as 'shared stiffness'. An example of the approach, applied to the CUF, has been presented by Carrera and Zappino (2013).

If the models in the second family have to be combined with those of the first, or have to be combined with each other, the shared stiffness approach cannot be used since the displacements do not appear as explicit unknowns. In this case, an appropriate technique has to be used in order to mix different FEs in the same model. A possible solution consists of introducing

Finite Element Analysis of Structures Through Unified Formulation, First Edition.
Erasmo Carrera, Maria Cinefra, Marco Petrolo and Enrico Zappino.

'Lagrange multipliers' (LMs), because they allow compatibility to be imposed between two generic points of the structure. An application of this method has been presented by Carrera *et al.* (2013).

A more 'smeared' approach is the 'Arlequin' method (Ben Dhia 1998; Biscani *et al.* 2011), which imposes an overlapping area in which the two models are merged together using some smooth functions.

14.1 Coupling Variable Kinematic Models via Shared Stiffness

The most common approach used to couple FEs with different kinematics consists of 'adding' or 'sharing' the stiffness of the nodes. Let us consider two structures, structure A and structure B, which 'share' (see Figure 14.1) a finite number of nodes. The displacement vector of structure A is $\boldsymbol{u}^A$, and that of structure B is $\boldsymbol{u}^B$. These vectors can be split into two contributions, a 'free' displacement, $\boldsymbol{u}_f$, and the displacements of the 'shared' nodes, $\boldsymbol{u}_s$. The shared nodes are the nodes that belong to both structures. The free nodes are those that belong to only one structure. Therefore, the following vectors can be written:

$$\boldsymbol{u}^{A^T} = \left(\boldsymbol{u}_f^A,\ \boldsymbol{u}_s^A\right) \tag{14.1}$$

$$\boldsymbol{u}^{B^T} = \left(\boldsymbol{u}_f^B,\ \boldsymbol{u}_s^B\right) \tag{14.2}$$

By utilizing these two vectors, the stiffness matrix can be split into four contributions and the virtual variation of the internal work of structure A becomes

$$\delta L_{int}^A = \left\{ \delta \boldsymbol{u}_f^A \quad \delta \boldsymbol{u}_s^A \right\} \begin{bmatrix} \boldsymbol{K}_{ff}^A & \boldsymbol{K}_{fs}^A \\ \boldsymbol{K}_{sf}^A & \boldsymbol{K}_{ss}^A \end{bmatrix} \left\{ \begin{matrix} \boldsymbol{u}_f^A \\ \boldsymbol{u}_s^A \end{matrix} \right\} \tag{14.3}$$

The same can be done for structure B. In explicit form, the virtual variation of the internal work of the two structures becomes

$$\delta L_{int}^A = \delta \boldsymbol{u}_f^A \boldsymbol{K}_{ff}^A \boldsymbol{u}_f^A + \delta \boldsymbol{u}_s^A \boldsymbol{K}_{sf}^A \boldsymbol{u}_f^A + \delta \boldsymbol{u}_f^A \boldsymbol{K}_{fs}^A \boldsymbol{u}_s^A + \delta \boldsymbol{u}_s^A \boldsymbol{K}_{ss}^A \boldsymbol{u}_s^A \tag{14.4}$$

$$\delta L_{int}^B = \delta \boldsymbol{u}_f^B \boldsymbol{K}_{ff}^B \boldsymbol{u}_f^B + \delta \boldsymbol{u}_s^B \boldsymbol{K}_{sf}^B \boldsymbol{u}_f^B + \delta \boldsymbol{u}_f^B \boldsymbol{K}_{fs}^B \boldsymbol{u}_s^B + \delta \boldsymbol{u}_s^B \boldsymbol{K}_{ss}^B \boldsymbol{u}_s^B \tag{14.5}$$

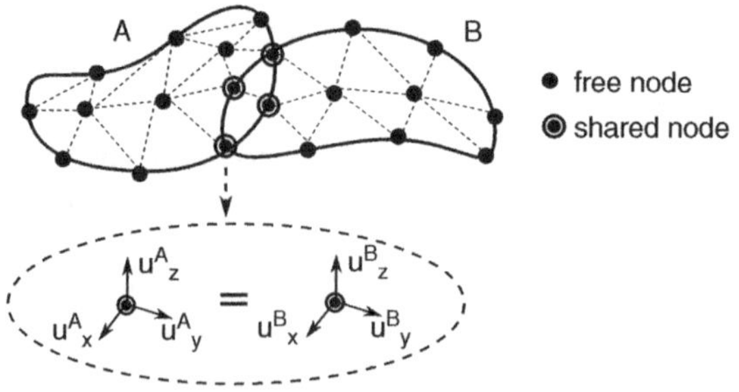

Figure 14.1 Example of two structures, A and B, which share some nodes in which the displacement compatibility is imposed

If compatibility is imposed over the shared nodes,

$$\boldsymbol{u}_s^A = \boldsymbol{u}_s^B = \boldsymbol{u}_s \tag{14.6}$$

the displacement vector of the whole structure can be written as

$$\boldsymbol{u}^T = \left(\boldsymbol{u}_f^A,\ \boldsymbol{u}_f^B,\ \boldsymbol{u}_s\right) \tag{14.7}$$

The virtual variation of the internal work of the two structures is

$$\delta L_{int} = \delta L_{int}^A + \delta L_{int}^B \tag{14.8}$$

which, in matrix form, becomes

$$\delta L_{int} = \left\{\delta \boldsymbol{u}_f^A \quad \delta \boldsymbol{u}_f^B \quad \delta \boldsymbol{u}_s\right\} \begin{bmatrix} \boldsymbol{K}_{ff}^A & \boldsymbol{0} & \boldsymbol{K}_{fs}^A \\ \boldsymbol{0} & \boldsymbol{K}_{ff}^B & \boldsymbol{K}_{fs}^B \\ \boldsymbol{K}_{sf}^A & \boldsymbol{K}_{sf}^B & \boldsymbol{K}_{ss}^A + \boldsymbol{K}_{ss}^B \end{bmatrix} \begin{Bmatrix} \boldsymbol{u}_f^A \\ \boldsymbol{u}_f^B \\ \boldsymbol{u}_s \end{Bmatrix} \tag{14.9}$$

The same approach can be used to derive the load vector. The loads can be split into $\boldsymbol{P}_f$ and $\boldsymbol{P}_s$; the first contributions are the loads on the free nodes, and the second ones are the loads on the shared nodes. The load vectors of the structures become

$$\boldsymbol{P}^{A^T} = \left\{\boldsymbol{P}_f^A,\ \boldsymbol{P}_s^A\right\} \tag{14.10}$$

$$\boldsymbol{P}^{B^T} = \left\{\boldsymbol{P}_f^B,\ \boldsymbol{P}_s^B\right\} \tag{14.11}$$

The virtual variation of the external work can be written as

$$\delta L_{ext}^A = \delta \boldsymbol{u}_f^A \boldsymbol{P}_f^A + \delta \boldsymbol{u}_s^A \boldsymbol{P}_s^A \tag{14.12}$$

$$\delta L_{ext}^B = \delta \boldsymbol{u}_f^B \boldsymbol{P}_f^B + \delta \boldsymbol{u}_s^B \boldsymbol{P}_s^B \tag{14.13}$$

The virtual variation of the external work of the whole structure becomes

$$\delta L_{ext} = \left\{\delta \boldsymbol{u}_f^A \quad \delta \boldsymbol{u}_f^B \quad \delta \boldsymbol{u}_s\right\} \begin{Bmatrix} \boldsymbol{P}_f^A \\ \boldsymbol{P}_f^B \\ \boldsymbol{P}_s^A + \boldsymbol{P}_s^B \end{Bmatrix} \tag{14.14}$$

By introducing the PVD ($\delta L_{int} = \delta L_{ext}$), it is possible to write

$$\begin{bmatrix} \boldsymbol{K}_{ff}^{A} & \boldsymbol{0} & \boldsymbol{K}_{fs}^{A} \\ \boldsymbol{0} & \boldsymbol{K}_{ff}^{B} & \boldsymbol{K}_{fs}^{B} \\ \boldsymbol{K}_{sf}^{A} & \boldsymbol{K}_{sf}^{B} & \boldsymbol{K}_{ss}^{A} + \boldsymbol{K}_{ss}^{B} \end{bmatrix} \begin{Bmatrix} \boldsymbol{u}_{f}^{A} \\ \boldsymbol{u}_{f}^{B} \\ \boldsymbol{u}_{s} \end{Bmatrix} = \begin{Bmatrix} \boldsymbol{P}_{f}^{A} \\ \boldsymbol{P}_{f}^{B} \\ \boldsymbol{P}_{s}^{A} + \boldsymbol{P}_{s}^{B} \end{Bmatrix} \tag{14.15}$$

which describes the problem as two parts of the structure that are modelled by two different kinematics, which share a given number of nodes.

14.1.1 Application of the Shared Stiffness Method

This section describes an example of the above technique. The structure to be investigated is shown in Figure 14.2. It is a simplified model of an aircraft. The geometry parameter a is considered to be equal to 0.5 m and the structure is made of an aluminium alloy.

The geometry of the structure drives the choice of the FE model that should be used in the analysis of the various parts of the aircraft. The wing, the fuselage and the horizontal stabilizer can be assumed to be slender bodies and the 1D FEs should, therefore, provide good accuracy. The structure between the fuselage and the wing and between the fuselage and the horizontal stabilizer could be conveniently modelled as a plate or a 3D element. Both cases are considered here. Figure 14.3 shows a model in which two plate elements are used to connect beams; this model is here called *Model* 1. The plate model uses a second-order layer-wise model, while the beam element is based on an LE model. *Model* 2 is presented in Figure 14.4. In this case, the beam elements are connected using two solid elements with 27 nodes each.

The free vibrations of the structure have been investigated. Table 14.1 gives the natural frequencies evaluated using the two models introduced previously and the reference value evaluated using a 2D plate model from the MSC Nastran commercial code. The first 10 frequencies match the reference values accurately. The DOFs of the models are reported at the head of the table. The method appears to be computationally efficient.

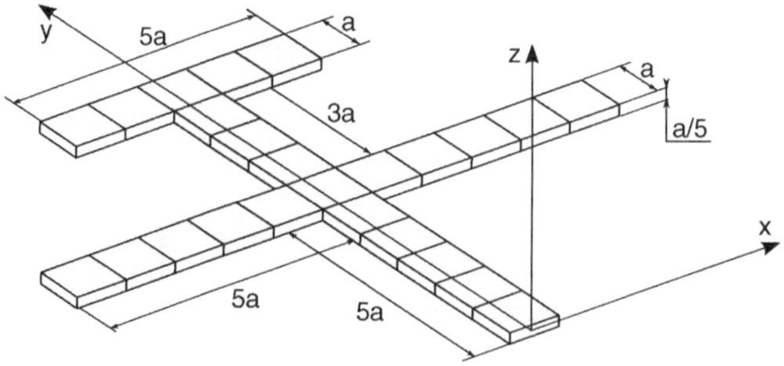

Figure 14.2 Reference system and geometry of the structure

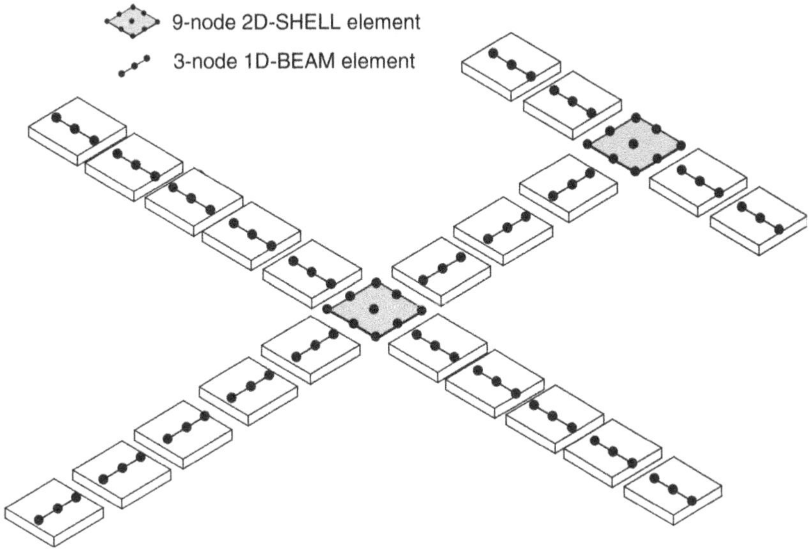

Figure 14.3 FEM discretization of model 1

14.2 Coupling Variable Kinematic Models via the LM Method

The LM method allows a generic relation to be imposed between the displacements of two (or more) arbitrary points. It provides the stationary conditions of a constrained functional. The calculation of the stationary conditions of a functional with n variables and k BCs is reduced to

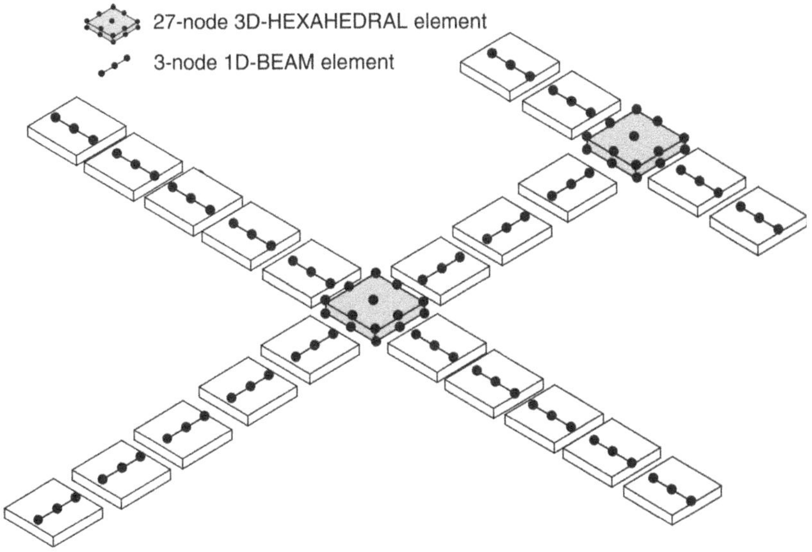

Figure 14.4 FEM discretization of model 2

Table 14.1 Dynamic response of a simplified aircraft (Carrera and Zappino 2013)

	Model 1	Model 2	MSC Nastran
DOFs	1323	1323	6120
I	10.92	10.93	10.80
II	18.18	18.18	17.26
III	20.20	20.21	20.05
IV	50.62	50.80	50.37
V	51.62	51.62	51.16
VI	52.98	52.98	51.31
VII	67.48	67.48	65.55
VIII	71.78	71.78	69.94
IX	76.69	76.69	75.77
X	88.59	88.59	87.44

This table shows the first 10 natural frequencies of the simplified aircraft model.

a stationary point problem of an unconstrained functional of $n + k$ variables. The LM approach can be applied to any kinematic model, while the method introduced in Section 14.1 can only be applied if the unknowns are displacements. A more generic discussion of the LM method can be found in the books by Courant (1936) and Strang (1991).

In this section, the LM method will be used to couple elements with variable kinematics. The formulation will be derived in the CUF. Let us consider two bodies, as in Figure 14.5, with a shared face. The following condition has to be imposed:

$$\mathbf{u}^1(x_k, y_k, z_k) = \mathbf{u}^2(x_k, y_k, z_k) \tag{14.16}$$

which is the same as writing

$$\mathbf{u}^1(x_k, y_k, z_k) - \mathbf{u}^2(x_k, y_k, z_k) = 0 \tag{14.17}$$

where $\mathbf{u}^1(x_k, y_k, z_k)$ is the displacement of a point on the interface cross-section of the first body, whereas $\mathbf{u}^2(x_k, y_k, z_k)$ is the displacement of the same point on the interface cross-section

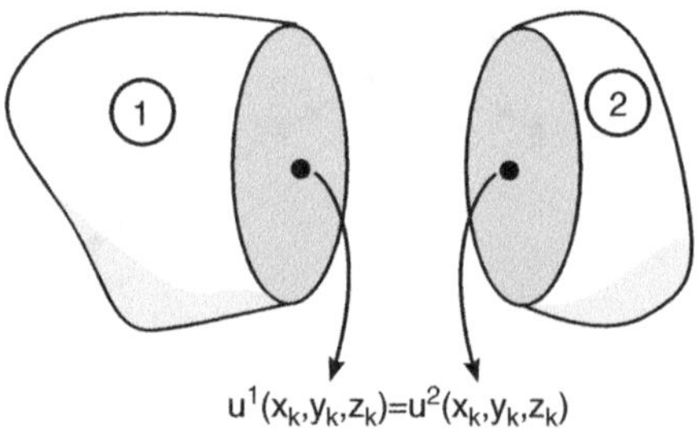

Figure 14.5 Two bodies coupled via LM at a generic point

of the second body. In order to impose the above condition, a new equation must be introduced into the system; this equation is called the 'Lagrangian' and it can be formulated as

$$\Pi = \boldsymbol{\lambda}^T \left(\mathbf{u}^1(x_k, y_k, z_k) - \mathbf{u}^2(x_k, y_k, z_k)\right) \tag{14.18}$$

Displacements can be formulated according to the CUF. Let us consider a beam-like approximation, as in Chapter 9,

$$\boldsymbol{u} = N_i(y) F_\tau(x, z) \boldsymbol{u}_{\tau i} \tag{14.19}$$

The *Lagrangian* can be rewritten as

$$\Pi = \boldsymbol{\lambda}^T \left(N_i^1(y_k) F_\tau^1(x_k, z_k) \boldsymbol{u}_{\tau i}^1 - N_i^2(y_k) F_\tau^1(x_k, z_k) \boldsymbol{u}_{\tau i}^2\right) \tag{14.20}$$

In matrix form, this equation becomes

$$\Pi = \boldsymbol{\lambda}^T \boldsymbol{B}_\gamma \boldsymbol{u} \tag{14.21}$$

where the FN of matrix $\boldsymbol{B}_\gamma$ is

$$\boldsymbol{B}_\gamma^{\tau i} = \left(N_i^1(y_k) F_\tau^1(x_k, z_k) \boldsymbol{u}_{\tau i}^1 - N_i^2(y_k) F_\tau^1(x_k, z_k) \boldsymbol{u}_{\tau i}^2\right) \boldsymbol{I} \tag{14.22}$$

and where $\boldsymbol{\lambda} = \{ \lambda_x \; \lambda_y \; \lambda_z \}^T$ is the vector that contains the LMs. The superscripts '1' and '2' also denote the fact that the bodies to be coupled are generally modelled with different order theories and with different types of beam elements along the beam axis.

The solution of the constrained problem is obtained by finding $\mathbf{u}$ and $\boldsymbol{\lambda}$ in the following linear system:

$$\begin{cases} \boldsymbol{K}\mathbf{u} + \dfrac{\partial \Pi}{\partial \mathbf{u}} = \mathbf{F} \\[2ex] \dfrac{\partial \Pi}{\partial \boldsymbol{\lambda}} \qquad\quad = \bar{\mathbf{u}} \end{cases} \tag{14.23}$$

Equation (14.23) can be rewritten in matrix form as

$$\begin{bmatrix} \boldsymbol{K} & \boldsymbol{B}_\gamma^T \\ \boldsymbol{B}_\gamma & 0 \end{bmatrix} \begin{bmatrix} \mathbf{u} \\ \boldsymbol{\lambda} \end{bmatrix} = \begin{bmatrix} \boldsymbol{F} \\ \bar{\mathbf{u}} \end{bmatrix} \tag{14.24}$$

Use of the multiplier method offers many advantages. However, the main disadvantage of its use in structural problems is that the matrix of Equation (14.24) is generally not positive definite.

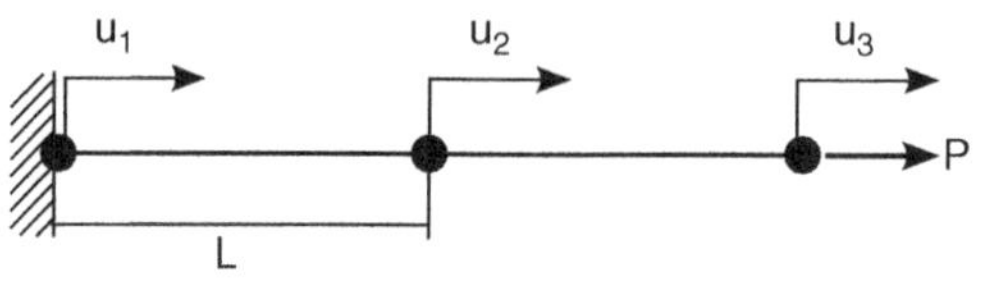

Figure 14.6 Classical FEM model

Example 14.2.1 *Let us consider a bar of length 2L. The FEM model is built from two linear bar elements of length L. The bar is clamped at one end and loaded with an axial load P at the other end.*

Figure 14.6 shows the structure in its conventional configuration. The solution can easily be obtained using the stiffness matrix introduced in Chapter 1. The problem that has to be solved becomes

$$\frac{EA}{L}\begin{bmatrix} 1 & -1 & 0 \\ -1 & 2 & -1 \\ 0 & -1 & 1 \end{bmatrix}\begin{Bmatrix} u_1 \\ u_2 \\ u_3 \end{Bmatrix} = \begin{Bmatrix} 0 \\ 0 \\ P \end{Bmatrix} \tag{14.25}$$

If the BC, $u_1 = 0$, is imposed, the constrained system assumes the form

$$\frac{EA}{L}\begin{bmatrix} 2 & -1 \\ -1 & 1 \end{bmatrix}\begin{Bmatrix} u_2 \\ u_3 \end{Bmatrix} = \begin{Bmatrix} 0 \\ P \end{Bmatrix} \tag{14.26}$$

and the solution is

$$u_2 = \frac{PL}{EA}, \quad u_3 = 2\frac{PL}{EA} \tag{14.27}$$

The same problem can be solved using the Lagrangian approach. Figure 14.7 shows the same problem in a non-conventional configuration. There are two different structures: the first is clamped at one end, while the second is loaded at one end but not constrained. Using LMs, it is necessary to impose

$$u_2^1 - u_1^2 = 0 \tag{14.28}$$

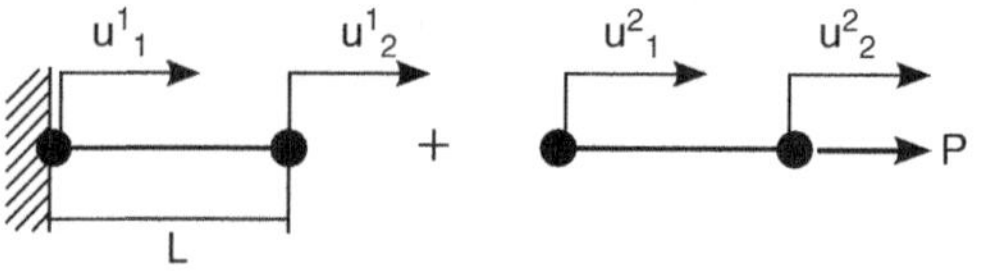

Figure 14.7 Coupling using an LM

The Lagrangian can be obtained by introducing the displacement formulation,

$$\Pi = \lambda\left(N_1^1(y_k)u_2^1 + N_1^2(y_k)u_2^1 - N_2^1(y_k)u_1^2 - N_2^2(y_k)u_2^2\right) \tag{14.29}$$

Substituting the values for the shape functions and using a matrix formulation, the equation becomes

$$\Pi = \lambda \underbrace{[0\ \ 1\ \ -1\ \ 0]}_{B_\gamma}\left[u_2^1\ \ u_2^1\ \ u_1^2\ \ u_2^2\right]^T \tag{14.30}$$

By using Equation (14.24), the problem can be written as

$$\frac{EA}{L}\begin{bmatrix} 1 & -1 & 0 & 0 & 0 \\ -1 & 1 & 0 & 0 & 1 \\ 0 & 0 & 1 & -1 & -1 \\ 0 & 0 & -1 & 1 & 0 \\ 0 & 1 & -1 & 0 & 0 \end{bmatrix}\begin{Bmatrix} u_1^1 \\ u_2^1 \\ u_1^2 \\ u_2^2 \\ \lambda \end{Bmatrix} = \begin{Bmatrix} 0 \\ 0 \\ 0 \\ P \\ 0 \end{Bmatrix} \tag{14.31}$$

If the BC $u_1^1 = 0$ *is imposed, this equation becomes*

$$\frac{EA}{L}\begin{bmatrix} 1 & 0 & 0 & 1 \\ 0 & 1 & -1 & -1 \\ 0 & -1 & 1 & 0 \\ 1 & -1 & 0 & 0 \end{bmatrix}\begin{Bmatrix} u_2^1 \\ u_1^2 \\ u_2^2 \\ \lambda \end{Bmatrix} = \begin{Bmatrix} 0 \\ 0 \\ P \\ 0 \end{Bmatrix} \tag{14.32}$$

and the solution of the problem is

$$u_2^1 = \frac{PL}{EA},\quad u_1^2 = \frac{PL}{EA},\quad u_2^2 = 2\frac{PL}{EA},\quad \lambda = -\frac{PL}{EA} \tag{14.33}$$

The solution fulfils the initial assumption, $u_1^2 = u_2^1$, *and is equal to the solution obtained with the classical approach.*

14.2.1 Application of the LM Method to Variable Kinematic Models

An application in which two models with different kinematics are coupled using the LM technique is described in this section. The following example, taken from Carrera *et al.* (2013), refers to a model that was proposed by Biscani *et al.* (2011), in which variable

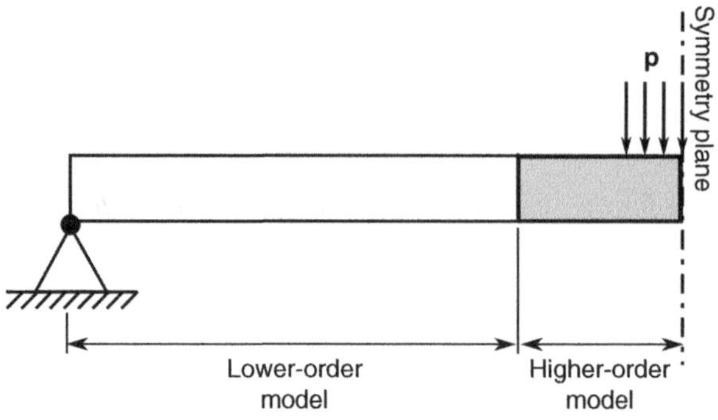

Figure 14.8 *LM* ^{a} model

kinematic models, based on the CUF, were obtained by exploiting the Arlequin method that is introduced in the next section. The structure has a square section and the material data are Young's modulus $E = 75$ GPa and Poisson's ratio $\nu = 0.3$. The beam was simply supported at both ends and subjected to a localized uniform pressure, p, equal to 1 Pa, which acted over 10% of the length and was centred at mid-span.

The results were computed for 20 B4 elements of the same length. The loading was applied to the top of the cross-section, as shown in Figures 14.8 and 14.9. Two different configurations were considered. The LM^a model uses a first-order beam theory in the area close to the constraint, whereas a fourth-order beam theory is used close to the load. The LM^b model uses the refined model close to the constraint and the lower-order theory in the loading area.

The two models were coupled using the LM approach. The interface was created by considering a different number of points in which the LMs were applied. Table 14.2 shows the results, in terms of vertical displacements. The first column gives the number of LMs used at the interface. When a large number of connection points are introduced, both models converge to a similar solution. Some discrepancies arise when a lower number of connection points is introduced; in this case, the LM^b model shows faster convergence than the LM^a model.

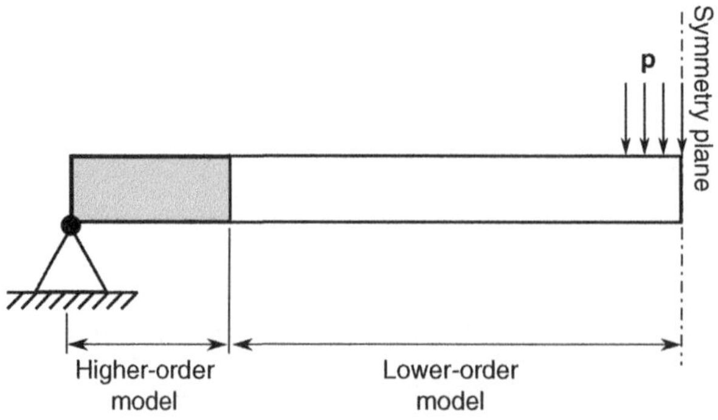

Figure 14.9 *LM* ^{b} model

Table 14.2 Displacements of a mixed kinematic beam

	u_z^*	
LM points	LM^a	LM^b
2	3.409	2.606
3	3.163	2.606
7	2.841	2.576
9	2.785	2.575
13	2.525	2.550
25	2.524	2.550

This table shows the vertical displacement vs the number of connection points for a square beam (Carrera *et al.* 2013).

14.3 Coupling Variable Kinematic Models via the Arlequin Method

The Arlequin method allows models with different kinematics to be coupled by overlapping a part of the domain and introducing the LM field. The method was first introduced by Ben Dhia (1998 1999) and further developed by Ben Dhia and Rateau (2005) and Ben Dhia (2008). The volume of a structure (V) is divided into two sub domains, V_1 and V_2, that are partially overlapped; see Figure 14.10 where V_s represents the overlapped volume. A different model is assumed for each sub domain. The Arlequin method merges the two sub domains, and the global problem can then be solved using a classical technique. The structural compatibility in the overlapped volume is ensured via the LM field (λ) and a coupling operator (C_ξ) that links the DOFs of each sub domain within the overlapped volume. The sub domains are indicated with the index ξ, and the PVD formulation for each domain becomes

$$\delta L^\xi_{int}(\mathbf{u}_\xi) + \delta L^\xi_c(\mathbf{u}_\xi) = \delta L^\xi_{ext} \tag{14.34}$$

where L^ξ_{int} is the internal work of the ξth volume, and L^ξ_c is the work due to the coupling. The virtual variation of the strain energy in each sub domain is

$$\delta L^\xi_{int} = \int_{A_\xi} \alpha_\xi \left(\delta \epsilon^T \sigma\right) dV \text{ with } \begin{cases} \alpha_\xi = 1 \text{ in } V_\xi \setminus V_s \\ \alpha_1 + \alpha_2 = 1 \text{ in } V_s \end{cases} \tag{14.35}$$

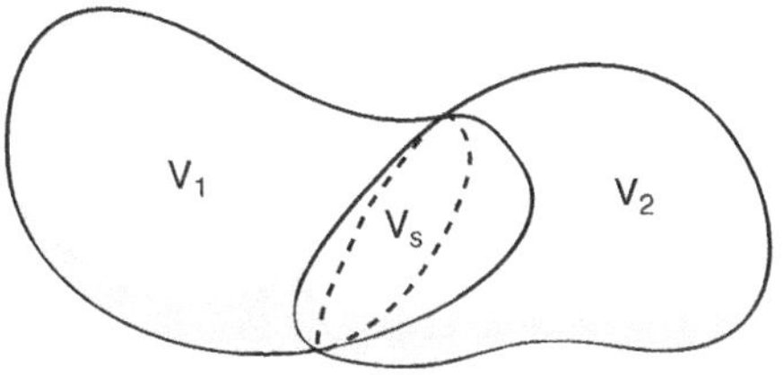

Figure 14.10 Example of the V domain, divided into two sub domains V_1 and V_2 with overlapping area S

where α_ξ are weighting functions which prevent the energy in the overlapped volume from being considered twice. The external virtual work is dealt with in a similar manner,

$$\delta L^{\xi}_{ext} = \int_{A_\xi} \alpha_\xi \left(\delta \boldsymbol{u}_\xi^T \boldsymbol{P}_\xi\right) dV \text{ with } \begin{cases} \alpha_\xi = 1 \text{ in } V_\xi \setminus V_s \\ \alpha_1 + \alpha_2 = 1 \text{ in } V_s \end{cases} \tag{14.36}$$

where δL^ξ_c is the virtual coupling work,

$$\delta L^\xi_c = (-1)^\xi\, C_\xi(\delta \boldsymbol{\lambda}, \mathbf{u}_\xi) \tag{14.37}$$

The coupling operator can be obtained using different formulations (see the papers by Ben Dhia and Rateau (2005) and by Guidault and Belytschko (2007)). The L^2 coupling has been considered here,

$$\delta C_\xi = \int_{S_\xi} \delta \boldsymbol{\lambda}^T \mathbf{u}_\xi \, dV \tag{14.38}$$

The above formulation can be used for any kinematic model. The following formulae are related to the beam model, but the same approach holds for any other structural model. The beam volume is axially divided into two sub domains that are partially overlapped. Different expansion orders are assumed for each sub domain,

$$\mathbf{u}_\xi = N_i F_{\tau_\xi} \mathbf{u}_{\tau_\xi i} \text{ with } \tau_\xi = 1, 2, \ldots, N_u^{A_\xi}, \quad \xi = 1, 2 \tag{14.39}$$

The LM can be expressed in the CUF as

$$\boldsymbol{\lambda} = N_i F_{\tau_\lambda} \boldsymbol{\Lambda}_{\tau_\lambda i} \tag{14.40}$$

where $\boldsymbol{\Lambda}_{\tau_\lambda i}$ is the unknown nodal vector. The FN of the coupling matrix $\mathbf{C}_\xi^{ij\tau_\xi s_\lambda}$ is derived by substituting Equation (14.40) in Equation (14.38),

$$\delta C_\xi = \delta \boldsymbol{\Lambda}^T_{\tau_\lambda i} \mathbf{C}_\xi^{\tau_\lambda s_\xi ij} \mathbf{q}_{s_\xi j} \tag{14.41}$$

In the case of the L^2 coupling, the FN is diagonal and its components are

$$C_{\xi mn}^{\tau_\lambda s_\xi ij} = \delta_{nm} I_{ij} J_{\tau_\lambda s_\xi} \text{ with } m, n = x, y, z \tag{14.42}$$

where δ_{nm} is Kronecker's delta and $J_{\tau_\xi s_\lambda}$ and I_{ij} are

$$J_{\tau_\xi s_\lambda} = \int_A F_{\tau_\xi} F_{s_\lambda} dA \tag{14.43}$$

$$I_{ij} = \int_l N_i N_j dy \tag{14.44}$$

According to Ben Dhia (1999) and Guidault and Belytschko (2007), the same approximation order should be assumed for a low-order model and the LM. This choice prevents the locking phenomenon arising when the approximation of the most refined model is adopted for the discretization of the LM field. Considering the whole structure and assuming that the refined model is adopted in sub domain V_2, the governing equations of the variable kinematic problem in the framework of the proposed unified formulation, coupled with the Arlequin method, are

$$\begin{bmatrix} \overline{\mathbf{K}}^{\tau_1 s_1 ij}_{V_1\backslash V_s} & \mathbf{0} & \mathbf{0} & \mathbf{0} & \mathbf{0} \\ \mathbf{0} & (1-\alpha)\,\overline{\mathbf{K}}^{\tau_1 s_1 ij}_{V_1\cap V_s} & \mathbf{0} & \mathbf{0} & \overline{\mathbf{C}}^{\tau_1 s_1 ijT}_1 \\ \mathbf{0} & \mathbf{0} & \overline{\mathbf{K}}^{\tau_2 s_2 ij}_{V_2\backslash V_s} & \mathbf{0} & \mathbf{0} \\ \mathbf{0} & \mathbf{0} & \mathbf{0} & \alpha\overline{\mathbf{K}}^{\tau_2 s_2 ij}_{V_2\cap V_s} & -\overline{\mathbf{C}}^{\tau_2 s_1 ijT}_2 \\ \mathbf{0} & \overline{\mathbf{C}}^{\tau_1 s_1 ij}_1 & \mathbf{0} & -\overline{\mathbf{C}}^{\tau_2 s_1 ij}_2 & \mathbf{0} \end{bmatrix} \begin{Bmatrix} \overline{\mathbf{q}}^{V_1\backslash V_s}_{s_1 j} \\ \overline{\mathbf{q}}^{V_1\cap V_s}_{s_1 j} \\ \overline{\mathbf{q}}^{V_2\backslash V_s}_{s_2 j} \\ \overline{\mathbf{q}}^{V_2\cap V_s}_{s_2 j} \\ \overline{\mathbf{\Lambda}}_{s_1 j} \end{Bmatrix} = \begin{Bmatrix} \overline{\mathbf{P}}^{V_1\backslash V_s}_{\tau_1 i} \\ (1-\alpha)\,\overline{\mathbf{P}}^{V_1\cap V_s}_{\tau_1 i} \\ \overline{\mathbf{P}}^{V_2\backslash V_s}_{\tau_2 i} \\ \alpha\overline{\mathbf{P}}^{V_2\cap V_s}_{\tau_2 i} \\ 0 \end{Bmatrix} \tag{14.45}$$

14.3.1 Application of the Arlequin Method

In this section, the Arlequin method is used to solve the problem already proposed in Section 14.2.1. A simply supported beam has a uniform pressure load, $\boldsymbol{p}$, located in the middle of its

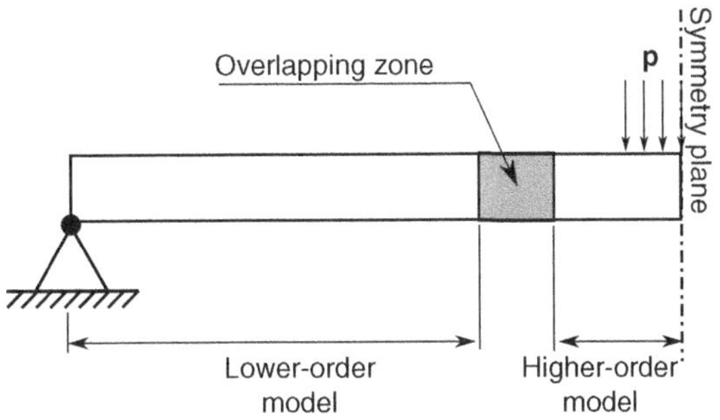

Figure 14.11 AR^a model

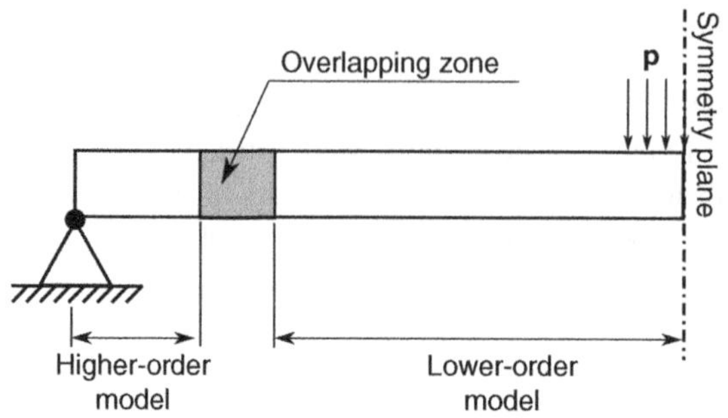

Figure 14.12 AR^b model

length, which covers 10% of the length of the structure. The structure has a square section and is made of aluminium alloy. The AR^a model, shown in Figure 14.11, has the refined structural model in the central part of the beam, while a lower-order model is used close to the simple support. Model AR^b (see Figure 14.12) uses a higher-order model near the constraint, while a lower-order model is used in the central part of the structure. In both cases, the lower-order theory is a first-order beam model, while the refined model is a fourth-order beam model.

Table 14.3 gives the vertical displacement in the middle of the beam using different models. The results show that the Arlequin method is able to connect efficiently two different models. The vertical displacements of the AR^a and AR^b models are, in fact, very close to the results obtained using the first- and the fourth-order models.

Table 14.3 Displacements of a mixed kinematic beam using the Arlequin method

Model	u_z	DOFs
$N = 1$	2.544	549
$N = 4$	2.522	2745
AR^a	2.537	1197
AR^b	2.547	1197

This table shows the vertical displacement of a beam using variable kinematic models (Biscani *et al.* 2011).

References

Ben Dhia H 1998 Multiscale mechanical problems: the Arlequin method. *Comptes Rendus de l'Academie des Sciences, Series IIB, Mechanics Physics Astronomy* **326**(12), 899–904.

Ben Dhia H 1999 Numerical modelling of multiscale problems: the Arlequin method. *CD Proceedings of ECCM'99*, Munich, Germany.

Ben Dhia H 2008 Further insights by theoretical investigations of the multiscale Arlequin method. *International Journal for Multiscale Computational Engineering* **6**(3), 215–232.

Ben Dhia H and Rateau H 2005 The Arlequin method as a flexible engineering tool. *International Journal for Numerical Methods in Engineering* **62**(11), 1442–1462.

Biscani F, Giunta G, Belouettar S, Carrera E and Hu H 2011 Variable kinematic beam elements coupled via Arlequin method. *Composite Structures* **93**(2), 697–708.

Carrera E, Pagani A and Petrolo M 2013 Use of Lagrange multipliers to combine 1D variable kinematic finite elements. *Computers & Structures* **129**, 194–206.

Carrera E and Zappino E 2013 Full aircraft dynamic response by simplified structural models. *54rd AIAA/ASME/ASCE/AHS/ASC Structures, Structural Dynamics, and Materials Conference (SDM)*, Boston, Massachusetts, USA.

Courant R 1936 *Differential and Integral Calculus*. Interscience.

Guidault PA and Belytschko T 2007 On the l2 and the h1 couplings for an overlapping domain decomposition method using Lagrange multipliers. *International Journal for Numerical Methods in Engineering* **70**(3), 322–350.

Strang G 1991 *Calculus*. Wellesley-Cambridge Press.

15

Extension to Multilayered Structures

The aim of this chapter is to offer some ideas about the capabilities of the CUF in the analysis of multilayered structures. The CUF was, in fact, conceived by the first author in order to provide a powerful tool for the study of advanced new materials. For the sake of brevity, only 2D multilayered theories will be discussed here. However, the CUF can also be used for multilayered beam structures, as shown in the previous chapters.

15.1 Multilayered Structures

The previous chapters dealt with traditional structures made of metallic materials and showing isotropic behaviour. Most structure theories have, in fact, been developed to obtain a better understanding of the mechanical behaviour of one-layered beams, plates and shells. Multilayered structures appear in many applications, some of these are listed below.

- Sandwich structures:
  Sandwich[1] beams, plates and shells (see Figure 15.1) are composite structures that consist of at least three different layers. Two or more high-strength stiff layers (faces) are bonded to one or more low-density flexible layers (core). The core, which is usually the cheapest material, mainly has the task of keeping the two faces away from the neutral axis, thus improving bending resistance. Sandwich structures can be defined as composite structures, since they are composed of at least two different materials at a microscopic level. There are many types of sandwich structures in aerospace construction that have not been dealt with in detail here.

[1] This word is named after the Earl of Sandwich, John Montagu, who lived in the eighteenth century. A very conversant gambler, Lord Sandwich did not take time to eat a meal during his many hours playing cards. Consequently, he would ask his servant to bring him slices of meat between two slices of bread, a habit well known to his gambling friends, who began to order 'the same as Sandwich!' – and thus the word 'sandwich' was created.

Finite Element Analysis of Structures Through Unified Formulation, First Edition.
Erasmo Carrera, Maria Cinefra, Marco Petrolo and Enrico Zappino.

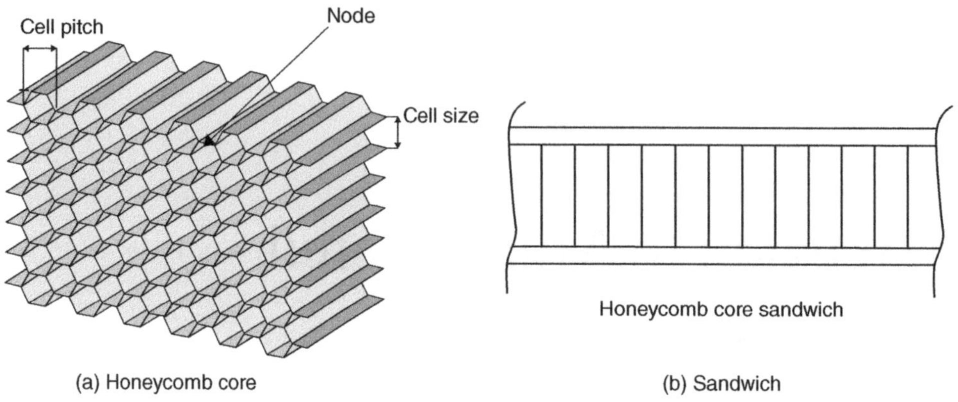

Figure 15.1 Honeycomb core sandwich

- Layered structures for thermal protection purposes:
  Engine components, re-entry vehicles and supersonic aircraft often require adequate thermal protection. In many cases, it is not possible to produce a material that can withstand both mechanical and thermal loadings at the same time. The solution, in this case, is to build a composite multilayered structure: the mechanical structure is protected by an additional layer that leads to high resistance with respect to thermal loadings, as shown in Figure 15.2.
- Piezo-layered materials for smart structures:
  The phenomenon of piezoelectricity is a particular feature of certain classes of crystalline materials. The piezoelectric effect consists of a linear energy conversion between the mechanical and electric fields in both directions that defines a direct or converse piezoelectric effect. The direct piezoelectric effect generates an electrical polarization by applying mechanical stresses. The converse piezoelectric effect instead induces mechanical stresses or strains by applying an electric field. Multilayered structures are also obtained when piezo-layers are bonded as sensors or actuators to a given structure, as can be seen in Figure 15.3.

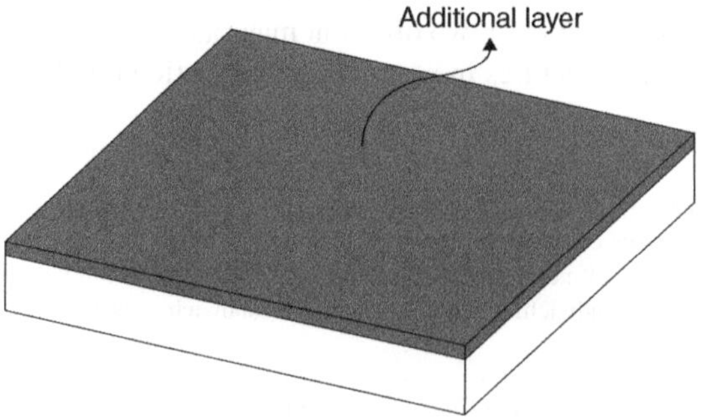

Figure 15.2 Layered plate made of isotropic layers for thermal protection

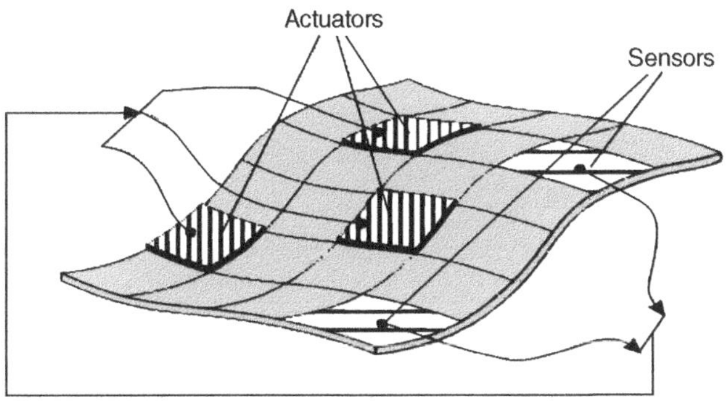

Figure 15.3 Example of a smart structure: sensor–actuator network for a plate

- Laminates:
  The laminate structure is the most common feature of composite materials (see Figure 15.4). Composites are multilayered structures (made mostly of flat and curved panels) comprising several layers or laminae that are bonded together perfectly. Each lamina is composed of fibres embedded in a matrix. These fibres are produced according to a specific technological process that confers high mechanical properties in the longitudinal direction (L) of the fibres (see Figure 15.4). The matrix has the role of holding them together. Carbon, boron and glass fibres are mostly used, along with organic products. The matrices are mostly of an epoxy type. There are several possible ways of putting the fibres and matrix together. Unidirectional laminae or laminates made of differently oriented laminae are used in most applications related to the construction of aerospace, automotive or marine vehicles. The laminae are placed one over the other, according to a given layout, as shown in Figure 15.4. Such a possibility, which is known as 'tailoring', permits the use of the material to be optimized for a given set of design requirements.

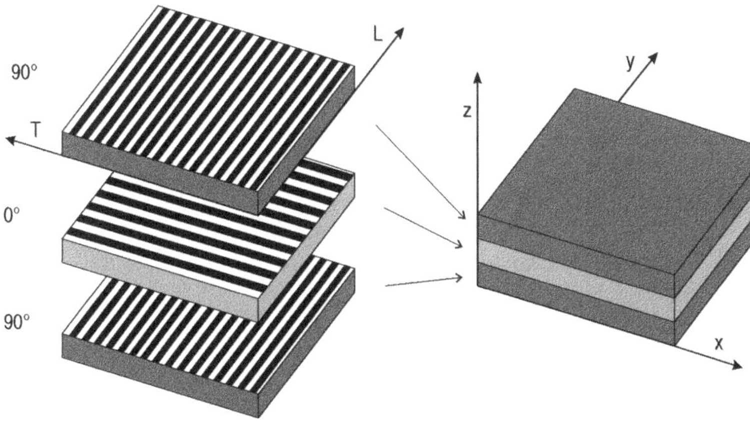

Figure 15.4 Layered plate made of unidirectional laminae

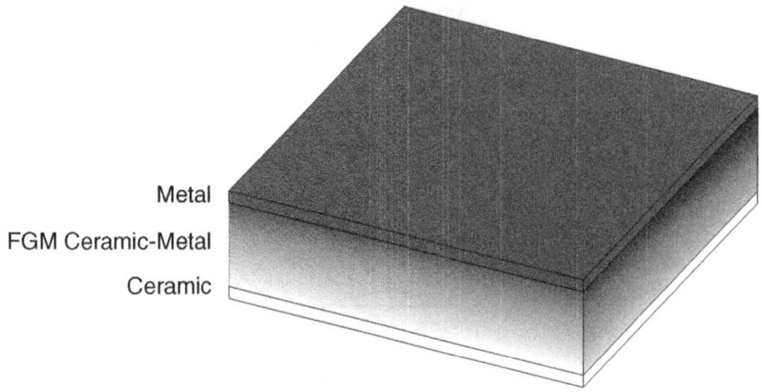

Figure 15.5 Multilayered plate embedding an FGM layer

- Functionally graded materials:
  FGMs are advanced composite materials, within which the composition of each constituent material varies gradually with respect to spatial coordinates. Therefore, the macroscopic material properties of FGMs vary continuously, distinguishing them from laminated composite materials in which an abrupt change in the material properties across the layer interfaces leads to large interlaminar stresses which can lead to the development of damage. As in the case of laminated composite materials, FGMs combine the desirable properties of the constituent phases to give a superior performance. To this end, functionally graded layers can be embedded in multilayered structures (see Figure 15.5).
- Multiwalled carbon nanotubes:
  Carbon nanotubes (CNTs) have exceptional mechanical properties (Young's modulus, tensile strength, toughness, etc.) due to their molecular structure, which consists of single or multiple sheets of graphite wrapped in seamless hollow cylinders (see Figure 15.6). Because of the great stiffness, strength and high aspect ratio of CNTs, it is expected that, by dispersing them evenly throughout a polymer matrix, it will be possible to produce composites with considerably improved and effective mechanical properties overall. Furthermore, CNTs have a relatively low density of about 1.75 g/cm^3 and nanotube reinforced polymers (NRPs), therefore, excel due to their extremely high specific stiffness, strength and toughness.

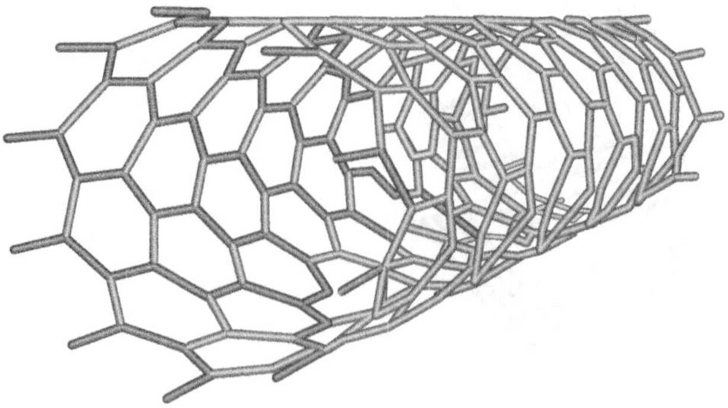

Figure 15.6 Example of single-walled CNT

15.2 Theories for Multilayered Structures

Theories for multilayered structures can be developed by making an appropriate choice concerning the following points.

1. Choice of the unknown variables:
   - displacement formulation;
   - mixed formulation.
2. Choice of the variable description:
   - equivalent single layer (ESL) models;
   - layer-wise (LW) models.

Many other choices which have already been adopted for traditional metallic structures can be made. Nevertheless, it will be clear, from reading this chapter, that the '*sic et simpliciter*' extension of the modellings that are already available for monocoque structures would not lead to satisfactory results. Complicating effects, such as the so-called C_z^0 requirements, should be taken into account for this purpose.

15.2.1 C_z^0 Requirements

Layered structures are said to be 'transversely anisotropic' because they exhibit different mechanical and physical properties in the thickness direction. Transverse discontinuous mechanical properties cause a displacement field, u, in the thickness direction, which can exhibit a rapid change in its slope corresponding to each layer interface. This is known as the zigzag, or ZZ, form of displacement field in the thickness direction (see Figure 15.7b). In-plane stresses $\sigma_p = (\sigma_{xx}, \sigma_{yy}, \sigma_{xy})$ can in general be discontinuous at each layer interface. Nevertheless, transverse stresses $\sigma_n = (\sigma_{xz}, \sigma_{yz}, \sigma_{zz})$, for equilibrium reasons, i.e. the Cauchy

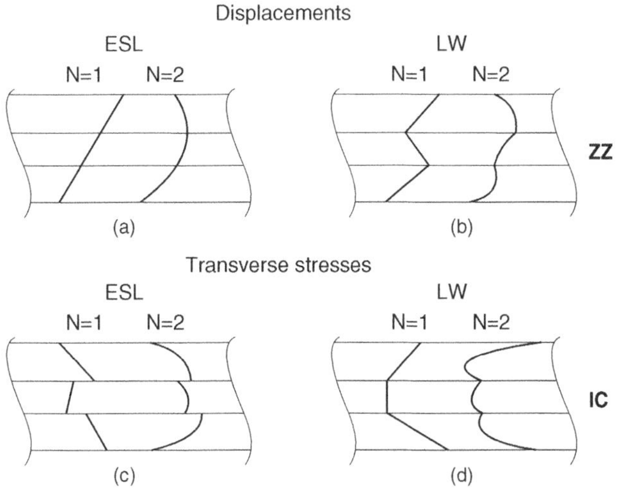

Figure 15.7 C_z^0 requirements

theorem, must be continuous at each layer interface. This is often referred to in the literature as interlaminar continuity (IC) of the transverse stresses. Figure 15.7 shows, from a qualitative point of view, the distribution scenarios of possible displacements and transverse stresses of a multilayered structure, which depend on the description of the variables. It appears that displacements and transverse stresses are C^0continuous functions in the thickness z direction, due to compatibility and equilibrium reasons, respectively. It is also evident that the displacement u has a discontinuous first derivative corresponding to each interface, where the mechanical properties change. Carrera (1995–1996, 1997) referred to ZZ and IC as the C_z^0 requirements. Fulfilment of the C_z^0 requirements is a crucial point in the development of any theory suitable for multilayered structures.

15.2.2 Refined Theories

Theories that were originally developed for single layer 'monocoque' structures made of traditional isotropic materials can conveniently be grouped into two cases: Love first approximation theories (LFATs) or Love second approximation theories (LSATs). Applications of LFAT to multilayered structures are often referred to as classical laminated theories (CLTs); see Reissner and Stavsky (1961). CLTs assume that the normals to the reference surface Ω (see Figure 15.8) remain normal in the deformed states and do not change in length; that is, transverse shear as well as normal strains are postulated to be negligible with respect to the other strains. Extensions of the so-called Reissner–Mindlin (after Reissner (1945) and Mindlin (1951)) LSAT-type model, which includes transverse shear strains, to layered structures are known as the shear deformation theory (SDT) (or First-order SDT, FSDT), see Yang *et al.* (1966).

A simple refinement of the Reissner–Mindlin theory was conducted by Vlasov (1957) for monocoque structures. Vlasov's FSDT-type theory permits the homogeneous conditions to be fulfilled for the transverse shear stresses corresponding to the top and bottom shell/plate surfaces $\sigma_{iz}(\pm h/2) = 0$. Reddy (1984) and Reddy and Phan (1985) have shown that such a simple inclusion can lead to significant improvements, compared with FSDT analysis, in tracing the static and dynamic response of thick laminated structures.

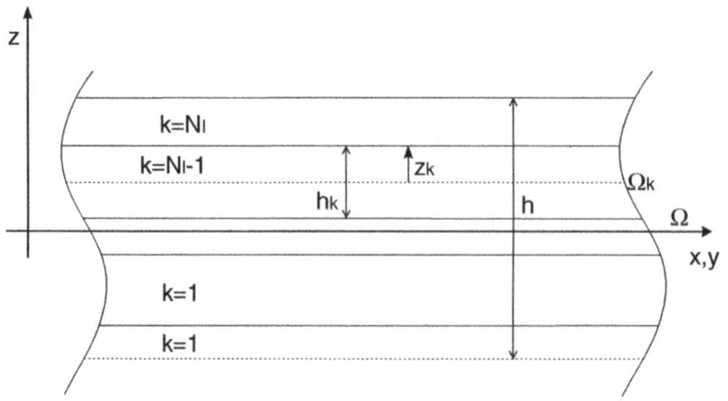

Figure 15.8 Notation for multilayered structures

Further refinements of FSDT are known as higher-order theories (HOTs). In general, HOTs are based on displacement models of the following type:

$$u_i(x, y, z) = u_0 + z\,u_1 + z^2\,u_2 + \ldots + z^N\,u_N \tag{15.1}$$

where N is the order of expansion used for the displacement variables. Displacement models related to HOT have been traced in Figures 15.7. Examples of the application of these types of models to flat and curved laminated structures can be found in Sun and Whitney (1973), Soldatos (1987), Librescu and Schmidt (1988), Touratier (1988), Dennis and Palazotto (1991), Gaudenzi (1992), Touratier (1992), Zenkour (1999) and Rabinovitch and Frosting (2001).

15.2.3 Zigzag Theories

The extension of CLT, FSDT and HOT to multilayered plates does not permit the C_z^0 requirements to be fulfilled. Refined theories have therefore been introduced to resolve this problem. Due to the form of the displacement field in the thickness direction, see Figure 15.7, these types of theories are referred to as zigzag (ZZ) theories.

The fundamental idea behind ZZ theories is that a certain displacement and/or stress model is assumed in each layer and then compatibility and equilibrium conditions are used at the interface to reduce the number of unknown variables.

The first, most significant contributions to ZZ theories came from the Russian school. The first ZZ theory was presented by Lekhnitskii (1935) for beam geometries. Other outstanding contributions to plates and shells were made by Ambartsumian (1958, 1961, 1969). An independent manner of formulating ZZ plate/shell theories was presented in the West by Reissner (1984, 1986a,b). Many other theories that are based on the original contributions of these three scientists have been introduced and are cited in the historical review by Carrera (2003a).

15.2.4 Layer-Wise Theories

The theories mentioned previously consider a number of unknown variables that are independent of the number of constitutive layers N_l. These all are known as equivalent single layer models or ESLMs. Although these kinematic theories can describe transverse shear and normal strains, including transverse warping of the cross-section, their approach is 'kinematically homogeneous' in the sense that the kinematics is insensitive to individual layers, unless ZZ models are used. If a detailed response from the individual layers is required, and if significant variations in the displacement gradients between layers exist, as is the case for a local phenomena description, this approach needs the use of special HOTs in each of the constitutive layers along with a concomitant increase in the number of unknowns in the solution process, which leads to an increase in the complexity of the analysis. In other words, a possible 'natural' way of including the ZZ effect in the framework of a classical model with only displacement variables could be to apply CLT, FSDT or HOT at a layer level; that is, each layer is considered as an independent plate, and compatibility of the displacement components, corresponding to each interface, is then imposed as a constraint (see Figure 15.7b). In these cases, layer-wise models or LWMs are obtained. Examples of these types of theories can be found in the articles by Hsu and Wang (1970), Cheung and Wu (1972), Srinivas (1973), Cho *et al.* (1991) and

Robbins and Reddy (1993). These displacement models require constraint conditions to be included in order to enforce the compatibility conditions at each interface. Generalizations of these types of theories were given by Nosier *et al.* (1993) and Reddy (1997), who expressed the displacement variables in the thickness direction in terms of Lagrange polynomials

$$u_i^k(x,y,z) = L_1(z_k)u_i^k|_{h/2} + L_2(z_k)u_i^k|_{-h/2} + L_3(z_k)u_{i3}^k + \ldots + L_N(z_k)u_{iN}^k \qquad (15.2)$$

with $k = 1, \ldots, N_l$. Interface values are used as the unknown variables (the first two terms of the expansion $u_i^k|_{h/2}$ and $u_i^k|_{-h/2}$) and this permits easy linkage of the compatibility conditions at each interface. $L1$, $L2$, in fact, coincide with linear Lagrangian polynomials, while $L3, \ldots, L_N$ should be an independent base of polynomials that start from the parabolic one ($L3$).

15.2.5 Mixed Theories

A third approach for laminated structures was presented in the two papers by Reissner (1984, 1986a), in which a mixed variational equation, namely the Reissner mixed variational theorem, or RMVT, was proposed. The displacement $\boldsymbol{u}$ and transverse stress $\boldsymbol{\sigma}_n$ variables are assumed to be independent in the framework of the RMVT, which states

$$\sum_{k=1}^{N_l} \int_{\Omega_k}\int_{A_k} \left\{ \delta\epsilon_{pG}^{k\ T} \sigma_{pH}^k + \delta\epsilon_{nG}^{k\ T} \sigma_{nM}^k + \delta\sigma_{nM}^{k\ T} \left(\epsilon_{nG}^k - \epsilon_{nH}^k\right) \right\} d\Omega_k\, dz = \sum_{k=1}^{N_l} \delta L_{ext}^k \qquad (15.3)$$

The subscript H indicates that stresses are computed by Hooke's law. The variation in the internal work has been split into in-plane and out-of-plane parts and involves stresses from Hooke's law and strains from geometrical relations (subscript G). δL_{ext} is the virtual variation of the work done by the external layer forces. The third 'mixed' term variationally enforces compatibility of the transverse strain components. Subscript M means that the transverse stresses are those of the assumed model, see the discussion reported in Carrera (2001).

Full mixed methods have also been developed, in which all six stress components and the three displacements are assumed to be the unknown variables. A detailed discussion has been reported by Pagano (1978). Applications of full mixed theories in the framework of an ESL description and plate and shell geometries have been proposed by Fares and Zenkour (1998) and Zenkour (1998, 2001). An interesting discussion on the possible improvement of FSDT-type models, using mixed and partially mixed formulations, has been presented by Auricchio and Sacco (2001). The Hellinger–Reissner mixed principle was employed in this work to determine the FSDT governing equations. Two FSDT-type models, both of which describe transverse shear stresses as independent variables, were discussed and implemented.

15.3 Unified Formulation for Multilayered Structures

As discussed in the previous section, the assumptions introduced for displacements and stresses can be made at layer or at multilayer level. An LW description is obtained in the first case, and an ESL description in the latter. If an LW description is employed, $\boldsymbol{u}_\tau$ and $\boldsymbol{\sigma}_{nM_\tau}$ are layer variables. These are different in each layer. If one refers to an ESL description, $\boldsymbol{u}_\tau$ and $\boldsymbol{\sigma}_{nM_\tau}$

are plate/shell variables. These are the same for the whole multilayer. Examples of ESL and LW assumptions are shown in Figure 15.7.

15.3.1 ESLMs

In the most general case, higher-order ESLMs appear in the following form:

$$\begin{aligned} u_x(x,y,z) &= u_{x0}(x,y) + z\, u_{x1}(x,y) + z^2\, u_{x1}(x,y) + \ldots + z^N\, u_{xN}(x,y) \\ u_y(x,y,z) &= u_{y0}(x,y) + z\, u_{y1}(x,y) + z^2\, u_{y1}(x,y) + \ldots + z^N\, u_{yN}(x,y) \\ u_z(x,y,z) &= u_{z0}(x,y) + z\, u_{z1}(x,y) + z^2\, u_{z1}(x,y) + \ldots + z^N\, u_{zN}(x,y) \end{aligned} \tag{15.4}$$

According to the unified formulation, these can be written in the following compact form:

$$\boldsymbol{u} = F_0\, \boldsymbol{u}_0 + F_1\, \boldsymbol{u}_1 + \ldots + F_N\, \boldsymbol{u}_N = F_\tau\, \boldsymbol{u}_\tau, \quad \tau = 0, 1, 2, \ldots, N \tag{15.5}$$

where N is the order of the expansion and

$$F_0 = 1, \quad F_1 = z, \quad F_0 = z^2, \ldots, F_N = z^N \tag{15.6}$$

These HOTs are denoted here by acronyms ED1, ED2, ED3, ..., EDN. Further, the letter 'd' means that the ED1d model is obtained by ED1 one by neglecting the linear term in u_z (ED1d, in fact, coincides with FSDT).

15.3.2 Inclusion of Murakami's ZZ Function

Murakami (1986) introduced in a first-order ESL displacement field a ZZ function able to describe the ZZ form for the displacements. He modified the FSDTs according to the following model:

$$\begin{aligned} u_x(x,y,z) &= u_{0x} + z\, u_{x1} + (-1)^k \zeta_k\, u_{xZ} \\ u_y(x,y,z) &= u_{0y} + z\, u_{x1} + (-1)^k \zeta_k\, u_{yZ} \\ u_z(x,y,z) &= u_{0z} \end{aligned} \tag{15.7}$$

Subscript Z refers to Murakami's introduced ZZ function. $\zeta_k = 2z_k/h_k$ is a non–dimensioned layer coordinate (z_k is the physical coordinate of the k layer whose thickness is h_k, as indicated in Figure 15.8). The exponent k changes the sign of the ZZ term in each layer. Such an artifice permits the discontinuity of the first derivative of the displacement variables to be reproduced in the z-direction, which physically comes from the intrinsic transverse anisotropy of multilayer structures.

Transverse normal strain/stress effects can be included as in the ED1 models, leading to

$$\begin{aligned} u_x(x,y,z) &= u_{0x} + z\, u_{x1} + (-1)^k \zeta_k\, u_{xZ} \\ u_y(x,y,z) &= u_{0y} + z\, u_{y1} + (-1)^k \zeta_k\, u_{yZ} \\ u_z(x,y,z) &= u_{0z} + z\, u_{z1} + (-1)^k \zeta_k\, u_{zZ} \end{aligned} \tag{15.8}$$

By introducing the notation

$$\tau = 0, 1, 2, \quad F_0 = 1, \quad F_1 = z, \quad F_2 = F_Z = (-1)^k \zeta_k \tag{15.9}$$

Equation (15.8) can be written in the following array form:

$$\boldsymbol{u} = \boldsymbol{u}_0 + (-1)^k \zeta_k\, \boldsymbol{u}_Z + z\, \boldsymbol{u}_1 = F_\tau\, \boldsymbol{u}_\tau, \quad \tau = 0, 1, Z \tag{15.10}$$

Such a model is denoted by the acronym EDZ1 in which Z indicates that a ZZ function has also been used for the expansion in z. Higher-order models appear in the following form:

$$\begin{aligned} u_x(x,y,z) &= u_{x0}(x,y) + z\, u_{x1}(x,y) + z^2\, u_{x1}(x,y) + \ldots + z^{N-1}\, u_{xN}(x,y) + (-1)^k \zeta_k\, u_{xZ} \\ u_y(x,y,z) &= u_{y0}(x,y) + z\, u_{y1}(x,y) + z^2\, u_{y1}(x,y) + \ldots + z^{N-1}\, u_{yN}(x,y) + (-1)^k \zeta_k\, u_{yZ} \\ u_z(x,y,z) &= u_{z0}(x,y) + z\, u_{z1}(x,y) + z^2\, u_{z1}(x,y) + \ldots + z^{N-1}\, u_{zN}(x,y) + (-1)^k \zeta_k\, u_{zZ} \end{aligned} \tag{15.11}$$

or in compact form

$$\boldsymbol{u} = \boldsymbol{u}_0 + (-1)^k \zeta_k\, \boldsymbol{u}_Z + z^r\, \boldsymbol{u}_r = F_\tau\, \boldsymbol{u}_\tau, \quad \tau = 0, 1, 2, \ldots, N \tag{15.12}$$

N is the order of the expansion and

$$F_0 = 1, \quad F_1 = z, \quad F_0 = z^2, \ldots, F_{N-1} = z^{N-1}, \quad F_N = F_Z = (-1)^k \zeta_k \tag{15.13}$$

These HOTs are denoted here by acronyms EDZ1, EDZ2, EDZ3, ..., EDZN.

15.3.3 LW Theory and Legendre Expansion

In the case of LWMs, each layer is seen as independent and the compatibility of displacement components at each interface is then imposed. To this end, Lagrange polynomials can be used, as in Reddy (1997). However, the use of thickness functions in terms of Legendre polynomials is preferred here. The following expansion is employed:

$$\begin{aligned} u_x^k &= F_t\, u_{xt}^k + F_b\, u_{xb}^k \\ u_y^k &= F_t\, u_{yt}^k + F_b\, u_{yb}^k \\ u_z^k &= F_t\, u_{zt}^k + F_b\, u_{zb}^k \end{aligned} \tag{15.14}$$

It is now intended that the subscripts t and b denote values related to the top and bottom layer surface, respectively. These two terms consist of the linear part of the expansion. The thickness functions $F_\tau(\zeta_k)$ are now defined at the k-layer level,

$$F_t = \frac{P_0 + P_1}{2}, \quad F_b = \frac{P_0 - P_1}{2} \tag{15.15}$$

in which $P_j = P_j(\zeta_k)$ is the Legendre polynomial of the jth order defined in the ζ_k domain: $\zeta_k = 2z_k/h_k$ and $-1 \leq \zeta_k \leq 1$. For instance, the first five Legendre polynomials are $P_0 = 1$, $P_1 = \zeta_k$, $P_2 = (3\zeta_k^2 - 1)/2$, $P_3 = \frac{5}{2}\zeta_k^3 - \frac{3}{2}\zeta_k$, $P_4 = \frac{35}{8}\zeta_k^4 - \frac{15}{4}\zeta_k^2 + \frac{3}{8}$. The chosen functions have the following interesting properties:

$$\zeta_k = \begin{cases} 1: & F_t = 1, \quad F_b = 0, \quad F_r = 0 \\ -1: & F_t = 0, \quad F_b = 1, \quad F_r = 0 \end{cases} \tag{15.16}$$

That allows one to have interface values as unknown variables, therefore avoiding the inclusion of constraint equations to impose C_z^0 requirements. In a unified form

$$\boldsymbol{u}^k = F_t\, \boldsymbol{u}_t^k + F_b\, \boldsymbol{u}_b^k = F_\tau\, \boldsymbol{u}_\tau^k, \quad \tau = t, b \tag{15.17}$$

This theory will be denoted as LD1: that is, LW theory with displacement unknowns and first-order expansion. Higher-order LW theories are written by adding higher-order terms,

$$\begin{aligned} u_x^k &= F_t\, u_{xt}^k + F_b\, u_{xb}^k + F_2\, u_{x2}^k + \ldots + F_N\, u_{xN}^k \\ u_y^k &= F_t\, u_{yt}^k + F_b\, u_{yb}^k + F_2\, u_{y2}^k + \ldots + F_N\, u_{yN}^k \\ u_z^k &= F_t\, u_{zt}^k + F_b\, u_{zb}^k + F_2\, u_{z2}^k + \ldots + F_N\, u_{zN}^k \end{aligned} \tag{15.18}$$

where

$$F_r = P_r - P_{r-2}, \quad r = 2, 3, \ldots, N \tag{15.19}$$

In a unified form

$$\boldsymbol{u}^k = F_t\, \boldsymbol{u}_t^k + F_b\, \boldsymbol{u}_b^k + F_r\, \boldsymbol{u}_r^k = F_\tau\, \boldsymbol{u}_\tau^k, \qquad \tau = t, b, \ r = 2, 3, \ldots, N \tag{15.20}$$

These higher-order expansions have been denoted by the acronyms LD2, LD3, ..., LDN.

15.3.4 Mixed Models with Displacement and Transverse Stress Variables

The ESLMs, described in the previous section, can still be used for displacement variables in the framework of RMVT applications. Besides, an appropriate transverse stress field is required. A first choice could be to use for $\boldsymbol{\sigma}_n$ the same expansions used for $\boldsymbol{u}$. Of course

IC should be fulfilled. To meet such a requirement, an LW description is required for the transverse stresses. The LE model already used for the displacement field seems to be very appropriate for this aim. For the generic higher-order case, one has

$$\begin{aligned}
\sigma^k_{xz} &= F_t\, \sigma^k_{txz} + F_b\, \sigma^k_{xzb} + F_2\, \sigma^k_{xz2} + \ldots + F_N\, \sigma^k_{xzN} \\
\sigma^k_{yz} &= F_t\, \sigma^k_{tyz} + F_b\, \sigma^k_{yzb} + F_2\, \sigma^k_{yz2} + \ldots + F_N\, \sigma^k_{yzN} \\
\sigma^k_{zz} &= F_t\, \sigma^k_{tzz} + F_b\, \sigma^k_{zzb} + F_2\, \sigma^k_{zz2} + \ldots + F_N\, \sigma^k_{zzN}
\end{aligned} \tag{15.21}$$

or in compact form,

$$\boldsymbol{\sigma}^k_n = F_t\, \boldsymbol{\sigma}^k_{nt} + F_r\, \boldsymbol{\sigma}^k_{nr} + F_b\, \boldsymbol{\sigma}^k_{nb} = F_\tau\, \boldsymbol{\sigma}^k_{n\tau}, \quad \tau = t, b, r, \quad r = 2, \ldots, N \tag{15.22}$$

Two possible choices can be made for displacements: the ESL or LW description.

In the first case, even though it is not appropriate, the mixed models are denoted as EMN. In case IC is imposed in the theories, i.e.

$$\boldsymbol{\sigma}^k_{nt} = \boldsymbol{\sigma}^{(k+1)}_{nb}, \quad k = 1 \ldots N_l - 1 \tag{15.23}$$

and/or top/bottom surface stress values are prescribed (zero or imposed values), the following additional equilibrium conditions should be accounted for:

$$\boldsymbol{\sigma}^1_{nb} = \bar{\boldsymbol{\sigma}}_{nb}, \quad \boldsymbol{\sigma}^{N_l}_{nt} = \bar{\boldsymbol{\sigma}}_{nt} \tag{15.24}$$

where the over-bar indicates the imposed values at the plate boundary surfaces.

The resulting models are denoted by the acronym EMCN. Letter C is included to indicate the fulfilment of IC.

The ZZ form of the displacement field can be included by using the EDZ-type model along with Murakami's related ZZ function. The resulting models are denoted by acronyms EMZN. Letter Z has been included to emphasize the description of ZZ effects.

A full LW description is acquired by combining the LW description for both the displacements and the transverse stresses. The resulting models are denoted by the LM1, LM2, LMN acronyms.

15.4 FE Formulation

Following the standard FEM, the unknown variables in the element domain are expressed in terms of their values at the element nodes. According to the isoparametric description, the displacements or stresses are expressed as follows:

$$\mathbf{u}^k_\tau = N_i \mathbf{u}^k_{\tau i} \quad (i = 1, 2, \ldots, N_n) \tag{15.25}$$

where

$$\mathbf{u}^k_{\tau i} = \left[{u_{x\tau i}}^k \quad {u_{y\tau i}}^k \quad {u_{z\tau i}}^k \right]^T \tag{15.26}$$

and

$$\boldsymbol{\sigma}_{n\tau}^{k} = N_i \mathbf{g}_{\tau i}^{k} \quad (i = 1, 2, \ldots, N_n) \tag{15.27}$$

where

$$\mathbf{g}_{\tau i}^{k} = \begin{bmatrix} g_{xz\tau i}^{k} & g_{yz\tau i}^{k} & g_{zz\tau i}^{k} \end{bmatrix}^T \tag{15.28}$$

N_n is the number of nodes of the element and it is taken as a free parameter of the model. N_i are the shape functions and $\mathbf{u}_{\tau i}^{k}$, $\mathbf{g}_{\tau i}^{k}$ are nodal variables. ξ, η are the natural coordinates.

By varying N and N_n, the FE matrices of the k layer, corresponding to the implemented 2D theories and number of nodes, are obtained.

By introducing the external work of applied loadings, the PVD leads for each k layer to the following equilibrium conditions:

$$\delta \mathbf{u}_{sj}^{kT} : \quad \mathbf{k}^{k\tau sij} \mathbf{u}_{\tau i}^{k} = \mathbf{p}_{sj}^{k} \tag{15.29}$$

Note that the matrix $\mathbf{k}^{k\tau sij}$ is a 3×3 array. Such an array consists of the FN of the FE matrix related to PVD applications.

The same steps taken in the PVD case can be extended to RMVT-formulated FEs. Transverse normal stress variables along with displacement ones will then lead to four 3×3 FNs. Three of them are related to equilibrium conditions. The others establish compatibility conditions. Therefore, the equilibrium and compatibility equations obtained by the RMVT (15.3) are

$$\begin{aligned} \delta \mathbf{u}_{sj}^{kT} &: \quad \mathbf{k}_{uu}^{k\tau sij} \mathbf{u}_{\tau i}^{k} + \mathbf{k}_{u\sigma}^{k\tau sij} \mathbf{g}_{\tau i}^{k} = \mathbf{p}_{sj}^{k} \\ \delta \mathbf{g}_{sj}^{kT} &: \quad \mathbf{k}_{\sigma u}^{k\tau sij} \mathbf{u}_{\tau i}^{k} + \mathbf{k}_{\sigma\sigma}^{k\tau sij} \mathbf{g}_{\tau i}^{k} = \mathbf{0} \end{aligned} \tag{15.30}$$

As for PVD case, by expanding the (τ, s) as well as (i, j) couples of indexes, the FE matrix is obtained for the given k layer.

15.4.1 Assembly at Multilayer Level

Both the PVD and RMVT formulations have been written for N_l independent layers. The C_z^0 requirements must be imposed to drive equations from the layer to multilayer level.

In general, multilayered equations can be obtained by summing stiffnesses and/or compliances related to the same variables. In the case describing the ESL, this means that stiffness matrices of different layers are summed to obtain the global stiffness matrix of the multilayer. In the case of LW approach, IC conditions are imposed by following the procedure shown in Figure 15.9.

Therefore, multilayered arrays are introduced at the very end of the assembly. Details of this procedure can be found in the paper by Carrera (2003b).

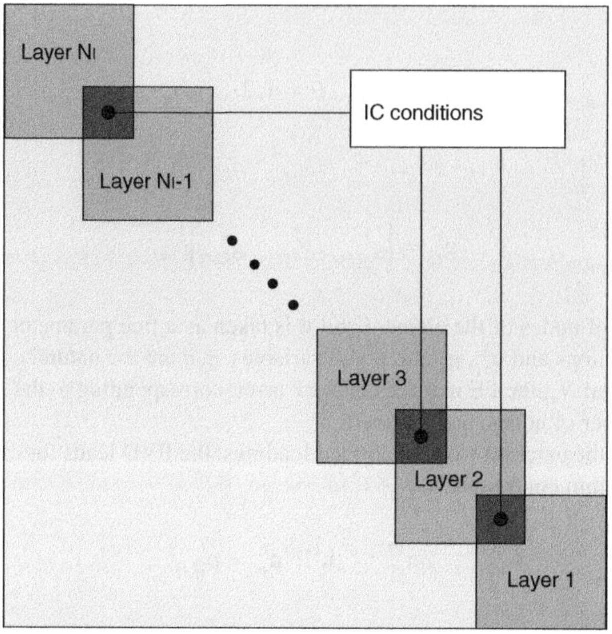

Figure 15.9 An example showing how IC conditions are imposed from the layer to multilayer level

15.4.2 Selected Results

A sandwich square plate bent by transverse bisinusoidal pressure applied to the top plate surface is considered. Exact 3D solutions have been provided by Pagano (1970). Results are presented for different values of the thickness parameter a/h. The faces have equal thickness $h_f = 0.1h$. The mechanical properties of the laminae, which are used as skins, are: $E_L = 25 \times 10^6$ psi, $E_T = 1 \times 10^6$ psi, $G_{LT} = 0.5 \times 10^6$ psi, $G_{TT} = 0.2 \times 10^6$ psi, $\nu_{LT} = \nu_{TT} = 0.25$. The core material used for the sandwich plate is transversely isotropic with respect the z axis and is characterized by the following elastic properties: $E_{xx} = E_{yy} = 0.04 \times 10^6$ psi, $E_{zz} = 0.5 \times 10^6$ psi, $G_{xz} = G_{yz} = 6 \times 10^4$ psi, $G_{xy} = 1.6 \times 10^4$ psi, $\nu_{xy} = \nu_{zy} = \nu_{zx} = 0.25$. The principal material directions of the core always coincide with the geometrical axes of the plate. Stress and displacement values are normalized according to the relations

$$\overline{\sigma}_{xx} = \frac{1}{p_z(a/h)^2}\sigma_{xx}, \quad \overline{\sigma}_{zx} = \frac{1}{p_z(a/h)}\sigma_{zx} \tag{15.31}$$

$$\overline{U_z} = u_z \frac{100 \cdot h^3 \cdot E_T}{p_z \cdot a^4} \tag{15.32}$$

where p_z is the amplitude of the transverse applied pressure.

A comprehensive comparison of different 2D theories has been conducted. FE solutions concerning displacement and stresses are given in Tables 15.1–15.3. Analytical closed-form solutions are also quoted in a few cases. Eight LW analyses are compared with 14 ESL results,

Table 15.1 Sandwich plate. Transverse displacement $\overline{U}_z$ $(a/2, a/2, 0)$. Comparison of different 2D theories (Carrera and Demasi 2002b)

a/h	2	4	10	20	50
3D	21.3	7.5	2.2	1.2	0.93
LM4	22.103	7.5947	2.2001	1.2260	0.934 33
LM3	22.103	7.5948	2.2001	1.2260	0.934 33
LM2	22.103	7.5943	2.2001	1.2260	0.934 33
LM1	22.462	7.6113	2.2003	1.2259	0.934 32
LD4	22.103	7.5948	2.2001	1.2264	0.934 33
LD3	22.103	7.5948	2.2001	1.2260	0.934 33
LD2	22.071	7.5931	2.2001	1.2260	0.934 33
LD1	22.450	7.6036	2.1976	1.2251	0.934 15
EMZC3	23.313	7.8723	2.2342	1.2343	0.935 40
EMZC2	23.298	7.8689	2.2335	1.2341	0.935 44
EMZC1	23.077	7.8177	2.2207	1.2264	0.928 91
EDZ3	23.305	7.8710	2.2338	1.2341	0.935 26
EDZ2	23.287	7.8660	2.2326	1.2337	0.935 20
EDZ1	23.063	7.8101	2.2180	1.2250	0.927 96
EMC4	23.345	7.5609	2.1341	1.2071	0.932 90
EMC3	22.189	7.4277	2.1287	1.2051	0.930 61
EMC2	15.625	4.6944	1.5098	1.0379	0.903 17
EMC1	14.408	4.6726	1.5407	1.0438	0.899 62
ED4	23.217	7.5175	2.1244	1.2044	0.932 33
ED3	21.960	7.3560	2.1132	1.2008	0.929 74
ED2	15.058	4.5144	1.4753	1.0287	0.901 36
ED1	12.960	4.2309	1.4550	1.0201	0.894 21
FSDT[a]	12.657	4.2259	1.4342	1.0120	0.890 98
CLT[a]	0.867 79	0.867 78	0.867 78	0.867 78	0.867 78

This table compares different 2D theories in terms of transverse displacement by considering various thickness ratios.

CLT and FSDT: 24 theories are therefore presented. Very thick ($a/h = 2$), moderately thick ($a/h = 10$) and thin ($a/h = 50$) sandwiches have been investigated. Transverse displacement as well as in-plane and transverse shear stresses have been evaluated. A mesh of 5×5 Q9 (nine nodes) plate elements have been used in the FE mathematical model. The following main comments can be made:

1. Results related to the different models merge in the thin-sandwich response.
2. Higher-order LW analyses furnish a complete 3D response for thin and very thick plate: displacement and both in-plane and transverse stresses are in excellent agreement with exact 3D solutions.
3. Formulations that are based on the RMVT lead to an improvement with respect to PVD applications. In particular, the RMVT becomes very effective as far as ESLM description is concerned: EMZC1–EMZC3 analyses lead to a much better description than ED1–ED3 ones.
4. LW descriptions are required for accurate evaluations of very thick sandwich plates.

Table 15.2 Sandwich plate. In-plane stress $\overline{\sigma}_{xx}$ $(a/2, a/2, \pm 1/2h)$. Comparison of different 2D theories (Carrera and Demasi 2002b)

	2		10		50	
a/h	$h/2$	$-h/2$	$h/2$	$-h/2$	$h/2$	$-h/2$
3D	3.278	−2.653	1.153	−1.152	1.099	−1.099
LM4	3.2426	−2.6233	1.1323	−1.1293	1.0786	−1.0786
LM3	3.2415	−2.6224	1.1324	−1.1293	1.0786	−1.0786
LM2	3.2352	−2.6172	1.1323	−1.1293	1.0785	−1.0785
LM1	3.0867	−2.4860	1.1332	−1.1304	1.0792	−1.0792
LD4	3.2420	−2.6228	1.1323	−1.1293	1.0785	−1.0785
LD3	3.2426	−2.6233	1.1324	−1.1293	1.0785	−1.0785
LD2	3.2259	−2.6091	1.1322	−1.1296	1.0785	−1.0785
LD1	3.0917	−2.5041	1.1297	−1.1270	1.0792	−1.0791
EMZC3	3.1594	−2.5612	1.1488	−1.1467	1.0714	−1.0714
EMZC2	3.1255	−2.5286	1.1588	−1.1568	1.0843	−1.0843
EMZC1	2.9179	−2.8960	1.1454	−1.1445	1.0690	−1.0689
EDZ3	3.1623	−2.5617	1.1484	−1.1462	1.0712	−1.0711
EDZ2	3.1331	−2.5323	1.1465	−1.1443	1.0710	−1.0710
EDZ1	2.9224	−2.9011	1.1427	−1.1418	1.0678	−1.0678
EMC4	2.9766	−2.4009	1.1435	−1.1424	1.0697	−1.0697
EMC3	3.0610	−2.4874	1.1444	−1.1427	1.0710	−1.0710
EMC2	1.0138	−0.472 31	1.0455	−1.0425	1.0642	−1.0641
EMC1	0.813 37	−0.767 10	1.0419	−1.0399	1.0629	−1.0628
ED4	2.9828	−2.4080	1.1440	−1.1430	1.0695	−1.0695
ED3	3.0752	−2.5015	1.1452	−1.1436	1.0706	−1.0706
ED2	1.0181	−0.479 43	1.0469	−1.0441	1.0635	−1.0634
ED1	0.830 25	−0.795 52	1.0446	−1.0431	1.0604	−1.0603
FSDT[a]	1.0921	−1.0921	1.0484	−1.0484	1.0902	−1.0902
CLT[a]	0.7555	−0.7555	1.0921	−1.0921	1.0921	−1.0921

This table compares different 2D theories in terms of in-plane stress by considering various thickness ratios.

5. Results related to ED1–ED3 theories are very much improved by the inclusion of ZZ functions.
6. The comparison between EMC1–EMC4 and EMZC1–EMZC4 results shows that RMVT applications require both the ZZ form of displacement field and IC. In other words, the use of the RMVT becomes ineffective if ZZ effects are not included.
7. Classical ESLM analyses based on the PVD comprise the worst description. In particular, theories with linear through-the-thickness displacement field ED1 are quite inaccurate even though moderately thick sandwich plates are considered.
8. The accuracy of the different theories is very much subordinate to the outputs considered. Better evaluations of in-plane stresses with respect to transverse ones are obtained by different modellings.
9. As far as FEs are concerned, note that the comparison between the FE and analytical solution further shows that RMVT-formulated elements have better convergence properties than those based on PVD applications.

Table 15.3 Sandwich plate. Transverse shear stress $\overline{\sigma}_{zx}(0, a/2, 0)$. Comparison of different 2D theories (Carrera and Demasi 2002b)

a/h	2	4	10	20	50
3D	0.186	0.239	0.3	0.317	0.323
LM4	0.189 76	0.245 93	0.304 25	0.322 25	0.328 72
LM3	0.159 57	0.245 83	0.304 25	0.322 25	0.323 72
LM2	0.185 33	0.243 52	0.303 23	0.321 48	0.328 10
LM1	0.190 39	0.245 33	0.303 50	0.321 60	0.328 10
LD4	0.178 46	0.227 78	0.280 19	0.295 95	0.301 77
LD3	0.178 46	0.227 78	0.280 19	0.295 95	0.301 77
LD2	0.178 34	0.227 78	0.280 19	0.295 95	0.301 77
LD1	0.178 55	0.227 75	0.280 07	0.295 79	0.301 62
FSDT[a]	0.250 63	0.283 51	0.313 39	0.320 24	0.322 34
CLT[a]	0.322 75	0.322 75	0.322 75	0.322 75	0.322 75

This table compares different 2D theories in terms of transverse shear stress by considering various thickness ratios.

15.5 Literature on the CUF Extended to Multilayered Structures

Many works by the authors of this book are devoted to the FEA of 1D and 2D multilayered structures embedding advanced materials.

The articles by Carrera and Petrolo (2012), Petrolo *et al.* (2012), Carrera *et al.* (2013h), Carrera *et al.* (2013a) and Carrera *et al.* (2013b) deal with refined beam elements for the static and dynamic analysis of laminated composite structures, while the buckling of composite thin-walled beams has been analysed by Ibrahim *et al.* (2012) and the static aeroelastic response of composite wings has been studied by Varello *et al.* (2013). A particular beam element scheme was introduced by Carrera *et al.* (2013c) for the structural analysis of single and multiple fibre/matrix cells. This so-called 'component-wise' scheme has been used for the analysis of laminated anisotropic composites by Carrera *et al.* (2012). The works by Carrera *et al.* (2013d), Carrera *et al.* (2013e), Carrera *et al.* (2013f) and Carrera *et al.* (2013g) employ both refined and CW beam models in the analysis of reinforced structures and wing structures. A further study on the effectiveness of higher-order terms in refined beam theories was carried out by Carrera and Petrolo (2011).

The article by Carrera (2003b) presents an overview of theories and FEs for multilayered plates and shells based on the unified formulation, with numerical assessments and benchmarks. An assessment of CUF plate elements on the analysis of composite and sandwich structures was also provided by Carrera *et al.* (2002) and Botshekanan Dehkordi *et al.* (2013), respectively. Plate elements, based on the CUF, have been used by Ferreira *et al.* (2013) for the analysis of viscoelastic sandwich laminates, and their performances in the failure diagnosis of laminated anisotropic plates were demonstrated by Carrera *et al.* (2009). Multilayered plate elements with various 2D assumptions have been compared by Carrera *et al.* (2010a) and their accuracy further discussed by Carrera *et al.* (2011b). In particular, the works by Carrera *et al.* (2011a) and Carrera and Miglioretti (2012) furnish useful recommendations, according to the problem considered, on the use of the best refined plate elements. More advanced multilayered plate elements for the analysis of multifield problems, including the electro mechanical problem in piezo-materials, have been presented by Carrera and Nali (2010). Shell FEs with different through-the-thickness kinematics for the linear analysis of cylindrical multilayered structures were introduced by Cinefra and Carrera (2013). Their extension to the study of functionally graded shells was performed by Cinefra *et al.* (2012).

Finally, some works exist concerning FEA by means of mixed models. Multilayered finite plate elements, based on the RMVT, have been presented and validated by Carrera and Demasi (2002a) and Carrera and Demasi (2002b), while enhanced mixed plate/shell elements, based on Mixed Interpolation of Tensorial Components (MITC) with four and nine nodes have been discussed by Carrera *et al.* (2010b) and Chinosi *et al.* (2013), respectively.

References

Ambartsumian SA 1958 On a theory of bending of anisotropic plates. *Investiya Akademii Nauk SSSR, Ot. Tekhnicheskikh Nauk.* No. 4.

Ambartsumian SA 1961 Theory of anisotropic shells. Fizmatzig, Moskwa; Translated from the Russian, NASA TTF118, 1964.

Ambartsumian SA 1969 *Theory of Anisotropic Plates*. Translated from the Russian by T. Cheron and Edited by J.E. Ashton. Technical Publishing Company.

Auricchio F and Sacco E 2001 Partial-mixed formulation and refined models for the analysis of composites laminated within FSDT. *Composite Structures* **46**, 103–113.

Botshekanan Dehkordi M, Cinefra M, Khalili SMR and Carrera E 2013 Mixed LW/ESL models for the analysis of sandwich plates with composite faces. *Composite Structures* **98**, 330–339.

Carrera E 1995–1996 A class of two–dimensional theories for anisotropic multilayered plates analysis. *Accademia delle Scienze di Torino, Memorie Scienze Fisiche* **19–20**, 1–39.

Carrera E 1997 C_z^0 requirements: Models for the two dimensional analysis of multilayered structures. *Composite Structures* **37**, 373–384.

Carrera E 2001 Developments, ideas and evaluations based upon Reissner's Mixed Variational Theorem in the modeling of multilayered plates and shells. *Applied Mechanics Review* **54**, 301–329.

Carrera E 2003a Historical review of zig-zag theories for multilayered plates and shell. *Applied Mechanics Review* **56**(3), 287–308.

Carrera E 2003b Theories and finite elements for multilayered plates and shells: a unified compact formulation with numerical assessment and benchmarking. *Archives of Computational Methods in Engineering* **10**(3), 215–296.

Carrera E, Büttner A, Nalif JP, Wallmerperge T and Kröplin B 2010a A comparison of various two-dimensional assumptions in finite element analysis of multilayered plates. *International Journal of Computational Engineering Science* **11**, 313–327.

Carrera E, Cinefra M and Nali P 2010b MITC technique extended to variable kinematic multilayered plate elements. *Composite Structures* **92**, 1888–1895.

Carrera E and Demasi L 2002a Classical and advanced multilayered plate element based upon PVD and RMVT. Part I: Derivation of finite element matrices. *International Journal for Numerical Methods in Engineering* **55**, 191–231.

Carrera E and Demasi L 2002b Classical and advanced multilayered plate element based upon PVD and RMVT. Part II: Numerical implementations. *International Journal for Numerical Methods in Engineering* **55**, 253–291.

Carrera E, Demasi L and Manganello M 2002 Assessment of plate elements on bending and vibrations of composite structures. *Mechanics of Advanced Materials and Structures* **9**, 333–358.

Carrera E, Filippi M and Zappino E 2013a Laminated beam analysis by polynomial, trigonometric, exponential and zig-zag theories. *European Journal of Mechanics – A/Solids* **41**, 58–69.

Carrera E, Filippi M and Zappino E 2013b Free vibration analysis of rotating composite blades via Carrera Unified Formulation. *Composite Structures* **106**, 317–325.

Carrera E, Filippi M and Zappino E 2013c Free vibration analysis of laminated beam by polynomial, trigonometric, exponential and zig-zag theories. *Journal of Composite Materials*. DOI:10.1177/0021998313497775.

Carrera E, Maiarù M and Petrolo M 2012 Component-wise analysis of laminated anisotropic composites. *International Journal of Solids and Structures* **49**, 1839–1851.

Carrera E, Maiarù M, Petrolo M and Giunta G 2013d A refined 1D element for the structural analysis of single and multiple fiber/matrix cells. *Composite Structures* **96**, 455–468.

Carrera E and Miglioretti F 2012 Selection of appropriate multilayered plate theories by using a genetic like algorithm. *Composite Structures* **94**, 1175–1186.

Carrera E, Miglioretti F and Petrolo M 2011a Guidelines and recommendations on the use of higher order finite elements for bending analysis of plates. *International Journal for Computational Methods in Engineering Science and Mechanics* **12**(6), 303–324.

Carrera E, Miglioretti F and Petrolo M 2011b Accuracy of refined finite elements for laminated plate analysis. *Composite Structures* **93**, 1311–1327.

Carrera E and Nali P 2010 Multilayered plate elements for the analysis of multifield problems. *Finite Elements in Analysis and Design* **46**, 732–742.

Carrera E, Nali P and Büttner A 2009 Mixed elements, stress fields and failure parameter in laminated anisotropic plates. *Aerotecnica Missili e Spazio* **88**, 93–104.

Carrera E, Pagani A and Petrolo M 2013e Classical, refined and component-wise analysis of reinforced-shell structures. *AIAA Journal* **51**(5), 1255–1268.

Carrera E, Pagani A and Petrolo M 2013f Component-wise method applied to vibration of wing structures. *Journal of Applied Mechanics* **80**(4), 041012.

Carrera E and Petrolo M 2011 On the effectiveness of higher-order terms in refined beam theories. *Journal of Applied Mechanics* **78**(2), 021013.

Carrera E and Petrolo M 2012 Refined one-dimensional formulations for laminated structure analysis. *AIAA Journal* **50**(1), 176–189.

Carrera E, Zappino E and Filippi M 2013g Free vibration analysis of thin-walled cylinders reinforced with longitudinal and transversal stiffeners. *Journal of Vibration and Acoustics* **135**(1), 011019.

Carrera E, Zappino E and Petrolo M 2013h Analysis of thin-walled structures with longitudinal and transversal stiffeners. *Journal of Applied Mechanics* **80**, 011006.

Cheung YK and Wu CI 1972 Free vibrations of thick, layered cylinders having finite length with various boundary conditions. *Journal of Sound and Vibration* **24**, 189–200.

Chinosi C, Cinefra M, Della Croce L and Carrera E 2013 Reissner's Mixed Variational Theorem toward MITC finite elements for multilayered plates. *Composite Structures* **99**, 443–452.

Cho KN, Bert CW and Striz AG 1991 Free vibrations of laminated rectangular plates analyzed by higher order individual layer theory. *Journal of Sound and Vibration* **145**, 429–442.

Cinefra M and Carrera E 2013 Shell finite elements with different through-the-thickness kinematics for the linear analysis of cylindrical multilayered structures. *International Journal for Numerical Methods in Engineering* **93**, 160–182.

Cinefra M, Carrera E, Della Croce L and Chinosi C 2012 Refined shell elements for the analysis of functionally graded structures. *Composite Structures* **94**, 415–422.

Dennis ST and Palazotto AN 1991 Laminated shell in cylindrical bending, two-dimensional approach vs exact. *American Institute of Aeronautics and Astronautics Journal* **29**, 647–650.

Fares ME and Zenkour AM 1998 Mixed variational formula for the thermal bending of laminated plates. *Journal of Thermal Stress* **22**, 347–365.

Ferreira AJM, Arajo AL, Neves AMA, Rodrigues JD, Carrera E, Cinefra M and Soares CMM 2013 A finite element model using a unified formulation for the analysis of viscoelastic sandwich laminates. *Composites Part B: Engineering* **45**, 1258–1264.

Gaudenzi P 1992 A general formulation of higher order theories for the analysis of laminated plates. *Composite Structures* **20**, 103–112.

Hsu T and Wang JT 1970 A theory of laminated cylindrical shells consisting of layers of orthotropic laminae. *American Institute of Aeronautics and Astronautics Journal* **8**, 2141–2146.

Ibrahim SM, Carrera E, Petrolo M and Zappino E 2012 Buckling of composite thin walled beams by refined theory. *Composite Structures* **94**(2), 563–570.

Lekhnitskii SG 1935 Strength calculation of composite beams. *Vestnik Inzhen i Tekhnikov* No. 9.

Librescu L and Schmidt R 1988 Refined theories of elastic anisotropic shells accounting for small strains and moderate rotations. *International Journal of Non-linear Mechanics* **23**, 217–229.

Mindlin R 1951 Influence of rotatory inertia and shear in flexural motions of isotropic elastic plates. *Journal of Applied Mechanics* **18**, 1031–1036.

Murakami H 1986 Laminated composite plate theory with improved in-plane responses. *Journal of Applied Mechanics* **53**, 661–666.

Nosier A, Kapania RK and Reddy JN 1993 Free vibration analysis of laminated plates using a layer-wise theory. *American Institute of Aeronautics and Astronautics Journal* **31**, 2335–2346.

Pagano NJ 1970 Exact solutions for rectangular bidirection composites and sandwich plates. *Journal of Composite Materials* **4**, 20–34.

Pagano NJ 1978 Stress fields in composite laminates. *International Journal of Solids and Structures* **14**, 385–400.

Petrolo M, Pagani A and Carrera E 2012 Multi-line-beam with variable kinematic models for the analysis of composite structures. *AIAA Journal* **50**(1), 176–189.

Rabinovitch O and Frosting Y 2001 Higher-order analysis of unidirectional sandwich panels with flat and generally curved faces and a 'soft' core. *Sandwich Structures and Materials* **3**, 89–116.

Reddy JN 1984 A simple higher order theory for laminated composites plates. *Journal of Applied Mechanics* **52**, 745–752.

Reddy JN 1997 *Mechanics of Laminated Composite Plates, Theory and Analysis*. CRC Press.

Reddy JN and Phan ND 1985 Stability and vibration of isotropic, orthotropic and laminated plates according to a higher order shear deformation theory. *Journal of Sound and Vibration* **98**, 157–170.

Reissner E 1945 The effect of transverse shear deformation on the bending of elastic plates. *Journal of Applied Mechanics* **12**, 69–76.

Reissner E 1984 On a certain mixed variational theory and a proposed application. *International Journal for Numerical Methods in Engineering* **20**, 1366–1368.

Reissner E 1986a On a certain mixed variational theorem and on laminated elastic shell theory. *Proceedings of the Euromech Colloquium* **219**, 17–27.

Reissner E 1986b On a mixed variational theorem and on a shear deformable plate theory. *International Journal for Numerical Methods in Engineering* **23**, 193–198.

Reissner E and Stavsky Y 1961 Bending and stretching of certain type of heterogeneous elastic plates. *Journal of Applied Mechanics* **9**, 402–408.

Robbins DH Jr and Reddy JN 1993 Modeling of thick composites using a layer-wise theory. *International Journal for Numerical Methods in Engineering* **36**, 655–677.

Soldatos KP 1987 *Cylindrical Bending of Cross-ply Laminated Plates: Refined 2D plate theories in comparison with the exact 3D elasticity solution*. Technical Report No. 140, University of Ioannina, Greece.

Srinivas S 1973 A refined analysis of composite laminates. *Journal of Sound and Vibration* **30**, 495–507.

Sun CT and Whitney JM 1973 On the theories for the dynamic response of laminated plates. *American Institute of Aeronautics and Astronautics Journal* **11**, 372–398.

Touratier M 1988 A refined theory for thick composites plates. *Mechanics Research Communications* **15**, 229–236.

Touratier M 1992 A refined theory of laminated shallow shells. *International Journal of Solids and Structures* **29**, 1401–1415.

Varello A, Carrera E and Demasi L 2013 A refined structural model for static aeroelastic response and divergence of metallic and composite wings. *CEAS Aeronautical Journal* **4**(2), 175–189.

Vlasov BF 1957 On the equations of bending of plates. *Dokla Akademii Nauk, Azerbeijanskoi-SSR* **3**, 955–979.

Yang PC, Norris CH and Stavsky Y 1966 Elastic wave propagation in hetereogenous plates. *International Journal of Solids and Structures* **2**, 665–684.

Zenkour AM 1998 Vibration of axisymmetric shear deformable cross-ply laminated cylindrical shells: a variational approach. *Composite Structures* **36**, 219–231.

Zenkour AM 1999 Transverse shear and normal deformation theory for bending analysis of laminated and sandwich elastic beams. *Mechanics of Composite Materials and Structures* **6**, 267–283.

Zenkour AM 2001 Bending, buckling and free vibration of nonhomogeneous composite laminated cylindrical shell using a refined first order theory. *Composites Part B* **32**, 237–247.

16

Extension to Multifield Problems

This chapter deals with classical and mixed variational statements for the analysis of layered structures under the effect of four different fields: mechanical, thermal, electric and magnetic. Constitutive equations, in terms of coupled mechanical–thermal–electric–magnetic field variables, are obtained on the basis of a thermodynamics approach. The PVD and the RMVT are employed. The latter permits interlaminar variables, such as transverse stresses, transverse electrical displacement, etc., to be assumed 'a priori'. A number of particular cases of the considered variational statements are proposed. A new condensed notation is introduced into the CUF and this leads to governing equations and FE matrices in terms of a few FNs. The FEA of multilayered plates has been addressed. Variable kinematics, as well as LW and ESL descriptions, have been considered for the FEs presented, according to the CUF. A few results are given to show the effectiveness of the proposed approach.

16.1 Mechanical vs Field Loadings

Over the last few centuries, scientists have identified four fundamental physical forces that act in nature: weak nuclear, strong nuclear, gravitational and electromagnetic. Each of these forces can be associated with a fundamental physical field. Both for computational reasons and the Heisenberg uncertainty principle, a deterministic modelling of the systems is not possible when referring to these four fields, as they refer to subatomic scales and relativistic quantities. In order to model these structures without reference to subatomic dimensions, other fields have been defined as substitutes for the four fundamental ones: mechanical, thermal, electric and magnetic; see the interaction processes in Figure 16.1. These are all based on measurable material properties (e.g. Young's modulus for the mechanical case). These properties describe the behaviour of the material at a suitable scale for engineering purposes. It is widely believed that most of the next generation aircraft and spacecraft will be manufactured as multilayered structures (MLSs) under the action of a combination of two or more of the four fields. Examples of the application of the multifield problem (MFP) are: the so-called 'smart structures' in which layers of piezoelectric or piezomagnetic materials are used as sensors or actuators to develop electromagnetic fields that are able to counteract

Finite Element Analysis of Structures Through Unified Formulation, First Edition.
Erasmo Carrera, Maria Cinefra, Marco Petrolo and Enrico Zappino.

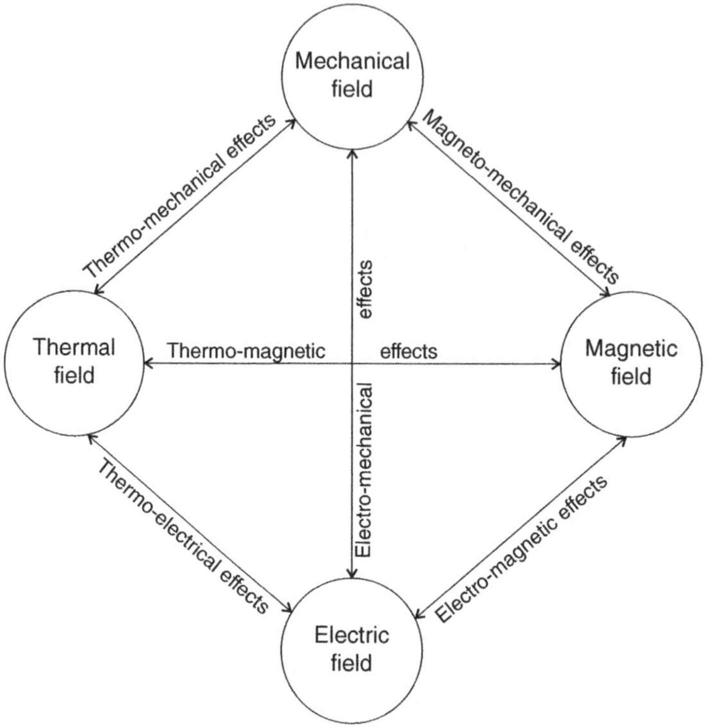

Figure 16.1 Interaction processes between the mechanical, thermal, electric and magnetic fields

thermomechanical deformations; and inflatable structures that have been planned for use in future space exploration missions with a special multilayer-made structure subjected to thermomechanical loadings and, in some cases, to electromagnetic loadings. These two examples also show that the structures that are usually employed in MFPs appear as assemblies of flat or curved MLSs.

16.2 The Need for Second-Generation FEs for Multifield Loadings

A number of requirements should be taken into account in order to obtain an accurate analysis of MFPs and MLSs. The following points have been dealt with in this chapter:

1. The constitutive equations must be derived in a consistent form.
2. The coupling between the various fields should be described accurately.
3. Inter layer continuity of the relevant variables should be guaranteed.
4. The kinematic model employed must be rich enough to describe the localized through-the-thickness-distribution of the variables involved in the various layers.

Reference to a thermodynamic basis is mandatory for point 1, see Ikeda (1996).

As far as point 2 is concerned, the coupling could be introduced in *partial* or *full* form. When the coupling is partial, the constitutive laws are uncoupled and the effect of coupling is only introduced as an external loading. If the coupling is full, the constitutive relations are coupled. For instance, if electromechanical coupling is fully included in the formulation of a piezoelectric problem, electromechanic stiffness appears in the governing equations.

As far as point 3 is concerned, it should be pointed out that a classical choice of primary variables for the various fields could violate some interlaminar continuities. This is the case of transverse shear and normal stresses, which, for equilibrium reasons, must be continuous at each layer interface, as mentioned in the previous chapter. Such continuity is not enforced in classical modelling, which only makes use of displacement variables (Carrera 2003). The same could be said for transverse electrical displacements or transverse magnetic inductance (Carrera and Brischetto 2007; Carrera and Fagiano 2007).

As far as point 4 is concerned, it is well known that the use of variable kinematic models is mandatory in MLSs subjected to MFP loadings. These loadings, in fact, have an *isotropic/anisotropic* and *localized* nature. For instance, the thermal field, being scalar, is isotropic by definition, while the electric and magnetic fields, being vectorial, can be isotropic or anisotropic. Loadings from these fields are completely different to those from mechanical cases. Consequently, kinematic models that were originally proposed for plate and shell structures subjected to mechanical loading could experience difficulties when used to analyse MFPs. Modellings that permit the use of both ESLMs and LWMs are mandatory in these cases.

A few pioneering works on the topic are mentioned below. Multilayered composite structures, made of orthotropic laminae with embedded piezoelectric patches/layers, were studied early by Mindlin in a displacement formulation context (Mindlin 1952). Coupled two-field formulations with generalized displacements, mechanical displacements and electric potential as independent variables have been proposed, among others, by Tiersten and Mindlin (1962), EerNisse (1967) and Tiersten (1969). The first FE model that relied on these formulations was proposed by Allik and Hughes (1970). Mindlin (1974) and Dokmeci (1978) proposed coupled variational principles for a vibration analysis of thermo-piezoelectric plates.

A contribution to the analysis of MLSs, in the case of MFPs, according to points 1–4 above, is presented here in the framework of the CUF. Constitutive equations are derived from the Gibbs free energy for fully coupled cases. Classical variational statements, e.g. the PVD, are extended to MFPs. Fully coupled and partially coupled cases are considered. The necessary continuity of secondary variables at the interface between two adjacent layers is imposed by extending the RMVT to MFPs (see the C_z^0 requirements in Carrera (2001), for the pure mechanical case). Variable kinematic models are derived according to the CUF proposed in earlier works by Carrera (1995, 2001, 2002). With respect to previous works such as those by Carrera and Brischetto (2007), Carrera and Fagiano (2007), Carrera (1995 and Carrera and Boscolo (2007), which were restricted to 2–3 fields, the derivation is presented here in compact form and for all four fields. The dimension of the FNs related to the CUF is given according to the number of variables involved in the variational statements. The whole notation (e.g. geometrical relations, constitutive equations, variational statements) has therefore been rearranged. In this way, a number of new, significant formulations can be proposed in the framework of both the PVD and RMVT. The subcases related to the RMVT, which can restrict the interface continuity to only those variables that are of particular significance for the particular problem, are of particular interest.

16.3 Constitutive Equations for MFPs

In this section, the constitutive equations are derived in the linear case for the MFPs considered. Standard tensor notation is used and Einstein's summation convention is implied over repeated indices. A set of intensive variables θ, ϵ, E and H, which are respectively the increment in temperature, strain, electric field and auxiliary magnetic field, are first assumed as independent variables. The relevant thermodynamic function is the Gibbs free energy per unit of volume G (Ikeda 1996) which is here extended to include the magnetic field

$$G = U - \theta\eta - \sigma_{ij}\epsilon_{ij} - E_iD_i - H_iB_i \tag{16.1}$$

where

U = internal energy per unit of volume (function of η, ϵ_{ij}, D_i and B_i);
η = variation of entropy per unit of volume;
σ_{ij} = stress tensor;
D_i = electrical displacement vector;
B_i = magnetic inductance vector.

The Gibbs free energy can be written in a quadratic form according to Ikeda (1996),

$$\begin{aligned} G = \frac{1}{2}\Bigg(& \theta^2\frac{\partial^2 G}{\partial\theta^2} + \epsilon_{ij}\epsilon_{lm}\frac{\partial^2 G}{\partial\epsilon_{ij}\partial\epsilon_{lm}} + E_iE_l\frac{\partial^2 G}{\partial E_i\partial E_l} + H_iH_l\frac{\partial^2 G}{\partial H_i\partial H_l} + \theta\epsilon_{lm}\frac{\partial^2 G}{\partial\theta\partial\epsilon_{lm}} \\ & + \theta E_l\frac{\partial^2 G}{\partial\theta\partial E_l} + \theta H_l\frac{\partial^2 G}{\partial\theta\partial H_l} + \epsilon_{ij}\theta\frac{\partial^2 G}{\partial\epsilon_{ij}\partial\theta} + \epsilon_{ij}E_l\frac{\partial^2 G}{\partial\epsilon_{ij}\partial E_l} + \epsilon_{ij}H_l\frac{\partial^2 G}{\partial\epsilon_{ij}\partial H_l} \\ & + E_i\theta\frac{\partial^2 G}{\partial E_i\partial\theta} + E_i\epsilon_{lm}\frac{\partial^2 G}{\partial E_i\partial\epsilon_{lm}} + E_iH_l\frac{\partial^2 G}{\partial E_i\partial H_l} + H_i\theta\frac{\partial^2 G}{\partial H_i\partial\theta} + H_i\epsilon_{lm}\frac{\partial^2 G}{\partial H_i\partial\epsilon_{lm}} \\ & + H_iE_l\frac{\partial^2 G}{\partial H_i\partial E_l}\Bigg) \end{aligned} \tag{16.2}$$

The exact differential of G is

$$dG = -\eta d\theta - \sigma_{ij}d\epsilon_{ij} - D_idE_i - B_idH_i \tag{16.3}$$

where

$$\eta = -\left[\frac{\partial G}{\partial\theta}\right]_{\epsilon,E,H}, \quad \sigma_{ij} = -\left[\frac{\partial G}{\partial\epsilon_{ij}}\right]_{\theta,E,H}, \quad D_i = -\left[\frac{\partial G}{\partial E_i}\right]_{\theta,\epsilon,H}, \quad B_i = -\left[\frac{\partial G}{\partial H_i}\right]_{\theta,\epsilon,E} \tag{16.4}$$

Subscripts refer to the quantities to be kept constant in the differentiation.

Substituting Equation (16.2) in Equation (16.4) one has

$$\eta = \theta\left[-\frac{\partial^2 G}{\partial\theta^2}\right]_{\epsilon,E,H} + \epsilon_{ij}\left[-\frac{\partial^2 G}{\partial\theta\partial\epsilon_{ij}}\right]_{E,H} + E_i\left[-\frac{\partial^2 G}{\partial\theta\partial E_i}\right]_{\epsilon,H} + H_i\left[-\frac{\partial^2 G}{\partial\theta\partial H_i}\right]_{\epsilon,E} \tag{16.5}$$

$$\sigma_{ij} = \theta\left[-\frac{\partial^2 G}{\partial\epsilon_{ij}\partial\theta}\right]_{E,H} + \epsilon_{ij}\left[-\frac{\partial^2 G}{\partial\epsilon_{ij}\partial\epsilon_{lm}}\right]_{\theta,E,H} + E_l\left[-\frac{\partial^2 G}{\partial\epsilon_{ij}\partial E_l}\right]_{\theta,H} + H_l\left[-\frac{\partial^2 G}{\partial\epsilon_{ij}\partial H_l}\right]_{\theta,E} \tag{16.6}$$

$$D_i = \theta\left[-\frac{\partial^2 G}{\partial E_i \partial \theta}\right]_{\epsilon,H} + \epsilon_{ij}\left[-\frac{\partial^2 G}{\partial E_l \partial \epsilon_{ij}}\right]_{\theta,H} + E_i\left[-\frac{\partial^2 G}{\partial E_i \partial E_j}\right]_{\theta,\epsilon,H} + H_i\left[-\frac{\partial^2 G}{\partial E_i \partial H_j}\right]_{\theta,\epsilon} \tag{16.7}$$

$$B_i = \theta\left[-\frac{\partial^2 G}{\partial H_i \partial \theta}\right]_{\epsilon,E} + \epsilon_{ij}\left[-\frac{\partial^2 G}{\partial H_l \partial \epsilon_{ij}}\right]_{\theta,E} + E_i\left[-\frac{\partial^2 G}{\partial H_i \partial E_j}\right]_{\theta,\epsilon} + H_i\left[-\frac{\partial^2 G}{\partial H_i \partial H_j}\right]_{\theta,\epsilon,E} \tag{16.8}$$

The following coefficients can be defined

$$\begin{aligned}
\frac{\rho C^{\epsilon,E,H}}{\theta_{ref}} &= -\left[\frac{\partial^2 G}{\partial \theta^2}\right]_{\epsilon,E,H} = \left[\frac{\partial \eta}{\partial \theta}\right]_{\epsilon,E,H} \\
C_{ijlm}^{\theta,E,H} &= -\left[\frac{\partial^2 G}{\partial \epsilon_{ij} \partial \epsilon_{lm}}\right]_{\theta,E,H} = \left[\frac{\partial \sigma_{ij}}{\partial \epsilon_{lm}}\right]_{\theta,E,H} \\
\varepsilon_{ij}^{\theta,\epsilon,H} &= -\left[\frac{\partial^2 G}{\partial E_i \partial E_j}\right]_{\theta,\epsilon,H} = \left[\frac{\partial D_i}{\partial E_j}\right]_{\theta,\epsilon,H} \\
\mu_{ij}^{\theta,\epsilon,E} &= -\left[\frac{\partial^2 G}{\partial H_i \partial H_j}\right]_{\theta,\epsilon,E} = \left[\frac{\partial B_i}{\partial H_j}\right]_{\theta,\epsilon,E} \\
\lambda_{ij}^{E,H} &= -\left[\frac{\partial^2 G}{\partial \theta \partial \epsilon_{ij}}\right]_{E,H} = \left[\frac{\partial \sigma_{ij}}{\partial \theta}\right]_{\epsilon,E,H} = \left[\frac{\partial \eta}{\partial \epsilon_{ij}}\right]_{\theta,E,H} \\
p_i^{\epsilon,H} &= -\left[\frac{\partial^2 G}{\partial \theta \partial E_i}\right]_{\epsilon,H} = \left[\frac{\partial D_i}{\partial \theta}\right]_{\epsilon,E,H} = \left[\frac{\partial \eta}{\partial E_i}\right]_{\theta,\epsilon,H} \\
e_{lij}^{\theta,H} &= -\left[\frac{\partial^2 G}{\partial \epsilon_{ij} \partial E_l}\right]_{\theta,H} = \left[\frac{\partial \sigma_{ij}}{\partial E_l}\right]_{\theta,\epsilon,H} = \left[\frac{\partial D_l}{\partial \epsilon_{ij}}\right]_{\theta,E,H} \\
r_i^{\epsilon,E} &= -\left[\frac{\partial^2 G}{\partial \theta \partial H_i}\right]_{\epsilon,E} = \left[\frac{\partial \eta}{\partial H_i}\right]_{\theta,\epsilon,E} = \left[\frac{\partial B_i}{\partial \theta}\right]_{\epsilon,E,H} \\
q_{lij}^{\theta,E} &= -\left[\frac{\partial^2 G}{\partial \epsilon_{ij} \partial H_l}\right]_{\theta,E} = \left[\frac{\partial \sigma_{ij}}{\partial H_l}\right]_{\theta,\epsilon,E} = \left[\frac{\partial B_l}{\partial \epsilon_{ij}}\right]_{\theta,E,H} \\
d_{ij}^{\theta,\epsilon} &= -\left[\frac{\partial^2 G}{\partial E_i \partial H_j}\right]_{\theta,\epsilon} = \left[\frac{\partial D_i}{\partial H_j}\right]_{\theta,\epsilon,E} = \left[\frac{\partial B_i}{\partial E_j}\right]_{\theta,\epsilon,H}
\end{aligned} \tag{16.9}$$

with:

- ρ = density;
- $C^{\epsilon,E,H}$ = specific heat per unit mass;
- θ_{ref} = reference temperature;
- C_{ijlm} = elastic coefficients – Hooke's law;
- ε_{ij} = permittivity coefficients;
- μ_{ij} = magnetic permeability coefficients;
- λ_{ij} = stress temperature coefficients;
- p_i = pyroelectric coefficients;

e_{lij} = piezoelectric coefficients;
r_i = pyromagnetic coefficients;
q_{lij} = piezomagnetic coefficients;
d_i = magnetoelectric coupling coefficients.

The first three constants in Equations (16.9) are principal constants in the respective individual systems; the latter six are coupling constants between two of the four fields considered. Upon introducing the constants of Equations (16.9), the following constitutive equations in the coupled four-field system are obtained from Equations (16.5)–(16.8):

$$\begin{aligned} \eta &= \frac{\rho}{\theta_{ref}} C^{\epsilon,E,H}\theta + \lambda_{ij}^{E,H}\epsilon_{ij} + p_i^{\epsilon,H}E_i + r_i^{\epsilon,E}H_i \\ \sigma_{ij} &= \lambda_{ij}^{E,H}\theta + C_{ijlm}^{\theta,E,H}\epsilon_{lm} + e_{ijl}^{\theta,H}E_l + q_{ijl}^{\theta,E}H_l \\ D_l &= p_l^{\epsilon,H}\theta + e_{lij}^{\theta,H}\epsilon_{ij} + \varepsilon_{lm}^{\theta,\epsilon,H}E_m + d_{lm}^{\theta,\epsilon}H_m \\ B_l &= r_l^{\epsilon,E}\theta + q_{lij}^{\theta,E}\epsilon_{ij} + d_{lm}^{\theta,\epsilon}E_m + \mu_{lm}^{\theta,\epsilon,E}H_m \end{aligned} \tag{16.10}$$

As emphasized by Ikeda (1996), physical constants are introduced by the second derivative of the relevant thermodynamic function. Each coupling constant is a second derivative with respect to two different variables, and is therefore considered to have a different meaning when interchanging the order of differentiation. For instance, from the definition given above, two kinds of piezoelectric constants

$$e_{lij}^{\theta,H} = \left[\frac{\partial D_l}{\partial \epsilon_{ij}}\right]_{\theta,E,H} \quad \text{and} \quad \widetilde{e}_{lij}^{\theta,H} = \left[\frac{\partial \sigma_{ij}}{\partial E_l}\right]_{\theta,\epsilon,H}$$

are derived. Here, e represents the electric flux density versus unit strain, whereas $\widetilde{e}$ represents stress versus unit electric field. It turns out that these correspond to the direct and converse piezoelectric effect, respectively. The piezoelectric coupling term in the third expression of Equations (16.10) indicates the direct effect and that in the second expression represents the converse effect. The equality of the direct and converse effect is thus self-evident. Similar relations are found for the other coupling constants (λ_{ij}, p_i, r_i, q_{lij} and d_{lm}).

Let us anticipate that, by using the RMVT, only extensive variables can be modelled in the through-the-thickness plate z direction (η, σ_{ij}, D_n, B_n).

16.4 Variational Statements for MFPs

As stated in Yang (1979), piezoelectricity is based on a quasi-static approximation (Nelson 1996). This approximation can be considered as the lowest-order approximation of a perturbation procedure, based on the fact that the speed of an acoustic wave is much lower than the speed of light. The same is true for magnetomechanical phenomena. As a result of this approximation, although the mechanical equations in the theory of piezoelectricity are dynamic, the electromagnetic equations are static, and the electric field and the magnetic field are not dynamically coupled. Therefore, the wave behaviour of the electromagnetic field is not described. The quasi-static theory is sufficient for many applications of piezoelectric acoustic wave devices, but there are situations in which full electromagnetic coupling needs to be considered. For example, electromagnetic waves generated by a mechanical field need to be studied in the calculation of radiated electromagnetic power from a piezoelectric device. Full Maxwell's equations also need to be considered in devices in which acoustic waves produce electromagnetic waves, or vice versa. When electromagnetic waves are involved, the complete set of Maxwell's equations needs to be used, coupled with the mechanical equations of motion. In this way, a fully dynamic theory can be obtained. An interesting discussion on this topic has been provided in the work by Yang (1979). However, in the present work, the electric field and magnetic field are not considered to be dynamically coupled.

16.4.1 PVD

In this section, the PVD is derived for the full case. This means that complete coupling among mechanical, thermal, electrical and magnetic variables is considered. It is convenient to start the derivation directly from Hamilton's principle

$$\delta \int_{t_0}^{t} (K - \Pi)\, dt = 0 \qquad \Rightarrow \qquad \delta \int_{t_0}^{t} K\, dt - \delta \int_{t_0}^{t} \Pi\, dt = 0 \tag{16.11}$$

where K is the kinetic energy and Π is the potential energy; δ is a variational symbol; t denotes time, t_0 and t_1 are the initial and generic instants. The kinetic energy variation can be treated as follows:

$$\begin{aligned} \delta \int_{t_0}^{t} Kdt &= \delta \int_{t_0}^{t} dt \int_V \left(\frac{1}{2}\rho\, \dot{u}_i\, \dot{u}_i\right) dV = \int_{t_0}^{t} \int_V \rho\, \dot{u}_i\, \delta\dot{u}_i\, dV\, dt \\ &= \int_V \rho\, \dot{u}_i\, \delta u_i\, dV\Big|_{t_0}^{t_1} - \int_{t_0}^{t_1} \int_V \rho\, \ddot{u}_i\, \delta u_i\, dV\, dt \end{aligned} \tag{16.12}$$

V is the plate volume, u_i is a displacement component and the dot denotes differentiation with respect to time. δu is equal to zero in $t = t_0$ and $t = t_1$, so that

$$\delta \int_{t_0}^{t} Kdt = - \int_{t_0}^{t_1} \int_V \rho\, \ddot{u}_i\, \delta u_i\, dV\, dt \tag{16.13}$$

It follows that

$$\delta \int_{t_0}^{t} Kdt = - \int_{t_0}^{t} \delta L_{ine}\, dt \tag{16.14}$$

in which δL_{ine} denotes the variation of the work done by inertial forces.

The variation of potential energy is written as the algebraic sum of the variation of the Gibbs free energy and the variation of the work done by applied mechanical, thermal and electrical loadings. Any hypothetical load due to the magnetic field is neglected here. For practical reasons, body forces and volumetric electrical charges are neglected too:

$$\begin{aligned}\delta \int_{t_0}^{t} \Pi \, dt &= \delta \int_{t_0}^{t} \left[\int_V G \, dV - \int_A \left(\bar{t}_j u_j - \bar{Q}\,\Phi \right) dA \right] dt \\ &= \delta \int_{t_0}^{t} \int_V G \, dV \, dt - \int_{t_0}^{t} \delta L_{ext} \, dt\end{aligned} \tag{16.15}$$

where:

A is the surface involved by loading;
$\bar{t}_j$ is the mechanical loading in j direction;
$\bar{Q}$ is the charge density on the plate surface;
ϕ is the electric potential;
δL_{ext} denotes the variation of the work done by external loads.

Upon substitution of Equations (16.14) and (16.15) in Equation (16.11), it follows that

$$\delta \int_{t_0}^{t} \int_V G \, dV \, dt = \int_{t_0}^{t} \delta L_{ext} \, dt - \int_{t_0}^{t} \delta L_{ine} \, dt \tag{16.16}$$

On differentiating the Gibbs free energy, Equation (16.16) takes the following form:

$$\int_{t_0}^{t} \int_V \left(-\frac{\partial G}{\partial T}\delta T - \frac{\partial G}{\partial \epsilon_{ij}}\delta \epsilon_{ij} - \frac{\partial G}{\partial E_l}\delta E_l - \frac{\partial G}{\partial H_l}\delta H_l \right) dV \, dt = \int_{t_0}^{t} \delta L_{ext} \, dt - \int_{t_0}^{t} \delta L_{ine} \, dt \tag{16.17}$$

Upon substitution of Equations (16.4) and by eliminating the time integral, the PVD for a 3D continuum is obtained,

$$\int_V (\eta \delta \theta + \sigma_{ij} \delta \epsilon_{ij} + D_l \delta E_l + B_l \delta H_l) \, dV = \delta L_{ext} - \delta L_{ine} \tag{16.18}$$

16.4.1.1 Use of Condensed Notation

Moving from indices to vectors, it is useful to collect vectors $\sigma, \mathbf{D}, \mathbf{B}, \eta$ and $\epsilon, \mathbf{E}, \mathbf{H}, \theta$ in $\boldsymbol{\mathcal{E}}$ and $\boldsymbol{S}$ respectively (bold script letters denote arrays); $\boldsymbol{S}$ is the vector of extensive variables while $\boldsymbol{\mathcal{E}}$ is the vector of intensive ones,

$$\boldsymbol{S}^T = \{\sigma_{11} \quad \sigma_{22} \quad \sigma_{12} \quad D_1 \quad D_2 \quad B_1 \quad B_2 \quad \eta \quad \sigma_{33} \quad \sigma_{13} \quad \sigma_{23} \quad D_3 \quad B_3\} \tag{16.19}$$

$$\boldsymbol{\mathcal{E}}^T = \{\epsilon_{11} \quad \epsilon_{22} \quad \epsilon_{12} \quad E_1 \quad E_2 \quad H_1 \quad H_2 \quad \theta \quad \epsilon_{33} \quad \epsilon_{13} \quad \epsilon_{23} \quad E_3 \quad H_3\} \tag{16.20}$$

Note that, in dealing with plates, subscript '3' indicates the through-the-thickness plate z direction while subscripts '1' and '2' are for the remaining two orthogonal in-plane directions.

By using these two vectors, the PVD statement for a multilayered plate results in the following compact form:

$$\int_V \left(\delta \boldsymbol{\mathcal{E}}_G^T \boldsymbol{S}_H\right) dV = \delta L_{ext} - \delta L_{ine} \tag{16.21}$$

where subscripts 'G' and 'H' indicate variables obtained from geometrical relations and by constitutive/Hooke's equations respectively. For MLSs, the volume integral is intended to be

$$\int_V (\ldots)\, dV = \sum_{k=1}^{N_l} \int_{\Omega_k} \int_{A_k} (\ldots)\, d\Omega_k\, dz \tag{16.22}$$

where Ω_k is the layer middle surface and A_k denotes the layer thickness domain.

The PVD can also be written in the following form:

$$\begin{aligned} \int_V & \delta \boldsymbol{\epsilon}_{pG}^T \boldsymbol{\sigma}_{pH} + \delta \boldsymbol{\epsilon}_{nG}^T \boldsymbol{\sigma}_{nH} + \boldsymbol{H}_{pG}^T \boldsymbol{B}_{pH} + \delta H_{nG} B_{nH} \\ & + \delta \boldsymbol{E}_{pG}^T \boldsymbol{D}_{pH} + \delta E_{nG} D_{nH} + \delta \theta_G \eta_H \, dV \; = \; \delta L_{ext} - \delta L_{ine} \end{aligned} \tag{16.23}$$

where the notation already used in the work by Carrera and Fagiano (2007) is referred to: subscript 'p' denotes in-plane unknowns and subscript 'n' denotes out-of-plane unknowns.

16.4.1.2 Particular Cases

In practical applications, not all the intensive variables (θ, ϵ, E, H) are taken to be independent. Considering all the possible combinations of virtual variations active in the model, many different governing equations can be obtained. For the PVD, virtual variations addressed here are six: δu_1, δu_2, δu_3, $\delta \phi$, $\delta \varphi$, $\delta \theta$; where $\delta \phi$ and $\delta \varphi$ indicate the variations of the electric and magnetic potential, respectively.[1] In general, a virtual variation can be considered alone or can be coupled with the others. A few examples of variational statements for different types of coupling are proposed below. Reference is made to the form of the PVD in Equation (16.23). Seven additional PVD forms are discussed.

PVD-1: Pure Mechanical Case. If only virtual variations of displacements δu_1, δu_2, δu_3 are considered, Equation (16.18) is reduced to

$$\int_V \delta \boldsymbol{\epsilon}_{pG}^T \boldsymbol{\sigma}_{pH} + \delta \boldsymbol{\epsilon}_{nG}^T \boldsymbol{\sigma}_{nH} \, dV \; = \; \delta L_{ext} - \delta L_{ine} \tag{16.24}$$

That is, a pure mechanical problem is described.

PVD-2: Coupled Thermomechanical Case. By adding the variation of the temperature, δu_1, δu_2, δu_3, $\delta \theta$, Equation (16.18) becomes

$$\int_V \delta \boldsymbol{\epsilon}_{pG}^T \boldsymbol{\sigma}_{pH} + \delta \boldsymbol{\epsilon}_{nG}^T \boldsymbol{\sigma}_{nH} + \delta \theta_G \eta_H dV = \delta L_{ext} - \delta L_{ine} \tag{16.25}$$

That is, the coupled thermomechanical problem is described.

[1] Note that δu_1, δu_2, δu_3 come from the strain $\delta \epsilon$, $\delta \phi$ from δE, and $\delta \varphi$ from δH.

PVD-3: Partially Coupled Thermomechanical Case. Saying that the thermal field is partially coupled with the mechanical field means that the extensive variable concerning the thermal field (η) is not considered in constitutive relations. The thermal field impacts the system in the form of thermal stresses σ_θ and the thermal load vector. The advantage of using a partially coupled system lies in the reduction of the number of system DOFs. The disadvantage is that the thermal effect due to strain is neglected and the temperature must be known at any point in the considered continuum. Correspondingly, Equation (16.18) is

$$\int \delta\boldsymbol{\epsilon}^T_{pG}(\boldsymbol{\sigma}_{pH} - \boldsymbol{\sigma}_{p\theta}) + \delta\boldsymbol{\epsilon}^T_{nG}(\boldsymbol{\sigma}_{nH} - \boldsymbol{\sigma}_{n\theta})dV = \delta L_{ext} - \delta L_{ine} \tag{16.26}$$

PVD-4: Coupled Electromechanical Case. The virtual variations here are δu_1, δu_2, δu_3, $\delta\phi$ and the PVD reduces to

$$\int_V \delta\boldsymbol{\epsilon}^T_{pG}\boldsymbol{\sigma}_{pH} + \delta\boldsymbol{\epsilon}^T_{nG}\boldsymbol{\sigma}_{nH} + \delta\boldsymbol{E}^T_{pG}\boldsymbol{D}_{pH} + \delta E_{nG}D_{nH}dV = \delta L_{ext} - \delta L_{ine} \tag{16.27}$$

PVD-5: Coupled Magnetomechanical Case. The virtual variations in this case are δu_1, δu_2, δu_3, $\delta\varphi$ and the PVD reduces to

$$\int_V \delta\boldsymbol{\epsilon}^T_{pG}\boldsymbol{\sigma}_{pH} + \delta\boldsymbol{\epsilon}^T_{nG}\boldsymbol{\sigma}_{nH} + \delta\boldsymbol{H}^T_{pG}\boldsymbol{B}_{pH} + \delta H_{nG}B_{nH}dV = \delta L_{ext} - \delta L_{ine} \tag{16.28}$$

PVD-6: Coupled Magneto-electromechanical Case. The virtual variations here are δu_1, δu_2, δu_3, $\delta\phi$, $\delta\varphi$ and the PVD is

$$\int_V \delta\boldsymbol{\epsilon}^T_{pG}\boldsymbol{\sigma}_{pH} + \delta\boldsymbol{\epsilon}^T_{nG}\boldsymbol{\sigma}_{nH} + \boldsymbol{H}^T_{pG}\boldsymbol{B}_{pH} + \delta H_{nG}B_{nH} + \delta\boldsymbol{E}^T_{pG}\boldsymbol{D}_{pH} + \delta E_{nG}D_{nH}dV = \delta L_{ext} - \delta L_{ine}$$

PVD-7: Pure Thermal Case. In this case only the virtual variation of the temperature $\delta\theta$ is considered and the PVD reduces to

$$\int_V \delta\theta_G\eta_H dV = \delta L_{ext} - \delta L_{ine} \tag{16.29}$$

A pure thermal problem is obtained that is equivalent to heat conduction problems.

16.4.2 RMVT

As stated in the introduction, the advantage of using the RMVT lies in the possibility of assuming two independent sets of variables: a set of primary unknowns and a set of extensive variables which are modelled in the thickness plate z direction. The main advantage of using the RMVT is the a priori and complete fulfilment of the C^0_z requirements for the modelled extensive mechanical variables. In this section, the RMVT is written for the multifield case. All the normal components (those in the z direction) of extensive variables are modelled in the thickness plate z direction.

The RMVT can be obtained by starting from the first term of Equation (16.16),

$$\delta \int_{t_0}^{t} \int_V G\, dV dt = \delta \int_{t_0}^{t} \int_V G\, dV dt \tag{16.30}$$

By referring to the condensed notation and considering that subscript 'a' indicates 'not-modelled quantities', while subscripts 'b' are related to 'modelled quantities', the following vectors can be introduced:

$\boldsymbol{S}_{aH}$ is the vector of not-modelled extensive variables, calculated from constitutive relations;
$\boldsymbol{S}_{bH}$ is the vector of modelled extensive variables, calculated from constitutive relations;
$\boldsymbol{\mathcal{E}}_{aG}$ is the vector of intensive variables associated to $\boldsymbol{S}_a$ and calculated from geometrical relations;
$\boldsymbol{\mathcal{E}}_{bG}$ is the vector of intensive variables associated to $\boldsymbol{S}_b$ and calculated from geometrical relations;
$\boldsymbol{\mathcal{E}}_{bH}$ is the vector of intensive variables associated to $\boldsymbol{S}_b$ and calculated from constitutive relations.

Since $\boldsymbol{S}_b$ are unknown variables, Lagrange multipliers associated to $\boldsymbol{S}_b$ should be introduced (see Reissner (1984) for the pure mechanical case)

$$\delta \int_{t_0}^{t} \int_V G\, dV\, dt = \delta \int_{t_0}^{t} \int_V \left[G + \delta \boldsymbol{S}_b^T \left(\boldsymbol{\mathcal{E}}_{bG} - \boldsymbol{\mathcal{E}}_{bH}\right)\right] dV\, dt \tag{16.31}$$

For our purpose it is convenient to differentiate as in the following:

$$\int_{t_0}^{t} \int_V \left[\delta \boldsymbol{\mathcal{E}}_{aG}^T \frac{\partial G}{\partial \boldsymbol{\mathcal{E}}_{aG}^T} + \delta \boldsymbol{\mathcal{E}}_{bG}^T \frac{\partial G}{\partial \boldsymbol{\mathcal{E}}_{bG}^T} + \delta \boldsymbol{S}_b^T (\boldsymbol{\mathcal{E}}_{bG} - \boldsymbol{\mathcal{E}}_{bH})\right] dV\, dt = \int_{t_0}^{t} \delta L_{ext}\, dt - \int_{t_0}^{t} \delta L_{ine}\, dt \tag{16.32}$$

By eliminating the time integral, the RMVT statement for an MLS in an MFP is obtained,

$$\int_V \left[\delta \boldsymbol{\mathcal{E}}_{aG}^T \boldsymbol{S}_{aH} + \delta \boldsymbol{\mathcal{E}}_{bG}^T \boldsymbol{S}_b^k + \delta \boldsymbol{S}_b^T (\boldsymbol{\mathcal{E}}_{bG} - \boldsymbol{\mathcal{E}}_{bH})\right] dV = \delta L_{ext} - \delta L_{ine} \tag{16.33}$$

If the through-the-thickness modelled variables are $\boldsymbol{\sigma}_n$, D_n and B_n, the RMVT can also be written in the following form:

$$\begin{aligned}\int_V & \delta \boldsymbol{\epsilon}_{pG}^T \boldsymbol{\sigma}_{pH} + \delta \boldsymbol{\epsilon}_{nG}^T \boldsymbol{\sigma}_n + \delta \boldsymbol{\sigma}_n^T (\boldsymbol{\epsilon}_{nG} - \boldsymbol{\epsilon}_{nH}) + \delta \theta_G \eta_H + \delta \boldsymbol{E}_{pG}^T \boldsymbol{D}_{pH} + \delta E_{nG} D_n \\ & + \delta D_n (E_{nG} - E_{nH}) + \delta \boldsymbol{H}_{pG}^T \boldsymbol{B}_{pH} + \delta H_{nG} B_n + \delta B_n (H_{nG} - H_{nH})\, dV = \delta L_{ext} - \delta L_{ine}\end{aligned} \tag{16.34}$$

16.4.2.1 Particular Cases

As for the PVD, several particular cases for the RMVT can be obtained. These can all be of particular interest in practical applications. By considering all the possible combinations of virtual variation active in the model, the problem unknowns can be thought as grouped in two sets: primary unknowns $(\delta u_1, \delta u_2, \delta u_3, \delta \phi, \delta \psi, \delta \vartheta)$ and unknowns modelled through the thickness $(\delta \sigma_{13}, \delta \sigma_{23}, \delta \sigma_{33}, \delta D_3, \delta B_3)$, which are intensive variables. Combinations of these two sets of virtual variations can be considered. If only variations associated to primary unknowns are chosen, the PVD is obtained, otherwise a particular case of the RMVT is given. In this section, a few examples of the latter cases are formulated, also considering a different number of involved fields. For the sake of clarity, the same notation as in the PVD cases is used.

RMVT-1: Pure Mechanical Case. The virtual variations in this case are δu_1, δu_2, δu_3, $\delta \sigma_{33}$, $\delta \sigma_{13}$, $\delta \sigma_{23}$ and the RMVT statement is

$$\int_V \delta \boldsymbol{\epsilon}_{pG}^T \boldsymbol{\sigma}_{pH} + \delta \boldsymbol{\epsilon}_{nG}^T \boldsymbol{\sigma}_n + \boldsymbol{\sigma}_n^T (\boldsymbol{\epsilon}_{nG} - \boldsymbol{\epsilon}_{nH}) \, dV = \delta L_{ext} - \delta L_{ine} \tag{16.35}$$

RMVT-2: Pure Mechanical Case (only normal transverse stress modelled). The virtual variations here are δu_1, δu_2, δu_3, $\delta \sigma_{33}$ and the RMVT statement is

$$\int_V \delta \boldsymbol{\epsilon}_{pG}^T \boldsymbol{\sigma}_{pH} + \delta \boldsymbol{\epsilon}_{nG}^T \boldsymbol{\sigma}_n + \sigma_{33} (\epsilon_{33G} - \epsilon_{33H}) \, dV = \delta L_{ext} - \delta L_{ine} \tag{16.36}$$

RMVT-3: Pure Mechanical Case (only shear transverse stresses modelled). The virtual variations are δu_1, δu_2, δu_3, $\delta \sigma_{13}$, $\delta \sigma_{13}$ and the RMVT reduces to

$$\begin{aligned} &\int_V \delta \boldsymbol{\epsilon}_{pG}^T \boldsymbol{\sigma}_{pH} + \delta \boldsymbol{\epsilon}_{nG}^T \boldsymbol{\sigma}_n + \sigma_{13} (\epsilon_{13G} - \epsilon_{13H}) + \sigma_{23} (\epsilon_{23G} - \epsilon_{23H}) \, dV \\ &= \delta L_{ext} - \delta L_{ine} \end{aligned} \tag{16.37}$$

RMVT-4: Coupled Thermomechanical Case. The virtual variations in this case are δu_1, δu_2, δu_3, $\delta \theta$, $\delta \sigma_{33}$, $\delta \sigma_{13}$, $\delta \sigma_{23}$ and the RMVT reduces to

$$\int_V \delta \boldsymbol{\epsilon}_{pG}^T \boldsymbol{\sigma}_{pH} + \delta \boldsymbol{\epsilon}_{nG}^T \boldsymbol{\sigma}_n + \delta \boldsymbol{\sigma}_n^T (\boldsymbol{\epsilon}_{nG} - \boldsymbol{\epsilon}_{nH}) + \delta \theta_G \eta_H dV = \delta L_{ext} - \delta L_{ine} \tag{16.38}$$

RMVT-5: Coupled Electromechanical Case. If the virtual variations are δu_1, δu_2, δu_3, $\delta \phi$, $\delta \sigma_{33}$, $\delta \sigma_{13}$, $\delta \sigma_{23}$, the RMVT statement is

$$\begin{aligned} &\int_V \delta \boldsymbol{\epsilon}_{pG}^T \boldsymbol{\sigma}_{pH} + \delta \boldsymbol{\epsilon}_{nG}^T \boldsymbol{\sigma}_n + \delta \boldsymbol{\sigma}_n^T (\boldsymbol{\epsilon}_{nG} - \boldsymbol{\epsilon}_{nH}) + \delta \boldsymbol{E}_{pG}^T \boldsymbol{D}_{pH} + \delta E_{nG} D_{nH} \, dV \\ &= \delta L_{ext} - \delta L_{ine} \end{aligned} \tag{16.39}$$

RMVT-6: Coupled Electromechanical Case. If the virtual variations are δu_1, δu_2, δu_3, $\delta\phi$, $\delta\sigma_{33}$, $\delta\sigma_{13}$, $\delta\sigma_{23}$, δD_n, the RMVT statement is

$$\begin{aligned}\int_V & \delta\boldsymbol{\epsilon}_{pG}^T\boldsymbol{\sigma}_{pH} + \delta\boldsymbol{\epsilon}_{nG}^T\boldsymbol{\sigma}_n + \delta\boldsymbol{\sigma}_n^T(\boldsymbol{\epsilon}_{nG} - \boldsymbol{\epsilon}_{nH}) + \delta\boldsymbol{E}_{pG}^T\boldsymbol{D}_{pH} + \delta E_{nG}D_n \\ & + \delta D_n(E_{nG} - E_{nH})\, dV = \delta L_{ext} - \delta L_{ine}\end{aligned} \tag{16.40}$$

RMVT-7: Coupled Magnetomechanical Case. If the virtual variations are δu_1, δu_2, δu_3, $\delta\varphi$, $\delta\sigma_{33}$, $\delta\sigma_{13}$, $\delta\sigma_{23}$, the RMVT reduces to

$$\begin{aligned}\int_V & \delta\boldsymbol{\epsilon}_{pG}^T\boldsymbol{\sigma}_{pH} + \delta\boldsymbol{\epsilon}_{nG}^T\boldsymbol{\sigma}_n + \delta\boldsymbol{\sigma}_n^T(\boldsymbol{\epsilon}_{nG} - \boldsymbol{\epsilon}_{nH}) + \delta\boldsymbol{H}_{pG}^T\boldsymbol{B}_{pH} + \delta H_{nG}B_{nH}\, dV \\ & = \delta L_{ext} - \delta L_{ine}\end{aligned} \tag{16.41}$$

RMVT-8: Coupled Magnetomechanical Case. If the virtual variations are δu_1, δu_2, δu_3, $\delta\varphi$, $\delta\sigma_{33}$, $\delta\sigma_{13}$, $\delta\sigma_{23}$, δB_n, the RMVT reduces to

$$\begin{aligned}\int_V & \delta\boldsymbol{\epsilon}_{pG}^T\boldsymbol{\sigma}_{pH} + \delta\boldsymbol{\epsilon}_{nG}^T\boldsymbol{\sigma}_n + \delta\boldsymbol{\sigma}_n^T(\boldsymbol{\epsilon}_{nG} - \boldsymbol{\epsilon}_{nH}) + \delta\boldsymbol{H}_{pG}^T\boldsymbol{B}_{pH} + \delta H_{nG}B_n \\ & + \delta B_n(H_{nG} - H_{nH})\, dV = \delta L_{ext} - \delta L_{ine}\end{aligned} \tag{16.42}$$

RMVT-9: Coupled Magneto-electromechanical Case. If the virtual variations are δu_1, δu_2, δu_3, $\delta\phi$, $\delta\varphi$, $\delta\sigma_{33}$, $\delta\sigma_{13}$, $\delta\sigma_{23}$, the RMVT statement is

$$\begin{aligned}\int_V & \delta\boldsymbol{\epsilon}_{pG}^T\boldsymbol{\sigma}_{pH} + \delta\boldsymbol{\epsilon}_{nG}^T\boldsymbol{\sigma}_n + \delta\boldsymbol{\sigma}_n^T(\boldsymbol{\epsilon}_{nG} - \boldsymbol{\epsilon}_{nH}) + \delta\boldsymbol{E}_{pG}^T\boldsymbol{D}_{pH} + \delta E_{nG}D_{nH} \\ & + \delta\boldsymbol{H}_{pG}^T\boldsymbol{B}_{pH} + \delta H_{nG}B_n\, dV = \delta L_{ext} - \delta L_{ine}\end{aligned} \tag{16.43}$$

RMVT-10: Coupled Magneto-electromechanical Case. If the virtual variations are δu_1, δu_2, δu_3, $\delta\phi$, $\delta\varphi$, $\delta\sigma_{33}$, $\delta\sigma_{13}$, $\delta\sigma_{23}$, δD_n, δB_n, the RMVT reduces to

$$\begin{aligned}\int_V & \delta\boldsymbol{\epsilon}_{pG}^T\boldsymbol{\sigma}_{pH} + \delta\boldsymbol{\epsilon}_{nG}^T\boldsymbol{\sigma}_n + \delta\boldsymbol{\sigma}_n^T(\boldsymbol{\epsilon}_{nG} - \boldsymbol{\epsilon}_{nH}) + \delta\boldsymbol{E}_{pG}^T\boldsymbol{D}_{pH} + \delta E_{nG}D_n \\ & + \delta D_n(E_{nG} - E_{nH}) + \delta\boldsymbol{H}_{pG}^T\boldsymbol{B}_{pH} + \delta H_{nG}B_n + \delta B_n(H_{nG} - H_{nH})\, dV \\ & = \delta L_{ext} - \delta L_{ine}\end{aligned} \tag{16.44}$$

16.5 Use of Variational Statements to Obtain FE equations in Terms of 'Fundamental Nuclei'

In the context of so-called axiomatic approaches, where a certain displacement, stress field or, more generally, electrical or magnetic variables are postulated in the thickness plate z direction, 2D theories are usually constructed according to the following four steps.

I. Material behaviour is assigned, i.e. constitutive relations are given (see Section 16.3).
II. An appropriate variational statement (e.g. the PVD or RMVT, Section 16.4) is used to establish governing equations and BCs which are variationally consistent with the hypotheses introduced at points I, III and IV.
III. Geometrical relations (e.g. strain–displacement relations) are assumed in the mechanical case.
IV. Displacement, stress distribution, electrical or magnetic variables in the thickness plate z direction are *postulated* by referring to a certain set of base functions.

The PVD and RMVT cases are discussed below in two different subsections.

16.5.1 PVD – Applications

16.5.1.1 Geometrical Relations

$\boldsymbol{U}^k$ denotes the vector containing the primary unknowns of the problem. Superscripts k and T indicates the kth layer of the plate and the array transposition, respectively. A suitable choice for PVD application is

$$\boldsymbol{U}^{kT} = \{u_1^k \quad u_2^k \quad u_3^k \quad \phi^k \quad \varphi^k \quad \theta^k\} \tag{16.45}$$

The intensive variables $\boldsymbol{\mathcal{E}}^k$ are linearly related to the unknowns $\boldsymbol{U}^k$ according to the following geometrical relations:

$$\boldsymbol{\mathcal{E}}_G^k = \boldsymbol{D}\boldsymbol{U}^k \tag{16.46}$$

where $\boldsymbol{D}$ denotes the following differential operator:

$$\boldsymbol{D} = \begin{bmatrix} \partial_x & 0 & 0 & 0 & 0 & 0 \\ 0 & \partial_y & 0 & 0 & 0 & 0 \\ \partial_y & \partial_x & 0 & 0 & 0 & 0 \\ 0 & 0 & 0 & -\partial_x & 0 & 0 \\ 0 & 0 & 0 & -\partial_y & 0 & 0 \\ 0 & 0 & 0 & 0 & -\partial_x & 0 \\ 0 & 0 & 0 & 0 & -\partial_y & 0 \\ 0 & 0 & 0 & 0 & 0 & 1 \\ 0 & 0 & \partial_z & 0 & 0 & 0 \\ \partial_z & 0 & \partial_x & 0 & 0 & 0 \\ 0 & \partial_z & \partial_y & 0 & 0 & 0 \\ 0 & 0 & 0 & -\partial_z & 0 & 0 \\ 0 & 0 & 0 & 0 & -\partial_z & 0 \end{bmatrix} \tag{16.47}$$

16.5.1.2 Constitutive Relations

The laminae of the multilayered plate are considered to be homogeneous and to operate in the linear elastic range. The material is assumed to be orthotropic. Considering the four-field linear coupling, constitutive coefficients can be organized in matrix $\boldsymbol{H}^k$ as

$$\boldsymbol{S}_H^k = \boldsymbol{H}^k \boldsymbol{\varepsilon}_G^k \tag{16.48}$$

where

$$\boldsymbol{H}^k = \begin{bmatrix}
C_{11}^k & C_{12}^k & C_{16}^k & 0 & 0 & 0 & 0 & \lambda_1^k & C_{13}^k & 0 & 0 & e_{31}^k & q_{31}^k \\
C_{12}^k & C_{22}^k & C_{26}^k & 0 & 0 & 0 & 0 & \lambda_2^k & C_{23}^k & 0 & 0 & e_{32}^k & q_{32}^k \\
C_{16}^k & C_{26}^k & C_{66}^k & 0 & 0 & 0 & 0 & 0 & C_{36}^k & 0 & 0 & e_{36}^k & q_{36}^k \\
0 & 0 & 0 & \varepsilon_{11}^k & \varepsilon_{12}^k & d_{11}^k & d_{12}^k & 0 & 0 & e_{15}^k & e_{14}^k & 0 & 0 \\
0 & 0 & 0 & \varepsilon_{12}^k & \varepsilon_{22}^k & d_{12}^k & d_{22}^k & 0 & 0 & e_{25}^k & e_{24}^k & 0 & 0 \\
0 & 0 & 0 & d_{11}^k & d_{12}^k & \mu_{11}^k & \mu_{12}^k & 0 & 0 & q_{15}^k & q_{14}^k & 0 & 0 \\
0 & 0 & 0 & d_{12}^k & d_{22}^k & \mu_{12}^k & \mu_{22}^k & 0 & 0 & q_{25}^k & q_{24}^k & 0 & 0 \\
\lambda_1^k & \lambda_2^k & 0 & 0 & 0 & 0 & 0 & (\rho C/\theta_{ref})^k & \lambda_3^k & 0 & 0 & p_3^k & r_3^k \\
C_{13}^k & C_{23}^k & C_{36}^k & 0 & 0 & 0 & 0 & \lambda_3^k & C_{33}^k & 0 & 0 & e_{33}^k & q_{33}^k \\
0 & 0 & 0 & e_{15}^k & e_{25}^k & q_{15}^k & q_{25}^k & 0 & 0 & C_{55}^k & C_{45}^k & 0 & 0 \\
0 & 0 & 0 & e_{14}^k & e_{24}^k & q_{14}^k & q_{24}^k & 0 & 0 & C_{45}^k & C_{44}^k & 0 & 0 \\
e_{31}^k & e_{32}^k & e_{36}^k & 0 & 0 & 0 & 0 & p_3^k & e_{33}^k & 0 & 0 & \varepsilon_{33}^k & d_{33}^k \\
q_{31}^k & q_{32}^k & q_{36}^k & 0 & 0 & 0 & 0 & r_3^k & q_{33}^k & 0 & 0 & d_{33}^k & \mu_{33}^k
\end{bmatrix} \tag{16.49}$$

16.5.1.3 Through-the-Thickness Assumptions of Primary Variables

The behaviour of primary unknowns $\boldsymbol{U}^k$ is postulated in the thickness plate z directions according to a given expansion,

$$\boldsymbol{U}^k(x, y, z) = F_\tau(z)\, \boldsymbol{U}_\tau^k(x, y), \quad \tau = 0, 1, \ldots, N \tag{16.50}$$

The repeated indexes are summed over their ranges. The polynomials $F_\tau(z)$ constitute a set of independent functions. Such a base is chosen arbitrarily: the power of z and combination of Lagrange polynomials can be considered. N denotes the order of the introduced expansion. Note that the variables concerning displacements, electric potential, magnetic potential and temperature are included in vector $\boldsymbol{U}^k$. The above expansion is done according to the CUF (Carrera 2001), which permits a large variety of plate theories to be written in unified form.

16.5.1.4 FE Discretization

In case of FEM implementation, unknowns can be expressed in terms of their nodal values via the shape functions N_i,

$$\boldsymbol{U}_\tau^k(x, y) = N_i(x, y) \boldsymbol{Q}_{\tau i}^k, \quad i = 1, 2, \ldots, N_n \tag{16.51}$$

where N_n denotes the number of nodes of the FE considered and $\boldsymbol{Q}^k_{\tau i}$ is the vector of nodal values of the primary unknowns

$$\boldsymbol{Q}^{kT}_{\tau i} = \left\{ Q^k_{u_1 \tau i} \quad Q^k_{u_2 \tau i} \quad Q^k_{u_3 \tau i} \quad Q^k_{\phi \tau i} \quad Q^k_{\varphi \tau i} \quad Q^k_{\theta \tau i} \right\} \tag{16.52}$$

Substituting Equation (16.51) in Equation (16.50), the final expression of primary unknowns can be obtained,

$$\boldsymbol{U}^k(x, y, z) = F_\tau N_i \boldsymbol{Q}^k_{\tau i} \tag{16.53}$$

16.5.1.5 FN

Upon substitution of Equations (16.46), (16.48), (16.50) and (16.51), the variational statement in Equation (16.21) leads to a set of equilibrium equations which can be formally put in the following compact form:

$$\delta \boldsymbol{Q}^k_{sj} : \boldsymbol{K}^{k\tau sij}_{PVD} \boldsymbol{Q}^k_{\tau i} \quad = \quad \boldsymbol{P}^k_{sj} \tag{16.54}$$

where $\boldsymbol{P}^k$ is the vector of external and inertial loads. The related BCs are

$$\boldsymbol{Q}^k_\tau = \overline{\boldsymbol{Q}}^k_\tau \tag{16.55}$$

The number of equations obtained coincides with the number of variables introduced: τ and s vary from 0 to N, i and j vary from 1 to N_n and k ranges from 1 to N_l. Matrix $\boldsymbol{K}^{k\tau sij}_{PVD}$ is the so-called 'fundamental nucleus' and, in this case, it is a 6×6 array.

FNs related to PVD–1÷PVD-7 can be obtained as particular cases of Equation (16.54). More details are given by Carrera and Nali (2007).

16.5.2 RMVT – Applications

16.5.2.1 Geometrical Relations

Unknown variables are collected in the vector $\boldsymbol{V}$, which includes the assumed interlaminar continuous variables

$$\boldsymbol{V}^{kT} = \left\{ u^k_1 \quad u^k_2 \quad u^k_3 \quad \phi^k \quad \varphi^k \quad \theta^k \quad \sigma^k_{33} \quad \sigma^k_{13} \quad \sigma^k_{23} \quad D^k_3 \quad B^k_3 \right\} \tag{16.56}$$

The explicit form of vectors introduced in Section 16.4.2 is

$$\boldsymbol{\varepsilon}^{kT}_a = \left\{ \epsilon^k_{11} \quad \epsilon^k_{22} \quad \epsilon^k_{12} \quad E^k_1 \quad E^k_2 \quad H^k_1 \quad H^k_2 \quad \theta^k \right\} \tag{16.57}$$

$$\boldsymbol{\varepsilon}^{kT}_b = \left\{ \epsilon^k_{33} \quad \epsilon^k_{13} \quad \epsilon^k_{23} \quad E^k_3 \quad H^k_3 \right\} \tag{16.58}$$

$$\boldsymbol{S}^{kT}_a = \left\{ \sigma^k_{11} \quad \sigma^k_{22} \quad \sigma^k_{12} \quad D^k_1 \quad D^k_2 \quad B^k_1 \quad B^k_2 \quad \eta^k \right\} \tag{16.59}$$

$$\boldsymbol{S}^{kT}_b = \left\{ \sigma^k_{33} \quad \sigma^k_{13} \quad \sigma^k_{23} \quad D^k_3 \quad B^k_3 \right\} \tag{16.60}$$

The following geometrical relations can be written:

$$\boldsymbol{\varepsilon}_{aG}^{k} = \boldsymbol{D}_a \boldsymbol{V}^k \tag{16.61}$$

$$\boldsymbol{\varepsilon}_{bG}^{k} = \boldsymbol{D}_b \boldsymbol{V}^k \tag{16.62}$$

$$\boldsymbol{S}_{bG}^{k} = \boldsymbol{D}_{b'} \boldsymbol{V}^k \tag{16.63}$$

In explicit form,

$$\boldsymbol{D}_a = \begin{bmatrix} \partial_x & 0 & 0 & 0 & 0 & 0 & 0 & 0 & 0 & 0 & 0 \\ 0 & \partial_y & 0 & 0 & 0 & 0 & 0 & 0 & 0 & 0 & 0 \\ \partial_y & \partial_x & 0 & 0 & 0 & 0 & 0 & 0 & 0 & 0 & 0 \\ 0 & 0 & 0 & -\partial_x & 0 & 0 & 0 & 0 & 0 & 0 & 0 \\ 0 & 0 & 0 & -\partial_y & 0 & 0 & 0 & 0 & 0 & 0 & 0 \\ 0 & 0 & 0 & 0 & -\partial_x & 0 & 0 & 0 & 0 & 0 & 0 \\ 0 & 0 & 0 & 0 & -\partial_y & 0 & 0 & 0 & 0 & 0 & 0 \\ 0 & 0 & 0 & 0 & 0 & 1 & 0 & 0 & 0 & 0 & 0 \end{bmatrix} \tag{16.64}$$

$$\boldsymbol{D}_b = \begin{bmatrix} 0 & 0 & \partial_z & 0 & 0 & 0 & 0 & 0 & 0 & 0 & 0 \\ \partial_z & 0 & \partial_x & 0 & 0 & 0 & 0 & 0 & 0 & 0 & 0 \\ 0 & \partial_z & \partial_y & 0 & 0 & 0 & 0 & 0 & 0 & 0 & 0 \\ 0 & 0 & 0 & -\partial_z & 0 & 0 & 0 & 0 & 0 & 0 & 0 \\ 0 & 0 & 0 & 0 & -\partial_z & 0 & 0 & 0 & 0 & 0 & 0 \end{bmatrix} \tag{16.65}$$

$$\boldsymbol{D}_{b'} = \begin{bmatrix} 0 & 0 & 0 & 0 & 0 & 0 & 1 & 0 & 0 & 0 & 0 \\ 0 & 0 & 0 & 0 & 0 & 0 & 0 & 1 & 0 & 0 & 0 \\ 0 & 0 & 0 & 0 & 0 & 0 & 0 & 0 & 1 & 0 & 0 \\ 0 & 0 & 0 & 0 & 0 & 0 & 0 & 0 & 0 & 1 & 0 \\ 0 & 0 & 0 & 0 & 0 & 0 & 0 & 0 & 0 & 0 & 1 \end{bmatrix} \tag{16.66}$$

Null rows or columns in matrices of Equations (16.64)–(16.66) are introduced for algebraic reasons in order to obtain the FN directly from a single matrix.

16.5.2.2 Constitutive Relations

Mixed variational statements require the rearrangement of constitutive relations for each layer. Equation (16.48) is rearranged in the following form:

$$\tilde{\boldsymbol{S}}_{H}^{k} = \tilde{\boldsymbol{H}}^{k} \tilde{\boldsymbol{\varepsilon}}_{G}^{k} \tag{16.67}$$

$\tilde{\boldsymbol{S}}_{H}^{k}$ comprises the vector of not-modelled extensive variables $\boldsymbol{S}_{aH}^{k}$ and the vector of intensive variables $\boldsymbol{\varepsilon}_{bH}$ (which is associated to $\boldsymbol{S}_{b}^{k}$); $\tilde{\boldsymbol{\varepsilon}}_{G}^{k}$ comprises the vector of intensive variables

$\boldsymbol{\mathcal{E}}^k_{aG}$ (which is associated to $\boldsymbol{S}^k_a$) and the vector of modelled extensive variables $\boldsymbol{S}^k_b$, which is considered a geometrical vector, by Equation (16.63)

$$\begin{aligned}\tilde{\boldsymbol{S}}^{kT}_H &= \left\{\boldsymbol{S}^{kT}_{aH} \quad \boldsymbol{\mathcal{E}}^{kT}_{bH}\right\} \\ \tilde{\boldsymbol{\mathcal{E}}}^{kT}_G &= \left\{\boldsymbol{\mathcal{E}}^{kT}_{aG} \quad \boldsymbol{S}^{kT}_{bG}\right\}\end{aligned} \tag{16.68}$$

Constitutive matrix $\boldsymbol{H}^k$ in Equation (16.48) can be partitioned by dividing cells related to modelled and not-modelled quantities,

$$\boldsymbol{H}^k = \begin{bmatrix} \boldsymbol{H}^k_{aa} & \boldsymbol{H}^k_{ab} \\ \boldsymbol{H}^k_{ba} & \boldsymbol{H}^k_{bb} \end{bmatrix} \tag{16.69}$$

where $\boldsymbol{H}^k_{ab} = \boldsymbol{H}^{kT}_{ba}$.

In explicit form,

$$\boldsymbol{H}^k_{aa} = \begin{bmatrix} C^k_{11} & C^k_{12} & C^k_{16} & 0 & 0 & 0 & 0 & \lambda^k_1 \\ C^k_{12} & C^k_{22} & C^k_{26} & 0 & 0 & 0 & 0 & \lambda^k_2 \\ C^k_{16} & C^k_{26} & C^k_{66} & 0 & 0 & 0 & 0 & 0 \\ 0 & 0 & 0 & \varepsilon^k_{11} & \varepsilon^k_{12} & d^k_{11} & d^k_{12} & 0 \\ 0 & 0 & 0 & \varepsilon^k_{12} & \varepsilon^k_{22} & d^k_{12} & d^k_{22} & 0 \\ 0 & 0 & 0 & d^k_{11} & d^k_{12} & \mu^k_{11} & \mu^k_{12} & 0 \\ 0 & 0 & 0 & d^k_{12} & d^k_{22} & \mu^k_{12} & \mu^k_{22} & 0 \\ \lambda^k_1 & \lambda^k_2 & 0 & 0 & 0 & 0 & 0 & (\rho C/\theta_{ref})^k \end{bmatrix} \tag{16.70}$$

$$\boldsymbol{H}^k_{ab} = \begin{bmatrix} C^k_{13} & 0 & 0 & e^k_{31} & q^k_{31} \\ C^k_{23} & 0 & 0 & e^k_{32} & q^k_{32} \\ C^k_{36} & 0 & 0 & e^k_{36} & q^k_{36} \\ 0 & e^k_{15} & e^k_{14} & 0 & 0 \\ 0 & e^k_{25} & e^k_{24} & 0 & 0 \\ 0 & q^k_{15} & q^k_{14} & 0 & 0 \\ 0 & q^k_{25} & q^k_{24} & 0 & 0 \\ \lambda^k_3 & 0 & 0 & p^k_3 & r^k_3 \end{bmatrix} \tag{16.71}$$

$$\boldsymbol{H}^k_{ba} = \begin{bmatrix} C^k_{13} & C^k_{23} & C^k_{36} & 0 & 0 & 0 & 0 & \lambda^k_3 \\ 0 & 0 & 0 & e^k_{15} & e^k_{25} & q^k_{15} & q^k_{25} & 0 \\ 0 & 0 & 0 & e^k_{14} & e^k_{24} & q^k_{14} & q^k_{24} & 0 \\ e^k_{31} & e^k_{32} & e^k_{36} & 0 & 0 & 0 & 0 & p^k_3 \\ q^k_{31} & q^k_{32} & q^k_{36} & 0 & 0 & 0 & 0 & r^k_3 \end{bmatrix} \tag{16.72}$$

$$\boldsymbol{H}^k_{bb} = \begin{bmatrix} C^k_{33} & 0 & 0 & e^k_{33} & q^k_{33} \\ 0 & C^k_{55} & C^k_{45} & 0 & 0 \\ 0 & C^k_{45} & C^k_{44} & 0 & 0 \\ e^k_{33} & 0 & 0 & \varepsilon^k_{33} & d^k_{33} \\ q^k_{33} & 0 & 0 & d^k_{33} & \mu^k_{33} \end{bmatrix} \tag{16.73}$$

Constitutive relations in Equation (16.48) can be arranged according to the above partitioning,

$$\begin{aligned} \boldsymbol{S}^k_{aH} &= \boldsymbol{H}^k_{aa}\boldsymbol{\varepsilon}^k_{aG} + \boldsymbol{H}^k_{ab}\boldsymbol{\varepsilon}^k_{bG} \\ \boldsymbol{S}^k_{bH} &= \boldsymbol{H}^k_{ba}\boldsymbol{\varepsilon}^k_{aG} + \boldsymbol{H}^k_{bb}\boldsymbol{\varepsilon}^k_{bG} \end{aligned} \tag{16.74}$$

From Equation (16.74) one has

$$\begin{aligned} \boldsymbol{S}^k_{aH} &= \widetilde{\boldsymbol{H}}^k_{aa}\boldsymbol{\varepsilon}^k_{aG} + \widetilde{\boldsymbol{H}}^k_{ab}\boldsymbol{S}^k_{bG} \\ \boldsymbol{\varepsilon}^k_{bH} &= \widetilde{\boldsymbol{H}}^k_{ba}\boldsymbol{\varepsilon}^k_{aG} + \widetilde{\boldsymbol{H}}^k_{bb}\boldsymbol{S}^k_{bG} \end{aligned} \tag{16.75}$$

with

$$\begin{aligned} \widetilde{\boldsymbol{H}}^k_{aa} &= \boldsymbol{H}^k_{aa} - \boldsymbol{H}^k_{ab}(\boldsymbol{H}^k_{bb})^{-1}\boldsymbol{H}^k_{ba} \\ \widetilde{\boldsymbol{H}}^k_{ab} &= \boldsymbol{H}^k_{ab}(\boldsymbol{H}^k_{bb})^{-1} \\ \widetilde{\boldsymbol{H}}^k_{ba} &= -(\widetilde{\boldsymbol{H}}^k_{ab})^T \\ \widetilde{\boldsymbol{H}}^k_{bb} &= (\boldsymbol{H}^k_{bb})^{-1} \end{aligned} \tag{16.76}$$

Matrix $\widetilde{\boldsymbol{H}}^k$ of Equation (16.67) is

$$\widetilde{\boldsymbol{H}}^k = \begin{bmatrix} \widetilde{\boldsymbol{H}}^k_{aa} & \widetilde{\boldsymbol{H}}^k_{ab} \\ \widetilde{\boldsymbol{H}}^k_{ba} & \widetilde{\boldsymbol{H}}^k_{bb} \end{bmatrix} \tag{16.77}$$

16.5.2.3 Through-the-Thickness Assumptions of Primary Variables

The behaviour of unknowns $\boldsymbol{V}^k$ is postulated in the thickness plate z direction according to the expansion described in the PVD case.

16.5.2.4 FE Discretization

In the case of FEM implementation, the unknowns can be expressed in terms of their nodal values, via the shape functions N_i, as in the PVD case,

$$V^k_\tau(x, y) = N_i(x, y)\boldsymbol{R}^k_{\tau i}, \quad i = 1, 2, \ldots, N_n \tag{16.78}$$

where N_n denotes the number of nodes concerning the FE and $\boldsymbol{R}^k_{\tau i}$ is the vector containing nodal values of the unknowns

$$\boldsymbol{R}^{kT}_{\tau i} = \left\{ R^k_{u_1 \tau i} \quad R^k_{u_2 \tau i} \quad R^k_{u_3 \tau i} \quad R^k_{\phi \tau i} \quad R^k_{\varphi \tau i} \quad R^k_{\theta \tau i} \quad R^k_{\sigma_{33} \tau i} \quad R^k_{\sigma_{31} \tau i} \quad R^k_{\sigma_{32} \tau i} \quad R^k_{D_3 \tau i} \quad R^k_{B_3 \tau i} \right\} \tag{16.79}$$

The final expression for the unknowns is

$$V^k(x, y, z) = F_\tau N_i \boldsymbol{R}^k_{\tau i} \tag{16.80}$$

16.5.2.5 FN

Upon substitution of Equations (16.61), (16.62), (16.63), (16.75) and (16.78), the variational statement in Equation (16.21) leads to a set of equilibrium equations which can be formally put in the following compact form:

$$\delta \boldsymbol{R}^k_{sj} : \boldsymbol{K}^{k\tau sij}_{RMVT} \boldsymbol{R}^k_{\tau i} = \boldsymbol{P}'^k_{sj} \tag{16.81}$$

where $\boldsymbol{P}'^k$ is the vector analogous to $\boldsymbol{P}^k$, in Section 16.5.1.

The related BCs are

$$\boldsymbol{V}^k_\tau = \overline{\boldsymbol{V}}^k_\tau \tag{16.82}$$

The number of equations obtained coincides with the number of introduced variables: τ and s vary from 0 to N, i and j vary from 1 to N_n and k ranges from 1 to N_l. Matrix $\boldsymbol{K}^{k\tau sij}_{RMVT}$ is the FN and, in this case, it is an 11×11 matrix. FNs related to RMVT-1–RMVT-10 can be obtained according to the procedure given by Carrera and Nali (2007).

16.6 Selected Results

Depending on which variational statement (PVD or RMVT) is used, the description of the through-the-thickness variables and the order of expansion N, a number of 2D theories and related FEs can be derived from the FN. Appropriate acronyms are introduced in order to identify the implemented FEs. The first field can be ‘E’ or ‘L’, according to the ESL or LW description, respectively; the second field can be ‘D’ or ‘M’, according to the PVD or RMVT application, respectively; the last field can assume the numbers 1–4, according to the order of the adopted expansion in the thickness direction; a third ‘Z’ and fourth ‘C’ field (which are optional in the ESL case) denote the use of the Murakami zigzag function and/or IC fulfilment (Interlaminar Continuity equilibrium), respectively. The results in the following sections represent a selection of previously published results, where the FNs were generated in a less general way compared with that in this chapter.

16.6.1 Mechanical–Electrical Coupling: Static Analysis of an Actuator Plate

In this section, the performances of the developed FEs are shown for the static response of a piezoelectric plate in an actuator configuration (the load consists of a bisinusoidal electric potential). In order to compare the analysis with closed-form exact solutions, attention has been restricted to simply supported plates. The reduced integration technique that was successfully applied in Carrera and Demasi (2002) has been retained in the present work. All the quoted results refer to a 6×6 mesh of Q9 elements (quadrilateral nine-node plate elements), for which a convergent solution was obtained. Multilayered plates consisting of four layers have been considered. The two inner layers coincide with cross-ply [0°/90°] carbon fibre and two external skins are made from the piezo-ceramic material, PZT-4. The properties of the material are given in Table 16.1.

The 3D solution was provided in Heyliger (1994). The two composite layers have a thickness $h_l = 0.4h$ and $h_l = 0.1h$ for the two skins. Unit value is assigned to the plate thickness. The results collected in Tables 16.2 and 16.3 show that the considered FEs in general provide a good approximation of the exact value of the electric potential ϕ. Moreover, better results for the displacement u_2 are obtained with FEs based on the RMVT, compared with the corresponding ones based on the PVD. This difference is more evident in an ESL analysis. As a final remark, it should be noted that, for an accurate transverse stress description, the use of the RMVT variational statement and LW analysis is mandatory.

16.6.2 Mechanical–Electrical Coupling: Comparison between RMVT Analyses

This section shows the performances of the FEs developed on the basis of the RMVT-6 application. For comparison purposes, the results of the RMVT-5 application are also given. Further comparisons are given in Carrera and Fagiano (2007) referring to the 3D solutions in

Table 16.1 Mechanical and electrical material properties

Properties	PZT-4	Gr/Ep	Properties	PZT-4	Gr/Ep
E_1 (GPa)	81.3	132.38	e_{15} (C/m^2)	12.72	0
E_2 (GPa)	81.3	10.756	e_{24} (C/m^2)	12.72	0
E_3 (GPa)	64.5	10.756	e_{31} (C/m^2)	−5.20	0
ν_{12}	0.329	0.24	e_{32} (C/m^2)	−5.20	0
ν_{13}	0.432	0.24	e_{33} (C/m^2)	15.08	0
ν_{23}	0.432	0.49	ϵ_{11}/ϵ_0	1475	3.5
G_{23} (GPa)	25.6	3.606	ϵ_{22}/ϵ_0	1475	3.0
G_{13} (GPa)	25.6	5.6537	ϵ_{33}/ϵ_0	1300	3.0
G_{12} (GPa)	30.6	5.6537	ρ (kg/m^3)	1	1

This table shows the properties of PZT-4 and Gr/Ep materials.

Table 16.2 Results for the actuator piezoelectric plate – comparison of various mechanical and electrical variables. Primary variables (Carrera *et al.* 2008)

FE type	$\phi(a/2, b/2, 0) \times 10^1$	$u_2(a/2, 0, -h/2) \times 10^{12}$
Exact 3D; Heyliger (1994)	4.476	−2.8625
LM4 (RMVT-5)	4.480 (0.089%)	−2.6590 (−7.109%)
LD4 (PVD-4)	4.480 (0.089%)	−2.6155 (−8.629%)
LM3 (RMVT-5)	4.480 (0.089%)	−2.6600 (−7.074%)
LD3 (PVD-4)	4.480 (0.089%)	−2.6158 (−8.618%)
LM2 (RMVT-5)	4.480 (0.089%)	−2.6559 (−7.217%)
LD2 (PVD-4)	4.481 (0.112%)	−2.6082 (−8.884%)
LM1 (RMVT-5)	4.470 (−0.134%)	−2.8652 (0.094%)
LD1 (PVD-4)	4.469 (−0.156%)	−3.0308 (5.879%)
EM4 (RMVT-5)	4.484 (0.179%)	−1.9461 (−32.01%)
ED4 (PVD-4)	4.484 (0.179%)	−1.2422 (−56.60%)
EM1 (RMVT-5)	4.461 (−0.335%)	−3.5838 (25.19%)
ED1 (PVD-4)	4.462 (−0.313%)	−4.3550 (52.14%)
EMZ3 (RMVT-5)	4.484 (0.179%)	−2.6814 (−6.327%)
EDZ3 (PVD-4)	4.484 (0.179%)	−2.1441 (−25.09%)

This table compares different 2D theories by considering primary electrical and mechanical variables.

Table 16.3 Results for the actuator piezoelectric plate – comparison of various variables; + denotes values corresponding to z positive. Shear stresses (Carrera *et al.* 2008)

FE type	$\sigma_{23}(a/2, 0, 0^+) \times 10^2$	$\sigma_{12}(0, 0, 0^+) \times 10^2$
Exact 3D; Heyliger (1994)	−2.3866	1.2286
LM4 (RMVT-5)	−2.0457 (−14.28%)	1.4770 (20.21%)
LD4 (PVD-4)	−1.9452 (−18.49%)	1.5325 (24.73%)
LM3 (RMVT-5)	−2.4082 (0.905%)	1.4832 (20.72%)
LD3 (PVD-4)	−7.0506 (195.4%)	1.4554 (18.46%)
LM2 (RMVT-5)	−0.9942 (−58.34%)	1.3512 (9.979%)
LD2 (PVD-4)	3.5829 (−250.1%)	1.6332 (32.93%)
LM1 (RMVT-5)	1.4902 (−162.4%)	1.4963 (21.78%)
LD1 (PVD-4)	−0.9403 (−60.60%)	0.9240 (−24.70%)
EM4 (RMVT-5)	12.048 (−604.8%)	2.9091 (136.7%)
ED4 (PVD-4)	13.119 (−649.6%)	3.0417 (147.5%)
EM1 (RMVT-5)	−40.630 (1602%)	−2.5329 (−306.1%)
ED1 (PVD-4)	56.882 (−2483%)	−3.0549 (−348.6%)
EMZ3 (RMVT-5)	11.168 (−567.9%)	−0.4796 (−139.0%)
EDZ3 (PVD-4)	9.2168 (−486.1%)	−0.8003 (−165.1%)

This table compares different 2D theories by considering transverse and in-plane shear stresses.

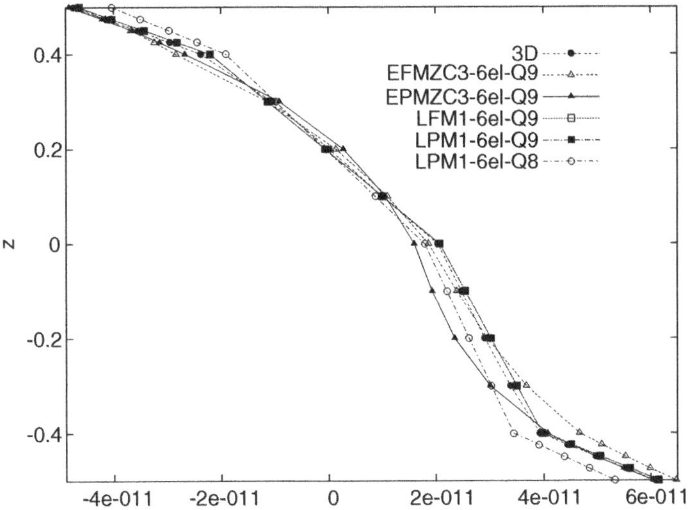

(a) Comparison between FEs based on RMVT-5 (results corresponding to acronym with 'P' as second letter) and RMVT-6 (results cor-tresponding to acronym with 'F' as second letter); '6el' means that six elements are employed in the analysis

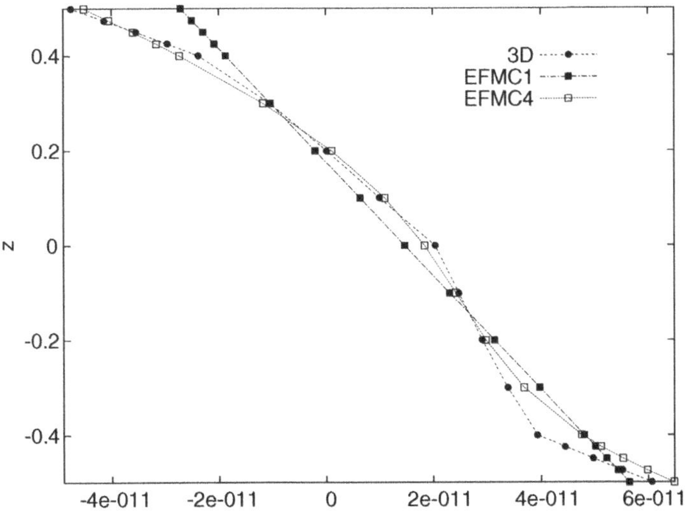

(b) FEs based on the ESL theory

Figure 16.2 Comparison of various FEs, used to predict displacement $u_2(\frac{a}{2}, 0)$ vs z, and the exact solution; $a/h = 4$. In acronyms, 'FM' and 'PM' indicate respectively RMVT-6 and RMVT-5

Heyliger (1994). In order to compare the analysis with closed-form exact solutions, attention has been restricted to simply supported square plates. LW as well as ESL analyses were performed for the Q8 and Q9 elements. Multilayered plates consisting of four layers were considered. The two inner layers coincide with cross-ply [0°/90°] carbon fibres and the two external skins are made from the piezo-ceramic material, PZT-4. The properties of the material are given in Table 16.1. The two composite layers have a thickness $h_l = 0.4h$ and $h_l = 0.1h$ for

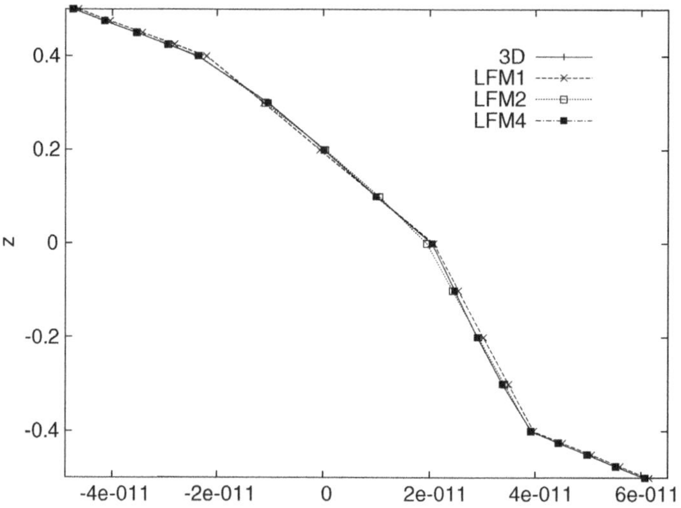

(a) FEs based on the LW theory

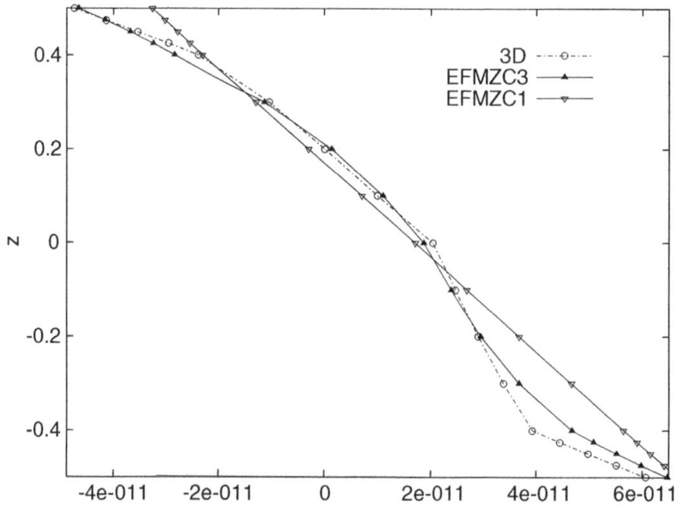

(b) ESL FEs taking into account Murakami's zigzag function

Figure 16.3 Comparison of various FEs, used to predict displacement $u_2(a/2, 0)$ vs z, and the exact solution; $a/h = 4$. In the acronyms, 'FM' and 'PM' indicate respectively RMVT-6 and RMVT-5

the two skins. Unit value is assigned to the plate thickness. A bisinusoidal distribution of the transverse pressure with amplitude $\hat{p}_z = 1$ is applied to the top surface. Figures 16.2 and 16.3 show the in-plane displacement u_2 in the through-the-thickness direction for selected plate elements. In particular, a comparison is reported in Figure 16.2a between the FEs based on RMVT-5 and the ones based on RMVT-6. Better results are obtained for the LM1 and EMZCI analyses based on RMVT-6, compared with the corresponding analyses based on RMVT-5.

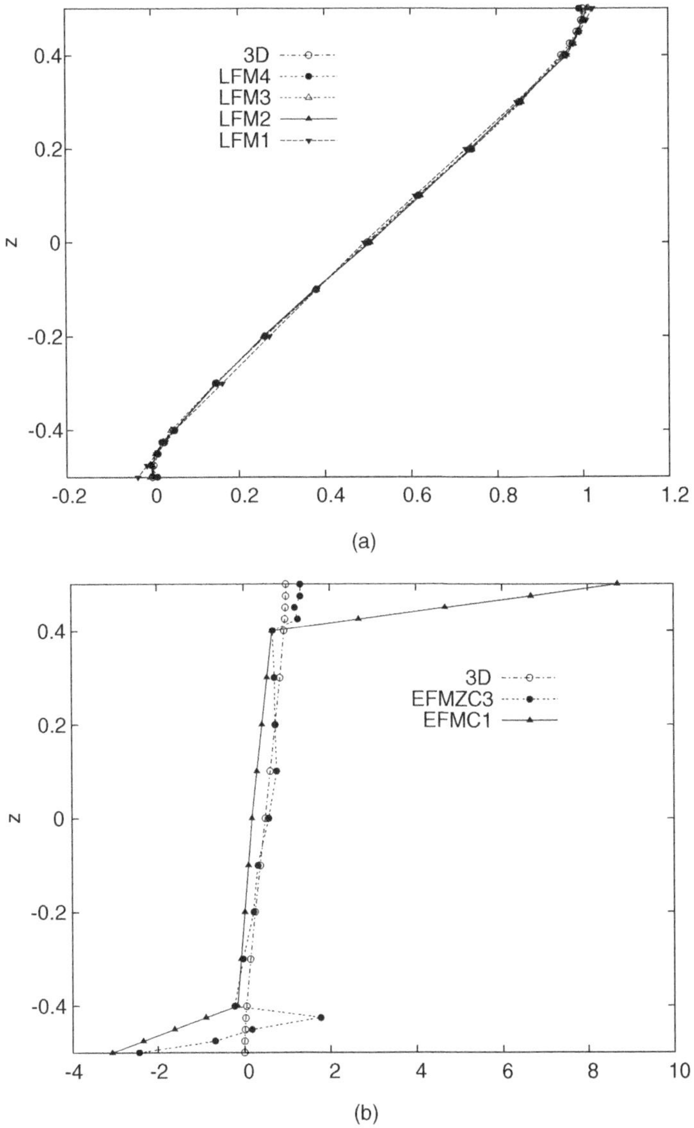

Figure 16.4 Comparison of various FEs used to predict transverse normal stress $\sigma_{33}(a/2, b/2)$ vs z; $a/h = 4$

The LW analyses lead to much better results than the ESL ones. The same conclusions can be made for the transverse normal stress evaluation in Figure 16.4: an LW theory is required with at least a parabolic distribution of the variables in each layer. Remarkable improvements are obtained when the Murakami zigzag function is used.

16.7 Literature on the CUF Extended to MFPs

This section presents some reference articles that deal with FEs based on the CUF for the analysis of MFPs. Most of them concern plate FEs, while beam and shell elements are still the subject of work in progress and were introduced in Miglioretti and Carrera (2012) and Cinefra *et al.* (2013), respectively.

Multilayered plate elements for the analysis of MFPs were presented by Carrera and Nali (2010a). In particular, hierarchic FEs for the study of multilayered piezoelectric plates have been described by Robaldo *et al.* (2006), Carrera and Robaldo (2010) and Carrera *et al.* (2007). Mixed elements, based on the extensions of the RMVT to the electromechanical problem, were introduced in Carrera and Nali (2010b). Using these elements, the static and dynamic analysis of piezoelectric plates was performed by Carrera and Boscolo (2007) and anisotropic piezoelectric plates have been analysed by Carrera *et al.* (2010). Works by Carrera and Nali (2009) and Carrera and Fagiano (2007) show the importance of directly evaluating the transverse electric displacement in layered smart structures in order to fulfil continuity conditions at the interface.

The work by Robaldo *et al.* (2005) is devoted to the thermoelastic problem in anisotropic composite plates, in which the thermal and mechanical fields are not considered coupled. Instead, the fully coupled thermomechanical problem and mixed thermomechanical elements based on the RMVT have been discussed by Nali *et al.* (2011) and Carrera and Robaldo (2007). Finally, refined and mixed multilayered plate elements were presented by Carrera *et al.* (2009b) and Carrera *et al.* (2009a) for the most general case of coupled magneto-electroelastic problems.

References

Allik H and Hughes TJR 1970 Finite element method for piezoelectric vibration. *International Journal for Numerical Methods in Engineering* **14**(4), 153–160.

Carrera E 1995 A class of two dimensional theories for anisotropic multilayered plates analysis. *Accademia delle Scienze di Torino* **19–20**, 49–87.

Carrera E 2001 Developments, ideas and evaluations based upon the Reissner's mixed theorem in the modeling of multilayered plates and shells. *Applied Mechanics Reviews* **54**, 301–329.

Carrera E 2002 Theories and finite elements for multilayered anisotropic, composite plates and shells. *Archives of Computational Methods in Engineering* **9**, 87–140.

Carrera E 2003 Historical review of zig-zag theories for multilayered plates and shells. *Applied Mechanics Review* **56**, 287–308.

Carrera E and Boscolo M 2007 Classical and mixed finite elements for static and dynamic analysis of piezoelectric plates. *International Journal for Numerical Methods in Engineering* **70**, 1135–1181.

Carrera E, Boscolo M and Robaldo A 2007 Hierarchic multilayered plate elements for coupled multifield problems of piezoelectric adaptive structures: formulation and numerical assessment. *Archives of Computational Methods in Engineering* **14**, 383–430.

Carrera E and Brischetto S 2007 Piezoelectric shell theories with a priori continuous transverse electromechanical variables. *Journal of Mechanics of Materials and Structures* **2**(2), 377–399.
Carrera E, Brischetto S, Fagiano C and Nali P 2009a Mixed multilayered plate elements for coupled magneto-electro-elastic problems. *Multidiscipline Modeling in Materials and Structures* **5**, 251–256.
Carrera E, Brischetto S and Nali P 2008 Variational statements and computational models for multi-field problems and multilayered structures. *Mechanics of Advanced Materials and Structures* **15**(3–4), 182–198.
Carrera E, Buttner A and Nali P 2010 Mixed elements for the analysis of anisotropic multilayered piezoelectric plates. *Journal of Intelligent Material Systems and Structures* **21**, 701–717.
Carrera E and Demasi L 2002 Classical and advanced multilayered plate elements based upon PVD and RMVT. Part 2: numerical implementations. *International Journal for Numerical Methods in Engineering* **55**, 253–291.
Carrera E, Digifico M, Nali P and Brischetto S 2009b Refined multilayered plate elements for coupled magneto-electro-elastic analysis. *Multidiscipline Modeling in Materials and Structures* **5**, 119–138.
Carrera E and Fagiano C 2007 Mixed piezoelectric plate elements with continuous transverse electric displacements. *Journal of Mechanics of Materials and Structures* **2**(3), 421–438.
Carrera E and Nali P 2007 *Description of Mathematica® software model for PVD and RMVT applications, in the framework of the Unified Formulation*. DIASP, Internal Report.
Carrera E and Nali P 2009 Mixed piezoelectric plate elements with direct evaluation of transverse electric displacement. *International Journal for Numerical Methods in Engineering* **80**(4), 403–424.
Carrera E and Nali P 2010a Multilayered plate elements for the analysis of multifield problems. *Finite Elements in Analysis and Design* **46**, 732–742.
Carrera E and Nali P 2010b Classical and mixed finite elements for the analysis of multifield problems and smart layered plates. *Acta Mechanica Solida Sinica* **23**, 115–121.
Carrera E and Robaldo A 2007 Mixed finite elements for thermoelastic analysis of multilayered anisotropic plates. *Journal of Thermal Stresses* **30**, 165–194.
Carrera E and Robaldo A 2010 Hierarchic finite elements based on the Carrera Unified Formulation for the static analysis of shear actuated multilayered piezoelectric plates. *Multidiscipline Modeling in Materials and Structures* **6**, 45–77.
Cinefra M, Carrera E and Valvano S 2013 Refined shell elements for the analysis of multilayered structures with piezoelectric layers. In: *6th ECCOMAS Thematic Conference on Smart Structure and Materials (SMART13)*, Turin, Italy.
Dokmeci MC 1978 Theory of vibration of coated, thermopiezoelectric laminae. *Journal of Mathematical Physics* **19**(1), 109–126.
EerNisse EP 1967 Variational method for electroelastic vibration analysis. *IEEE Transaction on Ultrasonic* **14**(4), 153–160.
Heyliger P 1994 Static behavior of laminated elastic-piezoelectric plates. *AIAA Journal* **32**(12), 2481–2484.
Ikeda T 1996 *Fundamentals of Piezoelectricity*. Oxford University Press.
Miglioretti F and Carrera E 2012 Refined 1D-Elements for multifields analysis. In: *DeMEASS V*, Ulrichsberg, Austria.
Mindlin RD 1952 Forced thickness-shear and flexural vibrations of piezoelectric crystal plates. *Journal of Applied Physics* **22**(1), 83–88.
Mindlin RD 1974 Equation of high frequency vibration of thermopiezoelectric crystal plates. *International Journal of Solids and Structures* **10**, 625–637.
Nali P, Carrera E and Calvi A 2011 Advanced fully coupled thermo-mechanical plate elements for multilayered structures subjected to mechanical and thermal loading. *International Journal for Numerical Methods in Engineering* **85**, 896–919.
Nelson DF 1996 *Fundamentals of Piezoelectricity*. Oxford University Press.
Reissner E 1984 On a certain mixed variational theorem and a proposed application. *International Journal for Numerical Methods in Engineering* **20**, 1366–1368.
Robaldo A, Carrera E and Benjeddou A 2005 Unified formulation for finite element thermoelastic analysis of multilayered anisotropic composite plates. *Journal of Thermal Stresses* **28**, 1031–1064.
Robaldo A, Carrera E and Benjeddou A 2006 A Unified Formulation for finite element analysis of piezoelectric plates. *Computers & Structures* **84**, 1494–1505.
Tiersten HF 1969 *Linear Piezoelectric Plate Vibrations*. Plenum.
Tiersten HF and Mindlin RD 1962 Forced vibrations of piezoelectric crystal plates. *Quarterly of Applied Mathematics* **20**(2), 107–119.
Yang J 1979 *Electric, Optic and Acoustic Interactions in Crystals*. John Wiley & Sons, Inc

Appendix A

Numerical Integration

The solution of a structural problem, through the FEM, requires the governing equations to be written in weak form on the basis of the variational principle, as shown in the previous chapters. Whichever model is used, 1D or 2D and 3D, the solution requires the integral of the shape functions to be evaluated over the physical domain.

Integration is one of the main topics of numerical analysis. The most common approach, which is widely used in FEM solvers, is the use of some interpolating functions defined over a standard domain. Legendre polynomials are used in the case of a finite domain, which is usually defined between -1 and 1, but there are many other solutions, such as Chebyshev or Jacobi polynomials. These polynomials can interpolate the original functions at a finite number of points. If the points are spaced equally, the method is called Newton–Cotes integration. If the points are placed arbitrarily, the process is called Gauss integration.

This appendix has the aim of showing the nature of the numerical integration on the basis of the Gauss–Legendre formula. A more exhaustive description of this approach can be found in classical FEM books such as the one by Zienkiewicz *et al.* (2005), or in the handbook by Abramowitz and Stegun (1964).

A.1 Gauss–Legendre Quadrature

The Gauss–Legendre formula approximates the integral of the function in the $[-1 \leq \xi \leq 1]$ domain with the sum of the values of the function at the ith Gauss point, $f(\xi_i)$, multiplied by a weight, w_i,

$$I_{1D} = \int_{-1}^{1} f(\xi) d\xi \approx \sum_{i=1}^{N} f(\xi_i) w_i \tag{A.1}$$

A 1D case is given in Equation (A.1). The positions and the weights of the Gaussian points are given in Table A.1. If more than seven points are needed, it is best to refer to the handbook

Finite Element Analysis of Structures Through Unified Formulation, First Edition.
Erasmo Carrera, Maria Cinefra, Marco Petrolo and Enrico Zappino.

Table A.1 Gauss points and weights

	Weight	Position
	1 point	
1	0. 000 000 000 000 000	2. 000 000 000 000 000
	2 points	
1	1. 000 000 000 000 000	−0. 577 350 269 189 626
2	1. 000 000 000 000 000	0. 577 350 269 189 626
	3 points	
1	0. 888 888 888 888 889	0. 000 000 000 000 000
2	0. 555 555 555 555 556	−0. 774 596 669 241 483
3	0. 555 555 555 555 556	0. 774 596 669 241 483
	4 points	
1	0. 652 145 154 862 546	−0. 339 981 043 584 856
2	0. 652 145 154 862 546	0. 339 981 043 584 856
3	0. 347 854 845 137 454	−0. 861 136 311 594 053
4	0. 347 854 845 137 454	0. 861 136 311 594 053
	5 points	
1	0. 568 888 888 888 889	0. 000 000 000 000 000
2	0. 478 628 670 499 367	−0. 538 469 310 105 683
3	0. 478 628 670 499 367	0. 538 469 310 105 683
4	0. 236 926 885 056 189	−0. 906 179 845 938 664
5	0. 236 926 885 056 189	0. 906 179 845 938 664
	6 points	
1	0. 360 761 573 048 139	0. 661 209 386 466 264
2	0. 360 761 573 048 139	−0. 661 209 386 466 264
3	0. 467 913 934 572 691	−0. 238 619 186 083 197
4	0. 467 913 934 572 691	0. 238 619 186 083 197
5	0. 171 324 492 379 170	−0. 932 469 514 203 152
6	0. 171 324 492 379 170	0. 932 469 514 203 152
	7 points	
1	0. 417 959 183 673 469	0. 000 000 000 000 000
2	0. 381 830 050 505 119	0. 405 845 151 377 397
3	0. 381 830 050 505 119	−0. 405 845 151 377 397
4	0. 279 705 391 489 277	−0. 741 531 185 599 394
5	0. 279 705 391 489 277	0. 741 531 185 599 394
6	0. 129 484 966 168 870	−0. 949 107 912 342 758
7	0. 129 484 966 168 870	0. 949 107 912 342 758

This table gives the weights and the positions for various Gauss point sets.

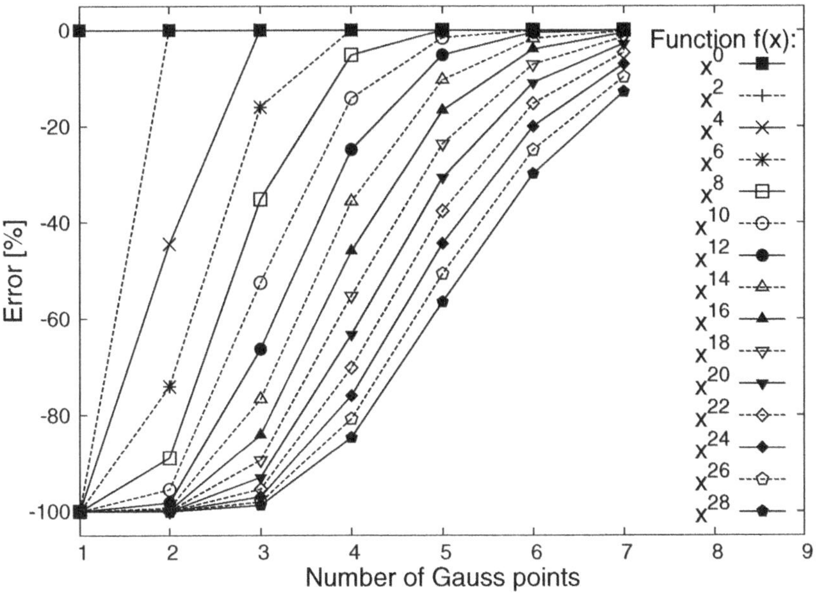

Figure A.1 Gauss quadrature convergence analysis

by Abramowitz and Stegun (1964). Otherwise, the Gauss points can be defined as the roots of the Legendre polynomials,

$$P_n(z) = \frac{1}{2\pi i} \oint (1 - 2tz + t^2)^{-1/2} t^{-n-1} dt \tag{A.2}$$

The weights can be derived using the following equation:

$$w_i = \frac{2}{\left(1 - x_i^2\right) \left[P'_n(x_i)\right]^2} \tag{A.3}$$

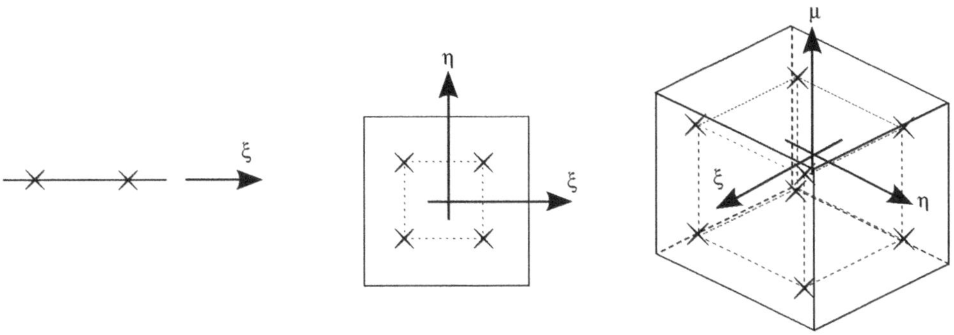

Figure A.2 Example of Gauss point distribution for 1D, 2D and 3D structures

where x_i is the coordinate of the Gauss point and P' is the derivative of the Legendre polynomials in t. The accuracy of the Gauss–Legendre quadrature formula depends on the number of Gauss points, N, used to integrate a function of order M. If N Gauss points are used, the Gauss–Legendre quadrature formula provides exact results for a function with an order of $M = 2N - 1$. Figure A.1 shows the error in the integration of functions with different orders considering different numbers of Gauss points. As an example, if three Gauss points are considered, the error is only zero for functions with an order lower than or equal to five. The Gauss–Legendre quadrature formula can also be used for integration over surfaces or volumes:

$$I_{2D} = \int_{-1}^{1} \int_{-1}^{1} f(\xi, \eta) d\xi \, d\eta \approx \sum_{i=1}^{N} \sum_{j=1}^{N} f(\xi_i, \eta_j) w_i w_j \tag{A.4}$$

$$I_{3D} = \int_{-1}^{1} \int_{-1}^{1} \int_{-1}^{1} f(\xi, \eta, \mu) d\xi \, d\eta d\mu \approx \sum_{i=1}^{N} \sum_{j=1}^{N} \sum_{k=1}^{N} f(\xi_i, \eta_j, \mu_k) w_i w_j w_k \tag{A.5}$$

Equations (A.4) and (A.5) show the 2D and 3D formulations. Index i refers to points in the ξ direction, while index j refers to points in the η direction. Index k instead refers to the point in the μ direction. An example of Gauss point distributions for the 1D, 2D and 3D domains is shown in Figure A.2.

References

Abramowitz M and Stegun IA 1964 *Handbook of Mathematical Functions With Formulas, Graphs, and Mathematical Tables*. NBS Applied Mathematics Series 55, National Bureau of Standards.

Zienkiewicz OC, Taylor RL and Zhu JZ 2005 *The Finite Element Method: Its Basis and Fundamentals*. Sixth Edition. Elsevier.

Appendix B

CUF FE Models: Programming and Implementation Guidelines

CUF models are particularly suitable for the implementation of FE codes. This is due to the hierarchical capabilities of the CUF which lead to the automatic implementation of arbitrary-order models. An FEM code based on the CUF has to deal with a number of specific issues. The main aim of this appendix is to provide some guidelines to deal with the most important programming and implementation issues related to the CUF. In particular, the following main topics will be discussed:

1. the preprocessing;
2. the FEA;
3. the postprocessing of the results.

It should be mentioned that this appendix does not present a detailed and comprehensive guide to implement FE codes; its purpose is instead to describe details related to the CUF FE models. Also, only 1D FE will be considered. Most of the guidelines provided in the following sections can be easily extended to 2D and 3D CUF models.

B.1 Preprocessing and Input Descriptions

Preprocessing includes all those operations required to set up the input data necessary for an FE. Attention is paid here to all the data needed for a CUF FE model analysis. Inputs are split into two categories: typical FE inputs and specific CUF inputs. The former indicate all those inputs commonly used in most FE codes. The latter indicate specific inputs needed for CUF models.

Finite Element Analysis of Structures Through Unified Formulation, First Edition.
Erasmo Carrera, Maria Cinefra, Marco Petrolo and Enrico Zappino.

B.1.1 General FE Inputs

B.1.1.1 Geometry and Mesh Data

The length of the beam (L) defines the geometry of the structure along the beam axis. The FE discretization is carried out along this axis. A mesh is usually defined according to the following:

- the type of beam element (i.e. the number of nodes of each beam element);
- the total number of beam elements (N_{BE}).

These two parameters are sufficient if homogeneous meshes are adopted; this means that constant length beam elements are considered. In this book, three beam elements are adopted:

- the two-node element (B2) based on linear shape functions;
- the three-node element (B3) based on parabolic shape functions;
- the four-node element (B4) based on cubic shape functions.

Node locations and their connectivity must be provided as inputs. This means that the spanwise coordinates of each node and the set of nodes of each beam element must be defined. Example B.1.1 presents different mesh definitions in CUF FEM models together with some comments concerning some critical issues. A convergence analysis is usually performed to define the mesh size of the problem.

Example B.1.1 *Let us consider a beam with $L = 1\,m$ and discretized through two beam elements. Three different meshes have to be used: two B2 elements, two B3s and two B4s. In the case of B2 elements the total number of nodes is equal to*

$$N_{nodes} = (\underbrace{2}_{number\ of\ nodes\ per\ element} - 1) \times \underbrace{2}_{N_{BE}} + 1 = 3 \tag{B.1}$$

Table B.1 shows the node location list. The first column contains the identification number of each node (ID), the second column shows the spanwise locations. Table B.2 presents the connectivity of each B2 element. The first column contains the ID of the element, the other two columns define the local nodes of an element. Figure B.1 explains the difference between global and local nodes; that is, between the global node numbering and the local one.

Table B.1 Node location list of the B2 mesh model

Node ID	y coordinate
1	0.0
2	0.5
3	1.0

This table presents the sample B2 mesh model node list related to Example B.1.1.

Table B.2 Connectivity of the B2 mesh model

Element ID	Node 1	Node 2
1	1	2
2	2	3

This table presents the sample B2 mesh model connectivity related to Example B.1.1.

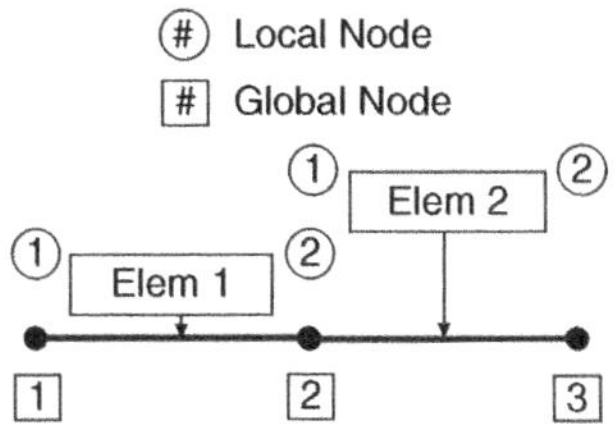

Figure B.1 Global and local node for a B2 element

If two B3 elements are used the total number of nodes is given by

$$N_{nodes} = (\underbrace{3}_{number\ of\ nodes\ per\ element} - 1) \times \underbrace{2}_{N_{BE}} + 1 = 5 \tag{B.2}$$

Tables B.3 and B.4 report the node locations and the connectivity, respectively. If two B4 elements are used, the total number of nodes is given by

$$N_{nodes} = (\underbrace{4}_{number\ of\ nodes\ per\ element} - 1) \times \underbrace{2}_{N_{BE}} + 1 = 7 \tag{B.3}$$

Tables B.5 and B.6 report the node locations and the connectivity, respectively, whereas Figures B.2 and B.3 show global and local nodes for B3 and B4 meshes, respectively.

Table B.3 Node location list of the B3 mesh model

Node ID	*y* coordinate
1	0.0
2	0.5
3	1.0
4	0.25
5	0.75

This table presents the sample B3 mesh model node list related to Example B.1.1.

Table B.4 Connectivity of the B3 mesh model

Element ID	Node 1	Node 2	Node 3
1	1	2	4
2	2	3	5

This table presents the sample B3 mesh model connectivity related to Example B.1.1.

Table B.5 Node location list of the B4 mesh model

Node ID	y coordinate
1	0.0
2	0.5
3	1.0
4	0.17
5	0.33
6	0.67
7	0.83

This table presents the sample B4 mesh model node list related to Example B.1.1.

Table B.6 Connectivity of the B4 mesh model

Element ID	Node 1	Node 2	Node 3	Node 4
1	1	2	4	5
2	2	3	6	7

This table presents the sample B4 mesh model connectivity related to Example B.1.1.

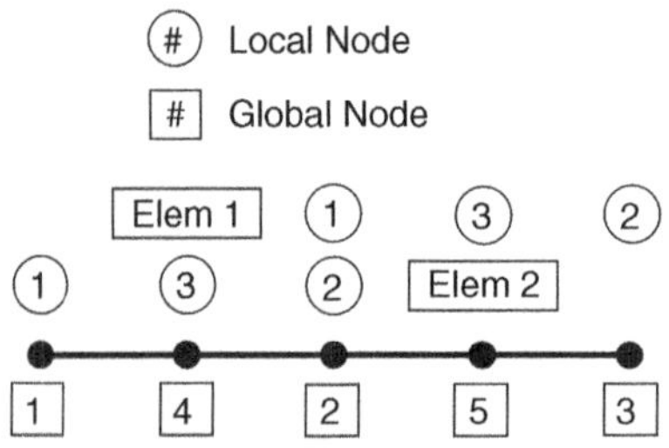

Figure B.2 Global and local node for a B3 element

B.1.1.2 Loads and BCs

Many different loading conditions can be adopted in the framework of an FEA, including mechanical, thermal and inertial loads, among others. A detailed analysis of loading conditions is not the aim of this book, but can be found in the excellent books by Bathe (1996) and Oñate (2009). If we consider a point load, we will need to define the following parameters:

- application point coordinates, $[x_p, y_p, z_p]$;
- load magnitude;
- load direction.

In the following sections, the procedure for computing the equivalent force nodal vector will be described.

Different types of constraints can be applied in the CUF beam model, the most important being clamped point, hinged, and hinged with free horizontal translation, which are graphically shown in Figures B.4, B.5 and B.6, respectively. In a TE model, a constraint is applied to the entire cross-section of the node; in an LE model, constraints are applied to the single Lagrange point above the cross-section. All the displacement components are locked in a clamped point;

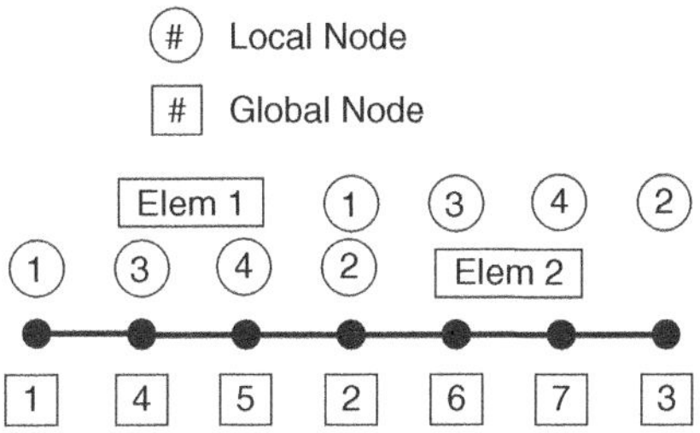

Figure B.3 Global and local node for a B4 element

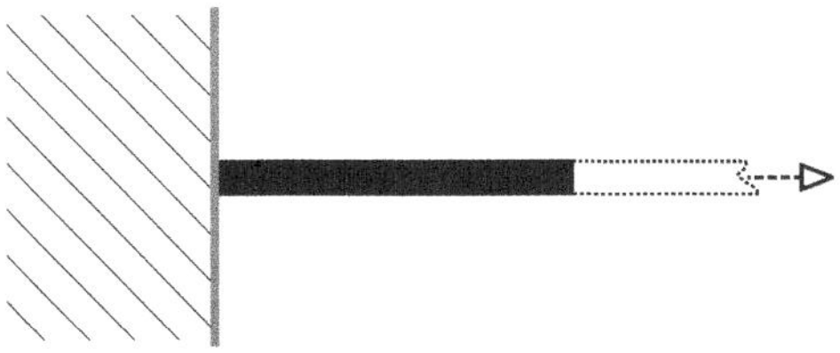

Figure B.4 A clamped node

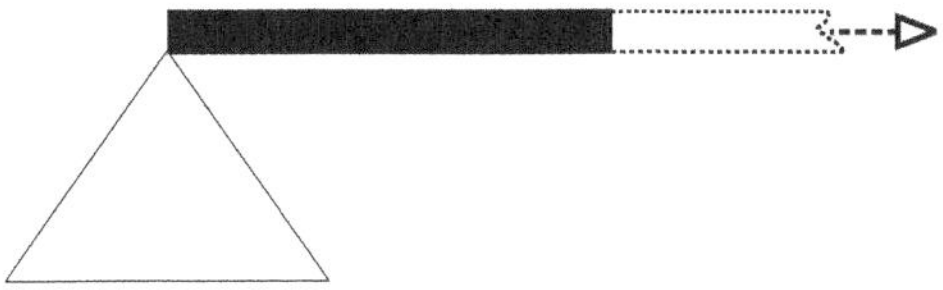

Figure B.5 A hinged node

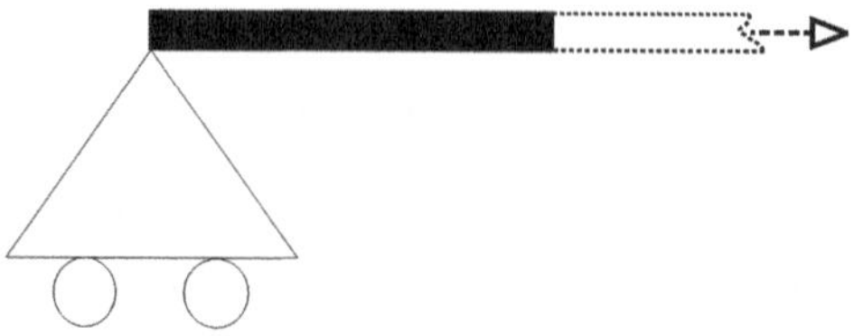

Figure B.6 A roll-hinged node

rotations are only allowed in a hinged point. If necessary, it is also possible to constrain given the DOFs of a certain point, as is usual in FEA. The constraining technique in the CUF will be discussed in the following sections.

B.1.1.3 Material Properties

The definition of the material properties depends to a great extent on what kind of material is adopted, e.g. isotropic, orthotropic, cross-ply laminates. In this book, the analyses were all conducted on isotropic structures. This makes the definition of the material characteristics particularly easy since only Young's modulus (E) and Poisson's ratio (ν) are needed.

B.1.1.4 Type of Analysis

This input defines the analysis that has to be conducted. In the present book two types of structural analysis were conducted, the linear static analysis and the linear free vibration analysis. The definition of the analysis determines the FE matrices that have to be computed and the output files to be considered. In the case of linear static analysis, the stiffness matrix and the loading vector will be computed and the displacement, strain and stress vectors are the typical outputs. As far as the free vibration analysis in concerned, the stiffness and mass matrices will be computed for the eigenvalue analysis. Natural frequencies and modal shapes will be provided as output data.

B.1.2 Specific CUF Inputs

This section deals with the description of the input data specifically needed for a CUF FE model.

B.1.2.1 Order of the Beam Model

The free choice of the order of the structural model is the most innovative feature of CUF models. The order N is set as an input and the analysis is then carried out by assembling all the FE matrices related to the order chosen. Classical models (EBBT and TBT) can also be adopted and they are implemented as particular cases of the $N = 1$ model. As an example, if N is set equal to two, the analysis will be conducted by means of the following beam model:

$$\begin{aligned}u_x &= u_{x_1} + x\,u_{x_2} + z\,u_{x_3} + x^2\,u_{x_4} + xz\,u_{x_5} + z^2\,u_{x_6}\\ u_y &= u_{y_1} + x\,u_{y_2} + z\,u_{y_3} + x^2\,u_{y_4} + xz\,u_{y_5} + z^2\,u_{y_6}\\ u_z &= u_{z_1} + x\,u_{z_2} + z\,u_{z_3} + x^2\,u_{z_4} + xz\,u_{z_5} + z^2\,u_{z_6}\end{aligned} \tag{B.4}$$

In the CUF, reduced higher-order models can be implemented. In other words, it is possible to choose which terms of a refined model have to be considered for the analysis. This means that the set of displacement variables that has to be retained can be defined as input. For instance, the following beam model can be chosen automatically:

$$\begin{aligned}u_x &= u_{x_1} + x\,u_{x_2} + x^2\,u_{x_4} + xz\,u_{x_5}\\ u_y &= z\,u_{y_3} + x^2\,u_{y_4} + xz\,u_{y_5} + z^2\,u_{y_6}\\ u_z &= u_{z_1} + x\,u_{z_2} + z\,u_{z_3} + z^2\,u_{z_6}\end{aligned} \tag{B.5}$$

The choice of the beam model implies the definition of the model variables and the total number of unknowns of the problem; that is, the computational cost of the analysis. The proper choice of the order is, in general, problem dependent. A convergence analysis is usually conducted to define the order of the beam model to be used, as commonly done to define the mesh size of the problem.

Example B.1.2 *Let us consider a cantilever beam having a square cross-section of edge length equal to* 0.2 *m made of isotropic material with $E = 75$ GPa and $v = 0.33$. The structure is loaded by a vertical force equal to* −50 *N. A convergence analysis on the vertical displacement of the loading point regarding both mesh and beam order has to be performed for two values of the beam length L, i.e.* 100 *and* 10*.*

Tables B.7 and B.8 show the results of the convergence study for the slender and the moderately thick beam, respectively. Rows are related to different meshes while columns are related to the expansion order. Figure B.7 shows the convergence study for B4 elements. Figure B.8 shows the effect of the beam element type on the convergence. This convergence study highlights some fundamental and typical aspects related to the use of refined models:

- *The refinement of the mesh and of the beam model leads to more flexible structures, i.e., to larger deflections.*
- *Slender structures are well modelled by lower-order models, while thick beams need refined theories.*
- *Linear and classical models tend to have their own convergence behaviour that is, in general, different from those of higher-order models. This is due to the Poisson locking correction that artificially enhance the flexibility of the beam.*
- *The Poisson locking correction can lead to linear models providing solutions that are different from those of higher-order models.*

In the case of free vibration analyses, convergence studies can be carried out on natural frequencies leading to qualitatively similar results (Carrera et al. *2010b).*

Table B.7 Doubly parametric convergence analysis for a slender beam (Carrera *et al.* 2010a)

No. elem.	EBBT	TBT	$N = 1$	$N = 2$
		$u_z \times 10^2$ m		
		B2		
1	−1.001	−1.001	−1.001	−0.893
3	−1.297	−1.297	−1.297	−1.165
5	−1.321	−1.321	−1.321	−1.236
10	−1.331	−1.331	−1.331	−1.287
40	−1.333	−1.333	−1.333	−1.323
		B3		
1	−1.333	−1.333	−1.333	−1.158
3	−1.333	−1.333	−1.333	−1.275
5	−1.333	−1.333	−1.333	−1.298
10	−1.333	−1.333	−1.333	−1.316
40	−1.333	−1.333	−1.333	−1.330
		B4		
1	−1.333	−1.333	−1.333	−1.239
3	−1.333	−1.333	−1.333	−1.302
5	−1.333	−1.333	−1.333	−1.315
10	−1.333	−1.333	−1.333	−1.325
40	−1.333	−1.333	−1.333	−1.332

This table shows the convergence analysis for a slender compact square beam by considering both the number of mesh elements and the order of the beam model.

Table B.8 Doubly parametric convergence analysis for a moderately thick beam (Carrera *et al.* 2010a)

No. elem.	EBBT	TBT	$N = 1$	$N = 2$	$N = 3$	$N = 4$
			$u_z \times 10^5$ m			
			B2			
1	−1.001	−1.010	−1.010	−0.902	−0.904	−0.904
3	−1.297	−1.306	−1.306	−1.173	−1.176	−1.176
5	−1.321	−1.330	−1.330	−1.244	−1.246	−1.246
10	−1.331	−1.340	−1.340	−1.293	−1.296	−1.296
40	−1.333	−1.343	−1.343	−1.325	−1.327	−1.328
			B3			
1	−1.333	−1.343	−1.343	−1.166	−1.168	−1.168
3	−1.333	−1.343	−1.343	−1.283	−1.285	−1.285
5	−1.333	−1.343	−1.343	−1.305	−1.307	−1.307
10	−1.333	−1.343	−1.343	−1.321	−1.323	−1.324
40	−1.333	−1.343	−1.343	−1.329	−1.331	−1.333
			B4			
1	−1.333	−1.343	−1.343	−1.248	−1.250	−1.250
3	−1.333	−1.343	−1.343	−1.309	−1.311	−1.311
5	−1.333	−1.343	−1.343	−1.320	−1.322	−1.323
10	−1.333	−1.343	−1.343	−1.327	−1.329	−1.330
40	−1.333	−1.343	−1.343	−1.330	−1.332	−1.333

This table shows the convergence analysis for a moderately thick compact square beam by considering both the number of mesh elements and the order of the beam model.

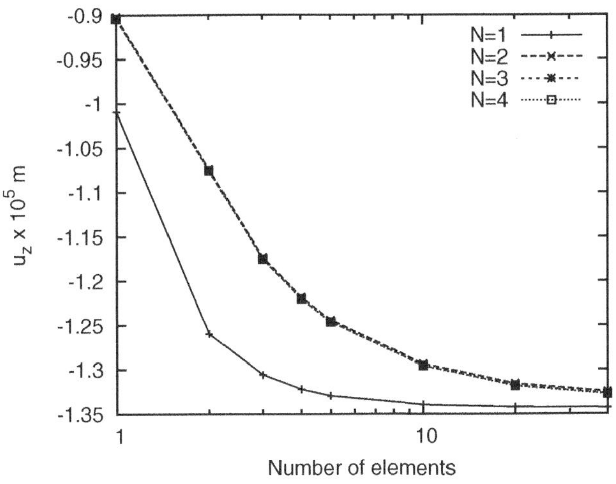

Figure B.7 Convergence study for different beam models and meshes via B4 elements for a moderately thick beam

B.1.2.2 Cross-section Geometry

The definition of the cross-section geometry is another important input in a CUF beam FE model. The present formulation is able to deal with arbitrary cross-section geometries. The cross-section geometry is directly involved in the computation of the FE matrix terms where surface integrals have to be computed,

$$\int_A F_\tau(x,z) \cdot F_s(x,z) dx\, dz$$

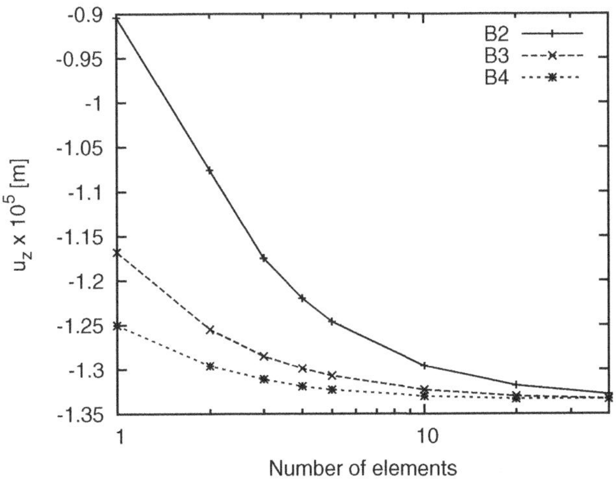

Figure B.8 Convergence study for different meshes via a $N = 4$ model for a moderately thick beam

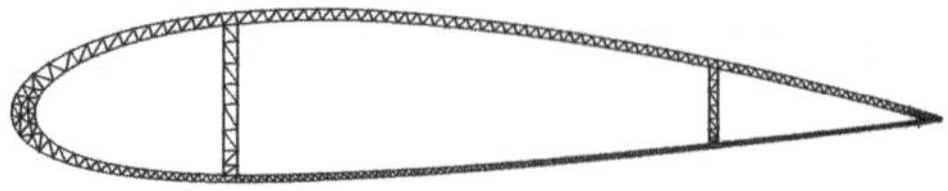

Figure B.9 Triangular mesh for an aerofoil cross-section having three cells

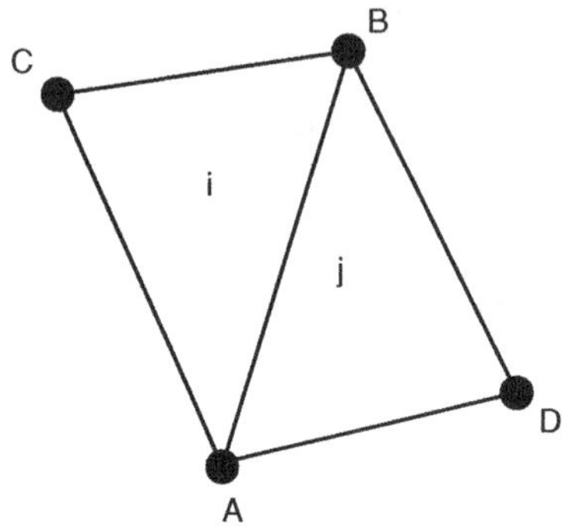

Figure B.10 Two triangular elements of a cross-section mesh

These integrals are computed numerically using a technique that will be described in the following sections. As far as the related input data are concerned, numerical computation of the surface integral requires the discretization of the cross-section in a number of triangular elements. Such a discretization can be performed using a common FE preprocessor. The data needed in the present FE code are the coordinates of each node above the cross-section and the connectivity of the triangular elements. An example of a numerical mesh for a three-cell aerofoil-shaped beam cross-section is shown in Figure B.9. The number of triangular elements has to be chosen via a convergence study of the integrals. In general, the higher the order of polynomials in the integral, the finer the mesh has to be. Examples of convergence studies on the numerical cross-section mesh will be given below. Figure B.10 shows two generic triangular elements (i and j). Table B.9 presents a list of the node coordinates and Table B.10 contains a connectivity list. These data have to be provided for all the elements of the cross-section numerical mesh.

Table B.9 Node location list of the triangular numerical cross-section mesh

Node ID	x coordinate	z coordinate
A	x_A	z_A
B	x_B	z_B
C	x_C	z_C
D	x_D	z_D

This table presents an example of input data to define the nodes above the cross-section numerical mesh, see also Figure B.10.

Table B.10 Connectivity of the triangular cross-section numerical mesh

Element ID	Node 1	Node 2	Node 3
i	A	B	C
j	A	D	B

This table presents an example of connectivity input for the triangular elements of the cross-section numerical mesh as in Figure B.10.

B.2 FEM Code

This section presents guidelines that can be used to build an FE code based on CUF beam models. As in the previous sections of this chapter, attention is focused on particular issues related to CUF implementation; the aim is not to give a comprehensive FE programming guide.

B.2.1 Stiffness and Mass Matrices

Computation of the stiffness and mass matrices is described by means of the nucleus-based approach. Critical issues related to the numerical computation of the integrals involved in the matrices are also discussed.

B.2.1.1 Nucleus-Based Implementation

The CUF beam model is based on a hierarchical implementation of the FE matrices which are expressed in terms of the so-called FNs. These nuclei are formally independent of the order of the beam model. The stiffness and mass matrix components are expressed as

$$k_{lk}^{\tau sij}, \quad l,k = x,y,z$$

$$m_{lk}^{\tau sij}, \quad l,k = x,y,z$$

It is clear that computation of the nodal matrices has to be performed by means of the six indexes, namely i, j, τ, s, l and k:

- i and j are the shape function indexes, and depend on the type of beam element adopted. They vary from one to the number of nodes per element (two for B2, three for B3 and four for B4).
- τ and s are the expansion function indexes, and depend on the beam model order (N). They vary from one to $(N + 1) \times (N + 2)/2$.
- l and k are the indexes needed to define all nine components of the nucleus, l, $k = x$, y, z.

The hierarchical structure of CUF FE models is described in Figure B.11. The outer cycles are related to the beam mesh characteristics, whereas the inner ones are related to the beam

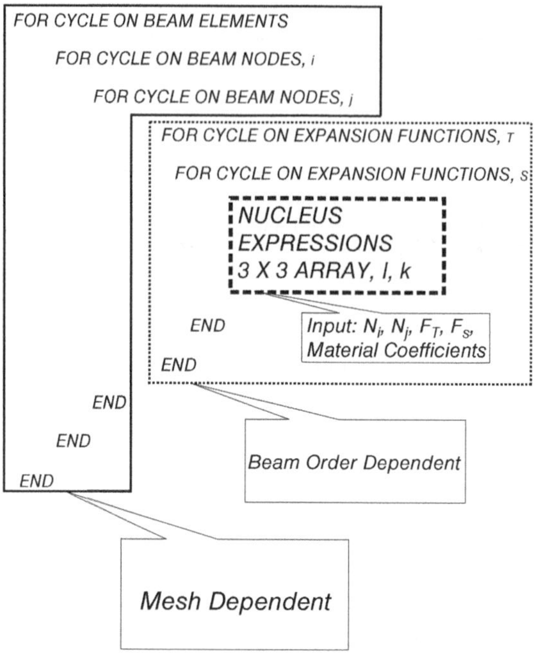

Figure B.11 Graphic description of the hierarchical structure of the nucleus-based formulation

expansion order. The core of the procedure is represented by the computation of the nucleus components, which involves line and surface integrals

$$\int_l N_i N_j dy$$

$$\int_A F_\tau(x, z) \cdot F_s(x, z) dx\, dz$$

These integrals are computed numerically; the adopted computational techniques are described in the following sections.

B.2.1.2 Line Integrals

Line integrals of shape functions are evaluated numerically by means of the Gauss quadrature technique shown in Appendix A. Line integrals are computed over each element; that is, each integral domain is defined by the element boundaries. It is convenient to integrate along the $-1 \leq \eta \leq +1$ domain (commonly referred to as the natural domain of the element). The Jacobian (J_E) of the element has to be computed,

$$\int_{l_E} f(y) dy = \int_{-1}^{+1} J_E(\eta) f(\eta) d\eta \simeq \sum_{i=1}^{k} J_E(\eta_i) f(\eta_i)\, W_i \tag{B.6}$$

In the FE isoparametric formulation, the Jacobian of the transformation is given by

$$J_E(\eta) = \frac{dy}{d\eta} = \sum_{i=1}^{N_{NE}} \frac{dN_i(\eta)}{d\eta}\, y_i \tag{B.7}$$

where N_{NE} is the number of nodes per element, N_i are the shape functions and y_i are the node locations. Another useful formula is related to the derivatives of the functions expressed in the global and natural domains,

$$\frac{d(f(y))}{dy} = \frac{1}{J_E}\frac{df(\eta)}{d\eta} \tag{B.8}$$

Example B.2.1 *Let us reconsider the data from Example 8.4.1, where the line integrals now have to be computed by means of the Gauss quadrature formula. The shape functions in natural coordinates are given by*

$$N_1 = 1 - \eta, \quad N_2 = \eta$$

The nodes of the B2 element are located at $y = 0, L$, and the Jacobian of the transformation is given by

$$J = \sum_{i=1}^{2} \frac{dN_i(\eta)}{d\eta}\, y_i = 1 \times 0 + 1 \times L = L$$

The integrals to be computed are related to the following stiffness matrix component:

$$\begin{aligned} k_{xx}^{2212} = {} & C_{22} \int_{-a}^{+a}\int_{-b}^{+b} F_{2,x}\, F_{2,x} dx\, dz \int_0^L N_1\, N_2 dy \\ & + C_{66} \int_{-a}^{+a}\int_{-b}^{+b} F_{2,z}\, F_{2,z} dx\, dz \int_0^L N_1\, N_2 dy \\ & + C_{44} \int_{-a}^{+a}\int_{-b}^{+b} F_2\, F_2 dxdz \int_0^L N_{1,y}\, N_{2,y} dy \end{aligned}$$

and in natural coordinates

$$\begin{aligned} k_{xx}^{2212} = {} & C_{22} \int_{-a}^{+a}\int_{-b}^{+b} 1 \cdot 1 dx\, dz \int_{-1}^{+1} L\ (1-\eta)\ \eta d\eta \\ & + C_{66} \int_{-a}^{+a}\int_{-b}^{+b} 0 \cdot 0 dx\, dz \int_{-1}^{+1} L\ (1-\eta)\ \eta d\eta \\ & + C_{44} \int_{-a}^{+a}\int_{-b}^{+b} x \cdot x dx\, dz \int_{-1}^{+1} \left(- \underbrace{\frac{1}{L}}_{1/J_E} \right) \underbrace{\frac{1}{L}}_{1/J_E} d\eta \end{aligned}$$

The polynomials to be integrated are of second order, therefore a second-order quadrature is needed

$$\int_{-1}^{+1} f(\eta) = f(0) \times 2.0 + f(+0.577\,350\,269\,2) \times 1.0 + f(-0.577\,350\,269\,2) \times 1.0$$

The final result is exactly the same as the analytical one,

$$k_{xx}^{2212} = \frac{2}{3}\, C_{22}\; a\; b\; L - \frac{4}{3}\, C_{44}\, \frac{b\, a^3}{L}$$

As mentioned in Chapter 8, shear locking has to be considered and the selective integration should be carried out to reduce its detrimental effects.

B.2.1.3 Surface Integrals

Surface integrals of F_τ are evaluated numerically by discretizing the integration area into a certain number of subdomains,

$$\int_A F_\tau(x,z)\; F_s(x,z) dx\, dz \simeq \sum_{m=1}^{M} [F_\tau(x_m, z_m)\; F_s(x_m, z_m)]\; A_m \tag{B.9}$$

where x_m and z_m are the coordinates of the centre of A_m. M is evaluated through a convergence study. This numerical technique is adopted in order to analyse arbitrary cross-section geometries through TE models since analytical solutions of the surface integrals might not be efficient from a numerical point of view. The subdomains can have arbitrary geometries. A triangular shape is usually preferred since it permits a better partitioning of irregular cross-sections (e.g. annular or aerofoil shaped).

The area of each triangular element (A_m) can be computed starting from the vertices of the triangle,

$$A_m = \frac{1}{2}\left[\det \begin{pmatrix} x_{1m} & x_{2m} & x_{3m} \\ z_{1m} & z_{2m} & z_{3m} \\ 1 & 1 & 1 \end{pmatrix}\right] \tag{B.10}$$

where the first two rows of the matrix contain the coordinates of the three vertices of the triangle. The centre point of each triangle is computed as

$$\begin{aligned} x_m &= \frac{x_{1m} + x_{2m} + x_{3m}}{3} \\ z_m &= \frac{z_{1m} + z_{2m} + z_{3m}}{3} \end{aligned} \tag{B.11}$$

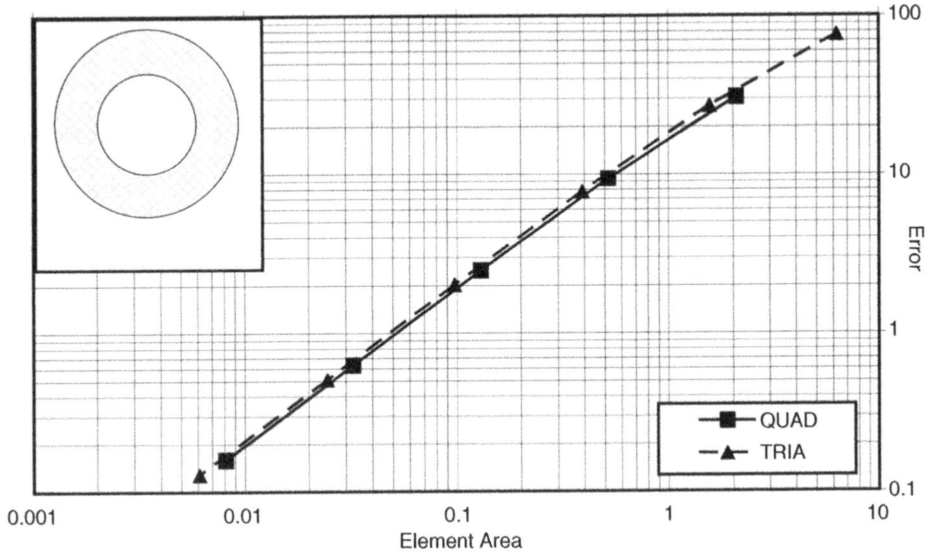

Figure B.12 Convergence study of $\int_A x^8 dx\, dz$ above an annular domain

Figure B.12 shows a typical convergence study on a numerical mesh of an annular cross-section. The integral computed is

$$\int_A x^8 dx\, dz \tag{B.12}$$

Both quadrilateral and triangular mesh elements are used. The horizontal axis shows the ratio between the mean area of the elements and the total area of the cross-section. The vertical axis shows the error with respect to the exact solution, which, in this case, can be easily obtained.

Example B.2.2 *This example aims to provide reference numerical integral data for comparison purposes. Two cross-section geometries are considered and different order integrals are computed. The first cross-section is shown in Figure B.13, where the annular cross-section has an outer radius equal to* 2 m *and a thickness equal to* 0.02 m*. Numerical integrals (Table B.11) are computed via a mesh composed of* 10^5 *triangular meshes. The second cross-section is shown in Figure B.14; the aerofoil-shaped cross-section was built using the NACA profile 2415 with unit chord and two vertical walls located at 25% and 75% chordwise. Numerical integrals (Table B.12) are computed via a mesh composed of* 1.5×10^5 *triangular meshes.*

B.2.2 Stiffness and Mass Matrix Numerical Examples

Some numerical examples of stiffness and mass matrices are provided in this section in order to generate reference data that can be used for comparison purposes.

The first data set is related to a simply supported beam with an annular cross-section. The outer diameter of the cylinder (d) is equal to 2 m, and $L/d = 10$. The thin-walled structure

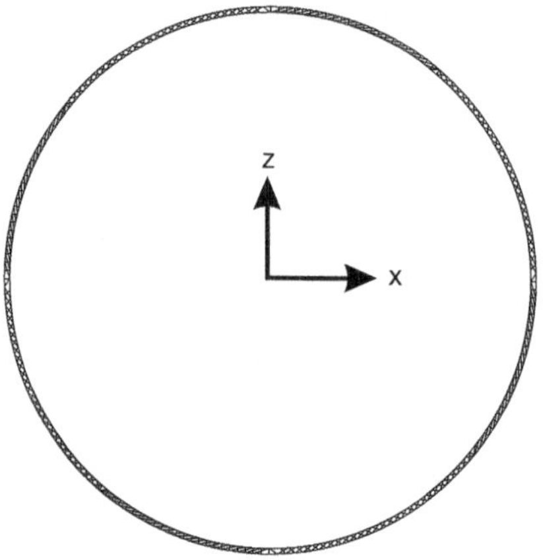

Figure B.13 Annular cross-section numerical mesh

Table B.11 Numerical integrals of the annular cross-section

$\int_A$	1	x	z	x^2
1	1.244×10^{-1}	0.0	0.0	6.097×10^{-2}
x	0.0	6.097×10^{-2}	0.0	0.0
z	0.0	0.0	6.097×10^{-2}	0.0
x^2	6.097×10^{-2}	0.0	0.0	4.483×10^{-2}

This table presents a set of cross-section integrals computed numerically, where each value is related to the integral of the product between the row and column functions.

Figure B.14 Aerofoil-shaped cross-section numerical mesh

Table B.12 Numerical integrals of the airfoil-shaped cross-section

$\int_A$	x^3	$x^2 z$	$x z^2$	z^3
x^3	5.371×10^{-5}	-1.027×10^{-6}	1.870×10^{-7}	1.856×10^{-9}
$x^2 z$	-1.027×10^{-6}	1.870×10^{-7}	1.856×10^{-9}	5.056×10^{-9}
$x z^2$	1.870×10^{-7}	1.856×10^{-9}	5.056×10^{-9}	4.734×10^{-10}
z^3	1.856×10^{-9}	5.056×10^{-9}	4.734×10^{-10}	7.518×10^{-10}

This table presents a set of cross-section integrals computed numerically, where each value is related to the integral of the product between the row and column functions.

is 0.02 m thick. A 10 B4 element mesh is adopted with an $N = 1$ model. The material is isotropic with $E = 75$ GPa, $\nu = 0.33$ and the density $\rho = 2700\,\text{kg/m}^3$. The surface integrals are computed numerically by means of a 10^5 triangular numerical mesh. Some components of the stiffness matrix are reported, in particular the terms related to the first B4 element is considered for $i = j$. The stiffness matrix components are then placed between the 10th and 18th rows and columns:

	$\tau = 1$			$\tau = 2$	$\tau = 3$
$s = 1$	1.894×10^{10}	0.0	0.0	...	...
	0.0	5.039×10^{10}	0.0		
	0.0	0.0	1.894×10^{10}		
$s = 2$		...		...	...
$s = 3$		...		...	...

$i = j = 2$, 1st *element*

The first bending frequency for this structural model is equal to 14.182 Hz (Carrera *et al.* 2010c).

The mass matrix is now considered for an $N = 2$ model. In this case, components related to second-order terms are considered (τ, $s = 4, 5, 6$). These terms are placed between the 19th and 36th rows and columns since the first B4 element is again considered with $i = j = 2$:

	$\tau = 4$	$\tau = 5$			$\tau = 6$
$s = 4$	...		...		...
$s = 5$	...	3.112×10^1	0.0	0.0	...
		0.0	$3.112 \times 10^1 \times 10^{10}$	0.0	
		0.0	0.0	3.112×10^1	
$s = 6$	...		...		...

$i = j = 2$, 1st *element*

The non-null values have the same value because of the symmetry of the structure; they are in fact given by the following integrals:

$\rho \int_l N_2 N_2 dl \int_A x^4 dx\, dz$	0.0	0.0
0.0	$\rho \int_l N_2 N_2 dl \int_A x^2 z^2 dx\, dz$	0.0
0.0	0.0	$\rho \int_l N_2 N_2 dl \int_A z^4 dx\, dz$

$i = j = 2$, 1st *element*

In this case the first bending frequency is equal to 14.185 Hz. A higher frequency value is found than for the $N = 1$ case; that is, a stiffer structure is modelled by the parabolic model.

This phenomenon is due to the aforementioned Poisson locking correction effect, which can make linear models less stiff than second-order ones.

An $N = 7$ model (Carrera *et al.*, 2010c) is considered with τ and s varying from 1 to 36 and with 108 DOFs per node. Some mass matrix values of this model are reported,

	$\tau = 34$	$\tau = 35$			$\tau = 36$
$s = 34$	...		...		...
$s = 35$	...	3.640	0.0	0.0	...
		0.0	3.640	0.0	
		0.0	0.0	3.640	
$s = 36$	...		...		...

$i = j = 2$, 1st *element*

which correspond to

$\rho \int_l N_2 N_2 dl \int_A x^4 z^{10} dx\, dz$	0.0	0.0
0.0	$\rho \int_l N_2 N_2 dl \int_A x^2 z^{12} dx\, dz$	0.0
0.0	0.0	$\rho \int_l N_2 N_2 dl \int_A z^{14} dx\, dz$

$i = j = 2$, 1st *element*

B.2.3 Constraints and Reduced Models

This section illustrates the technique used to impose constraints on the structure and to create reduced higher-order beam models. The methodology is based on a penalty technique (Bathe, 1996), which acts on the stiffness matrix. The procedure is based on the following steps:

- The DOFs to be constrained have to be chosen.
- The nodes to be constrained have to be selected.
- All the *diagonal terms* of the selected node stiffness matrix related to the chosen DOFs have to be penalized with a penalty value (Π).

The value assigned to Π should be chosen via a convergence study. However, one reliable way of choosing it is to employ the maximum value of the stiffness matrix, K_{max}:

$$\Pi \geq 10^3 \times K_{max} \tag{B.13}$$

In each diagonal position K_{ii}, which has to be penalized, the following substitution has to be made:

$$K_{ii} \rightarrow \Pi$$

If reduced models have to be used, all the nodes will be affected by the penalty (Carrera and Petrolo 2011). This means that, to obtain a displacement model, such as

$$
\begin{aligned}
u_x &= x\, u_{x_2} + x^2\, u_{x_4} + z^2\, u_{x_6} \\
u_y &= u_{y_1} + x\, u_{y_2} + xz\, u_{y_5} \\
u_z &= x\, u_{z_2} + z\, u_{z_3} + xz\, u_{z_5}
\end{aligned}
$$

the following procedure is needed:

1. A second-order beam model has to be implemented.
2. The diagonal terms corresponding to the generalized variables u_{x_1}, u_{x_3}, u_{x_5}, u_{y_3}, u_{y_4}, u_{y_6}, u_{z_1}, u_{z_4} and u_{z_6} have to be penalized in all the nodes of the FE model.

Example B.2.3 *Let us consider a node of an $N = 1$ model and three different constraint options: clamped, hinged, and hinged with horizontal translation allowed. The proper positions of the penalty value (Π) within the nodal stiffness matrix have to be calculated for all three constraint cases. A full linear model has nine generalized displacement variables per node; that is, the nodal stiffness matrix is a 9×9 array. Penalties have to be inserted on the diagonal terms only, thus $i = j$ and $\tau = s$. In the case of the clamped node, all the displacement variables have to be penalized:*

	$\tau = 1$			$\tau = 2$			$\tau = 3$			
	Π									
$s = 1$		Π			…			…		
			Π							
				Π						
$s = 2$		…			Π			…		
						Π				
							Π			
$s = 3$		…			…			Π		
									Π	$i=j$

If the node has to be hinged, the rotations with respect to the x- and z-axes have to be unconstrained:

	$\tau = 1$			$\tau = 2$			$\tau = 3$			
	Π									
$s = 1$		Π			…			…		
			Π							
				Π						
$s = 2$		…						…		
						Π				
							Π			
$s = 3$		…			…					
									Π	$i=j$

If a rolling supported node has to be implemented, the translation along the axial direction has to be left free as well:

	$\tau = 1$	$\tau = 2$	$\tau = 3$
$s = 1$	Π Π	...	...
$s = 2$	...	Π Π	...
$s = 3$	...	...	Π Π $i=j$

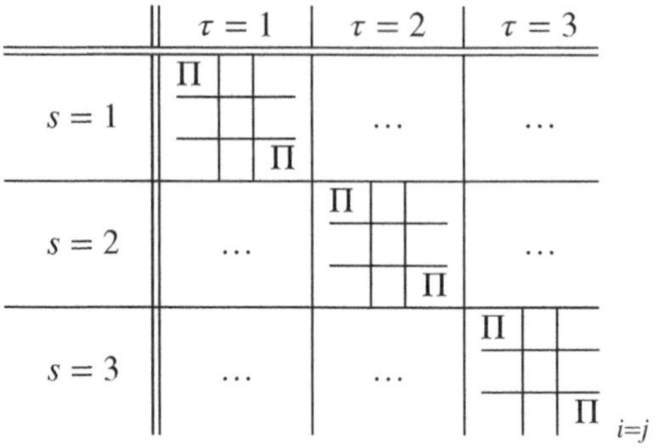

Example B.2.4   *Let us consider the following reduced beam model:*

$$u_x = x\, u_{x_2} + z\, u_{x_3}$$
$$u_y = u_{y_1} + x\, u_{y_2} + z\, u_{y_3}$$
$$u_z = x\, u_{z_2} + z\, u_{z_3}$$

This model has to be implemented via the penalty technique. A first-order model has to be chosen, therefore the nodal stiffness matrix is a 9×9 array. All the diagonal terms related to the variables u_{x_1} and u_{z_1} have to be penalized for all the nodes

	$s = 1$	$s = 2$	$s = 3$
$\tau = 1$	Π	...	...
$\tau = 2$	...		...
$\tau = 3$	...	...	Π $i=j$, $\forall$ nodes

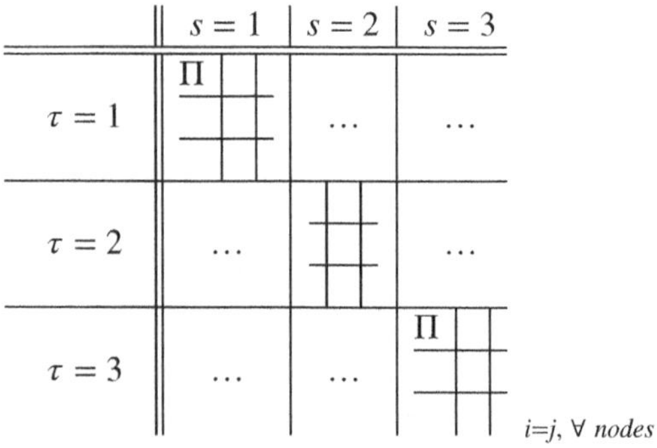

This operation can be easily carried out by rewriting the FE code where an 'if' structure similar to the following has to be inserted:

```
IF (i == j) .AND. ( tau == s) .AND.
& ( tau == 1 .OR. tau == 3) THEN
    K(1,1) = PI
ENDIF
```

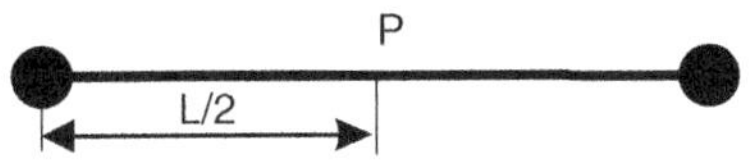

Figure B.15 B2 element loaded at $L/2$

B.2.4 Load Vector

Computation of the equivalent nodal load vector is another task where CUF models lead to specific procedures. Higher-order terms play a fundamental role in determining this vector. As in a general FEM model, some steps have to be taken in order to construct the force vector:

1. The type of load has to be chosen (e.g. concentrated, surface, inertial load, etc.).
2. The portion of the structure to be loaded has to be indicated.
3. The elements in the loading region have to be identified.
4. The loading vector is then built by means of the PVD.

The aim of this section is to portray the role of higher-order terms in the load vector. Example B.2.5 presents a comprehensive set of loading cases that are able to emphasize the importance of refined theories.

Example B.2.5 *Let us consider a B2 element having a square cross-section and loaded by two opposite forces of equal magnitude as shown in Figures B.15 and B.16. The equivalent nodal vector has to be computed for TBT and N* = 1 *models. The load vector is computed by exploiting the PVD,*

$$\delta L_{ext} = \boldsymbol{P}\,\delta\boldsymbol{u} = \boldsymbol{P}\,F_\tau\,N_i\,\boldsymbol{u}_{\tau i} \tag{B.14}$$

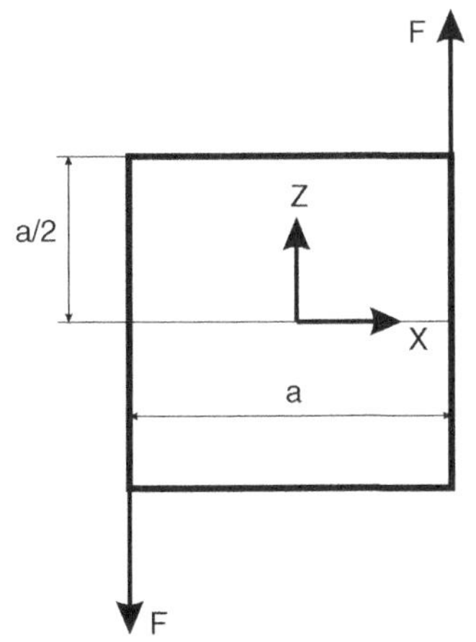

Figure B.16 Square cross-section torsion load for the B2 element

The shape functions at the loading point are equal to

$$N_1\left(\frac{1}{2}\right) = N_2\left(\frac{1}{2}\right) = \frac{1}{2}$$

P acts in the z direction; that is, only the u_z components of the load vector are involved. In the case of TBT, the virtual variation of the external work is given by

$$\begin{aligned}
\delta L_{ext} &= P\,N_1\left(\frac{1}{2}\right) \underbrace{F_1}_{F_1=1}(a/2)\;u_{z_{11}} + P\,N_2\left(\frac{1}{2}\right) F_1\,(a/2)\;u_{z_{12}} \\
&\quad - P\,N_1\left(\frac{1}{2}\right) F_1\,(-a/2)\;u_{z_{11}} - P\,N_2\left(\frac{1}{2}\right) F_1\,(-a/2)\;u_{z_{12}} \\
&= \frac{1}{2}\,P\,u_{z_{11}} + \frac{1}{2}\,P\,u_{z_{12}} - \frac{1}{2}\,P\,u_{z_{11}} - \frac{1}{2}\,P\,u_{z_{12}} \\
&= 0
\end{aligned}$$

It is clear that in TBT (and EBBT as well) such a loading condition does not involve a virtual variation of the external load. In the case of $N = 1$, the virtual variation of the external work is given by

$$\begin{aligned}
\delta L_{ext} &= P\,N_1\left(\frac{1}{2}\right) \underbrace{F_1}_{F_1=1}(a/2)\;u_{z_{11}} + P\,N_1\left(\frac{1}{2}\right) \underbrace{F_2}_{F_2=x}(a/2)\;u_{z_{21}} \\
&\quad + P\,N_2\left(\frac{1}{2}\right) F_1\,(a/2)\;u_{z_{12}} + P\,N_2\left(\frac{1}{2}\right) F_2\,(a/2)\;u_{z_{22}} \\
&\quad - P\,N_1\left(-\frac{1}{2}\right) F_1\,(-a/2)\;u_{z_{11}} - P\,N_1\left(-\frac{1}{2}\right) F_2\,(-a/2)\;u_{z_{21}} \\
&\quad - P\,N_2\left(-\frac{1}{2}\right) F_1\,(-a/2)\;u_{z_{12}} - P\,N_2\left(-\frac{1}{2}\right) F_2\,(-a/2)\;u_{z_{22}} \\
&= \underbrace{\frac{1}{2}\,P\,a}_{\text{First node load}} \times u_{z_{21}} + \underbrace{\frac{1}{2}\,P\,a}_{\text{Second node load}} \times u_{z_{22}}
\end{aligned}$$

and, according to Equation (8.67), load components have to be placed in the sixth position of each nodal load vector.

B.3 Postprocessing

The postprocessing of results represents the final step of an FEA. In a displacement-based model, the postprocessing input is the vector of the generalized nodal displacement variables, in the case of a static analysis, or the eigenvectors and eigenvalues, in the case of a modal analysis. A method to postprocess the results requires that these vectors are expanded over all the points where postprocessing data are needed. The expansion process involves shape functions ($N_i(y)$) along the beam axis and expansion functions (F_τ) above the cross-section,

$$\begin{cases} u_x(x_p, y_p, z_p) = N_i(y_p)\; F_\tau(x_p, z_p)\; u_{x_{\tau i}} \\ u_y(x_p, y_p, z_p) = N_i(y_p)\; F_\tau(x_p, z_p)\; u_{y_{\tau i}} \\ u_z(x_p, y_p, z_p) = N_i(y_p)\; F_\tau(x_p, z_p)\; u_{x_{\tau i}} \end{cases} \tag{B.15}$$

where $[x_p, y_p, z_p]$ is a generic point of the structure where the displacement components have to be computed. The terms $u_{x_{\tau i}}$ have to be extracted from the nodal unknown vector; the extraction process requires that the FE to which the postprocessing point belongs is identified.

Example B.3.1 *Let us consider the B2 element in Figure B.15. The displacement component u_x has to be computed in P at a generic cross-section point of coordinates $[x_p, z_p]$. An $N = 2$ model is assumed:*

$$\begin{aligned} u_x(x_p, y_p, z_p) &= N_1(y_p)\; F_\tau(x_p, z_p)\; u_{x_{\tau 1}} \\ &\quad + N_2(y_p)\; F_\tau(x_p, z_p)\; u_{x_{\tau 2}} \end{aligned}$$

where $u_{x_{\tau 1}}$ and $u_{x_{\tau 2}}$ are the x components of the nodal variables of the first and second node of the element, respectively. Shape functions in P are equal to $1/2$,

$$\begin{aligned} u_x(x_p, y_p, z_p) &= \frac{1}{2}\; \big[F_1(x_p, z_p)\; u_{x_{11}} + F_2(x_p, z_p)\; u_{x_{21}} \\ &\quad + F_3(x_p, z_p)\; u_{x_{31}} + F_4(x_p, z_p)\; u_{x_{41}} \\ &\quad + F_5(x_p, z_p)\; u_{x_{51}} + F_6(x_p, z_p)\; u_{x_{61}}\big] \\ &\quad + \frac{1}{2}\; \big[F_1(x_p, z_p)\; u_{x_{12}} + F_2(x_p, z_p)\; u_{x_{22}} \\ &\quad + F_3(x_p, z_p)\; u_{x_{32}} + F_4(x_p, z_p)\; u_{x_{42}} \\ &\quad + F_5(x_p, z_p)\; u_{x_{52}} + F_6(x_p, z_p)\; u_{x_{62}}\big] \end{aligned}$$

and on substituting the F_τ expressions the displacement component becomes

$$\begin{aligned} u_x(x_p, y_p, z_p) &= \frac{1}{2}\big[u_{x_{11}} + x_p\; u_{x_{21}} + z_p\; u_{x_{31}} \\ &\quad + x_p^2\; u_{x_{41}} + x_p\; z_p\; u_{x_{51}} + z_p^2\; u_{x_{61}}\big] \\ &\quad + \frac{1}{2}\big[u_{x_{12}} + x_p\; u_{x_{22}} + z_p\; u_{x_{32}} \\ &\quad + x_p^2\; u_{x_{42}} + x_p\; z_p\; u_{x_{52}} + z_p^2\; u_{x_{62}}\big] \end{aligned}$$

Modal shapes from free vibration analyses have to be computed by considering each eigenvector as equivalent to a nodal displacement vector and then expanding it as in Example B.3.1.

B.3.1 Stresses and Strains

Stress and strain components can be straightforwardly computed from the nodal displacement vector. Strains, in particular, require that the partial derivatives are calculated by means of shape and expansion functions,

$$\begin{cases} \mathbf{u}_{,x} = (F_\tau\, N_i\, \mathbf{u}_{\tau i})_{,x} = F_{\tau,x}\, N_i\, \mathbf{u}_{\tau i} \\ \mathbf{u}_{,y} = (F_\tau\, N_i\, \mathbf{u}_{\tau i})_{,y} = F_\tau\, N_{i,y}\, \mathbf{u}_{\tau i} \\ \mathbf{u}_{,z} = (F_\tau\, N_i\, \mathbf{u}_{\tau i})_{,z} = F_{\tau,z}\, N_i\, \mathbf{u}_{\tau i} \end{cases} \tag{B.16}$$

As the strain field is computed, stress components are obtained through the constitutive laws,

$$\{\sigma\} = [\mathbf{C}]\,\{\varepsilon\}$$

Example B.3.2 *Let us consider Example B.3.1 to compute ε_{xx}.*

$$\begin{aligned} \varepsilon_{xx}(x_p, y_p, z_p) = u_x(x_p, y_p, z_p)_{,x} = N_1(y_p)\, F_{\tau,x}(x_p, z_p)\, u_{x_{\tau 1}} \\ + N_2(y_p)\, F_{\tau,x}(x_p, z_p)\, u_{x_{\tau 2}} \end{aligned}$$

thus

$$\begin{aligned} \varepsilon_{xx}(x_p, y_p, z_p) = \frac{1}{2}\big[& F_{1,x}(x_p, z_p)\, u_{x_{11}} + F_{2,x}(x_p, z_p)\, u_{x_{21}} \\ &+ F_{3,x}(x_p, z_p)\, u_{x_{31}} + F_{4,x}(x_p, z_p)\, u_{x_{41}} \\ &+ F_{5,x}(x_p, z_p)\, u_{x_{51}} + F_{6,x}(x_p, z_p)\, u_{x_{61}}\big] \\ &+ \frac{1}{2}\big[F_{1,x}(x_p, z_p)\, u_{x_{12}} + F_{2,x}(x_p, z_p)\, u_{x_{22}} \\ &+ F_{3,x}(x_p, z_p)\, u_{x_{32}} + F_{4,x}(x_p, z_p)\, u_{x_{42}} \\ &+ F_{5,x}(x_p, z_p)\, u_{x_{52}} + F_{6,x}(x_p, z_p)\, u_{x_{62}}\big] \end{aligned}$$

and on substituting the $F_{\tau,x}$ expressions the displacement component becomes

$$\begin{aligned} u_x(x_p, y_p, z_p) = \frac{1}{2}\,&[u_{x_{21}} + 2\, x_p\, u_{x_{41}} + z_p\, u_{x_{51}}] \\ &+ \frac{1}{2}\,[u_{x_{22}} + 2\, x_p\, u_{x_{42}} + z_p\, u_{x_{52}}] \end{aligned}$$

References

Bathe K 1996 *Finite Element Procedure*. Prentice Hall.

Carrera E, Giunta G, Nali P and Petrolo M 2010a Refined beam elements with arbitrary cross-section geometries. *Computers and Structures* **88**(5–6), 283–293.

Carrera E, Giunta G and Petrolo M 2010b A modern and compact way to formulate classical and advanced beam theories. In: *Developments and Applications in Computational Structures Technology*, Ch. 4. Saxe-Coburg Publications.

Carrera E and Petrolo M 2011 On the effectiveness of higher-order terms in refined beam theories. *Journal of Applied Mechanics* **78**. DOI: 10.1115/1.4002207.

Carrera E, Petrolo M and Nali P 2010c Unified formulation applied to free vibrations finite element analysis of beams with arbitrary section. *Shock and Vibrations* **18**. DOI: 10.3233/SAV-2010-0528.

Oñate E 2009 *Structural Analysis with the Finite Element Method: Linear Statics*, Volume 1. Springer.

Index

Finite Element Analysis of Structures Through Unified Formulation, First Edition.
Erasmo Carrera, Maria Cinefra, Marco Petrolo and Enrico Zappino.

Printed and bound by CPI Group (UK) Ltd, Croydon, CR0 4YY

16/04/2025

14658473-0003